D1566484

DEEP IMPURITIES IN SEMICONDUCTORS

DEEP IMPURITIES IN SEMICONDUCTORS

A. G. MILNES
Carnegie-Mellon University

A WILEY-INTERSCIENCE PUBLICATION

JOHN WILEY & SONS, New York . London . Sydney . Toronto

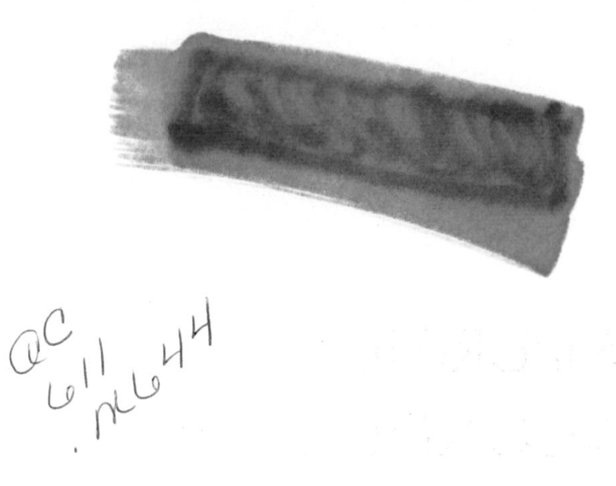

Library of Congress Cataloging in Publication Data:

Milnes, Arthur George.
Deep impurities in semiconductors.

"A Wiley-Interscience publication."
Bibliography: p.
1. Semiconductors. I. Title.

QC611.M644 537.6'22 73-5844
ISBN 0-471-60670-7

Printed in the United States of America

10 9 8 7 6 5 4 3 2 1

To

Mary and Sheila

Brian and John

Preface

Impurities that have energy levels lying deep in the band gap of semiconductors may create desirable effects in devices. The use of gold to give fast recombination in silicon junctions that require nanosecond switching times is well known. The high quantum gain of a band-gap photoconductor is usually achieved by the trapping action of one carrier by a deep impurity level. Impurity photoconductivity is a further important application of deep impurity effects in semiconductors. Gold, zinc, mercury, cadmium, and other centers in silicon and germanium have been explored for such uses. Impurities that have energy levels of less than 0.1 eV from the nearest band edge act as deep centers at very low temperatures and provide impurity photoconductivity detectors in the wavelength range from 10 to 100 μm. Semiconductors containing deep impurities are also of use as temperature sensors in thermistor-type applications.

In light-emitting diodes the deep centers may be the factors that determine the wavelength of the emitted radiation. In the red-light emission of gallium phosphide diodes zinc–oxygen pairs are the impurities responsible for the very high efficiencies (over 8% at 300°K) that are observed. In semi-insulating gallium arsenide the deep centers oxygen and chromium are responsible for the high resistances observed. There are some special uses for this material such as dielectric substrates for microwave integrated circuits.

Deep impurities also produce a wide range of effects in semiconductors that are fascinating, but are not yet of proven usefulness. There are the counting effects observed when a number of pulses are applied to bulk high-resistance gallium arsenide or silicon. Oscillations of mega-hertz frequencies are observed in silicon; low-frequency field-dependent trap-emptying domain oscillations are observed in oxygen–copper-doped GaAs;

and a wide variety of other space-charge-limited current effects and two-terminal negative-resistance effects have been seen that have still to find practical application. Deep impurities also have been found to have large effects on the sensitivity of junction and bulk semiconductors to stress and so to have potential strain gauge uses.

Aside from the valuable effects of deep impurities there are other aspects which are less agreeable. For example, deep impurities are not readily avoided in large-band-gap compound semiconductors, and they can result in undesired trapping, oscillation, and negative-resistance effects that can be a source of considerable frustration to the would be user of these materials. By the study of deep impurity effects in well-controlled semiconductors such as silicon and germanium, a reasonable understanding of the physics involved has been obtained. Such understanding has then been helpful in interpreting the more involved effects that may be seen in compound semiconductors where more than one deep impurity level may be present.

Deep impurity effects in semiconductors have been the subject of study for about 20 years, and the number of papers published is large and widely scattered. Apart from a few review papers, a summary of the subject has not previously been attempted. To keep the present treatment within reasonable bounds it has been necessary to confine it to deep impurity effects in silicon, germanium, and the III–V compound semiconductors. The present state of knowledge on deep impurity effects in II–VI compound semiconductors is provided by the review volumes on photoconductivity, electroluminesence, and II–VI technology that are available. Defect centers created by radiation or ion implantation are excluded from this book since conference proceedings and other reviews are available in these special areas.

The present monograph discusses the following.

1. Energy levels and capture cross sections of deep impurities.
2. Lifetime, recombination, and trapping effects in bulk and junction structures.
3. Capacitance effects.
4. Space-charge-limited flow effects.
5. Recombination wave and slow domain effects.
6. Thermally stimulated emission studies.
7. Impact ionization of deep impurities.
8. Hopping and impurity band conduction effects.

In general extensive mathematical developments are avoided and results of such developments are presented with references to the original papers. In this approach the overall picture is less obscured by the detailed steps.

I am deeply indebted to my previous students, K. L. Ashley, A. M. Barnett, K. P. Lisiak, L. J. Kroko, A. E. McCombs, I. Melngailis, H. K.

Sacks, E. Schibli, P. C. Smith, and J. J. Wagener for the use of material that their studies have generated. From 1961 these investigations have been supported by grants from the National Science Foundation.

Many other investigators and their publishers have kindly given me permission to quote from their papers. These are acknowledged in the appropriate places and in the references in the extensive bibliography.

Special thanks are due to Effie Lipanovich for extensive editorial, typing, and drafting services.

A. G. MILNES

Pittsburgh, Pennsylvania
April 1973

Contents

Tables

CHAPTER ONE

Energy Level Concepts

This book is concerned primarily with the effects of deep impurities on the recombination and generation properties of semiconductors under a variety of transport and injection conditions. However, we begin with a brief review of the state of knowledge on the calculation of the energy levels of impurities in semiconductor band gaps. There is no need for our purposes to examine in great detail the various models that have been proposed. Indeed, modeling of very deep impurities has not been particularly successful in a quantitative way. However, an overview is desirable in providing a setting for the topics treated in later chapters. The ionization energies of shallow impurities are examined first and then the problems of revising such treatments for deeper lying impurity levels are considered.

1.1 Shallow Impurities

For the group IV semiconductors, Si and Ge, shallow donors are produced by impurities such as P, As, and Sb which substitute for the host atoms. The group V elements have five electrons in the outer shell. Four of these are used in the covalent substitutional bonds, leaving one electron per atom available for conduction. At very low temperatures, such as 4°K, this electron still remains bound to the impurity atom. Ionization energy E_i is a measure of the energy required to release the electron to the conduction energy band; E_i may be determined by Hall measurements which give the number of thermally liberated carriers as a function of temperature. The dependence observed for the electron density n is usually of the form $\exp(-E_i/kT)$ where k is Boltzmann's constant and T is the absolute temperature. Another method to determine E_i is to measure the infrared photon energy required to liberate the electrons from the impurity atoms.

Shallow acceptors in Si and Ge are produced by substitution impurities from group III of the periodic table such as B, Al, and Ga. These impurities require an extra electron to complete the valence bonding requirements and therefore at all except the lowest temperatures take an electron from a neighboring host atom and thereby create a hole. The activation energy with respect to the valence band can be determined by Hall measurements or optically as before.

Some experimental determinations of ionization energies for Si and Ge are presented in Table 1.1.

The ionization energies of all shallow donors and acceptors in Ge are seen to be very similar, about 0.011 eV. For Si the shallow donors and

TABLE 1.1

Ionization Energies of Shallow Impurities in Silicon and Germanium
(After Kohn, 1957)

	Impurity Element	E_i (eV) Thermal	E_i (eV) Optical
In silicon	Donors		
	Li	0.033	
	P	0.044	0.045
	As	0.049	0.053
	Sb	0.039	0.043
	Bi	0.069	
	Acceptors		
	B	0.045	0.046
	Al	0.057	0.067
	Ga	0.065	0.071
	In	0.16	0.154
In germanium	Donors		
	P	0.0120	
	As	0.0127	
	Sb	0.0096	
	B	0.0104	
	Acceptors		
	Al	0.0102	
	Ga	0.0108	
	In	0.0112	

acceptors tend to be in the range 0.04 to 0.07 eV with the exception of indium, which is much deeper than the rest. The optical measurements are in fair agreement with the thermal results.

Transitions of carriers from the impurity ground state to excited levels that are still bound may be inferred from infrared absorption studies. These are of interest to users of shallow-doped semiconductors as long-wavelength infrared sensors and to theorists for information against which to check the degree of success of their models.

1.2 Calculation of Ionization Energies of Shallow Impurities

A simple effective mass model is able to provide an approximate calculation of the binding energy of an electron to a donor impurity such as As^+. Following Mott and Gurney (1940) and Smith (1958) the force F acting on a distant free electron in the crystal will be very nearly equal to that due to a single positive charge located at the As^+ ion. This is given by

$$F = \frac{e^2}{4\pi\varepsilon r^2} \tag{1.1}$$

where r is the distance of the electron from the As^+ ion and ε the permittivity of the crystal. This will be valid only if r is greater than two or three lattice spacings. When r is of the order of the nearest neighbor distance the extra electron will perturb the bonding electrons appreciably and (1.1) will no longer hold. However, such small values of r have very little effect when ε is large. The force F may now be regarded as an "external" force acting on the electron, regarded as moving in a perfect crystal. Thus we consider an electron of effective mass m^* moving in the field F. The problem is then reduced to that of a hydrogen atom, and we obtain a series of energy levels with binding energy E_n given by

$$E_n = \frac{m^* {\varepsilon_0}^2 E_{\mathrm{H}}}{m_0 n^2 \varepsilon^2} = \frac{(m^*/m_0){\varepsilon_0}^2}{n^2 \varepsilon^2} \times 13.5 \text{ eV} \tag{1.2}$$

where n is an integer and ε_0 is the permittivity of free space.

This simple model requires modification to take account of the departure of the energy bands of Si and Ge from spherical symmetry. This introduces considerable complication which we do not wish to consider here. However, if we take an effective mass for electrons in silicon, from mobility considerations, of $m^* = 0.4m_0$ and $\varepsilon/\varepsilon_0 = 12$, we obtain $E_n = 0.04/n^2$ eV, from which for the lowest state $E_1 = 0.04$ eV. The experimental values, as we see from Table 1.1, range from 0.033 to 0.069 eV, depending on the specific donor impurity.

For Ge the observed values are about 0.011 eV, and to obtain this from (1.2) it is necessary to assume $m^* = 0.2m_0$, which is less acceptable from mobility considerations.

It is interesting to estimate the extent of the spread of the wave functions of electrons or holes bound to the impurity centers. A measure of this is the "radius of the first Bohr orbit" a_n, corresponding to the hydrogen-like wave functions. This is given by

$$a_n = \frac{(m_0/m^*)\varepsilon a_0 n^2}{\varepsilon_0} \tag{1.3}$$

where a_0 is the radius of the first Bohr orbit for hydrogen (equal to 0.53×10^{-8} cm). For group V donors in Ge, $a_1 = 80a_0$, so we see that the wave function is spread over many lattice spacings. In this case the contribution from the field at values of r of the order of a_0 may be shown to be small. This is the justification of the use of a coulomb field in the approximate calculation of E_n. For group V donors in Si, $a_1 = 30a_0$, and although the wave function covers many lattice sites the agreement with the simple theory may be expected to be not quite so good as for germanium. For excited states the agreement should be better because of the factor n^2.

However, we cannot hope for much precision from such a simple model and therefore the topic has had the attention of many theorists. Their progress has been summarized by Kohn (1957). It is interesting to quote Kohn as he becomes enthusiastic about the subject.

It may seem very surprising at first sight that the orbit of a trapped electron or hole wending its way through hundreds of crystal cells can be described at all simply. The theory shows, however, that the larger the orbit the more accurately it can be understood!

To appreciate this point let us consider a donor state for a moment. Such a state can be pictured roughly as follows. Assume a neutral impurity atom is introduced into the lattice and imagine that one electron is removed from it, leaving a positive ion behind. This ion will polarize the semiconductor so that at large distances it produces an electrostatic potential

$$u = \frac{e}{\varepsilon r} \tag{1.4}$$

Here ε is the static dielectric constant. When the electron is brought back, one of two things may happen. It may be energetically favorable for this electron to occupy an orbit rather near the impurity ion. In this case (1.4) does not apply and we have a rather complicated state of affairs in the immediate vicinity of the impurity ion. In brief, we obtain a "deep" impurity state. On the other hand, the lowest energy state may have a large orbit over most of which (1.4) applies. One would then guess that such a state can be described by the Schrödinger equation

$$\left(-\frac{\hbar^2}{2m^*}\nabla^2 - \frac{e^2}{\varepsilon r}\right)F(r) = EF(r) \tag{1.5}$$

where m^* is an appropriate effective mass. (The actual band structures of silicon and germanium are rather complex, which requires a modification of the kinetic energy term.)

It is clear that (1.5) will apply more accurately the larger the orbit, since then the region near the impurity ion, where (1.4) does not hold, plays an insignificant role. Here we see one of the main reasons why we can form an accurate picture of the shallow impurity states: The potential is basically Coulomb-like and hence the impurity states are very closely related to the states of the hydrogen atom. A second reason is slightly more subtle. We shall see later that the concept of an effective mass holds only for a state whose wave function $F(r)$ has a slow spatial variation. This principle also favors the description of shallow impurity states with large orbits.

The analogy with the hydrogen atom suggests the existence not only of a lowest bound state, but of a whole spectrum of excited bound states. Such states have been clearly identified experimentally. Since their orbits are even larger than that of the ground state, equations of the type (1.5) are particularly applicable to them.

Our understanding of the physical properties of impurity states is still being rapidly expanded. However, a great deal of insight into these fascinating "atoms" has already been gained. They are very delicate structures. They may be disrupted by an energy of the order of 0.01 eV and their orbits have the relatively immense dimensions in the range from 10^{-7} to 10^{-6} cm. Nevertheless, their structure is determined by the same quantum-mechanical laws and the same Coulomb attraction which govern their robust counterparts, the mesic atoms, with binding energies of about 10^6 eV and orbital diameters of about 10^{-12} cm. The range of energies, dimensions, and physical conditions in which hydrogen-like atoms occur in nature is truly remarkable.

From Kohn's review we conclude that the subject, notwithstanding his remarks, is one of considerable complexity although reasonably good understanding has been achieved. However there is still room for progress, as can be seen by inspecting Table 1.2.

Multivalley effects with respect to shallow donor states in silicon have

TABLE 1.2

Comparison of Experimental and Calculated Ground-State Shallow Donor Energies for Silicon and Germanium

(From P. Csavinszky, 1963)

	Silicon (eV)		Germanium (eV)	
Donor	Experimental	Calculated	Experimental	Calculated
P	0.045	0.031	0.0128	0.0089
As	0.053	0.035	0.0140	0.0093
Sb	0.043	0.039	0.0098	0.0096
Bi	0.069	0.045	0.0125	0.0010

been studied by Ning and Sah (1972) in an effective mass approximation that includes intervalley overlap terms and provides closer agreement between calculated and experimental energies.

1.3 The Problem of Deep Impurity Levels

When impurities that produce deep energy levels are considered, we encounter the problem that the levels may be associated with impurity complexes or impurity-vacancy complexes and that the impurity may be interstitial rather than substitutional.

Even if attention is confined to substitutional-type impurities there are considerable difficulties in applying the effective mass theory successfully. Bebb and Chapman (1971) in a study of the wave functions to be expected for substitutional impurities, with the experimental energy levels accepted as starting information, comment as follows.

Changes in ground state binding energies of substitutional impurities arise from deviations of the impurity core potential from that of a simple point charge. Attempts to account for observed chemical shifts of impurity level ground states usually proceed by dividing space around the impurity into inner and outer regions. In the outer regions it is argued that the usual effective mass equations are valid and can be used to obtain wave functions for $r > r_c$ where r_c is some radius to be determined dividing the inner and outer regions. Nara and Morita have shown in some detail that for $r > r_c$ where r_c is of the order of the nearest neighbor distance, the potential energy could be replaced by its asymptotic form $-e^2/\varepsilon r$ in the effective mass equations. Numerous approximations have been considered for describing the core region $r < r_c$ with varying degrees of success ranging from poor to fair. Wave functions describing all space are then obtained by requiring the core and exterior functions to satisfy certain continuity conditions. The success of the theory is normally measured by its ability to predict the empirical ground state binding energy. The energy integrals computed obtain a major contribution from the region of small r and thus depend critically on the accuracy of the core functions.

However, in many applications the integrals to be computed obtain their major contribution from the region of large r where the effective mass equations are valid. In particular it is known that optical integrals are very insensitive to the core region of the wave functions and, consequently, can be accurately evaluated from only a knowledge of the wave functions in the region of large r. Similar assertions are true for other integrals, e.g., phonon coupling strengths, etc., to varying degrees of accuracy. Therefore, in many (and probably most) applications, we only require wave functions valid in the region of large r where the effective mass equations hold.

Attempts to solve the effective mass equation in the region of large r in a way that reflects the influence of the core potential usually involve replacing the eigenenergy by the observed energy. Three such semiempirical solutions have been published. The simplest and most widely used approximation is to assume that the potential is coulombic over all space. The resulting hydrogenic wave functions are then scaled by adjusting the effective Bohr radius, a^*, to reproduce the observed impurity binding energy, $E_{obs} = -(\hbar^2/2m^*)/a^{*2}$. This procedure is partially successful for reasonably shallow impurity levels. It is philosophically objectionable because it requires scaling parameters formally determined by the host lattice.

Another approach valid in the limit of very deep impurities has been proposed by Lucovsky. He assumed that the increased binding energy of the ground state was due entirely to the core potential. He accordingly approximated the ion potential by a delta-function-well with the depth adjusted to reproduce the empirical binding energy. While Lucovsky's approach does at least take cognizance of the physics of the problem, its validity is limited to very deep impurity levels. It is philosophically objectionable because it requires specifying the potential near $r = 0$ where the general effective mass equations are of uncertain validity. However, the wave function lies outside the delta-function-well and is assumed valid for large r.

A more general solution completely compatible with the assumptions inherent to effective mass theory is provided by the quantum defect method (QDM), which has been successfully applied in atomic physics. The QDM relies on the observation that the potential becomes coulombic, $-e^2/\varepsilon r$, for large r ($r > r_c$); the solutions to the effective mass equation in the exterior region are therefore also coulombic (but not necessarily hydrogenic). In the quantum defect method, the "allowed values of the principal quantum number, ν, are determined from the observed binding energy rather than from the boundary conditions at $r = 0$. Consequently, the values of ν are not constrained to integer numbers in this procedure, and the wave functions may diverge at $r = 0$. This does not, however, affect their validity away from $r = 0$."

From the quantum defect model Bebb and Chapman proceed to calculate quantities such as bound–bound transition rates, phonon coupling strengths, and band-impurity recombination lifetimes. The reader is referred to their paper for details of the calculations.

1.4 Deep Impurity Levels

A direct approach at calculating deep impurity levels has been made by Glodeanu (1967). He considers bivalent substitutional impurities that generate two localized energy levels in the forbidden band and uses a helium-like model in a single-band approximation and calculates the deep donor and acceptor levels of a number of impurities in GaAs, Si, and Ge.

In considering a donor with two supplementary electrons he writes the following Hamiltonian

$$H = -\frac{\hbar^2}{2m^*}(\nabla_1{}^2 + \nabla_2{}^2) + V(r_1) + V(r_2) + U_{\text{eff}}(r_1, r_2) \quad (1.6)$$

and the corresponding Schrödinger equation:

$$H\Psi(r_1, r_2) = E\Psi(r_1, r_2) \tag{1.7}$$

where $V(r_1)$ is a periodic potential and represents the interaction of electron 1 with the effective crystal field and $V(r_2)$ is a similar potential for electron 2. The last term in (1.6) is given by

$$U_{\text{eff}}(r_1, r_2) = -\frac{Ze^2}{\varepsilon r_1} - \frac{Ze^2}{\varepsilon r_2} + \frac{e^2}{\varepsilon |r_1 - r_2|} \tag{1.8}$$

where the first two terms represent the interaction of the two electrons with a supplementary positive charge Ze that has been introduced with the purpose of conserving the crystal neutrality.

The wave function in (1.7) is taken to be of the form

$$\Psi(r_1, r_2) = \frac{1}{N\Omega} \sum_{k_1,k_2} C(k_1, k_2) U_{c,0}(r_1) U_{c,0}(r_2) \times \exp(ik_1 r_1) \exp(ik_2 r_2) \tag{1.9}$$

where $U_{c,0}$ is the Bloch function at the minimum of the conduction band satisfying the equation

$$\left[-\frac{\hbar^2}{2m^*} \nabla_1{}^2 + V(r_1)\right] U_{c,0}(r_1) = E_{c,0}(r_1) U_{c,0}(r_1) \tag{1.10}$$

Ω is the volume of the elementary cell, and N is the number of cells.

If we introduce in (1.7) the expression of Ψ given by (1.9), we obtain the following helium-like equation:

$$\left[-\frac{\hbar^2}{2m^*} (\nabla_1{}^2 + \nabla_2{}^2) - \frac{Ze^2}{\varepsilon r_1} - \frac{Ze^2}{\varepsilon r_2} + \frac{e^2}{\varepsilon |r_1 - r_2|}\right] F_n(r_1, r_2) = EF_n(r_1, r_2) \tag{1.11}$$

From (1.11) Glodeanu determines the energy by a variational treatment, choosing for $F_n(\bar{r}_1, \bar{r}_2)$ a trial function of the form

$$F_n(r_1, r_2) = \frac{Z'^3}{\pi r_0{}^3} \exp\left[-\frac{Z'}{r_0}(r_1 + r_2)\right]; \qquad r_0 = \frac{\hbar^2 \varepsilon}{m^* e^2} \tag{1.12}$$

and obtains for the two ionization energies the expressions

$$E_1 = a\left[Z_{\text{eff}}^2 - \frac{5}{4} Z_{\text{eff}} + \frac{25}{128}\right]; \qquad E_2 = aZ_{\text{eff}}^2 \tag{1.13}$$

where a is a parameter, $a \sim m^*/\varepsilon$, and $Z' = Z_{\text{eff}} - \frac{5}{16}$.

Because a correct considering of the core difference is a difficult problem, Z_{eff} is treated as a parameter in (1.13) which is fitted to the experimental results in a manner such that the sum of the squares of the relative errors becomes minimum.

The calculated and experimental ionization energies are given in Table 1.3. The effective parameter Z_{eff} is also shown in Table 1.3. The depth of the donor levels is counted from the conduction band and that of the acceptor levels from the valence band.

TABLE 1.3

Calculated and Observed Values of the Ionization Energy for Deep Impurities in GaAs, Si, and Ge

				E_1 (eV)		E_2 (eV)	
Crystal	Impurity	Type[b]	Z_{eff}	Calculated	Observed	Calculated	Observed
GaAs	Cu	*A*	2	0.171	0.15	0.407	0.47
Si	S	*D*	2.145	0.206	0.18	0.447	0.52
	Ni	*A*	2.380	0.280	0.23	0.550	0.70
	Co	*A*	2.515	0.328	0.35	0.614	0.58
	Zn	*A*	2.428	0.297	0.31	0.573	0.55
Ge	Se	*D*	2.275	0.138	0.14	0.285	0.28
	Te	*D*	2.195	0.124	0.11	0.262	0.30
	Mn	*A*	2.475	0.177	0.16	0.322	0.37
	Co	*A*	2.815	0.249	0.25	0.431	0.43
	Fe	*A*	3.060	0.312	0.34	0.509	0.47
	Ni	*A*	2.753	0.234	0.22	0.413	0.44
	Cr[a]	*A*	2.580	0.068	0.07	0.123	0.12
	Cd[a]	*A*	2.735	0.079	0.06	0.138	0.20
	Zn[a]	*A*	2.050	0.034	0.03	0.078	0.09

[a] For Cr, Cd, and Zn impurities the effective mass $m^* = 0.34\, m_0$ has been used.
[b] *A* is acceptor; *D* is donor.

The determined values for Z_{eff} lie in a reasonable range, $2 \leq Z_{eff} < 3$; and for Cu in GaAs, S in Si, Se and Zn in Ge the values are very close to 2, as is expected from the hypothesis of covalent bonds. For impurities for which the core differs greatly from the core of the host crystal atom the values of Z_{eff} have corresponding magnitudes.

The parameter Z_{eff} could be determined separately for the E_1 and E_2 levels. Then the effective radii r_1 and r_2 may be calculated. The obtained values of Z_{eff}, r_1, and r_2 usually do not differ much from those determined by the procedure above. Unfortunately for some impurities it leads to Z_{eff} values which contradict the hypothesis of covalent bonds. Also in some cases the incorrect result $r_1 < r_2$ is obtained in this way.

Glodeanu remarks on the numerous difficulties and oversimplifications in his approach. For instance, for the Cr, Cd, and Zn impurities in Ge he must use the effective mass for calculating the parameter a. If the free-electron mass is used for these impurities, he finds $Z_{eff} < 2$, in contradiction with the hypothesis of covalent bonds. Thus his results show that for ionization energies larger than 0.1 eV the free-electron mass has to be used, whereas for energies below 0.1 eV the effective mass is necessary.

However, in spite of the limitations of Glodeanu's model, it is one of the most direct attacks on the theoretical problem that has yet been reported.

In another approach Watkins and Messmer (1970) simulate a deep defect level in a semiconductor by a large cluster of host atoms surrounding the defect. The electronic states of the entire system are then computed by LCAO–MO techniques. They apply this approach to boron and nitrogen in diamond.

Ryabokon' and Svidzinskii (1972) discuss an approach for calculating the deep levels of substitutional acceptor impurities in which allowance is made for the penetration of a hole from the valence band into the inner shell of an impurity. Mn, Fe, Co, Ni and Zn in germanium are considered.

Electron spin resonance studies also provide some wave-function information on deep energy levels (Lancaster, 1967).

As we have commented earlier, however, there are in practice so many factors that may contribute to the energy levels of deep impurities that there seems no hope for a unifying theory. We must therefore turn to experimental determination of the levels.

In the two chapters that follow we discuss deep levels observed in Si and Ge and in GaAs and other III–V semiconductors.

CHAPTER TWO

Deep Impurity Levels in Silicon and Germanium

The energy levels of impurities depend on how well these impurities fit the electronic and geometric periodic pattern of the host lattice, and are thus determined by the lattice position, the size, and the electronic shell structure of the impurities. Closely related to these factors is the maximum solid solubility. However, the tendency to form compounds or precipitates upon cooling, or the presence of the same impurity in different lattice sites, may result in varying proportions of electrically active and inactive species. Therefore even careful investigations have sometimes led to seemingly contradictory conclusions.

To reveal some of the multiple levels of a deep impurity it is necessary to provide compensation by shallow impurities. This requires great care in material preparation and deep impurity studies are not lightly undertaken.

Experimental values that have been determined for the energy levels of deep impurities in silicon and germanium are discussed in this chapter. References are provided to the original literature since the values determined are not always independent of techniques used and of the material preparation. The order of presentation for silicon is impurities of groups I, II, and III; the transition-metal impurities; group VI impurities; group IV impurities; and then other impurities such as Li, N, and the rare earths. In Section 2.3 there is a discussion of the Fischler relationship between solubilities and distribution coefficients. Diffusion coefficients and resistivity are also discussed.

The presentation for impurities in germanium follows a somewhat similar order. Useful data summaries covering most aspects of the properties of Si,

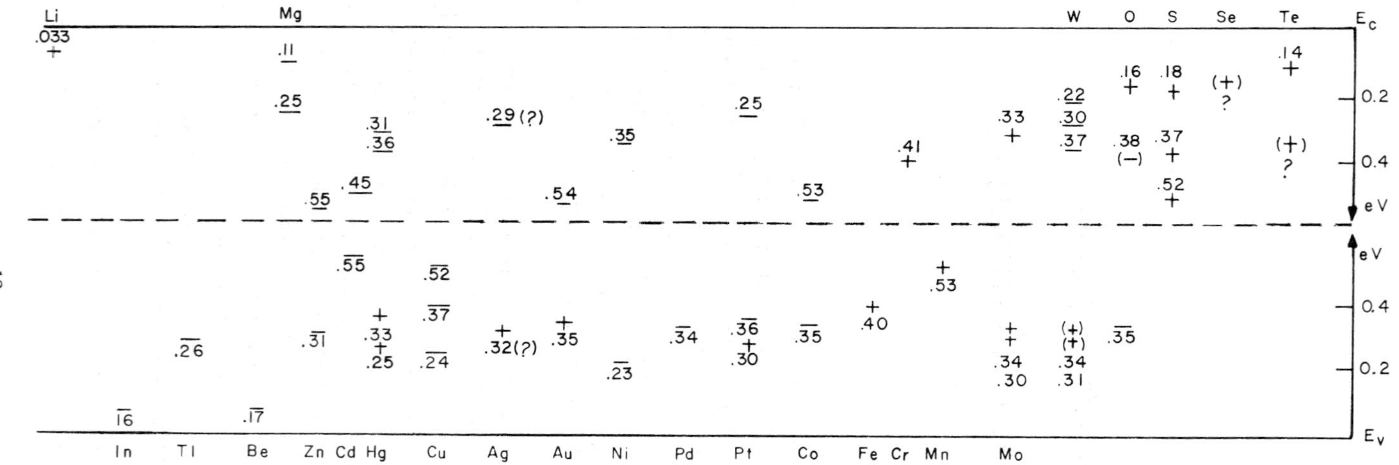

FIG. 2.1 Deep energy levels in silicon.

The energy values are measured from the nearest band edge. The symbols + and − indicate donor and acceptor levels, respectively. Parentheses indicate uncertain levels. See also Table 2.1. For Pt the 0.36eV acceptor level is uncertain: a recent study (K. P. Lisiak) places it at $E_c - 0.49$eV.

Ge, and other semiconductors, are available from the Electronic Properties Information Center of Hughes Aircraft Company, Culver City, California.

Energy level data for deep impurities in silicon are shown in Fig. 2.1, and are discussed for the most part later in the chapter. In general the levels were determined in many different laboratories and by several different techniques, although Hall effect versus temperature studies predominated. The deep doping was normally achieved by impurity addition during crystal growth or by diffusion.

Recently Fahrner and Goetzberger (1972) have described an interesting surface capacitance technique for determining deep impurity energy levels in silicon. The element to be investigated is ion implanted into an Si–SiO_2 interface such that the peak of the ion distribution coincides with the interface. If the ion concentration profile is narrow compared to the width of the space-charge layer, the variation of the potential across the ion distribution is small and all the ions behave as if they were at the surface potential. However, if the ion distribution is several hundred angstroms wide, most ions will presumably have bulk energy levels and not be influenced by the presence of the SiO_2 (with its different dielectric constant). In effect then, bulk levels have been obtained that behave like surface states in MOS measurements. From high- and low-frequency capacitance-voltage curves, plots can be derived of the surface-state density versus voltage. With deep impurities present, surface-state density peaks are observed which are characteristic of the impurities and give their energy levels.

In general the energy levels determined by surface capacitance studies are in good agreement with the levels obtained by more conventional methods. This can be seen from Table 2.1 where the values obtained so far are summarized and compared, where possible, with those shown in Fig. 2.1 and those given by Sze (1969). Although the agreement is excellent for some elements, there remains the prospect that the ion-implantation process itself may be affecting a few of the levels observed, even though an annealing step (typically 500°C for 30 minutes) is applied before the MOS capacitance measurements are made. The following detailed discussions of levels therefore concentrate on the values of Fig. 2.1 as being conservative, even though there may now be need for reexamination of some earlier work in the light of the surface capacitance studies.

2.1 Silicon: Impurities of Groups I, II, and III

The number and type of most impurity energy levels follow a few simple rules (Conwell, 1958). Substitutional impurities of groups II and III tend to complete the tetrahedral bonding, need electrons to do so, and therefore

TABLE 2.1

Deep Impurity Levels in Silicon: Measurements by Conventional Techniques and by Surface Capacitance Studies[a]

Impurity	Level(s) from Conventional Techniques: From Fig. 2.1 (eV)	Level(s) from Conventional Techniques: From Sze (1969) (eV)	Levels from Surface Capacitance Studies, Fahrner and Goetzberger (1972) (eV)
Ag	0.29 (*c*, *a*)	0.33 (*c*, *a*)	0.36 (*c*)
	0.32 (*v*, *d*)	0.34 (*v*, *d*)	0.33 (*v*)
Au	0.54 (*c*, *a*)	0.54 (*c*, *a*)	0.30 (*c*)[b]
	0.35 (*v*, *d*)	0.35 (*v*, *d*)	0.83 (*c*)
	0.033 (*v*, *a*)?		
Ba			0.32 (*c*)
			0.50 (*v*)
Be	0.17 (*v*, *a*)		0.42 (*v*)
Cd	0.45 (*c*, *a*)		0.20 (*c*)
	0.55 (*v*, *a*)		0.30 (*v*)
Co	0.53 (*c*, *a*)		0.63 (*c*)
	0.35 (*v*, *a*)	0.39 (*v*, *a*)	
Cr	0.41 (*c*, *d*)		
Cs			0.30 (*c*)
Cu	0.52 (*v*, *a*)	0.52 (*v*, *a*)	0.26 (*c*)
	0.37 (*v*, *a*)	0.37 (*v*, *a*)	0.53 (*v*)
	0.24 (*v*, *a*)	0.24 (*v*, *a*)	0.40 (*v*)
Fe			0.14 (*c*)
		0.55 (*c*, *d*)	0.51 (*c*)
	0.40 (*v*, *d*)	0.40 (*v*, *d*)	0.40 (*v*)
Ge			0.27 (*c*)
			0.50 (*v*)
Hg	0.31 (*c*, *a*)		
	0.36 (*c*, *a*)	0.33 (*c*)	0.35 (*c*)
	0.33 (*v*, *d*)	0.36 (*v*)	0.39 (*v*)
	0.25 (*v*, *d*)		
In	0.16 (*v*, *a*)	0.16 (*v*, *a*)	
Li	0.033 (*c*, *d*)	0.033 (*c*, *d*)	
Mg	0.11 (*c*, *d*)		
	0.25 (*c*, *d*)		
Mn	0.53 (*v*, *d*)	0.53 (*v*, *d*)	0.43 (*c*)
			0.45 (*v*)

[a] The letters *c*, *v* indicate whether the energy has been given with respect to the conduction or valence band edges. The letters *a*, *d* show whether the level is generally considered to be an acceptor or a donor.

[b] Some instability of these levels was observed.

TABLE 2.1 (continued)

Impurity	Level(s) from Conventional Techniques: From Fig. 2.1 (eV)	Level(s) from Conventional Techniques: From Sze (1969) (eV)	Levels from Surface Capacitance Studies, Fahrner and Goetzberger (1972) (eV)
Mo	0.33 (*c*, *d*)		
	0.34 (*v*, *d*)		
	0.30 (*v*, *d*)		
N	0.14 (*c*, *d*)		
Ni	0.35 (*c*, *a*)	0.35 (*c*, *a*)	
	0.23 (*v*, *a*)	0.22 (*v*, *a*)	
O	0.16 (*c*, *d*)		0.51 (*c*)[c]
	0.38 (*c*, *a*)		0.41 (*v*)
		0.03 (*v*, *a*)	
Pb			0.17 (*c*)
			0.37 (*v*)
Pd	0.34 (*v*, *a*)		
Pt	0.25 (*c*, *a*)		
	0.36 (*v*, *a*) ?	0.37 (*c*, *a*)	
	0.30 (*v*, *d*)		
S	0.18 (*c*, *d*)	0.18 (*c*, *d*)	0.26 (*c*)
	0.37 (*c*, *d*)	0.37 (*c*, *d*)	0.48 (*v*)
	0.52 (*c*, *d*)		
Se	? (*c*, *d*)		0.25 (*c*)
			0.40 (*c*)
Sn			0.17 (*c*)
			0.37 (*v*)
Ta			<0.14 (*c*)
			0.43 (*c*)
Te	0.14 (*c*, *d*)		
	? (*c*, *d*)		
Ti			0.21 (*c*)
Tl	0.26 (*v*, *a*)	0.26 (*v*, *a*)	0.30 (*v*)
V			0.49 (*c*)
			0.40 (*v*)
W	0.22 (*c*, *a*)		
	0.30 (*c*, *a*)		
	0.37 (*c*, *a*)		
	0.34 (*v*, *d*)		
	0.31 (*v*, *d*)		
Zn	0.55 (*c*, *a*)	0.55 (*c*, *a*)	0.55 (*c*)
	0.31 (*v*, *a*)	0.31 (*v*, *a*)	0.26 (*v*)

[c] These oxygen levels broaden at room temperature and disappear after some weeks.

act as acceptors. The depth of the acceptor level appears to increase within group III as the row number increases (B, 0.045 eV; Al, 0.057 eV; Ga, 0.065 eV). This can be thought of as repulsion between the electronic shells, making it less favorable for electrons to complete the tetrahedral bonding. Similar trends are observed for Zn, Cd, and Hg and for Cu, Ag, and Au (groups IIB and IB) in germanium. Such regularities can be expected for substitutional impurities, or for interstitial impurities in a common state.

Of the group III elements, indium and thallium are deep substitutional acceptors, with energy levels at $E_v + 0.16$ and $E_v + 0.26$ eV, respectively. These are the only known stable single-level deep impurities in silicon.

However, alloyed contacts of In are ohmic on *n*-type silicon, and rectifying on *p*-type silicon. The interpretation is not certain but might indicate donor action of interstitial indium in the recrystallized region (Migitaka, 1965). The existence of interstitial In (In^+) has also been suggested during investigations of diffusion mechanisms. Similarly, interstitial Al is believed to act as a donor in silicon (Millea, 1966). However, McCaldin and Mayer (1970) attribute donor behavior in indium-doped silicon to phosphorus contamination from quartz.

Figure 2.1 shows some of the reported energy levels in silicon. Zinc, as a substitutional acceptor, introduces two levels (0.31 and 0.55 eV) which are much deeper than the corresponding two levels in germanium (0.03 and 0.09 eV) (Carlson, 1957). This is perhaps related to the smaller lattice constant of silicon (5.43 versus 5.66 Å for germanium). Preliminary work on mercury at densities of 10^{14} cm^{-3} indicates two deep acceptors, 0.31 and 0.36 from E_c, and two deep donors, 0.25 and 0.33 from E_v (Zibuts et al., 1964).

Mercury lies between Au and Tl in the periodic table and since both of the latter elements are substitutional, it may be expected that Hg is substitutional also, although this has not been explicitly reported. The presence of two acceptor and two donor levels is compatible with the tetrahedral bonding model, and the size effect presumably accounts for the Hg acceptor levels being nearer the conduction band than those for Zn. The size effect presumably also accounts for the absence of Zn donor levels in the energy gap. Mg, however, shows two deep donor levels, according to Franks and Robertson (1967).*

Silicon containing Cd has been examined recently by Gulamova et al., 1971; Avak'yants et al., 1971; and Grinberg et al., 1972. Cd is a double acceptor with levels at $E_v + 0.55$ eV (singly charged) and at $E_c - 0.45$ eV (doubly charged). The Cd in the Gulamova et al. studies was introduced by a 1200°C diffusion and the electrically active density was 5×10^{13} cm^{-3}.

Copper in silicon appears to be a triple acceptor with levels at $E_v + 0.24$, $+0.37$, and $+0.52$ eV. Theoretical considerations by Weiser (1962) and

* *Solid State Com.*, **5**, 479.

Millea (1966) explain the extremely high diffusion coefficient of interstitial copper. Electrical measurements on copper-doped samples are poorly reproducible. Hall and Racette (1964) have indirect evidence for interstitial copper (a donor) in p-type and in moderately n-type silicon, and for substitutional copper (a triple acceptor) in heavily doped n-type silicon. They speculate that earlier determined energy levels (acceptor at $E_v + 0.49$ eV and donor at $E_v + 0.24$ eV) are properties of precipitates rather than of the dissolved copper.

For silver in silicon, Boltaks and Shih-yin (1961) find one donor level at 0.32 eV above the valence band in agreement with Irmler (1958) and one acceptor level 0.22 eV from the conduction band. Thiel and Ghandhi (1970) report the donor level as $E_v + 0.26$ and the acceptor level as $E_c - 0.29$ eV. Recently Yau and Sah (1972) have measured the donor level as $E_v + 0.405$ eV and the acceptor level as $E_c - 0.593$ eV. The surface capacitance studies of Fahrner and Goetzberger (1972), however, show values of $E_v + 0.33$ and $E_c - 0.36$ eV. Clearly there are problems to be solved and values to be reconciled in connection with silver doped silicon.

Silver, like gold, is a fast diffusant in silicon ($D = 10^{-8}$ cm^2 sec^{-1} at 1200°C). The diffusion appears to be interstitial followed by reversion to substitutional sites. Although the solubility from tracer studies is greater than 10^{17} cm^{-3} it appears difficult to obtain electrically active densities above 10^{15} cm^{-3}.

Gold is the deep impurity that has been the subject of most study in silicon because it has considerable technological importance as a lifetime control impurity. It has a donor level at 0.35 eV above the valence band and an acceptor level at 0.54 eV from the conduction band. Gold diffuses somewhat faster than does silver in silicon and a variety of diffusion mechanisms have been identified. The electrically active solubility is limited to about 10^{17} cm^{-3}. The properties of gold in silicon have been reviewed by Bullis (1966). Although the tetrahedral radii of Cu, Ag ,and Au (1.35, 1.53, and 1.50 Å, respectively) do not follow their position in the periodic table, the energy levels of these substitutional impurities obey a reasonable sequence, namely: Cu, three acceptor levels; Ag, one acceptor plus one donor level; Au, one acceptor plus one donor level, as in Fig. 2.1.

Figure 2.2 shows the levels produced by gold in silicon with and without shallow impurities, as understood in 1966.

At low temperatures, the levels are many kT apart and they may be treated independently of each other. As a result the activation energies can be determined directly from the slope of low-temperature Hall coefficient data. In n-type material, with $N_D > N_A$ and $N_{\mathrm{Au}} > N_D - N_A$, the electron concentration is

$$n \approx \frac{N_D - N_A}{N_{\mathrm{Au}} - (N_D - N_A)} g_A N_c \exp\left(-\frac{E_c - E_A}{kT}\right) \tag{2.1}$$

for

$$n \gg p, \qquad n \ll N_D - N_A, \qquad \text{and} \qquad n \ll N_{\text{Au}} - (N_D - N_A)$$

In (2.1) g_A is the degeneracy factor associated with the gold acceptor level. In p-type material, with $N_A > N_D$ and $N_{\text{Au}} > N_A - N_D$, the hole concentration is

$$p \approx \frac{N_A - N_D}{N_{\text{Au}} - (N_A - N_D)} \frac{N_v}{g_D} \exp\left(-\frac{E_D - E_v}{kT}\right) \tag{2.2}$$

for

$$p \gg n, \qquad p \ll N_A - N_D, \qquad \text{and} \qquad p \ll N_{\text{Au}} - (N_A - N_D)$$

where g_D is the degeneracy factor associated with the gold donor level.

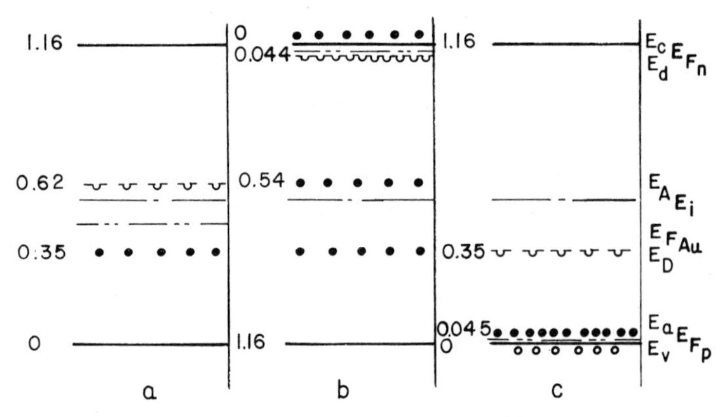

FIG. 2.2 Energy band diagram for silicon doped with gold:
(a) no other impurity present; (b) shallow donor present; (c) shallow acceptor present. In cases (b) and (c) it is assumed that the shallow impurity concentration is much greater than the gold concentration. Note that in case (c) the acceptor level does not exist since the donor levels are empty. (After Bullis, 1966.)

The Hall coefficient and electrical resistivity of p-type gold-doped silicon are shown as functions of temperature in Fig. 2.3, from which the donor level of $E_v + 0.35 \pm 0.02$ eV may be inferred.

The resistivity of n-type silicon (300°K) as a function of the gold concentration for this simple model is shown calculated in Fig. 2.4. It is seen that the resistivity becomes high when the gold concentration exceeds the shallow donor concentration. As the gold concentration is further increased, the resistivity becomes that typical of intrinsic silicon. With a further increase in gold concentration the resistivity reaches a maximum and the material becomes distinctly p-type. When gold becomes the dominant impurity,

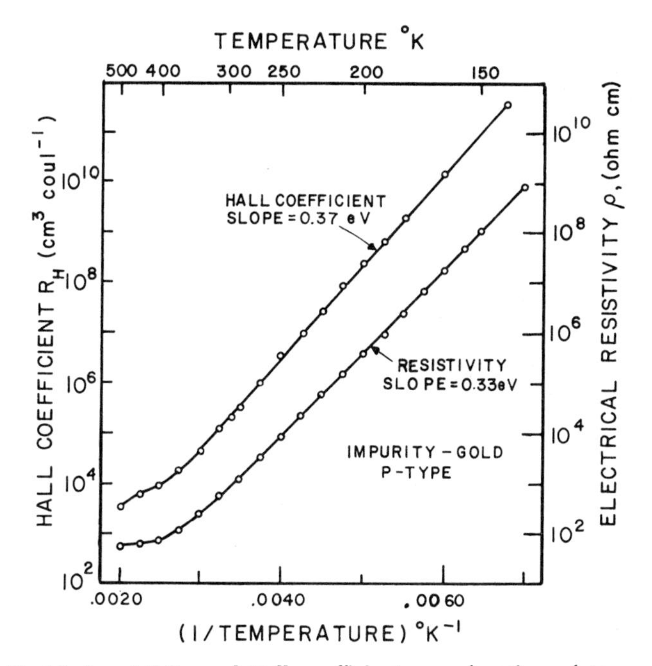

FIG. 2.3 Electrical resistivity and Hall coefficient as a function of temperature for *p*-type gold-doped silicon.

The concentration of gold is about 10^{16} cm^{-3} and the concentration of net residual acceptors (boron) is about 2×10^{14} cm^{-3}. (After Collins et al., 1957.)

the position of the Fermi level is determined principally by the relative occupancy of the gold levels, and the resistivity reaches a limiting value. Then, if $p \ll N_D - N_A$ and $p \ll N_{\text{Au}} - (N_D - N_A)$, the hole concentration is given by

$$p \approx \frac{N_{\text{Au}} - (N_D - N_A)}{N_D - N_A} \frac{N_v}{g_A} \exp\left(-\frac{E_A - E_v}{kT}\right) \tag{2.3}$$

and the energy difference between the upper gold level and the valence band edge can be determined directly.

For p-type silicon, the resistivity also begins to increase as the gold concentration approaches the shallow acceptor concentration; see Bullis (1966). However, in this case the increase is monotonic, reaching the limiting value when the gold concentration significantly exceeds the shallow acceptor concentration.

Experimental evidence, however, is in only fair agreement with the results of these calculations. For n-type silicon Bullis and Strieter (Bullis, 1966) find that the sharp increase in resistivity occurs at a gold concentration 1.5 to

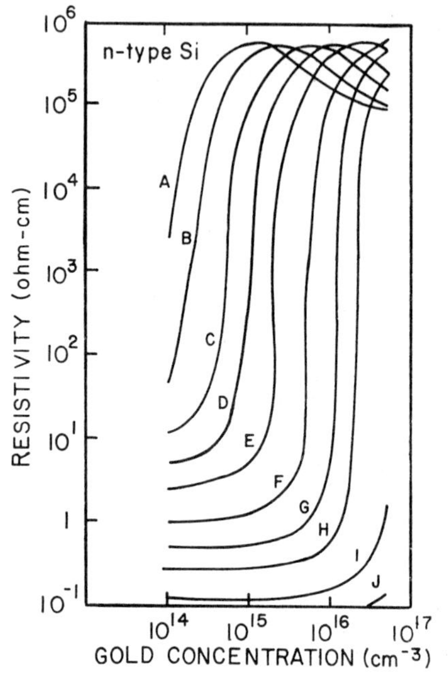

FIG. 2.4 Calculated value of resistivity versus gold concentration in *n*-type silicon at 300°K.

Phosphorus concentrations (cm^{-3}) assumed are A, 1×10^{14}; B, 2×10^{14}; C, 5×10^{14}; D, 1×10^{15}; E, 2×10^{15}; F, 5×10^{15}; G, 1×10^{16}; H, 2×10^{16}; I, 5×10^{16}; J, 1×10^{17}. The degeneracy factors were taken as $g_D = 0.5$ and $g_A = 2$.

2 times larger than predicted and the maximum resistivity obtained is somewhat lower than that calculated. For *p*-type silicon the measured and computed resistivities are even further apart than those for *n* type.

Brüchner (1971) has recently concluded that there is an impurity state in gold-doped silicon that is located 0.033 eV above the valence band and has suggested that it is an electrically active complex of gold with other defects. The concentration of these acceptors increases rapidly as the gold concentration increases. For *p*-type material, analysis indicates that the introduction of an energy state between the valence band edge and the gold donor state, with a concentration that depends on the gold concentration, causes a decrease of resistivity as the gold concentration increases that is in reasonable agreement with that observed experimentally.

Figure 2.5*a* and 2.5*b* (supplied by courtesy of W. M. Bullis and his coworkers) gives calculated resistivities of *n*- and *p*-type silicon with the Brüchner center included.

The calculations have been made for a temperature of 300°K. The phosphorus concentration ranges from 1×10^{13} (curve *A*) to 1×10^{17} cm^{-3} (curve *M*) in steps 1, 2, 5, 10, 20, and so on, for the initially *n*-type silicon. In this case as more gold is added the hole concentration increases so that to the right of the vertical bar, $p > n$, and to the right of the dot, the Hall

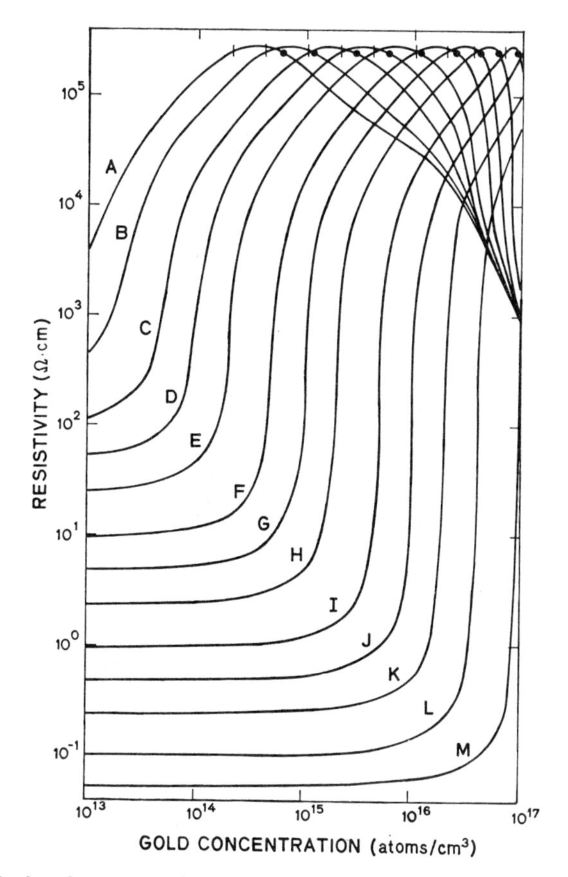

FIG. 2.5a Calculated curves of resistivity versus gold concentration for initially n-type silicon at 300°K.

Phosphorus concentrations (cm^{-3}) *are A,* 1×10^{13}; *B,* 2×10^{13}; *C,* 5×10^{13}; *D,* 1×10^{14}; *E,* 2×10^{14}; *F,* 5×10^{14}; *G,* 1×10^{15}; *H,* 2×10^{15}; *I,* 5×10^{15}; *J,* 1×10^{16}; *K,* 2×10^{16}; *L,* 5×10^{16}; *M,* 1×10^{17}. *At the bar on each curve,* $n = p = n_i$. *To the right of the bar,* $p > n$. *At the dot on each curve, the Hall coefficient,* R_H, *is zero. To the left of the dot* $R_H < 0$. *The modeling includes a gold-coupled acceptor state at 0.033 eV above* E_v.

coefficient becomes positive. For the p-type silicon the boron concentration and steps are the same.

In these calculations the energies of the gold donor and acceptor states were 0.35 and 0.58 eV, respectively, above the valence band; the energy gap at 300°K was 1.111 eV; and the intrinsic concentration was 1.5×10^{10} cm^{-3}. The effective masses were taken from the work of Barber (1967).

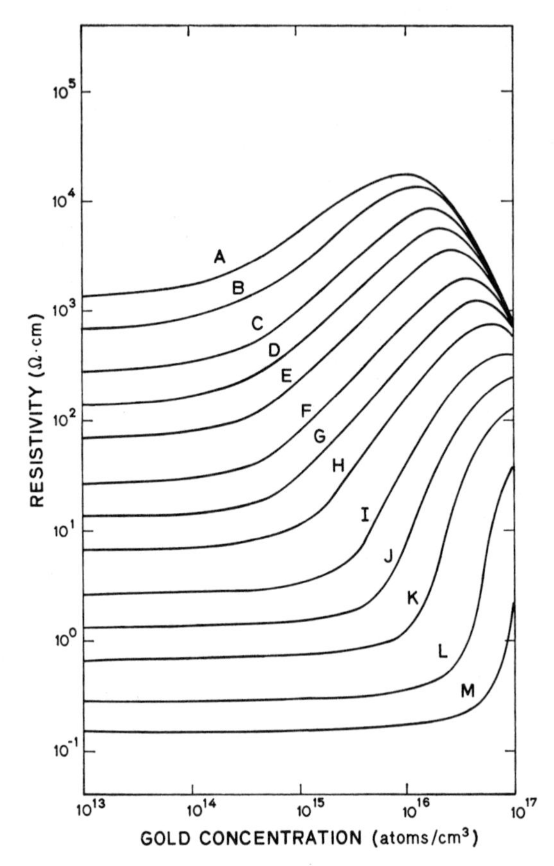

FIG. 2.5*b* Calculated curves of resistivity versus gold concentration for *p*-type silicon at 300°K.
Boron concentrations (cm^{-3}) are A, 1×10^{13}; B, 2×10^{13}; C, 5×10^{13}; D, 1×10^{14}; E, 2×10^{14}; F, 5×10^{14}; G, 1×10^{15}; H, 2×10^{15}; I, 5×10^{15}; J, 1×10^{16}; K, 2×10^{16}; L, 5×10^{16}; M, 1×10^{17}. The modeling includes a gold-coupled acceptor state at 0.033 eV above E_v. (After Bullis, 1972, private communication.)

The degeneracy factor used was 0.125 for both the gold donor and the gold acceptor levels and the lattice mobility and impurity mobility were combined reciprocally to obtain the carrier mobility used in the calculation of the resistivity. Also included was the gold-coupled acceptor state located at 0.033 eV above the valence band. Its dependence on gold concentration is not well established, but based on Bullis' experimental work, the acceptor concentration increases approximately as the third power of the gold concentration with a value of 4.5×10^{15} cm^{-3} at a gold concentration of 1×10^{17} cm^{-3}.

These curves represent a step forward in the agreement between calculated and experimental results, particularly for *p*-type material. However, for *n*-type material there remains the problem that the calculated curves predict a sharp increase in resistivity when the gold and donor concentrations are equal, whereas experimental data show that the rise does not occur until the gold concentration is about twice that of the shallow donor. The position of the sharp rise is affected neither by choice of degeneracy factor for either gold energy level nor by the shallow acceptor concentration (within reasonable limits). Although there is evidence that the model on which these curves are based represents the behavior of initially *n*-type silicon at large gold concentrations, it must be applied with caution until a modified one can be developed from which the position of the sharp increase is predicted successfully.

From diffusion and other studies of gold in silicon Kendall and De Vries (1969) have speculated that the electrically active gold state may be a "digold" formed by the reaction of two gold interstitials with a divacancy. This is unproven as yet, although it is curious to note that there is some similarity between the $E_c - 0.54$ and $E_v + 0.35$ eV levels of gold and two of the energy levels of a divacancy ($E_c - 0.54 \pm 0.02$ and $E_v + 0.28 \pm 0.05$ eV). Lappo and Tkachev (1972), however, have suggested that the negatively charged divacancy in silicon corresponds to an acceptor state at $E_c - 0.42$ eV. Thus there is at present no uniformity of agreement on the divacancy levels themselves.

A recent interesting contribution to the problem of gold in silicon is that of Parrillo and Johnson (1972), who show that certain inconsistencies in the interpretation of hole and electron emission data disappear if the acceptor level for gold is taken as $E_c - 0.45$ eV at 300°K.

Some preliminary work has been done with other group I impurities in silicon: in particular sodium and potassium have been implanted by ion bombardment (Zorin et al., 1972). Potassium appears to be a donor with an activation energy of less than 0.04 eV and a solubility of about 10^{16} cm^{-3}.

2.2 Silicon with Ni, Pd, Pt, Co, Fe, Mn, Cr, and W

Electrically active nickel in silicon appears to be substitutional (Yoshida and Furusho, 1964) and interstitial nickel is electrically inactive, perhaps because of precipitation to inactive sites on cooling. The substitutional solubility is reported to be 7×10^{17} cm^{-3}. The acceptor levels observed are $E_v + 0.23$ and $E_c - 0.35$ eV (Ghandhi and Thiel, 1969; Tokumaru, 1963).

Palladium and platinum in silicon in the low 10^{16} cm^{-3} concentration have been studied by spin resonance methods and shown to occupy distorted substitutional sites in the silicon lattice (Woodbury and Ludwig, 1962).

Acceptor levels at 0.34 to 0.36 eV from the valence band have been reported for both impurities. However, there appears to be considerable uncertainty about the Pt level. Zibuts and coworkers (1964, 1967) report two donor levels in place of the 0.34-eV Pt acceptor level. An apparent Pt level observed by Carlson (Woodbury and Ludwig, 1960) at 0.45 eV above the valence band may be due to a defect pair (Ludwig and Woodbury, 1961) but it is not clearly understood; poor reproducibility of levels in platinum-doped silicon had been reported earlier. Glinchuk and Litovchenko (1965) find that the electrical activity of Pt ($E_v + 0.34$ eV acceptor level), as well as of Au, Zn, Fe, and S in silicon, is reduced by prolonged heat treatment of samples, but can be restored by suitable quenching procedures.

Carchano and Jund (1970) have found two acceptor levels for Pt, $E_v + 0.36$ and $E_c - 0.25$ eV, and a donor level at $E_v + 0.30$ eV. The solid solubility of Pt in silicon has been studied by Conti and Panchieri (1970) and by Smith and Milnes (1970) in recent work. The solubility from neutron-activation measurements is 10^{17} cm^{-3} after diffusion at 1150°C. The fraction of the platinum that is electrically active is probably high, although this is still not well-resolved information. However, further studies are in progress on Pt, since interest is developing in its possible use as a lifetime controller in silicon.

The energy levels of Co in silicon have been studied in some detail by Penchina et al. (1966) and also reported in some earlier works; Ghandhi et al., 1965; Holonyak, 1962). The two acceptor levels, 0.35 eV from E_v and 0.53 from E_c, were determined by Hall, resistivity, and photoconductivity measurements. Recent measurements by Yau and Sah (1972) give the values as 0.377 and 0.593 eV, respectively. Double-acceptor behavior might be taken as indicative of substitutional sites, as in the case of Ni.

Iron (Collins and Carlson, 1957) and manganese (Carlson, 1956) are reportedly donors; the high diffusion coefficients of Fe, Mn, Cr, and V suggest interstitial species, but other measurements (Woodbury and Ludwig, 1960a) are inconclusive as to the site of these impurity elements. Apparently these impurities go into stable substitutional positions (e.g., as Mn^+ or Mn^{2-}) if vacancies are available; otherwise, they remain mobile interstitials (Mn^- or Mn^{2+}). Limited studies (Zibuts et al., 1964, 1967) of tungsten in silicon (10^{15} cm^{-3}) indicate three acceptor levels as shown in Fig. 2.1 and other possible levels. Molybdenum apparently has three donor levels in silicon.

Chromium in silicon forms a donor level at $E_c - 0.41$ eV (Lebedev and Sultanov, 1971). The electrically active solubility was about 10^{14} cm^{-3} when the chromium was introduced by a 1200 to 1250°C diffusion.

Iron in silicon has been shown to influence Si(111) surface structures in homoepitaxial growth (Thomas and Francombe, 1970).

A great many impurity pairs have been observed (Ludwig and Woodbury, 1962) involving an interstitial donor (Fe, Mn, Cr) and a substitutional acceptor (B, Al, In). Such pairing may result in enhanced solubility of the impurities involved, and may contribute to frequency-dependent hopping conductivity (Uchinokura and Tanaka, 1967). Although most of the pairs are believed to be electrically active, only a few of the energy levels are known. This is to be expected in view of the difficulties of identifying the pairs, which may form even at room temperature (Woodbury and Ludwig, 1960; Shepherd and Turner, 1962).

Manganese tends to form clusters of four Mn atoms. Rearrangement of such Mn groups seems possible, even at room temperature, under high mechanical stress (Bullough and Newman, 1963), as indicated by long-time hysteresis effects observed in pressure experiments on manganese-doped silicon.

Other pairs involve two substitutional acceptors (Zn and B, Ga, Al) (Fuller and Morin, 1957), and pairing of Zn and P, but not As, has been suspected recently.

2.3 Relations of Solid Solubilities to Distribution Coefficients

Fischler (1962) has observed, both for silicon and germanium, that the experimental maximum molar solid solubility x_m is for most impurities about one-tenth of the distribution coefficient k° at the melting point. For impurities in silicon the expected solid solubility should be 5.2×10^{21} k° cm^{-3}. This empirical relationship is given by the straight line in Fig. 2.6, and the points represent the known values of solid solubility and distribution coefficient for specified impurities. Statz (1963) has offered some basis for this dependence from thermodynamical considerations.

The solubility shown in Fig. 2.6 for Ni is from tracer information and includes precipitated Ni. The electrical activity is reported as about three orders of magnitude lower. The indium solubility has been assigned the value at which Backenstoss (1957) inferred the onset of precipitation. However, Kozlovskaya and Rubinshtein (1962), from thermodynamical considerations, arrive at a solubility value for In of 2×10^{18} cm^{-3}, which would conform with the Fischler line.

Table 2.2 lists the reported solid solubilities of a number of impurities arranged according to their positions in rows of the periodic table. The table tends to show decreasing solubilities in the sequences Ge, Ga, Cu, Ni and Sn, In, Ag, Pd. On this basis the solubilities of Zn and Cd appear to be unexpectedly low. Although Boltaks (1963) quotes 6×10^{16} cm^{-3} as the Zn solubility, it may be noted that the Boltaks distribution coefficient for

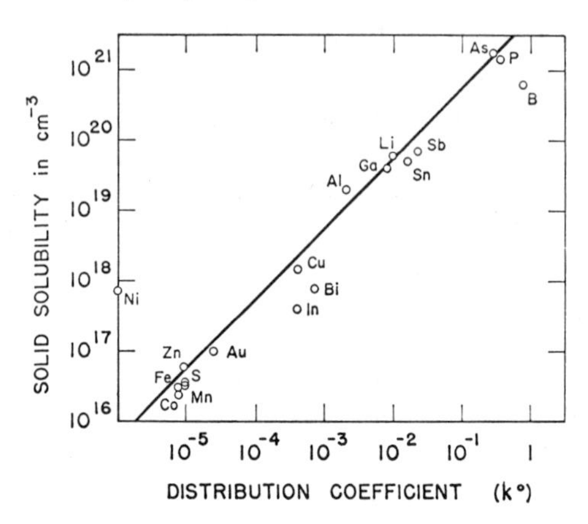

FIG. 2.6 Maximum solid solubilities of impurities in silicon versus distribution coefficient k° at the melting point. The straight line is Fischler's relation. (After Schibli and Milnes, 1967.)

Zn ($k^\circ = 4 \times 10^{-4}$) taken in conjunction with the Fischler relationship would conform with the apparent pattern of Table 2.1. However, the evidence from electrical measurements, and radiotracer studies, for 6×10^{16} cm^{-3} as the maximum solubility of Zn in Si is substantial. This would be consistent with Table 2.2 if it is accepted that Cu is dominantly interstitial whereas Ga and Ge are substitutional.

Although no distribution coefficient determination is available for Cd in silicon, one may speculate similarly that the solubility of Cd should be a few times 10^{17} cm^{-3} if it were not for the interstitial-substitutional argument.

If the distribution coefficients of impurities at the melting point of silicon or germanium are plotted as functions of their tetrahedral radii as in Fig.

TABLE 2.2

Partial Table of Measured Solid Solubilities of Impurities in Silicon (atoms cm^{-3})

Mn, Fe, Co	Ni	Cu	Zn	Ga	Ge
2×10^{16}	7×10^{17}	2×10^{18}	6×10^{16}	4×10^{19}	Isomorph
	Pd	Ag	Cd	In	Sn
	3×10^{16}	5×10^{17}	$0.1\text{–}3 \times 10^{16}$	$0.4\text{–}2 \times 10^{18}$	5×10^{19}

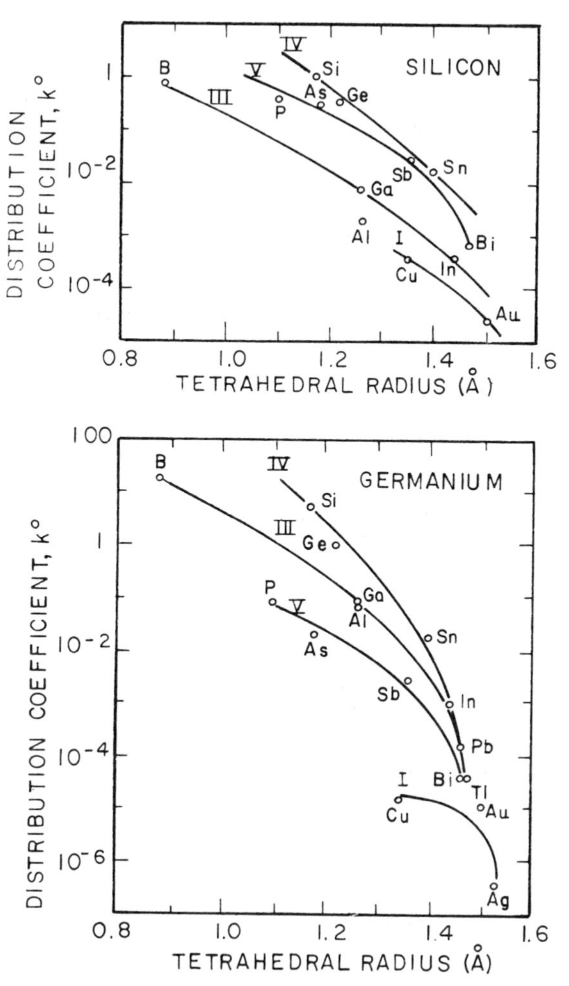

FIG. 2.7 Distribution coefficients of impurities at the melting point of silicon or germanium as a function of the tetrahedral radii and the column of the atomic table. (After Trumbore, 1960.)

2.7, there is evidence of the effect of size. Also there are indications of the role played by the column of the periodic table from which the element comes.

The solid solubilities of various impurity elements in silicon as a function of temperature are given in Fig. 2.8. Most of the deep impurity elements are seen to be low in solubility (10^{18} to 10^{16} cm^{-3}) compared with the

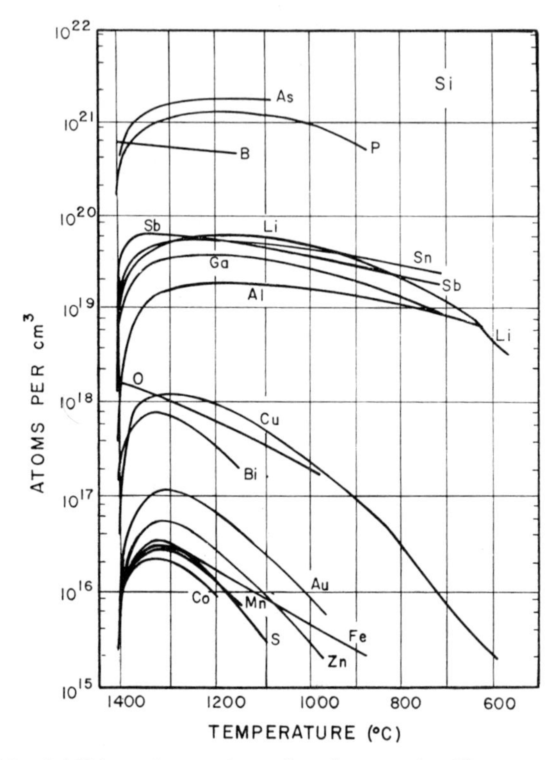

FIG. 2.8 Solid solubilities of some impurity elements in silicon versus temperature (After Trumbore, 1960.)

substitutional impurities. Impurity diffusion coefficients are shown in Fig. 2.9 where the usual deep impurity elements are seen to be fast diffusers. Their diffusion is normally interstitial which accounts for the high diffusion coefficients. Sometimes the atoms after interstitial diffusion may occupy both interstitial and substitutional sites, and the diffusion behavior and electrical activity comprise a complicated function of the lattice perfection and the diffusion conditions. Three lines are included for Au in Fig. 2.9 to remind us of this kind of problem. The data of Fig. 2.9 are helpful in providing an impression of overall diffusion trends, but the original experimental studies should be consulted in planning any use of the information. Such studies are referenced in Hannay (1959), Boltaks (1963), Burger and Donovan (1967), Bullis (1966), and in Kendall and De Vries (1969).

The resistivity of p- and n-type Si and p- and n-type Ge as a function of shallow impurity concentration is given in Fig. 2.10. These curves are experimental and of course include the effects of impurity scattering on the

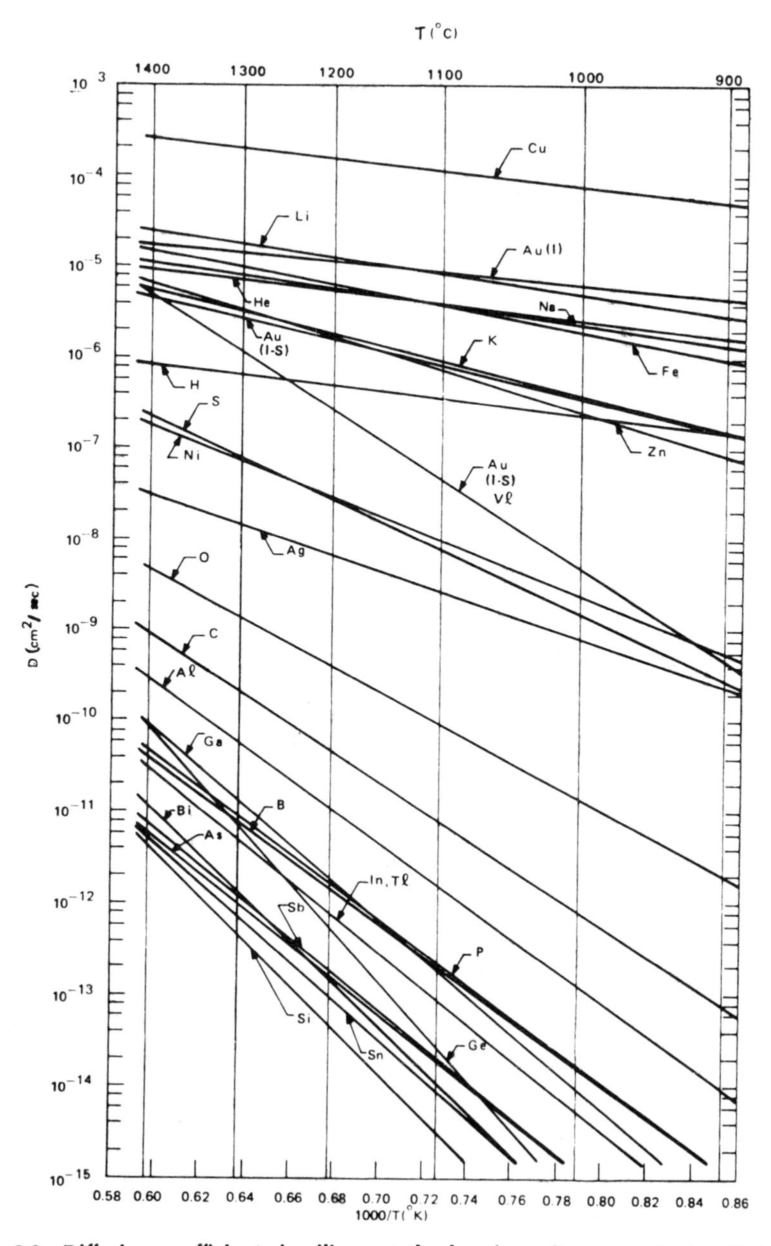

FIG. 2.9 Diffusion coefficients in silicon at the low impurity concentration limit. *For gold diffusion, the three curves denote the interstitial (I), the interstitial-substitutional (I–S) with unlimited vacancy supply through dislocations, and the vacancy-limited interstitial-substitutional process (I–S, Vl). (After Burger and Donovan, 1967.)*

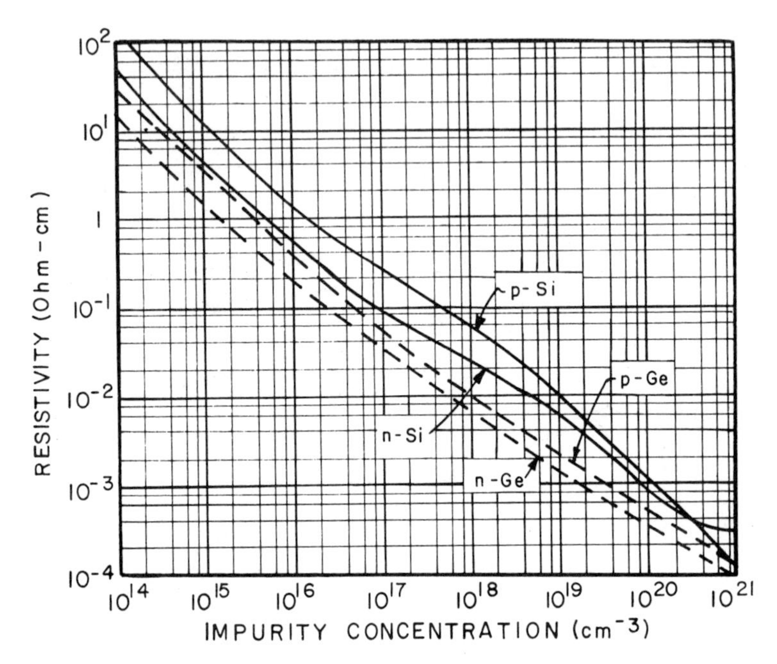

FIG. 2.10 Resistivity versus impurity concentration for Si and Ge at 300°K. (After Cuttriss, 1961, and Irvin, 1962.)

mobilities and assume that no significant compensation is present either by deep or shallow impurities. If a deep impurity is the dominant impurity present and if its degree of ionization is known, the curves of Fig. 2.10 may be used to provide first order-estimates of the resistivity. However, this is only justifiable if there is an absence of specific information on the effective mobilities observed with deep impurity doping. The scattering of carriers from Zn impurity sites in germanium has been studied by Tyler and Woodbury (1959). The relative magnitude of hole mobility in zinc-doped and gallium-doped germanium, at 300°K, indicates that the scattering cross section for doubly charged Zn sites is four times that for Ga sites.

2.4 Silicon Containing Group VI Impurities O, S, Se, and Te

Oxygen is present in large quantities in silicon, the amount depending on the crystal growth method. Ingots pulled with fast rotation may contain about 10^{18} cm^{-3} of oxygen from the silica crucible, while slow rotation gives less

(Kaiser et al., 1956, 1957). "Oxygen-free" ingots are obtained through float zoning, and repeated passes under vacuum reduce the oxygen content to less than 10^{16} cm^{-3}.

In pulled ingots, most oxygen is dissolved interstitially and forms the electrically inactive Si_2O group, although some oxygen is also present in higher complexes (Matukura, 1959). The interstitial oxygen is fairly mobile above 300°C and forms groups with an increasing number of oxygen atoms. Thermal donors (probably the SiO_4 complex) appear from 300 to 500°C, but disappear at higher temperatures. At 1000°C, silicon oxide particles of 0.1 to 1 μm in size have been observed in pulled ingots, and these grow within a few hours to disk-shaped precipitates of 10 to 40 μm diameter at 1200°C if the original oxygen content is sufficient (Furusho, 1964). Vacancy-oxygen complexes have been studied by de Kock (1971).

Infrared absorption is a major tool for investigation of oxygen in silicon (Gross et al., 1972). The Si_2O group has a strong absorption around 9 μm, due to stretching vibrations of the Si–O bond. Its intensity serves to determine the oxygen content. (The silicon lattice itself has only a small absorption peak at 9 μm.) At various stages during the heat treatment, absorptions at other than the 9-μm band appear and disappear again, due to the formation of intermediate silicon–oxygen complexes (Corbett et al., 1964). The 9-μm peak resolves into seven individual peaks (Pajot, 1967), and 30 further absorption peaks have been resolved in the range from 20 to 30 μm. Many of these investigations were prior to 1958 and have been summarized by Kaiser et al. (1958).

The question of energy levels of oxygen in silicon, however, is not quite solved yet. A donor level at $E_c - 0.16$ eV is found consistently, and is due to the above-mentioned SiO_4 thermal complex. An acceptor level at $E_c - 0.17$ eV is due to the *A* center, which consists of an oxygen atom trapped at a radiation-induced vacancy and could in principle be viewed as substitutional oxygen. Other levels have been found at $E_c - 0.38$ eV (acceptor), $E_c - 0.31$ eV (donor), and $E_v + 0.35$ eV (acceptor). Trapping by oxygen complexes has been reported at the $E_c - 0.38$ eV acceptor level and for the $E_c - 0.16$ eV thermal (SiO_4) donor level (Mordkovich, 1963, 1964, 1965). The observed $E_v + 0.35$ eV acceptor level acts as an efficient recombination center.

This multiplicity of energy levels is not surprising because oxygen interacts with vacancies, radiation defects, dislocations, with deep impurities like Cu or Ni in the presence of vacancies, with shallow acceptors (B, Al, Ga) to form shallow donors around 600°C, and with Li (Fuller and Doleiden, 1958; Fuller and Wolfstirn, 1965; Gilmer et al., 1965). Oxygen reacts with Al and leads to precipitation of both (Bullough and Newman, 1963). No mention of an indium–oxygen complex has been found.

The diffusion coefficient of oxygen in silicon has been calculated from diffusion experiments, from internal friction, and from stress relaxation along a 111 axis. The results agree surprisingly well and indicate a diffusion coefficient $D = 0.23 \exp(-2.561 \pm 0.005 \text{ eV}/kT)$; that is, oxygen diffuses about ten times as fast as Al (Fig. 2.9) but much slower than other interstitial diffusants (Corbett et al., 1964).

The information on sulfur in silicon is less complete than that on oxygen in silicon. Sulfur can be introduced from hydrogen sulfide gas during the growth of the crystal or by diffusion. In the latter case, erosion of the silicon surface may occur, involving probably SiS or SiS_2. The solubility is much lower (3×10^{16} cm^{-3}) than that of oxygen, but the diffusion coefficient is two orders of magnitude higher (Carlson et al., 1959). Most of the sulfur seems to be electrically active, depending on heat treatment however (like other retrograde dopants), with donor levels observed at $E_c - 0.18$ and $E_c - 0.52$ eV. A previously reported second level at $E_c - 0.37$ eV seems to be associated with sulfur pairs. The lattice position of sulfur is not known; indications other than the fast diffusion are for a substitutional site. Three different sulfur–iron pairs were also observed in spin resonance measurements. Recent infrared absorption experiments on sulfur-doped silicon under uniaxial stress indicate two centers, at $E_c - 0.61$ and $E_c - 0.37$ eV, associated with singly ionized S, and two centers, at $E_c - 0.19$ and $E_c - 0.11$ eV, involving neutral S (Krag et al., 1966). Many more shallow excited levels due to these four S centers are observed near the conduction band. However, no definitive model for the configurations of S in silicon has been established. Diodes doped with S have been studied by Lebedev and coworkers, 1971.

Sulfur, selenium and tellurium have been tried as transport media in the vapor growth of silicon single crystals (Fischler, 1962a). While all three elements are reasonably efficient in the transport of germanium (Ignatkov and Kosenko, 1961), only Te works well with silicon. Such vapor transport is the only practical way to date to obtain tellurium-doped silicon. Tellurium has one donor level at $E_c - 0.14$ eV, and a second donor level might be expected deeper in the energy gap, in analogy with the behavior of group VI elements in germanium. However, this simple reasoning is weakened in view of the complexity of the more recent determinations of the S levels in silicon. Tellurium has an electrically active solubility of at least 3×10^{16} cm^{-3}. Recently, the Si–Te system has been studied in more detail, in particular with respect to $SiTe_2$ and Si_2Te_3 semiconductors (Bailey, 1966; Peart, 1966).

Selenium may be expected to have properties between those of S and Te. Early observations indicating a very low electrically active solubility of 10^{15} cm^{-3} of selenium-diffused silicon probably require reexamination since similar early determinations have been revised upward for other dopants (Co, Ni, Te, Ag). Diffusion of Se into silicon is expected to be difficult because of possible transport effects, which are very weak for oxygen, noticeable for sulfur, and quite strong for tellurium.

2.5 Silicon Containing Group IV Impurities: C, Ge, and Sn

These impurities have high solubilities in silicon, indeed Ge forms continuous solid solutions with Si. No electrical activity has been reported so far [except for a statement by Peart (1966) that Ge may be a donor in silicon and the surface capacitance levels of Table 2.1]. Their diffusion coefficients are rather small (Table 2.3).

TABLE 2.3
Solid Solubilities and Diffusion Coefficients of Group IV Impurities in Silicon

	Diffusion Coefficient ($cm^2\ sec^{-1}$)	Temperature (°C)	Solubility ($g\ cm^{-3}$)
C	$0.33 \exp[-(2.95 \pm 0.25\ eV)/kT]$	1070–1400	4×10^{18}
Si	$9000 \exp(-5\ eV/kT)$	1100–1300	—
Ge	$6 \times 10^5 \exp(-5.3\ eV/kT)$	1150–1350	Isomorph
Sn	$1.2–2.5 \times 10^{-13}$	1217	5×10^{19}

Carbon probably occupies substitutional sites in silicon, as suggested by the activation energy of 3 eV for diffusion (Bullough and Newman, 1963; Newman and Wakefield, 1961). Vibrational absorptions at 16.6 μm due to the C–Si bond, and near 9 μm due to the C–O bond, have been observed (Newman and Willis, 1965). Another absorption peak at 12.2 μm in polycrystalline samples could be due to silicon carbide precipitates (Balkanski et al., 1965). Carbon is believed to precipitate with Al and O. Oxidation rates of silicon at low oxygen pressures (less than 0.1 mmHg) depend critically on C contamination, which forms transparent, nonoxidizing films of SiC (Gulbransen et al., 1966; Henderson, 1972; Assour, 1972). The activation energy of the self-diffusion coefficient of silicon has been estimated as 4.75, 4.9, and 3.94 $\pm$ 0.33 eV, in reasonable agreement with values of 4.78 and 5.51 $\pm$ 0.06 eV obtained from tracer studies (Masters and Fairfield, 1966; Peart, 1966). Conventional measurements and implanted surface capacitance measurements indicate that C is electrically inactive in silicon.

The diffusion coefficient of Ge in silicon is, from Fig. 2.9, a little greater than that of Sn and Si in silicon. Interdiffusion of thin (10^{-6} cm) evaporated layers of Ge and Si seems to indicate a much faster diffusion (Pines and Grebennik, 1963), but this might be caused by the high defect densities of these films. An infrared absorption peak due to vibrations of the Ge–Si bond has been seen at 19.7 μm in germanium-doped silicon, but occurs at 46.7 μm in the germanium-rich alloy (Braunstein, 1963). Investigations of the phonon spectra indicate that Ge atoms in the silicon lattice do not form clusters, but rather occupy next nearest sites in the host lattice (Logan et al., 1964).

2.6 Silicon with Other Impurities: Li, N, Rare Earths, and Radiation-Induced Levels

Lithium has a high solubility in silicon (Trumbore, 1960) and is a shallow (0.033 eV) interstitial donor. It has a high diffusion coefficient, $D = 0.0025 \exp(-0.65 \text{ eV}/kT)$, from 25 to 1350°C as determined by several methods (Pratt and Friedman, 1966). Lithium drift-compensated material may have high resistivity, as needed for nuclear radiation detectors or for double-injection diodes. Lithium drift continues near room temperature, particularly in the presence of the high fields in *pn* junctions (Antonov, 1966). Radiation damage, however, reduces the drift mobility of Li, and in certain circumstances interactions of Li with radiation defects may reduce the degradation of silicon solar cells (Wysocki, 1965 and Berman, 1972). Five infrared absorptions from 15 to 20 μm are due to B–Li pairs (Waldner et al., 1965). Prolonged annealing at higher temperatures (16 h at 850°C) has led to precipitation of Li, thus increasing the absorption at 10 μm (Spitzer and Waldner, 1965). Gallium–lithium and Al–Li pairs also cause absorptions near 20 μm. Disk-shaped Li precipitates form preferentially at lattice defects (Weltzin et al., 1965).

Nitrogen is difficult to introduce into silicon because of its low segregation coefficient (10^{-7}) and the tendency to form silicon nitride upon cooling. Silicon nitride films, like oxide films, mask selectively during diffusion or they serve as dielectrics in metal-nitride-semiconductor type structures. Nitrogen may also be introduced by ion bombardment and subsequent annealing. In recent studies, it was concluded that nitrogen is substitutional in silicon with an energy level 0.14 eV below the conduction band (Kleinfelder, 1967). Nitrogen in the silicon lattice causes infrared absorption at 10.6 μm (Balkanski et al., 1965).

Argon has been detected in silicon by neutron activation methods after being introduced by bombarding processes (Comas and Wolicki, 1970).

Diffusion of Sm in either silicon (at 1240°C) or germanium (at 930°C) was unsuccessful (Gusev et al., 1964). Two rare earth elements, Nd and Th, have been introduced by ion bombardment into silicon (Gibbons and Moll, 1965). Deep donor levels, possibly due to Nd and Th, were observed at 0.33 ± 0.07 and 0.29 eV, respectively, below the conduction band. Diffusion of rare earth elements in germanium is further discussed in Section 2.12.

2.6.1 Radiation-Induced Levels in Silicon

Incident particles of sufficient energy can displace silicon atoms from lattice positions, thus creating interstitial-vacancy complexes. Such a damaged site, for example, the *A* center, is capable of capturing free carriers. At not too high radiation dosages discrete levels appear in the energy gap, some of which will anneal at elevated temperatures. As with other vacancies or dislocations, many impurities will aggregate at these damage sites. The phenomena are therefore very involved and several reviews have been published recently (Vavilov, 1965; Corbett, 1966; Hasiguti, 1968; Corbett and Watkins, 1971).

2.7 Doping Impurities in Germanium

The group III elements B, Al, Ga, In, and Te and the group V elements P, As, and Sb are shallow impurities in germanium. The levels are in the range 0.0096 to 0.013 eV and are considerably shallower than the levels of the same impurities in silicon. The energy levels of shallow and deep impurities in germanium are shown in Fig. 2.11. There is some tendency for deep impurities to exhibit more energy levels in germanium than in silicon. This is true of Cu, Ag, and Au for example. However, in general the pattern in Fig. 2.11 is similar to that in Fig. 2.1.

The solid solubility curves for various impurity elements in germanium as functions of temperature are given in Fig. 2.12. Comparison with similar curves for silicon (Fig. 2.8) shows that solubilities tend to be less in germanium than in silicon, both for shallow and deep impurities, with the exception of Zn. Bismuth in germanium has been studied by Kozlovskaya and Rubinshtein (1962) and Korol'kov and Romanenko (1964).

Impurity diffusion coefficients in germanium as functions of temperature are shown in Fig. 2.13. Most of the deep impurities are faster diffusers than the shallow impurities by many orders of magnitude. However, it is seen that both Au and Zn are much slower diffusers in germanium than in silicon, presumably because substitutional action is present in germanium to a

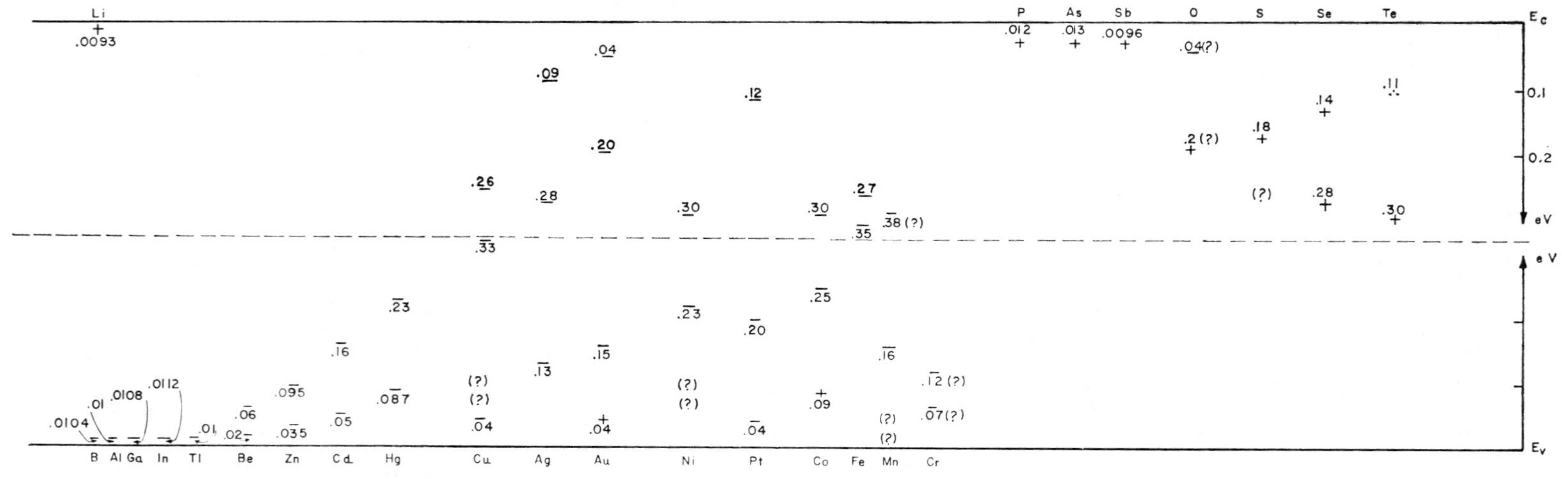

FIG. 2.11 Deep energy levels in germanium.

The energy values are measured from the nearest band edge. The symbols + and − indicate donor and acceptor levels, respectively. Question marks indicate uncertain energy levels.

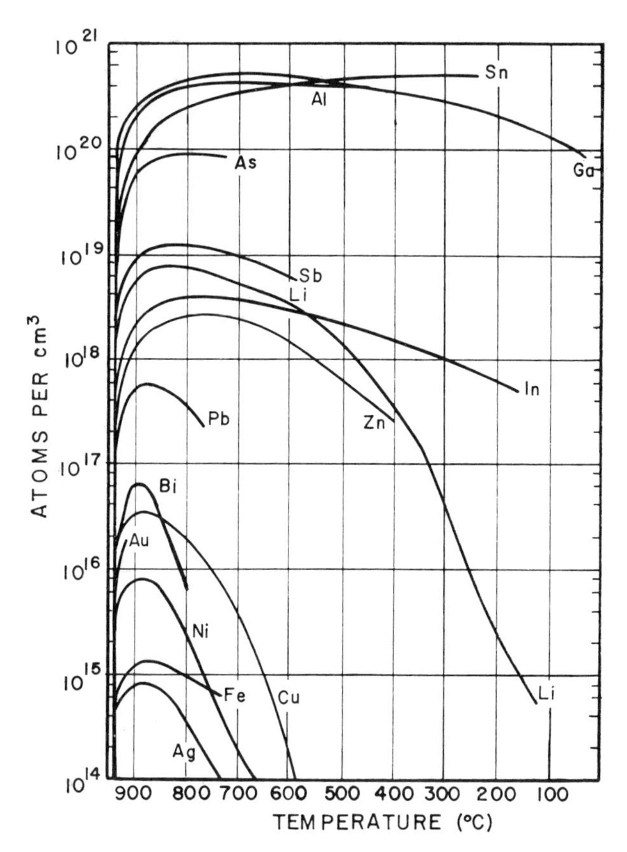

FIG. 2.12 Solid solubilities of various impurity elements in germanium versus temperature. [After Trumbore, 1960, and *J. Electrochem. Soc.*, **109, 734 (1962).]**

greater extent than in silicon. The diffusion of Tl in germanium, which is not shown in Fig. 2.13, has been studied by Ibragimov et al. (1963).

Deep impurity levels in germanium have been reviewed in papers by Burton (1954), Dunlap (1956), Tyler (1956), Newman and Tyler (1959), and others. The Tyler papers are still important sources of information and of references to studies in the 1950s. During the 1960s there has been continued work with germanium, including a considerable amount from Russian sources. In the sections that follow in this chapter the work of the last decade is added to the previous information to provide a more complete overview of the behavior of deep impurities in germanium.

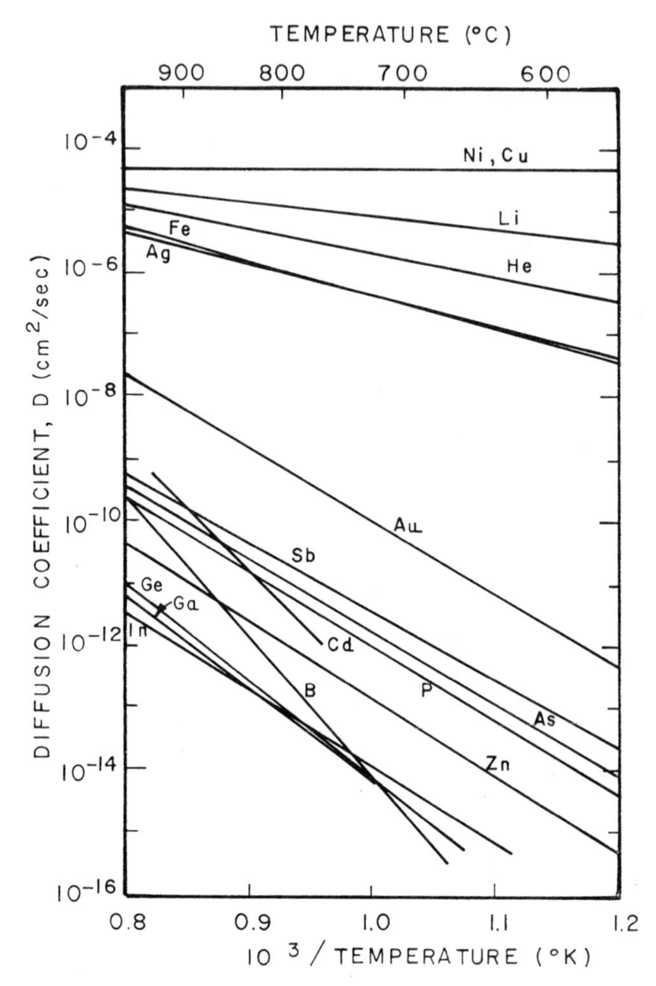

FIG. 2.13 Impurity diffusion coefficients in germanium. (After Sze, 1969.)

2.8 Germanium with Group II Impurities: Be, Zn, Cd, and Hg

The solubility of Be in germanium is high, about 1×10^{19} cm^{-3}, and it introduces two acceptor levels at 0.02 and 0.06 eV from the valence band edge (Penin et al., 1969; Besfamil'naya and Ostroborodova, 1969a). The diffusion coefficient of Be in germanium is about 1.5×10^{-11} cm^2 sec^{-1} at 900°C and the activation energy for diffusion is about 2.5 eV (Belyaev and

Zhidkov, 1961). The 0.06-eV acceptor level of Be decreases with compensation doping (Tyapkina and Vavilov, 1965).

As the periodic table sequence of group II elements Be, Zn, Cd, and Hg is followed, there is a progressive increase in acceptor level depth. The double-acceptor pattern is retained, as would be expected for group II impurities in a group IV semiconductor. The energy levels are Zn: 0.035, 0.095; Cd: 0.05, 0.16; and Hg: 0.087, 0.23 eV. The solubility of Zn is 7×10^{18} cm^{-3}. The solubility of Cd has been reported as 6×10^{14} cm^{-3} by Woodbury and Tyler (1956). This is compatible with the studies of L'vova (1964) where 3×10^{14} cm^{-3} was the electrical activity obtained. On the other hand, Kosenko (1960) reports a solubility of 2×10^{18} cm^{-3} from tracer studies with Cd^{115}. A diffusion coefficient of 2×10^{-10} cm^2 sec^{-1} was observed and a vacancy mechanism was considered to be involved. The solubility of Hg in germanium is uncertain; 5×10^{14} cm^{-3} seems to be the upper limit of electrical activity that is readily obtained (Blinov, 1965).

Doping with Zn and Cd has been studied primarily with respect to germanium detectors of infrared radiation (Newman and Tyler, 1959; Picus, 1961; Morozov and Kalashnikov, 1962; Sidorov, 1964). Mercury-doped germanium detectors of 10.6-μm laser radiation have been examined by Picus and Buczek (1968).

2.9 Germanium Doped with Li, Cu, Ag, Au, and Other Group I Elements

Hydrogen is an electrically inactive impurity in bulk germanium and presumably is interstitial in the form of neutral atoms. Even though hydrogen in the bulk is not active, the presence of atomic hydrogen on germanium surfaces results in fast recombination surface states (Sedgwick, 1968; Romanov et al., 1969). There are of course a number of instances of adsorbed gas affecting surface states. Water adsorbed on the surface of germanium produces recombination centers. These can be reduced by the presence of ozone (Novototskii–Vlasov, 1965).

Interstitial Li is a shallow donor in germanium (0.0093 eV) (Kurova and Tyapkina, 1961) and also in silicon (0.033 eV). The fact that Li ionizes and interstitial H does not has been found to be in accord with a quantum-mechanical study. The diffusion coefficient of Li is very fast in both germanium and silicon, as can be seen from Figs. 2.13 and 2.9. Because of its high mobility at moderate temperatures, and under electric field conditions, Li is a very suitable impurity for producing compensation of acceptor impurities. This allows high-resistance Ge to be obtained for wide-depletion region high-energy particle detectors. Such detectors are not considered further here since the subject has been reviewed at conferences on their applications and books are available such as those of Taylor (1963) and Vavilov (1965). Recent work on Ge crystal growth has shown that very high purity Ge can now be obtained and this may reduce the need for lithium-drifted structures.

Copper is an impurity that has been examined extensively in germanium. It produces three acceptor levels, Cu^- at $E_v + 0.04$ eV, Cu^{2-} at $E_v + 0.33$ eV, and Cu^{3-} at $E_c - 0.26$ eV. The diffusion coefficient is very high, 2.8×10^{-5} cm^2 sec^{-1} (500°C), and the maximum solubility is about 3×10^{16} cm^{-3}. The behavior of copper as an impurity in germanium is complicated by the fact that both the substitutional (Cu^s) and interstitial (Cu^i) species must be considered and so must pairing actions between Cu^s and donors. The classic papers on interactions among impurities in germanium and silicon include those of Reiss et al. (1956), and Hall and Racette (1964). The latter paper shows that the ratio of substitutional to interstitial copper in intrinsic germanium at 700°C is 6 : 1. The solubility of interstitial copper is enhanced by the presence of a p-type impurity such as Ga. This is illustrated in Fig. 2.14. Williams (1969) has shown that such an effect occurring in and near a gallium-diffused layer in germanium can result in a $p^+ - \pi - p$ structure which is usually undesirable. Copper can be gettered from germanium by elements such as gallium, iron, rhodium (Tissen, 1959) and lead (Kalashnikov and Mednikov, 1961). Copper can be used to decorate dislocation lines in germanium and silicon which can then be made visible by infrared microscopy.

Copper-doped germanium is important for infrared detectors. These operate in the 8 to 12 μm region when cooled to 15°K or below (Teich et al., 1966; Quist, 1968).

Silver in germanium is a triple acceptor with levels at $E_v + 0.13$, $E_c - 0.28$, and $E_c - 0.09$ eV (Tyler, 1959). The solubility is 9×10^{14} cm^{-3} and the diffusion coefficient has fast and slow components (Kosenko, 1962). The diffusion of silver at the surface of germanium has been studied by Kosenko and Khomenko (1962).

Gold in germanium has a very deep donor level at $E_v + 0.04$ eV and three acceptor levels at $E_v + 0.15$, $E_c - 0.20$, and $E_c - 0.04$ eV. Gold, with an electronic structure of $5s^2 5p^6 5d^{10} 6s^1$, is mainly substitutional in the germanium with the $6s$ electron behaving as the 0.04-eV donor. This implies that the strength of bonding of this electron to an Au–Ge pair is less by 0.04 eV than the strength of bonding of a valence electron to a Ge–Ge pair. The three acceptor states represent the three unsaturated Au–Ge bonds. The first acceptor state, $E_v + 0.15$ eV, is the energy required to remove an electron from a nearby Ge–Ge bond and place it on a neutral gold atom,

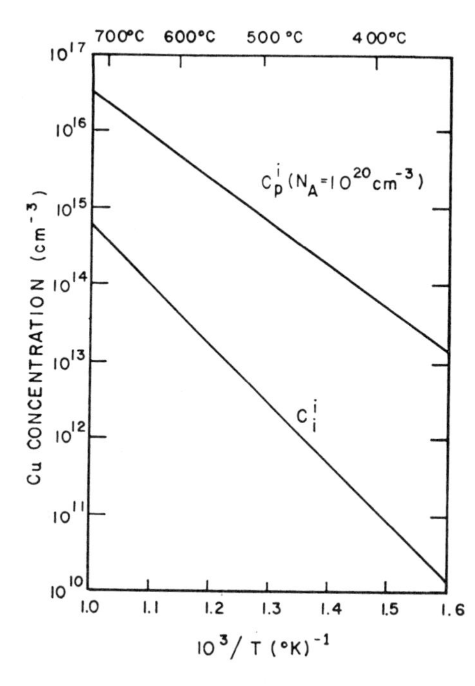

FIG. 2.14 Solubility of interstitial copper in germanium.

The solubility of interstitial copper is much higher in germanium containing a p-type impurity at a concentration of 10^{20} *cm*$^{-3}$. *The solubility of copper in both pure and doped material is a strong function of temperature. (After Williams, 1969.)*

thereby creating a hole. The next acceptor state appears only by partial compensation of the material by donor impurities which must be present in sufficient density to fill all the gold $E_v + 0.15$ eV states with electrons and to begin the filling of the $E_c - 0.20$ eV second acceptor level Au^{2-}. Similarly the high-lying acceptor level $E_c - 0.04$ eV appears only when the two acceptor levels have been filled.

The energy levels of Au in germanium therefore conform in a simple way with our expectations from the electronic structure. This is in contrast to Au in silicon where only the donor level ($E_v + 0.35$ eV) and the first acceptor level ($E_c - 0.54$ eV) are seen. The higher acceptor levels in silicon are presumably lost in the conduction band.

The solubility of Au in germanium is 3×10^{16} cm^{-3}, which is less than the 10^{17} cm^{-3} value found in silicon. The diffusion coefficient is 1.5×10^{-9} cm^2 sec^{-1} at 850°C. This is faster than for the shallow impurities, which tend to be in the range 10^{-10} to 10^{-12} cm^2 sec^{-1}. However, Au is not as fast a diffuser in germanium as it is in silicon (about 10^{-6} cm^2 sec^{-1}), presumably because substitutional action is more controlling in germanium.

The response of gold-doped germanium to infrared radiation has been studied by Dunlap (1958), Johnson and Levinstein (1960), and others.

2.10 Germanium Doped with Ni, Pt, Co, Fe, Mn, and Cr

Nickel produces two acceptor levels in germanium, at $E_v + 0.23$ eV and at $E_c - 0.30$ eV. The maximum solid solubility is about 8×10^{15} cm^{-3} and it is a fast diffuser, 5×10^{-5} cm^2 sec^{-1}, indicating interstitial behavior. Acceptor levels at $E_v + 0.07$ eV and at $E_v + 0.10$ eV form when supersaturated solutions of Ni or Cu in arsenic-doped germanium are annealed in the temperature range 160 to 400°C or slow quenched. The 0.07-eV acceptor is observed after a short anneal of nickel-doped specimens or after a slow quench of copper-doped specimens. The 0.10-eV acceptor is found after fast quench in the case of Cu or after long anneals in both the Ni and Cu specimens. The increase in hole mobility during annealing indicates that similar vacancy-donor associates are produced in both kinds of specimens. The constitutions of the acceptors are not known with certainty (Fuller and Wolfstirn, 1966a). The behavior of Ni in germanium has also been studied by Ostroborodova and Ivanova (1965) and they find the lower acceptor level to be at $E_v + 0.20$ eV and not at 0.23 eV.

The piezoresistance of nickel-doped p-germanium has been studied by Traum et al. (1969), and the π coefficient is found to be almost temperature independent in the range 160 to 300°K. This unusual behavior is interpreted in terms of a stress-dependent deionization of the lower acceptor level (taken as $E_v + 0.22$ eV).

Newman and Tyler (1959) report that Pt introduces acceptor levels at 0.04 and 0.20 eV from E_v and at 0.12 eV from E_c. However, the 0.04-eV level occurs in appreciably higher concentrations than the other levels and interpretation therefore is not simple.

Cobalt introduces a deep donor at about $E_v + 0.09$ eV and two acceptor levels, at $E_v + 0.25$ and $E_c - 0.30$ eV. The maximum solubility observed is about 2×10^{15} cm^3 (Tyler, 1959). The activation energy of the cobalt donor has been reported as 0.081 eV by Schultz et al. (1962), and recently as 0.083 ± 0.003 eV by Barnik et al. (1971) in work that also determined photon and hole-capture cross sections.

Iron creates two acceptor levels, at $E_v + 0.25$ and $E_c - 0.30$ eV, and its solubility is approximately 1.5×10^{15} cm^3. The diffusion coefficient is high, 8×10^{-6} cm^2 sec^{-1}. Bistable space-charge injection is readily obtained in high-resistance iron-doped germanium (Tyler and Newman, 1959).

Manganese in germanium produces acceptor levels at $E_v + 0.16$ eV (Tyler, 1959) and at $E_c - 0.37$ eV (Newman and Tyler, 1959). These authors comment that acceptor levels are also observed at 0.05 eV and at about 0.01 eV in concentrations equal to about 20 to 30% of that of the deeper levels. They present also the photoconductivity of manganese-doped germanium and compare it with Fe, Co, and Ni doping. Relatively little has been done with Cr doping of germanium but the levels reported by the same authors are $E_v + 0.07$ and $E_v + 0.12$ eV.

2.11 Germanium Doped with Group VI Elements: O, S, Se, and Te

Oxygen is quite soluble in germanium with densities of about 4×10^{17} atoms cm^{-3} reported (Whan and Stein, 1963; Adachi, 1967). Oxygen-vacancy complexes tend to occur. Donor levels at $E_c - 0.04$ and $E_c - 0.20$ eV cause trapping actions but are not well understood.

The properties of Te in germanium have been studied by Tyler (1959), Glinchuk et al. (1963), and others. It is a double donor with levels at $E_c - 0.11$ and $E_c - 0.30$ eV. The maximum solubility has been reported to be as high as 6×10^{18} cm^{-3} by radioactive isotope studies (Ignatkov and Kosenko, 1962). The diffusion behavior is quite complex with slow and fast components: the fast component is 5×10^{-7} $cm^2\ sec^{-1}$ at 800°C. Tellurium is an effective transport agent for germanium (Ignatkov and Kosenko, 1961).

For S and Se the donor levels nearest the conduction band edge are at 0.18 and 0.14 eV, respectively. The deeper donor levels are less well established. Tyler (1959) reports that the 0.28-eV level for Se is present in concentrations which are several times greater than that of the 0.14-eV level in both diffused and melt-doped crystals. It may be that S and Se can be present in both substitutional and interstitial forms that are electrically active.

2.12 Other Impurities in Germanium

The solubility of Sn in germanium is high, approximately 5×10^{20} atoms cm^{-3}. Assuming that less energy is required to take an electron from a Sn–Ge bond than from a Ge–Ge bond, it might be expected that Sn would introduce a donor level near the valence band which would show up in Ge doped with comparable concentrations of both Ga and Sn. Such a level was not found by Tyler (1959) for either Sn or Pb doping. It is interesting to note that strong radiative recombination with approximately 0.50 eV energy has been observed at 77°K in germanium containing Sn. Hergenrother and Feldman (1969) propose that this radiation is due to excitons bound to an isoelectronic Sn center. Similar emission has been seen from Pb-doped germanium (Feldman and Hergenrother, 1971).

The diffusion of rare earth elements (Ce, Nd, Tb, Lu, Yb) in germanium has been studied by Gusev et al. (1964). The values obtained were approximately 5×10^{-13} cm^2 sec^{-1} at 910°C with surface concentrations in the 10^{19} to 10^{21} cm^3 range. Electrical activity was apparently not examined. The magnetoresistance of germanium doped with neodymium and europium has been studied by Lashkarev et al. (1972).

Grain boundary states in germanium are acceptor-like according to Matukura (1963). They have also been studied recently by Mataré and Laakso (1969). Recombination at dislocation lines in germanium has been examined by Kolesnik (1962) and Gippius and Vavilov (1963), and activation energies at $E_c - 0.22$ and $E_c - 0.14$ eV are reported.

Radiation effects are regarded as outside the scope of the present book. The complex nature of radiation damage effects is discussed in the books of Vavilov (1965); Corbett (1966); Hasiguti (1968); and Corbett and Watkins (1971). Defect electronics in semiconductors have been treated recently by Wertheim et al. (1971) and by Mataré.*

The effects of pressure on the energy levels of impurities in germanium and silicon have been studied by Holland and Paul (1962). The coefficient of separation from the conduction band edge is normally greater than that from the valence band edge. The effect of hydrostatic pressure on the properties of Cu in germanium has been studied recently by Kanda et al. (1972).

* H. F. Mataré, *Defect Electronics in Semiconductors*, Wiley, New York, (1971).

CHAPTER THREE

Deep Impurity Levels in Gallium Arsenide and Other III–V Compounds

The techniques of preparation of single-crystal gallium arsenide, gallium phosphide, and some other III–V compound semiconductors have been improved greatly in recent years. Good-quality crystals are now available with reasonable uniformity of properties such as doping density and mobility. In the course of these developments, doping has been studied for a wide spectrum of impurities. Most of these impurities produce energy levels deep in the energy gap. In some instances impurity pairing, or interaction with native defects, occurs and produces unexpected additions to the energy level structure.

Epitaxial growth processes with GaAs are becoming increasingly important for creating device structures that cannot be achieved by conventional diffusion and alloying processes. Some of these processes involve the possibility of undesired doping by deep impurities or cause levels associated with native defects. Deep impurity doping is also sometimes deliberately attempted to obtain high-resistance semiinsulating layers.

Gallium arsenide and gallium phosphide structures and devices after fabrication usually show trapping effects, caused by deep levels and defect levels. They are more prone to this than are comparable silicon or germanium devices. Transistors and diodes and light-emitting devices of GaAs all show trapping effects in the response times and often characteristics that loop when retraced. Such effects can become dominant at temperatures greatly below

300°K. The problems are more severe in GaAs than in Si or Ge devices because it is a wider energy gap material and a compound semiconductor.

The first part of this chapter summarizes known impurity levels in GaAs. The detail is sufficient to lead directly to the relevant literature when needed. Doping in GaP and InP is then discussed and the final section reviews the rather incomplete information available on doping effects in InSb, InAs, GaSb, and AlSb.

3.1 GaAs with Impurities of Group I: Li, Na, Cu, Ag, and Au

Lithium

The solubility of lithium at 800°C is 1.6×10^{19} cm^{-3} in undoped GaAs and about 2.3×10^{19} in GaAs doped heavily with Te or Zn. Lithium diffuses rapidly into GaAs with an interstitial diffusion coefficient $D = 0.53 \exp(-1.0/kT)$, where kT is in electron volts (Fuller and Wolfstirn, 1962).

When nominally pure crystals of GaAs are saturated with lithium by diffusion at temperatures greater than 500°C and cooled to room temperature, they are compensated to a high p-type resistivity (greater than 10 Ω-cm). The compensation phenomenon is caused by the action of Li^+ which permeates the crystal and tends to make it n type. As a result a large increase in gallium vacancies occurs according to the law of mass action relation

$$V^- = V_i^- \frac{(Li^+)}{(n_i)} \tag{3.1}$$

where V_i^- is the concentration of vacancies (assumed singly charged) in undoped GaAs, and n_i is the intrinsic electron concentration at the temperature considered. The presence of the Li^+ interstitials increases the gallium vacancy concentration by orders of magnitude. The possible reactions that follow result in the species Li^+V^-, Li^-, Li^{2-}, Li^+Li^{2-}, and $(Li^+)_2Li^{2-}$, where Li^+ is interstitial and Li^- and Li^{2-} are substituted ions. On cooling to room temperature after a high-temperature lithium diffusion, excess Li^+ readily precipitates because of its high mobility. This leaves relatively immobile excess acceptors which give rise to the observed p-type compensation. The activation energy observed is 0.023 eV from the valence band edge (Fig. 3.1).

Lithium is also found to compensate n-type GaAs that has tellurium concentrations as high as 5×10^{18} cm^{-3}.

Chemical interaction among defects can result in ion-pair formation in semiconductors. The evidence for pairing is mostly indirect. However, when

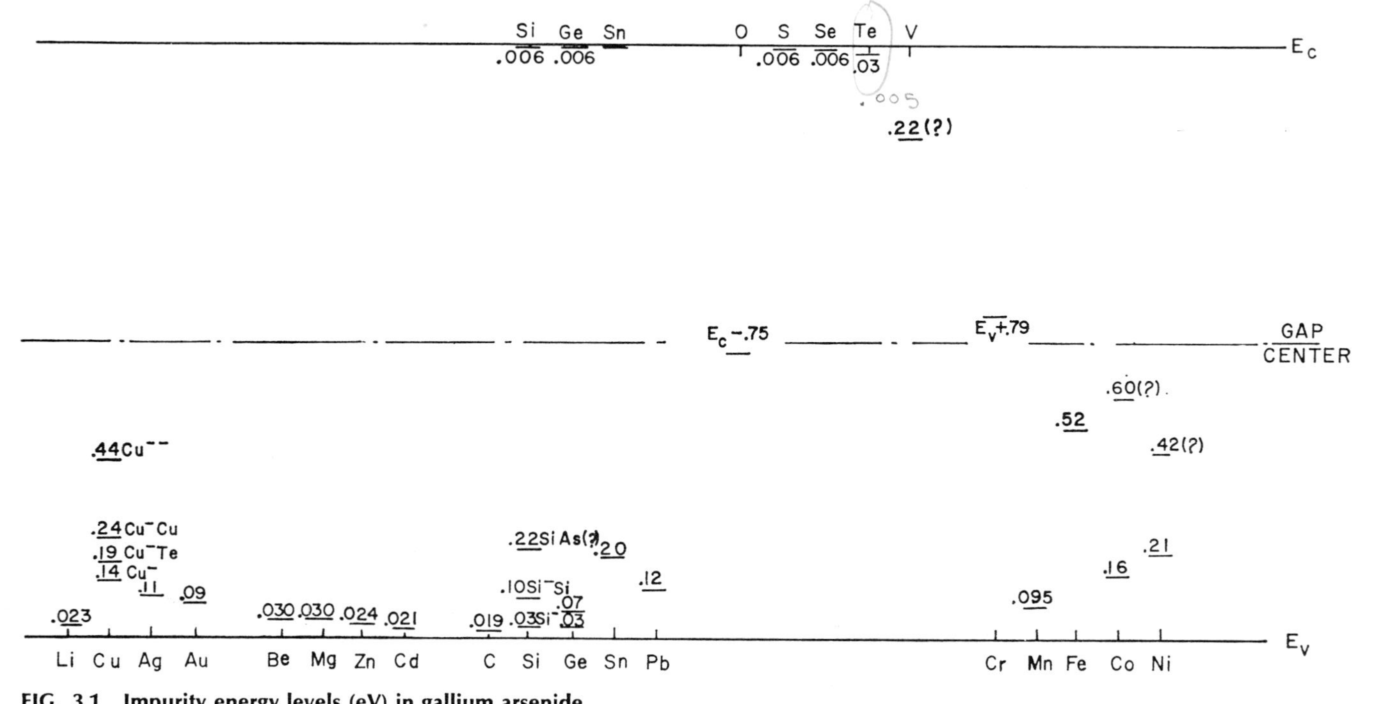

FIG. 3.1 Impurity energy levels (eV) in gallium arsenide.
The oxygen level, E_c − .75eV, and the vanadium level are donors.

ion pairs have an infrared-active local vibrational mode, infrared spectroscopy provides a sensitive tool for studying the pairing reactions and pair structure. Pairing is recognized by the reduction of site symmetry and the consequent lifting of degeneracies of vibrational modes. Local mode frequencies of unpaired impurities must be known in order to recognize site symmetry reduction. If more than one stable isotope is available for one of the impurity species, the change in the absorption band frequency due to changing the isotope can be used to study impurities which are heavier than host lattice atoms, provided they are paired with a light atom.

When the behavior of electrically active impurities is studied in GaAs by local mode infrared spectroscopy, the free-carrier concentration introduced by the impurities must be reduced by compensation with another electrically active defect. The behavior of Li has made it a good choice for the compensating impurity: it is a rapid diffuser, and two stable isotopes are readily available. Usually it is the local mode of the Li paired with the original dopant which is experimentally observed. The bandwidth range of interest is typically 300 to 460 cm^{-1}. In GaAs, the pair systems studied include Te–Li, Mg–Li, Cd–Li, Zn–Li, Mn–Li, and Si–Li (Lorimor and Spitzer, 1966; Allred et al., 1968; Spitzer and Allred, 1968; Leung et al., 1972). In addition to the local modes observed in GaAs doped during growth, absorption bands have also been reported in pure GaAs into which Li had been diffused; the corresponding absorption centers are believed to be Li complexed with native defects.

Acceptor actions associated with Ga vacancies and with As vacancies in GaAs have been studied by Muñoz et al.* For Ga-vacancy–associated acceptors the energy levels observed were 0.01 and 0.18 eV above E_v. For As-vacancy–associated acceptors the level observed was 0.12 eV above E_v.

Sodium

This has been reported as an acceptor by Hilsum and Rose-Innes (1961). However, there appear to be no recent studies with Na as a dopant.

Copper

Copper is a fast interstitial diffuser in GaAs, and may be present as a contaminant in crystals. It has been the subject of numerous studies (Hall and Racette, 1964; Hwang, 1968a–c; Vasil'ev et al., 1969; Blakeslee and Lewis, 1969; Gansauge and Hoffmeister, 1966; Mil'vidskii et al., 1966, 1969; Fistul', 1965; Furukawa et al., 1966, 1967; Sugiyama, 1967; Shirafuji, 1968; Fuller et al., 1967; and Lyubchenko et al., 1968).

* E. Muñoz, W. L. Snyder, and J. L. Moll, (1970), Effect of Arsenic Pressure on Heat Treatment of Liquid Epitaxial GaAs, *Appl. Phys. Lett.*, **16**, 262.

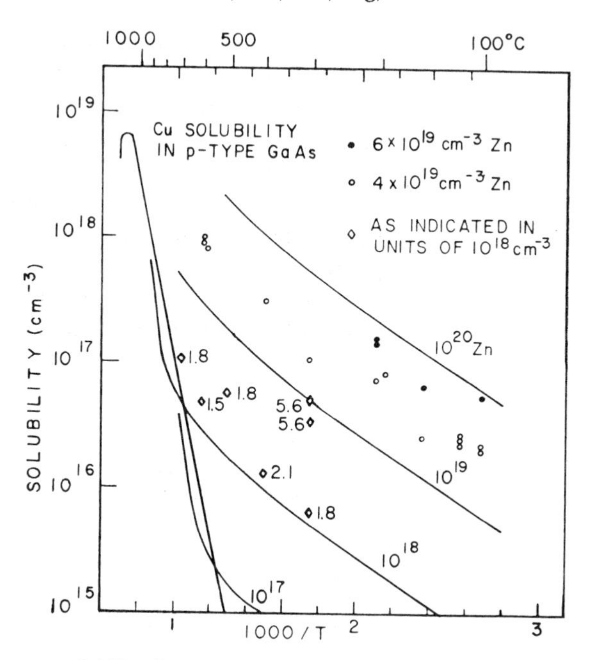

FIG. 3.2 Copper solubility in *p*-type GaAs versus temperature for various acceptor concentrations.

Data points are measured using Cu^{64}. Curves for 10^{17} to 10^{20} Zn are calculated. (After Hall and Racette, 1964.)

The interstitial species Cu^i is a single donor with the value $E_c - 0.07$ eV, assigned by Morgan et al. (1965). Its presence is normally masked by substitutional copper Cu^s, which is a double acceptor. The substitutional to interstitial ratio at 700°C, for example, has been estimated to be about 30. The interstitial diffusion coefficient D^i is 1.0×10^{-5} cm^2 sec^{-1} at 500°C and the diffusion activation energy is 0.53 eV. The solubility of copper in GaAs depends on the other impurities present. It is shown in Fig. 3.2 for various doping levels and temperatures. The distribution coefficient k (concentration in solid to concentration in liquid) is 2×10^{-3} or less. This is much lower than might be expected from tetrahedral radii considerations.

The primary energy levels associated with copper in GaAs are 0.14 to 0.15 eV above E_v, which is the first acceptor level (Cu^-), and 0.44 eV, which is the second acceptor level (Cu^{2-}). A level is also seen at $E_v + 0.24$ eV which is related to a pair of copper atoms. A Cu_{Ga}–Te complex has been observed at 0.19 eV, and there are unidentified levels believed to be associated with copper at 0.023, 0.123, 0.166, and 0.51 eV. At times donor levels, notably one around $E_c - 0.6$ eV, have been attributed to copper interacting with defects or other impurities.

Photoluminescent studies show a broad band at 1.37 eV which has been ascribed to $V_{As}Cu_{Ga}$ pairs or $D_{As}Cu_{Ga}$ pairs where V_{As} and D_{As} denote, respectively, an arsenic vacancy and a donor impurity occupying an arsenic site.

The characteristics of GaAs diodes doped with copper have been examined by Morgan et al. (1965) and by Furukawa et al. (1966). Copper does not appear to be the impurity responsible for the degradation of GaAs tunnel diodes. Instead there is some evidence that this degradation is related to oxygen (Kessler and Winogradoff, 1966). Degradation in III–V electroluminescent diodes has been discussed by Weisberg (1971).

Silver

Diffusion and distribution coefficient studies of silver in GaAs have been made. The diffusion studies of Boltaks and Shishiyanu (1964) are not in

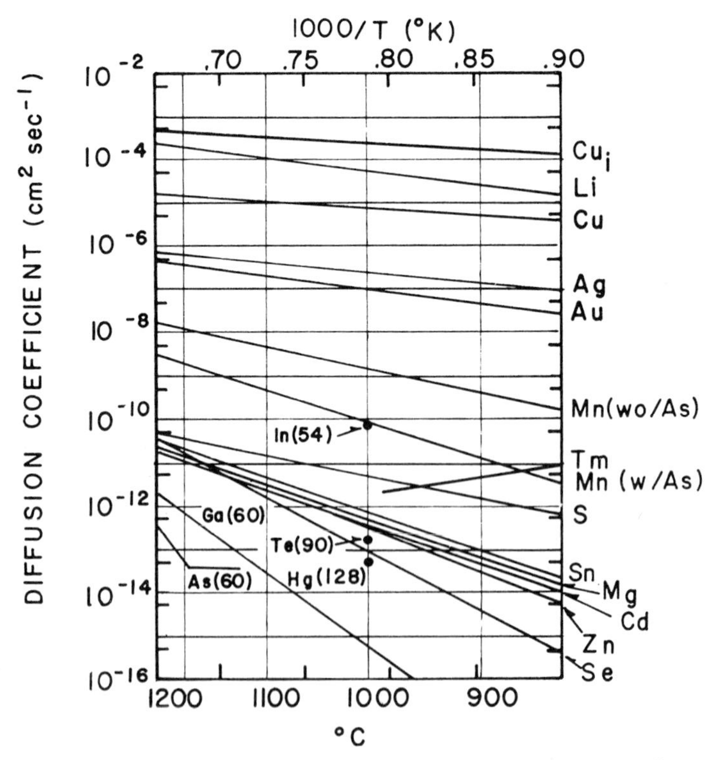

FIG. 3.3 Diffusion coefficients in GaAs at low concentration limit. (After Kendall, 1968.)

good agreement with those of Rybka et al. (1962). The diffusion is interstitial and the line marked Ag in Fig. 3.3 roughly represents the behavior. This illustration, taken from Kendall's review (1968) of diffusion in III–V compounds, is helpful in visualizing the overall diffusion pattern for GaAs.

The diffusion studies suggest that the bulk solubility of Ag in the temperature range 500 to 1160°C is 2 to 8 $\times$ 10^{17} cm^3. In other work, 3 $\times$ 10^{16} cm^3 has been reported for the solubility at 800°C (Shishiyanu and Boltaks, 1966). The distribution coefficient is low, less than 4 $\times$ 10^{-3}. Distribution measurements for impurities in GaAs have been summarized by Willardson and Allred (1966) and are given in Table 3.1.

TABLE 3.1

Distribution Coefficients of Impurities in Gallium Arsenide[a]

Doping Element	(1)	(2)	(3)	(4)	(5)
Al			3		0.2
Ag			0.1		4×10^{-3}
Bi					5×10^{-3}
Be					3
Ca			<0.02		2×10^{-3}
C			0.8		
Cr	6.4×10^{-4}				5.7×10^{-4}
Co	8.0×10^{-5}				4×10^{-4}
Cu				2×10^{-3}	$<2 \times 10^{-3}$
Ge			0.03	0.018	0.01
In			0.1		7×10^{-3}
Fe	2.0×10^{-3}			3×10^{-3}	1×10^{-3}
Pb			<0.02		$<1 \times 10^{-5}$
Mg		0.047	0.3		0.1
Mn	0.021				0.02
Ni	6.0×10^{-4}		<0.02		4×10^{-5}
P			2	3	3
Sb			<0.02		0.016
Se				0.44–0.55	0.30
Si		0.11	0.1	0.14	0.14
S		0.17	0.3	0.5–1.0	0.30
Te		0.025	0.3	0.054–0.16	0.059
Sn		0.048	0.03		0.08
Zn		0.36	0.1	0.27–0.9	0.40

[a] (1) Haisty and Cronin (1964); (2) Edmond (1959); (3) Weisberg (1961 a, b); (4) Whelan et al, (1960); (5) Willardson and Allred (1966).

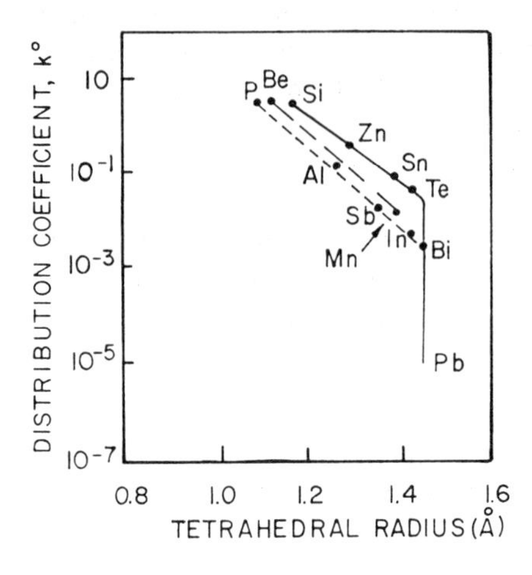

FIG. 3.4 Distribution coefficients of impurities at the melting point of GaAs as a function of tetrahedral radii. (After Willardson and Allred, 1966.)

The overall pattern that emerges (Fig. 3.4) is a trend to lower values for the distribution coefficients as the tetrahedral radii of the impurities increase. Beyond about 1.45 Å a sharp decrease corresponding to a very limited solubility is observed.

The electrical activity of Ag in GaAs shows an acceptor level at $E_v + 0.11$ eV (Shishiyanu and Boltaks, 1966). In this study it was not practical to look for deeper acceptor levels since the solubility was low (3×10^{16} cm^3) and copper contamination was a problem. In more recent studies an acceptor level at $E_v + 0.238$ eV (4°K) has been reported (Blätte et al., 1970).

Gold

The diffusion of Au in GaAs shows a high-concentration surface branch and a lower concentration deep branch as in Fig. 3.5. The deep branch can be represented approximately by an erfc function. The deep diffusion coefficient varies from 6×10^{-9} to 2×10^{-7} cm^2 sec^{-1} in the temperature range 740 to 1025°C. The solubility varies from 2.5×10^{16} to 1.6×10^{18} cm^{-3} for the temperature range 900 to 1140°C.

From Hall measurements, Au gives an acceptor level at 0.09 eV (Shishiyanu and Boltaks, 1966). In the presence of copper and gold an acceptor level of $E_v + 0.04$ eV has been observed (Krivov et al., 1970).

A complex of Au–Ge is found in degenerate germanium-doped *n*-type layers of GaAs grown from a gold-rich melt and produces an energy level at $E_v + 0.16$ eV (Andrews and Holonyak, 1972).

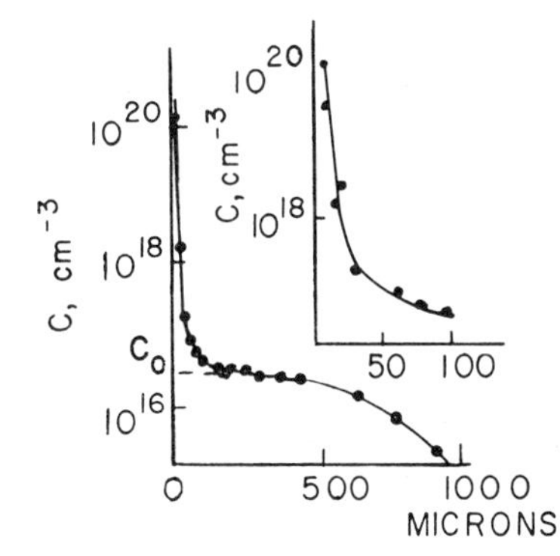

FIG. 3.5 Concentration distribution for the diffusion of Au in GaAs.

The first part of the concentration curve is shown on an enlarged scale on the right. (After Sokolov and Shishiyanu, 1964.)

3.2 GaAs with Impurities of Group II: Be, Mg, Zn, Cd, and Hg

Beryllium

Beryllium is an acceptor impurity in GaAs (Yunovich et al., 1964; Ormont et al., 1966) and its solubility limit is 1 to 3 × 10^{19} cm^{-3}. The diffusion coefficient takes the form $D = 7.3 \times 10^{-6} \exp(-1.2/kT)$ (Poltoratskii and Stuchebnikov, 1966). Luminescence studies suggest that Be is a shallow acceptor, 0.030 eV, but the structure observed is not well understood. This value is confirmed by the studies of Kressel and Hawrylo (1970).

Magnesium

Magnesium creates a shallow acceptor level originally proposed as 0.012 eV but more recently set at 0.030 eV (Kressel and Hawrylo, 1970) and has been used as a dopant in GaAs diodes and transistors (Doerbeck et al., 1968). Deep levels (e.g., $E_v + 0.22$, $E_v + 0.38$, and $E_v + 0.57$ eV) have been observed in magnesium-doped GaAs but their origin is unknown. As Mg is a reactive element, it is possible that these levels represent contaminants.

Zinc

Zinc is the most commonly used *p*-type dopant in GaAs. The activation energy is 0.024 to 0.031 eV (apparently depending on the measurement technique used) and the solubility is high. Zinc diffusion behavior is complicated and has been discussed by Kendall (1968) and others.

Since Zn is a shallow impurity its effects are not further discussed here. Some recent references are the studies of Hunsperger and Marsh (1969),

Hwang (1968a–c, 1969a), Imenkov et al. (1966, 1967), Boltaks et al. (1964), Blashku et al. (1971). A deep level, $E_v + 0.65$ eV, possibly associated with Zn doping, has been observed by Su et al. (1971) in liquid-phase epitaxially grown GaAs.

Cadmium

Cadmium is the second most frequently used *p*-type shallow dopant in GaAs. The activation energy is 0.021 to 0.030 eV. This dopant has been discussed by Doerbeck et al. (1968), Hunsperger and Marsh (1969), Kogan et al. (1964), and Ivanova et al. (1964). A deep acceptor level at about 0.36 eV above the valence band has also been observed with Cd doping and is attributed to lattice defects or to Cd lattice defect complexes (Huth, 1970).

Mercury

Kanz (according to Kendall, 1968) in a diffusion of Hg^{203} at 1000°C found low surface concentration (5×10^{17} cm^{-3}) and low diffusion coefficient (5×10^{-14} cm^2 sec^{-1}). Mercury, therefore, does not appear to be an interesting dopant in GaAs.

3.3 GaAs with Impurities of Group IV: C, Si, Ge, and Sn

Carbon

Early studies in the preparation of GaAs in graphite boats showed that the crystal could contain as much as 8×10^{16} atoms cm^{-3} of carbon (Rossi et al., 1969b; Weisberg, 1961a, b). An activation energy of $E_v + 0.019$ eV for carbon is reported by Sze and Irvin (1968).

Silicon

Silicon is an interesting amphoteric dopant in GaAs. When GaAs is grown from a melt of stoichiometric composition, silicon, if present, is incorporated on gallium sites as a shallow (0.002 to 0.0058 eV) donor. However, if GaAs is solution-grown from gallium, under conditions where the As vapor pressure is low, the silicon enters mainly on As sites and is then an acceptor at 0.025 to 0.030 eV.

In solution growth, by careful control of the temperature range and cooling cycle, it is possible to grow *np* or *pn* amphoterically doped diodes (Kressel et al., 1968a; Moriizumi and Takahashi, 1969). Usually the *p* region is highly compensated and it is desirable to add a zinc-doped contact to create a p^+pn structure. These diodes then emit infrared light

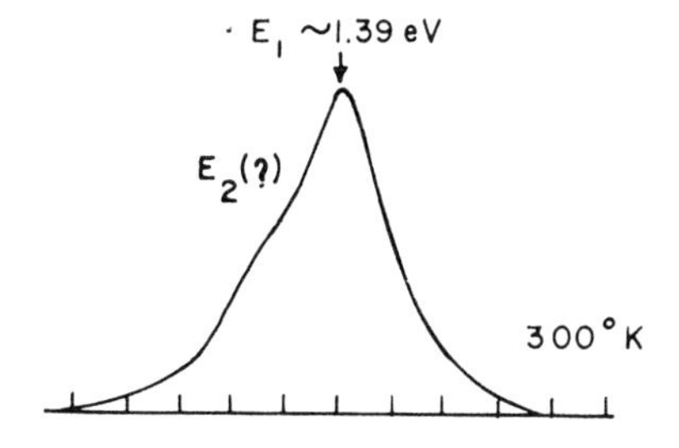

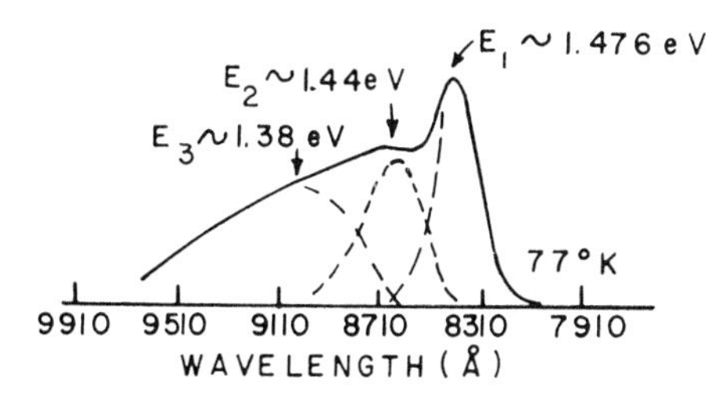

FIG. 3.6 Typical photoluminescence spectra of silicon-doped gallium arsenide at 300 and 77°K (After Kressel et al., 1968.)

under forward bias conditions with external quantum efficiencies of up to 6% at 300°K. The active light-emitting region is of the order of 50 μm wide and is mainly the π-doped region of the diode.

The bandwidth of the emission is fairly broad, as shown in Fig. 3.6. The variation of the peak with temperature, Fig. 3.7a, suggests transitions to three levels, $E_v + 0.03$, $E_v + 0.10$, and $E_v + 0.22$ eV, as shown in Fig. 3.7b. Kressel et al. (1969b) suggest that the 0.03 level is due to Si occupying random As sites and that the 0.10-eV level is a complex involving Si. This might be (Si_{Ga}–Si_{As}) or (Si_{Ga}–V_{Ga}) (Spitzer and Allred, 1968). The 0.22-eV level is attributed to an arsenic-vacancy complex that is related to the presence of silicon.

Recently Weisberg (1968), from a study of Auger concepts, has suggested that the high electroluminescence efficiency of amphoteric silicon *p-n* junctions in GaAs could be due to a silicon–silicon transition and not a band to impurity transition.

In silicon-doped GaAs prepared by vapor-phase growth, from arsine and HCl-transported Ga, a deep acceptor center has been found when the Si doping was in excess of 10^{18} cm^{-3}. A broad-band photoluminescence (77°K) centered at 1.8 eV was obtained. Indirect evidence suggested that the acceptor consisted of a complex of Si with an unknown impurity (from the arsine), rather than with a native defect such as a vacancy (Kressel and von Philipsborn, 1969). Localized concentrations of Si have been observed

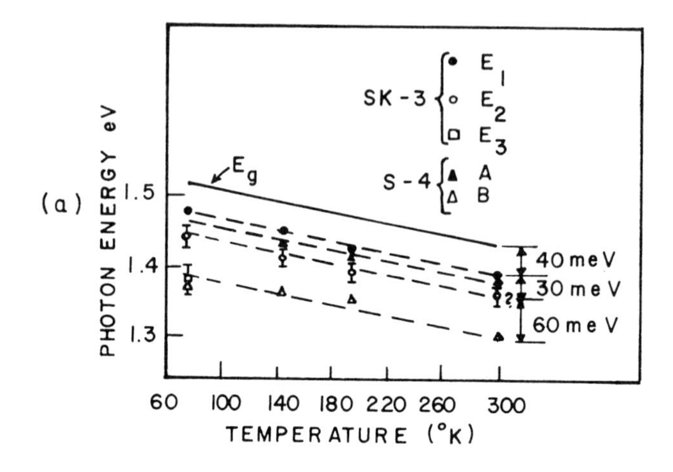

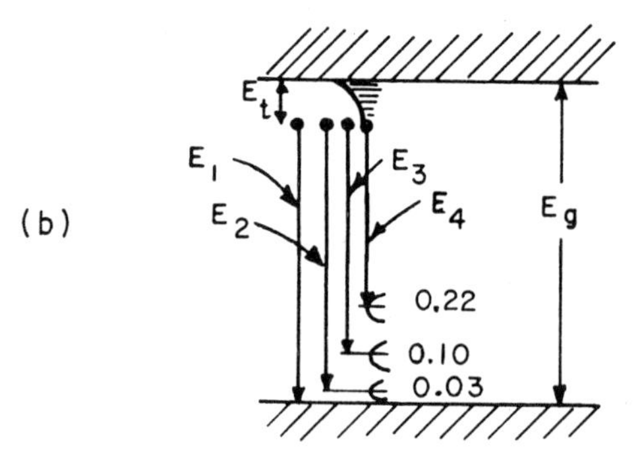

FIG. 3.7 Photoluminescence peaks and transitions in p-type silicon-doped material. *(a) Peak positions between 77 and 300°K compared to the variation of the band-gap energy. (b) Possible radiative transitions in silicon-doped p-type GaAs. The transitions are assumed to originate at donor states in the empty conduction band "tail" at an energy* E_t *below the unperturbed band edge (low injection level). (After Kressel et al., 1968.)*

at nn^+ interfaces in GaAs vapor phase and liquid phase epitaxy (DiLorenzo, 1971).

Silicon may be introduced into GaAs by interaction with quartz that is in contact with the molten semiconductor during growth (Fertin et al., 1966; Weiner, 1972). The carrier mobility may suffer on this account. Also, there is evidence that Si may contribute to thermal conversion of GaAs when processing is at temperatures above 900°C.

If GaAs is grown in the presence of silicon, SiO_2 and/or oxygen complexes of silicon and oxygen may form. In particular, Weiner and Jordan (1972) infer that silicon atoms on gallium sites pair with interstitial oxygen atoms, forming a complex which behaves as an acceptor with energy levels near 0.2 and 0.4 eV below the conduction band.

Germanium

Germanium like silicon is an amphoteric dopant in GaAs. On a Ga site germanium introduces a shallow donor, and on an As site it becomes an acceptor. At 77°K the acceptor levels observed are $E_v + 0.03$ and $E_v + 0.07$ eV. The 0.03-eV level is attributed to isolated Ge atoms on As sites and the 0.07-eV level to an atomic complex of Ge. This 0.07-eV level was 10^{-2} or 10^{-3} of the density of the shallower level.*

In the photoluminescence studies of Moriizumi and Takahashi (1969), germanium-doped *p–n* GaAs diodes gave an external efficiency lower than that of silicon-doped diodes. One reason was larger absorption in the *n* region because the emission peak energy of the germanium-doped diode was quite near the band-gap energy. As the Ge content is increased above 5×10^{17} cm^{-3} the total luminescence may decrease. This is presumably due to the increase in the number of nonradiative centers caused by increasing precipitation and vacancy formation at high doping levels. Similar effects are observed in silicon-doped GaAs diodes.

Laser transitions have been induced between the conduction band and the shallow acceptor level of Ge (Burnham et al., 1969).

Tin or Lead

In tin-doped GaAs grown by liquid-phase epitaxy under arsenic-deficient conditions, compensation due to the possible occupation of As sites by Sn may exist (Kressel et al., 1968c, d). A broad photoluminescence band is observed and suggests that the Sn acceptor level is at $E_v + 0.2$ eV. Evidence is inconclusive for the existence of a shallow acceptor as observed for Si or Ge in GaAs. Thus the deep level may be due to Sn on isolated As sites rather than to a complex involving Sn atoms.

The behavior of Pb is similar, with an acceptor level observed at $E_v + 0.12$ eV (Kressel et al., 1968c, d).

3.4 GaAs with Group VI Impurities: O, S, Se, and Te

Oxygen is a deep donor, at $E_c - 0.75$ eV, in GaAs. The solubility is in excess of 10^{17} cm^{-3}. If shallow acceptors are also present in the crystal,

* See also *Phys. Rev. B.*, **1**, 1603 (1970) for a report of a 6.08-meV level.

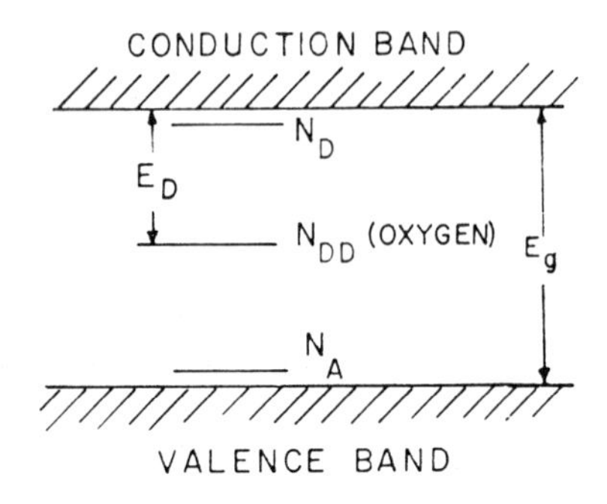

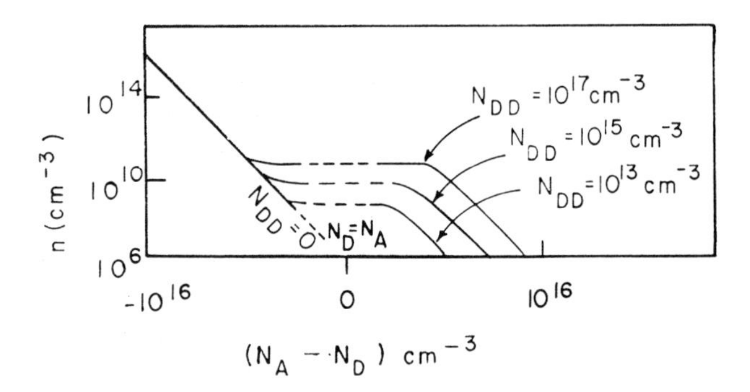

FIG. 3.8 High-resistivity GaAs from oxygen doping.
(a) Model with oxygen as deep donor, N_{DD}, near center of band gap. (b) Variation of electron concentration, n, with $(N_A - N_D)$ at room temperature for various values of N_{DD} with $E_D = 0.76$ eV. (After Blanc and Weisberg, 1961,)

the oxygen provides the compensating electrons needed and the GaAs has high resistance (more than 10^7 Ω-cm at 300°K) (Blanc and Weisberg, 1961). This requires no critical balance between the oxygen and the net shallow acceptor concentrations to achieve high resistance. The electron-carrier concentration expected under various doping conditions is shown in Fig. 3.8. Providing N_{DD} exceeds $(N_A - N_D)$ the free-carrier concentration is small because of the great depth of the oxygen donor, Ostroborodova and Kandidova (1970). Two deep levels have been observed in oxygen-doped GaAs by Vorob'ev et al. (1971); one at $E_c - 0.80$ eV and the other at $E_c - 1.2$ eV.

Semiinsulating GaAs exhibits a variety of photoconductive trapping effects (Shah and Yacoby, 1968) and space-charge-limited injection effects

(Holonyak, 1962; Saunders, 1968). Low-frequency current oscillations are also observed in the presence of photoexcitation (Tokumaru, 1969, 1970; Viehmann, 1969). In gallium arsenide transistors trapping effects have been observed and attributed to oxygen (Antell and White, 1968). The depletion region capacitance of Schottky barrier junctions on GaAs also shows trapping effects due to the oxygen level (Williams, 1966). The amount of oxygen in the crystal is of course dependent on the growth system and subsequent processing. Solution growth of epitaxial GaAs from Ga results in a much lower oxygen content than most vapor-growth processes (Carbellès et al., 1968).

Oxygen is also found to have undesirable effects on the performance of gallium arsenide structures such as tunnel diodes and injection lasers through a surface effect (Kessler and Winogradoff, 1966; Arthur, 1967).

Sulfur, Selenium, and Tellurium

Sulfur, selenium, and tellurium are shallow *n*-type dopants in GaAs (Weisberg, 1961; Madelung, 1964; Willardson and Goering, 1962; Strack, 1966). The activation energies reported for Se and Te are 0.005 and 0.03 eV, respectively. More recently the values 5.89 and 6.10 $\pm$ 0.025 meV have been determined for Se and S, respectively (Summers et al. 1970).

The variation of mobility and resistance of GaAs with shallow doping is shown in Fig. 3.9 for both *n*- and *p*-type material at 300°K. Field impact ionization of shallow donors at temperatures below 20°K has been studied by Hughes and Tree (1970).

Recently it has been recognized that the doping behavior of Se and Te impurities in GaAs is unusual at high dopant concentrations (Winteler and Steinemann, 1968; Kressel et al., 1968c; Williams, 1968; Vieland and Kudman, 1963; Rashevskaya and Fistul', 1967, 1968; Iizuka, 1968; Hwang and Dyment, 1968; Vul et al., 1964; Bagaev et al., 1964; Yurova et al., 1972; Mil'vidskii et al., 1972). The problem is one of impurity complexing or precipitation and shows itself in the radiative efficiency and deep-level luminescence of *n*-type GaAs.

A broad emission band centered at 1.2 eV is observed in GaAs doped with group VI elements, whether melt grown, vapor grown, or grown by liquid-phase epitaxy with gallium as the solvent. This photoluminescence increases in intensity, relative to the band-gap radiation, with increasing dopant concentration in the 10^{18} cm^{-3} range. The luminescing combination centers responsible for this 1.2-eV band may be complexes such as V_{Ga} + 3Se, or V_{Ga} + 3Te. These represent the solid solution of Ga_2Se_3 or Ga_2Te_3 in GaAs; $V_{Ga}Te_{As}$ complexes also have been reported (Logan, 1971). With

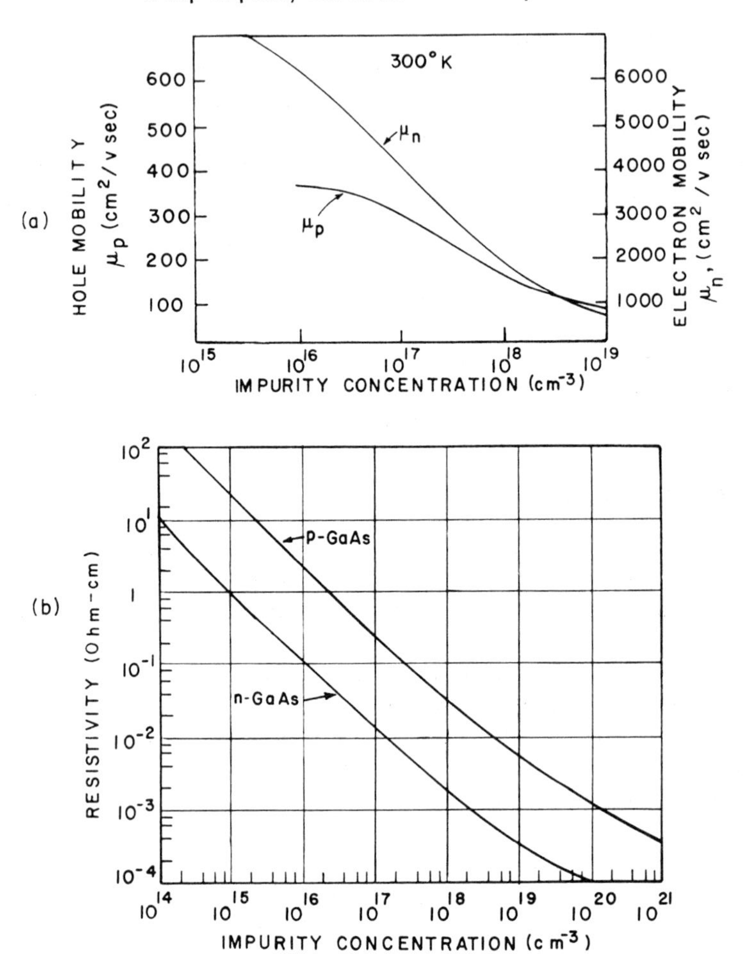

FIG. 3.9 Mobility and resistivity variation with impurity concentration in *p*- **and** *n*-**type GaAs.**
(a) Mobility μ_p, μ_n *at 300°K. (b) Resistivity of p- and n-GaAs at 300°K.*

increasing dopant content the solubility limit is eventually exceeded and precipitates of Ga_2Se_3 or Ga_2Te_3 are formed. With the dopant concentration increased beyond 2 to 3 × 10^{18} cm^{-3} the radiative efficiency falls off sharply. Kressel et al. (1968c) suggest that the falloff is partly due to nonradiative recombination via small precipitates, each containing on the order of 1000 dopant atoms.

The radiative efficiency of *n*-type silicon-doped GaAs prepared by liquid-phase epitaxy does not fall off up to 4.2 × 10^{18} cm^{-3}, the highest concentration studied by Kressel and his coworkers. They conclude that Si may be preferable to either Se or Te doping in optical-device applications requiring electron concentrations in the mid 10^{18} range. Falloff of luminescence efficiency of Si amphoterically doped GaAs diodes, however, has been noticed for high Si concentration levels (Moriizumi and Takahashi, 1969).

3.5 GaAs with Transition Elements: Cr, Mn, Fe, Co, Ni, and V

Transition-element doping of GaAs has been studied by Haisty and Cronin (1964); by Kolchanova et al. (1970); by Omel'yanovskii et al. (1970); and others. Transition element dopants, with the exception of V, are deep acceptors, as can be seen from Table 3.2.

TABLE 3.2

GaAs with Transition Metal Dopants

Dopant	Energy Level (eV)	Solubility (atoms cm^{-3})	Distribution Coefficient	Resistivity (typical, 300°K) (Ω-cm)
Cr	0.79 (A)	$>1.6 \times 10^{17}$	6.4×10^{-4}	3×10^{8}
Mn	0.095 (A)	$>10^{17}$	0.021	0.15
Fe	0.52 (A)	$>10^{17}$	2×10^{-3}	4×10^{4}
Co	0.16 (A)	$>10^{16}$	8×10^{-5}	2.2
Ni	0.21 (A)	$>10^{17}$	6×10^{-4}	17
V	$E_c - 0.22$ (D)	—	—	—

Chromium doping is seen to be a convenient method for preparing semi-insulating GaAs. This differs from oxygen-doped GaAs since chromium is a deep acceptor and oxygen is a deep donor. Recombination levels at 0.8 and 0.6 eV from the conduction band of GaAs: Cr have been reported from luminescence studies (Peka and Karkhanin, 1972). Photothermoelectric studies show the Cr level at $E_c - 0.56$ eV (Harper et al. 1970).

Iron doping also produces high-resistance material, up to about 10^5 Ω-cm at room temperature. Figure 3.10 shows the dependence of resistivity on temperature for typical doping densities.

Although a 0.37-eV level has been reported for iron in GaAs, this was not seen by Haisty and Cronin (1964). Copper together with iron produced only the 0.15-eV level characteristic of copper. Iron and tin, however, were found to interact and gave a donor level of 0.1 eV.

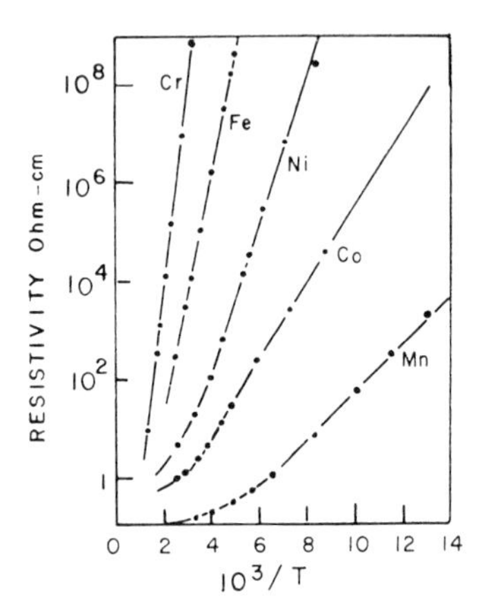

FIG. 3.10 Variation of resistivity with temperature for transition-metal-doped GaAs. (After Haisty and Cronin, 1964.)

The results summarized in Table 3.2 are qualitatively consistent with simple valency considerations. Manganese, cobalt, and nickel with stable oxidation states of +2 should introduce single-acceptor levels in GaAs by replacing gallium. Although chromium and iron have stable oxidation states of +3, they also exist in the +2 state and therefore could introduce single-acceptor levels in GaAs. Vanadium commonly shows an oxidation state of +4 and this is compatible with the donor action reported.

The energy levels of deep transition-metal centers in gallium arsenide have been considered by Bazhenov and Solov'ev (1972). The values they regarded as most reliable were 0.85 eV for Cr, 0.5 eV for Fe, 0.6 eV for Co, 0.42 eV for Ni, and 0.145 eV for Cu. The Hall effect and optical studies of Brown and Blakemore (1972) however, give values of 0.16 eV for Co, 0.20 eV for Ni and 0.09 eV for Mn in agreement with Haisty and Cronin (1964).

The photoluminescence of manganese-doped GaAs has been studied by Chang (1964). The capture cross section for electrons by a neutral Mn atom was found to be 10^{-16} cm^2 at 300°K and 3×10^{-16} cm^2 at 77°K. The capture cross section for holes by an ionized Mn atom was estimated to be 7×10^{-14} cm^2 at 77°K. Avalanching and negative resistance in manganese-doped GaAs have been studied by Weiser et al. (1967) and Dumke (1964).

Iron doping of GaAs has been studied by a number of workers (Strack, 1966; Hoyt and Haisty, 1966; Doerbeck et al., 1968; Pervova et al., 1969; Bazhenov et al., 1968; Lukicheva et al., 1971). Perhaps the most interesting discovery has been that iron doping of GaAs transistors results in an improvement in cutoff frequency and in operating temperature range (4.2 to 500°K). GaAs bipolar transistors have also been studied by Nuese et al. (1972).

3.6 Rare Earth Studies in GaAs: Tm, and Nd

Rare earth elements are of interest because their incomplete $4f$ subshells give spectral properties that are useful, such as stimulated emission when incorporated in various crystal hosts. Relatively little work has been done on the incorporation of rare earths in covalent semiconductors such as Si, Ge, and GaAs since the solubilities are expected to be low.

The solubility and diffusion of thulium in GaAs have been studied by Casey and Pearson (1964). Thulium was chosen because of its relatively small atomic size (1.746 Å) and the availability of a suitable radioactive isotope, ^{170}Tm. The maximum solubility was found to be 4×10^{17} cm^{-3} at 1150°C. This low solubility was attributed to the fact that the Tm atom is 25% larger than the Ga atom that it replaces substitutionally.

Richman (1964) has made initial studies of Nd doping of GaAs. The pumping band of Nd closely matches the emission obtained from GaAs luminescent diodes. Vapor-phase epitaxy was attempted and some inclusion of Nd in the grown crystal was observed.

3.7 Deep Impurities in GaP and InP

Gallium phosphide is a semiconductor with an indirect energy gap of 2.26 (300°K) or 2.338 eV (0°K). It is readily doped p-type by group II impurities Be, Mg, Zn, and Cd occupying gallium sites, and n-type by group VI impurities S, Se, and Te on phosphorus sites. Although an indirect gap material, GaP is found to be an efficient emitter of light from forward-biased pn junctions. The light is red if the crystal contains oxygen together with Zn or Cd, and the photoluminescence spectrum has a great deal of structure, as shown in Fig. 3.11. The light is green if the crystal contains N on P sites and if the other impurities in the GaP are carefully controlled. The red luminescence has been understood from the work of Gershenzon et al. (1965), Morgan et al. (1968), Henry et al. (1968), and Thomas et al. (1964).

Pair recombination occurs for electrons bound to donors and holes bound to acceptors for both close and remote impurity pairs. For a discrete pair separation r, the emission energy is

$$h\nu = E_g - (E_A + E_D) + \frac{e^2}{\varepsilon r} \tag{3.2}$$

where E_A and E_D are the acceptor and donor ionization energies, ε is the dielectric constant, and $e^2/\varepsilon r$ is the coulomb energy. At low temperature,

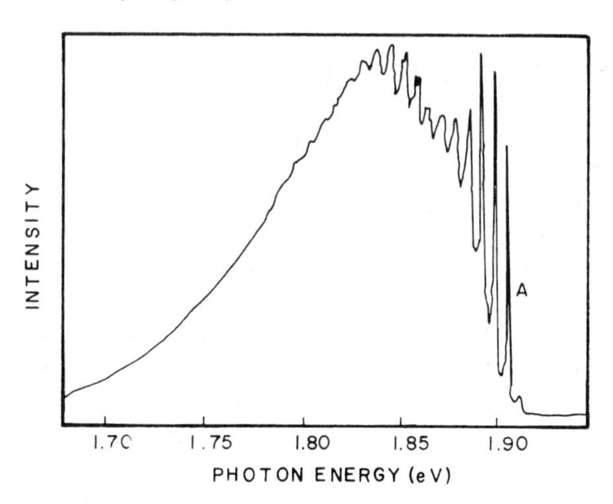

FIG. 3.11 Photoluminescence spectrum of gallium phosphide doped with cadmium and oxygen, at 20°K.
The emitted light is red. The line marked A *is a sharp electronic transition. The structure of the emission at lower energies consists of phonon or vibrational side bands. (After Thomas, 1968.)*

many discrete lines are observed in the emission spectra. These lines differ in energy by discrete values of $e^2/\varepsilon r$ for various separations of the impurities on lattice sites. The sharp lines correspond to pair separations between about 10 and 40 Å. For larger separations the discrete lines tend to merge into a broad band. The low-temperature emission lines may show phonon replicas involving the longitudinal optical phonon. When suitably examined they permit precise determination of E_A and E_D from transitions between pairs, such as Zn and S.

For Zn in GaP the isolated Zn acceptor level is 0.064 eV and the isolated O level is a donor about 0.90 eV below the conduction band edge. When Zn and O occupy adjacent lattice sites, an electrically neutral or isoelectric complex results, since the Zn and O with a total of eight valence electrons substitute for Ga and P also with eight valence electrons. However, the Zn–O complex has an attraction for an electron, which is assumed to be primarily due to the less effective shielding of the positive nuclear charge in O compared with P because of the smaller number of core electrons in the oxygen atom. Hence the Zn–O complex acts as an electron trap and the level found for it is about 0.3 eV below the conduction band edge. The trapped electron can combine radiatively with a bound hole at the 0.036-eV exciton level or as pair emission with a hole bound to the Zn acceptor (thought to be less important for room-temperature emission). These

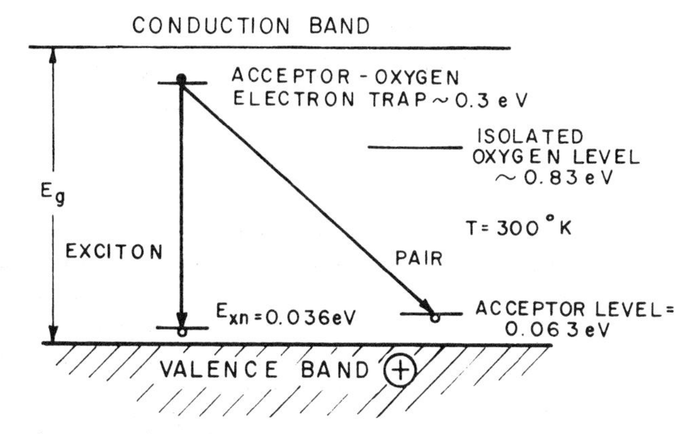

FIG. 3.12 The energy level diagram of the Zn–O levels in GaP that lead to the bound exciton red emission.

Electrons are trapped at Zn–O complexes, and holes are trapped at the negatively charged Zn–O complexes. The exciton hole level is E_{xn}. *(After Casey and Trumbore, 1970.)*

processes are illustrated in Fig. 3.12. The Zn–O complex concentrations for efficient operation should be in the range 0.5 to 5 $\times$ 10^{16} pairs cm^{-3} with Zn doping of the order of 10^{18} cm^{-3}. By very careful control of the doping in *pn* junctions grown by liquid-phase epitaxy from gallium, Saul et al. (1969) have obtained 7% for the external quantum efficiency of the red luminescence at room temperature. Oxygen incorporation in GaP liquid phase epitaxy layers has been studied by several workers including Saul and Hackett (1970) and Kowalchik et al. (1972).

Green luminescence is associated with an isoelectronic trap formed by nitrogen substituting for phosphorus in the *p* side of the junction. Since the cores of the N and P atoms are different, a short-range perturbation exists which acts as a shallow electron trap. Some of the electrons injected into the *p* side of the junction are trapped by the isoelectronic N, and a hole is bound by coulombic attraction to form an exciton. The radiative decay of this bound exciton results in green emission, peaking at about 2.22 eV. External efficiencies of 0.6% (300°K) have been reported for N concentrations of 10^{19} cm^{-3} in *pn* junctions with Zn, 2 to 5 $\times$ 10^{17} cm^{-3}, and S, 3 to 6 $\times$ 10^{16} cm^{-3}. As the eye is about 30 times more sensitive to the green band than to the red Zn–O band, the luminous effect is very satisfactory.

Yellow emission from GaP is obtained by Mg–O doping (Bhargava et al., 1972).

The ionization energies of a number of impurities in GaP are given in Table 3.3 (after Casey and Trumbore, 1970). Silicon is an amphoteric

TABLE 3.3

Ionization Energies of Impurities in GaP

Impurity	Donor (eV)	Acceptor (eV)	Ref.
Sn	$E_c -$ 0.065		Dean et al., 1970.
Si	$E_c -$ 0.082		Casey and Trumbore, 1970, review paper
Te	$E_c -$ 0.0895		Casey and Trumbore, 1970, review paper
Se	$E_c -$ 0.102		Casey and Trumbore, 1970, review paper
S	$E_c -$ 0.104		Casey and Trumbore, 1970, review paper
Unknown	$E_c - \leq 0.165$		Gershenzon et al., 1965
Unknown	$E_c -$ 0.24		Gloriozova et al., 1969
Unknown	$E_c -$ 0.6	Electron trap	Goldstein and Perlman, 1966
O	$E_c -$ 0.896		Dean and Henry, 1968
Cu		$E_v + 0.68$	Bowman, 1967
Co		$E_v + 0.41$	Loescher et al., 1966
Ge		$E_v + 0.30$	Dean 1970
Si		$E_v + 0.203$	Dean 1970
Cd		$E_v + 0.097$	Dean 1970
Zn		$E_v + 0.064$	Dean 1970
Be		$E_v + 0.056$	Dierschke and Pearson, 1970
Mg		$E_v + 0.054$	Dean et al., 1970
C		$E_v + 0.048$	Dean et al., 1970
C		$E_v + 0.041$	Bortfield et al., 1972
N	Isoelectronic trap	$E_c - 0.008$	Dean 1970
Bi	Isoelectronic trap	$E_v + 0.038$	Dean et al., 1969
Zn–O	Isoelectronic trap	$E_c - 0.30$	Cuthbert et al., 1968
Cd–O	Isoelectronic trap	$E_c - 0.40$	Cuthbert et al., 1968

dopant, tin is a shallow donor and germanium is a deep acceptor. By compensating *n*-type GaP with a deep acceptor such as Cu it is possible to obtain material with high dark resistivity (more than 10^8 Ω-cm) (Bowman, 1967) and considerable photosensitivity (Goldstein and Perlman, 1966; Grimmeiss and Olofsson, 1969). Cobalt in GaP has been studied by Loescher et al. (1966) and Ni by Baranowski et al. (1968). Both are acceptors with the impurity substituting for Ga. Gold- and silver-doped GaP have been studied by Ikizli et al. (1971). Iron in GaP acts as an electron trap near the con-dunction band edge according to Suto and Nishizawa (1972). The ionized sulfur donor in GaP has been found to have a large cross section (Dishman, 1972).

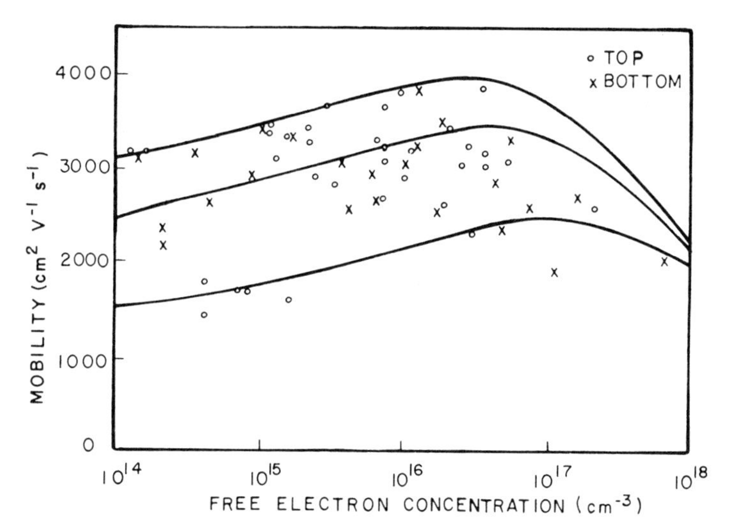

FIG. 3.13 Variation of mobility of InP with free-electron concentration. *Circles and crosses represent measurements at the top and bottom of crystals, respectively. Full lines are theoretical curves using acceptor levels of* 6.2×10^{16}, 8.8×10^{16}, *and* 1.8×10^{17} cm^3, *reading from top to bottom. (After Mullin, Royle, and Straughan, 1971.)*

Recently, Kukimoto et al. (1972) and Henry et al.* have shown that photocapacitance studies in GaP *p-n* junctions provide new insight to the deep states associated with the O donor. In addition to the well-known neutral state in which O binds an electron by 0.9 eV, a new state in which O deeply binds a second electron was discovered. This has been seen also by Jayson et al. (1972). The second electron is captured into a level approximately 0.45 eV deep after which lattice relaxation increases the average optical ionization energy to about 2.0 eV. The role of lattice relaxation was established by studying the temperature dependence of the photoionization of the trapped electrons. Optical absorption by impurities in GaP has been studied by Dishman and DiDomenico (1971).

Turning our attention now to InP, we know that this is a semiconductor with a direct energy gap of 1.35 (300°K) or 1.421 eV (0°K). The shallow impurities that have been studied in InP are the acceptors Zn and Cd and the donors S, Se, and Te. The B_2O_3 liquid encapsulation technique with a silica crucible and an overpressure of 27 atm has produced excellent single crystals. Figure 3.13 shows the 300°K mobility for a large number of *n*-type crystals (undoped). The theoretical lines were calculated (Mullin et al.,

* "Photocapacitance Studies of the Oxygen Donor in GaP," to appear in *Phys. Rev.* (1973).

1971) assuming the effective mobility to be the reciprocal addition of impurity and lattice mobilities. The two dominant scattering mechanisms were taken to be polar optical mode scattering by the lattice and ionized impurity scattering. The lattice mobility was assumed to be 7000 $cm^2\ V^{-1}$ sec^{-1} and the ionized impurity mobility was calculated using the Brooks Herring equation (Hilsum and Rose-Innes, 1961)

$$\mu_I = \frac{3.2 \times 10^{15}(m_0/m^*)^{1/2}\varepsilon^2 T^{3/2}/(N_D + N_A)}{\ln\,[1.3 \times 10^{14}T^2\varepsilon(m^*/m_0)/(N_D - N_A)]} \tag{3.3}$$

with the relative dielectric constant taken as 10.9 and the effective mass ratio $m^*/m_0 = 0.062$. Compensating acceptor densities between 6.2×10^{16} and 1.8×10^{17} cm^{-3} were assumed to obtain theoretical lines that reflected the spread of the experimental points. The undoped crystals therefore were heavily compensated. The total impurity content $(N_D + N_A)$ was in the range 1 to 5×10^{17} cm^{-3} and the variation in the acceptor level was only a factor of 3, although the net electron concentration covered a range of $2\frac{1}{2}$ decades; $N_D - N_A$ was found to remain approximately the same over the temperature range 77 to 300°K, which suggests that shallow impurities are involved.*

Analysis of lightly silicon-doped crystals revealed (in parts per million atomic) the following: C, less than 5 to 13; O_2, 1.0; S, 0.15; Cu, 0.02; Zn, 0.3 to 0.6; and the added silicon. The range 1 to 5×10^{17} cm^{-3} would represent 5 to 25 ppm atomic of impurities. The role of C in InP has not been established.

With the degree of compensation that exists in InP crystals at present, it is clear that the deep impurity studies that can be made usefully are limited. Chromium has been studied and found to produce resistivities of 10^3 to 10^4 Ω-cm and preliminary work suggests a donor level 0.45 eV below the conduction band. When zinc is added in addition to the chromium, to compensate possibly shallow donors, resistivities of 10^6 Ω-cm are obtained.

Germanium and tin are shallow donors in InP when crystals are grown from In solutions at 650 to 600°C. The free-carrier concentrations vary linearly with the impurity concentrations in the melt up to 1 to 1.6×10^{19} carriers cm^{-3}. The distribution coefficient is 0.011 for Ge, and 0.002 for Sn (Rosztoczy et al., 1971).

InP crystals have been made photosensitive by the diffusion of copper into *n*-type material which then becomes high in resistance (Bube et al., 1961). InP has been grown epitaxially by the reaction of PCl_3 with high-purity indium. The carrier concentrations observed without intentional doping

* Further progress in the preparation and examination of InP is reported by Mullin in the *Proceedings of the 4th International Symposium on Gallium Arsenide and Related Compounds, 1972, Boulder*, The Institute of Physics, London.

were 10^{14} to 10^{18} cm^{-3} and it was suspected that oxygen was affecting the impurity level in this material (Joyce and Williams, 1971).

Manganese is an acceptor in InP with an activation energy 0.4 eV with respect to E_v and a solubility at least equal to 3×10^{18} cm^{-3} (Starosel'tseva et al., 1972).

3.8 Deep Impurities in InSb, InAs, GaSb, and AlSb

Impurity studies in the semiconductors InSb and InAs have arisen out of their use as photosensitive detectors and Hall and related magneto devices. The uses of GaSb and AlSb are fewer and the impurity studies are more limited.

The distribution coefficients of some impurities are given in Table 3.4 and the diffusion coefficients are given in Table 3.5. Some of the values are

TABLE 3.4

Distribution Coefficients $k^0 = C_s/C_l$ in Some III–V Compounds[a]

Impurity	InSb	InAs	GaSb	AlSb
Ag	4.9×10^{-5}			
As	5.4		2–4	
Au	1.9×10^{-6}			
B				0.01–0.02
C				0.6
Cd	0.26	0.13	0.02	0.002
Co				0.002
Cu	6.6×10^{-4}			0.01
Fe	0.04			0.02
Ge	0.045	0.07	0.08–0.32	0.026
In			~1	~1
Mg		0.7		0.1
Ni	6.0×10^{-5}			0.01
P	0.16			
Pb				0.01
S	0.1	1.0	0.06	0.003
Se	0.17–0.5	0.93	0.18–0.4	0.003
Si		0.40	~1	~1
Sn	0.057	0.09	0.01	$(2–8) \times 10^{-4}$
Te	0.54–3.5	0.44	0.4	0.01
Zn	2.3–4.13	0.77	0.16–0.3	0.02

[a] Literature references to the values in this table may be found in Madelung (1964). Distribution coefficients for GaAs are given in Table 3.1.

TABLE 3.5

Diffusion Coefficient D_0 and Energy ΔE [from $D = D_0 \exp(-\Delta E/kT)$] for Diffusion of Elements in Some III–V Compounds[a]

Diffusing Element	InSb		InAs		GaSb		AlSb	
	D_0 ($cm^2\ sec^{-1}$)	ΔE (eV)	D_0 ($cm^2\ sec^{-1}$)	ΔE (eV)	D_0 ($cm^2\ sec^{-1}$)	ΔE (eV)	D_0 ($cm^2\ sec^{-1}$)	ΔE (eV)
Al								1.8
Au	7×10^{-4}	0.32						
Cd	1.3×10^{-4}	1.2	4.35×10^{-4}	1.17				
	1.0×10^{-5}	1.1						
	1.23×10^{-9}	0.52						
Co	10^{-7}	0.25						
Cu	See Fig. 3.14						3.5×10^{-3}	0.36
Fe	10^{-7}	0.25						
Ga					3.2×10^{3}	3.15		
Ge			3.74×10^{-6}	1.17				
In	0.05	1.81			1.2×10^{-7}	0.53		
	1.8×10^{-9}	0.28						
Mg			1.98×10^{-6}	1.17				
Ni	10^{-7}	0.25						
S			6.78	2.20				
Sb	0.05	1.94			3.4×10^{4}	3.44		
	1.4×10^{-6}	0.75			8.7×10^{-3}	1.13		1.5
Se			12.55	2.20				
Sn	5.5×10^{-8}	0.75	1.49×10^{-6}	1.17	2.4×10^{-5}	0.80		
Te	1.7×10^{-7}	0.57	3.43×10^{-5}	1.28	3.8×10^{-4}	1.2		
Zn	1.6×10^{6}	2.3						
	5.5	1.6	3.11	1.17			0.33	1.93
	0.5	1.35						

[a] Literature references to the values in this table may be found in Madelung (1964). See also Yarbrough (1968).

widely scattered because the results observed depend on the perfection of the semiconductor material and the experimental techniques used. The diffusion values for InSb are shown graphically in Fig. 3.14.

The preparation, properties, and device applications of InSb have been reviewed by Hulme and Mullin (1962), and their conclusions on the electrical properties of impurities are shown in Table 3.6. The ionization energies for S, Se, and Te are negligibly small in InSb and InAs. The low effective electron mass and the hydrogen-like model of an impurity lead to large electron "orbits" and an impurity band for small donor concentrations that coalesces with the conduction band.

Copper is seen to be an acceptor in InSb as it is in GaSb and AlSb. However, in InAs copper is a donor and is the cause of carrier density changes with heat treatment which have been studied by Dixon and Enright (1959) and by Hilsum (1959). Heat treatments and quenching from temperatures in the range 450 to 850°C were found to result in increases in electron concentrations of the order of 10^{16} to 10^{17} cm^{-3}. The effects could be reversed almost entirely by annealing at temperatures of 250 to 300°C.

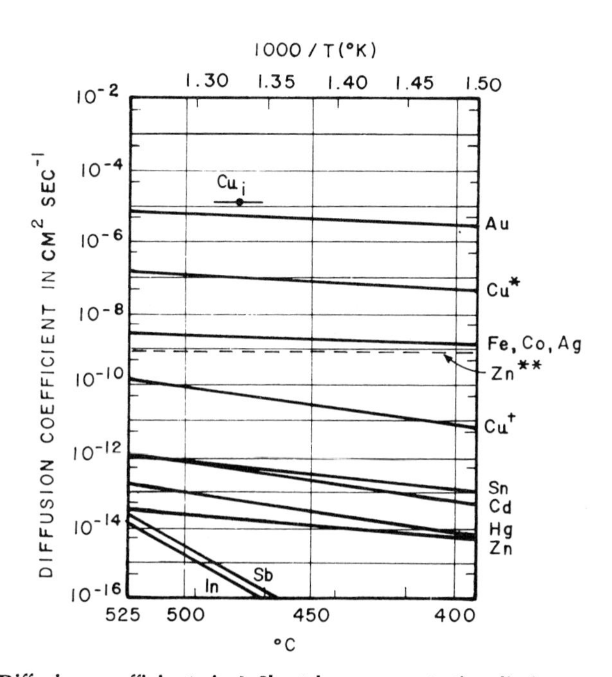

FIG. 3.14 Diffusion coefficients in InSb at low concentration limit except as noted. ***at 2 × 10^{20} cm^{-3}; †dislocation free; *10^4 dislocations cm^{-2}. (After Kendall, 1968, Chap. 3.)*

TABLE 3.6

Electrical Effects of Impurities in InSb

Element	Electrical Effect	Energy Levels where Known
Na, Li?	Donors	
Cu	Double acceptor	0.023 eV from valence band; deeper level 0.056 eV
Ag	Double acceptor	0.022 eV from valence band; deeper level 0.039 eV
Au	Double acceptor	0.32 eV from valence band; deeper level 0.066 eV
Mg, Zn, Cd	Acceptors	0.0075 eV Zn?
Al	Acceptor?	
Ga, As	Neutral	
Si, Ge[a]	Acceptors	
Sn	Donor	
Pb	Donor or neutral?	
S, Se, Te	Donors	Apparently zero ionization energy (see text)
Mn[b]	Acceptor	
F	Donor?	

[a] Activation energy of Ge, 0.925×10^{-2} eV, Murzin et al., 1972.
[b] Activation energy 0.95×10^{-2} eV, Dashevskii et al., 1971.

Copper atoms at dislocations in InAs are apparently electrically inactive, but on heating they diffuse away and can be frozen into the lattice as electrically active centers by subsequent quenching. They return to the dislocations during the low-temperature anneal. A similar effect is observed with Cu in InSb (Madelung, 1964), but is not seen in GaAs.

Several deep donors of undetermined origin have been observed in InSb studies (Trifonov and Yaremenko, 1971; Ismailov et al., 1971; Mirianashvili and Nanobashvili, 1971).

It is interesting to consider the doping behavior of group IV atoms (Si, Ge, etc.) in the III–V semiconductors:

1. A group IV atom is a donor on a III sublattice and an acceptor on a V sublattice.
2. The group IV atoms may be distributed statistically between the two lattices and whether the predominant effect is as a donor or acceptor depends on the nature and concentration of the impurities and on the external conditions during doping. There also may be some tendency for the atoms to be located on neighboring lattice III, IV sites as pairs, which would then be an electrically neutral condition.

TABLE 3.7

Electrical Effects of Group IV Atoms in III–V Semiconductors[a]

Doping Element	AlSb	GaP	GaAs	GaSb	InP	InAs	InSb
C	*A*		*A* (0.019)				
Si	*A*	*D* (0.082)	*D* (0.02)	*A*	*D*	*D*	*A*
		A (0.203)	*A* (0.03)				
			A (complexes)				
Ge	*A*	*A* (0.30)	*D* (shallow)	*A*	*D* (shallow)	*D*	*A*
			A (0.03)				
			A (complex)				
Sn	*D* + *A*	*D* (0.058)	*D* (shallow)	*A*	*D* (shallow)	*D*	*D*
			A (0.20)				
Pb	*A*		*A* (0.12)	*N*	*N*	*N*	*D* (?)

[a] The doping behavior is indicated by *A*, acceptor; *D*, donor; and *N* for no effect.

Covalent radii of the group IV elements are C, 0.77; Si, 1.17; Ge, 1.22; and Sn, 1.40 Å. The covalent radii for the group III elements are Al, 1.26; Ga, 1.58; and In, 1.65 Å. For the group V elements they are P, 1.10; As, 1.18; and Sb, 1.36 Å. On the assumption that the group IV impurity atom prefers to occupy the site that offers the largest radius, Si or Ge may be expected to be on the group III sublattice and therefore to be donors. This is true in GaAs and InAs. For instance, in GaAs prepared by vapor-phase epitaxy Si is a donor contributing one electron per atom up to a density of 10^{18} atoms cm^{-3}. Thereafter the effective doping rate decreases so that only 2 to 3 × 10^{18} carriers cm^{-3} are obtained for doping levels in the 10^{19} atoms cm^{-3} range. However, if GaAs is prepared by solution growth from gallium under conditions when gallium vacancies might be suppressed and the arsenic pressure is low, Si is an acceptor. Therefore consideration of site locations in terms of covalent radii sizes and vacancy concentrations results in some correct predictions. However, it is not entirely successful, as can be seen by inspection of Table 3.7, which summarizes the behavior of group IV dopants.

The behavior of Sn in GaSb has been studied by Fredericks et al. (1963), and the ternary phase diagram Ga–Sb–Sn was examined. Lithium in GaSb has been studied by Beer et al. (1965), and Si in GaSb by Burdiyan et al. (1972). Radiative recombination effects in various III–V semiconductors have been reviewed by Varshni (1967).

In certain ternary semiconductors, such as $Al_xGa_{1-x}As$, in which a transition from direct to indirect band gap occurs as x is increased, the conduction bands may have minima that are closely the same in energy. It then appears that one impurity can have several different levels associated with different extrema of the allowed bands. This concept has been developed by Alferov et al. (1972).

Luminescence studies of indirect band gap $Al_xGa_{1-x}As$ have been made by Kressel et al. (1970) and the optical ionization energies of Sn and Te donors determined to be in the range 59 ± 7 meV and to be 56 ± 5 meV for Zn. The theory of luminescent efficiency of ternary semiconductors has been discussed by Hakki (1971).

CHAPTER FOUR

Steady-State Statistics of Deep Impurities

The application of Fermi-Dirac statistics to semiconductors with shallow impurities (see, for instance, Blakemore, 1962) is assumed to be familiar to the reader. The statistics of deep impurities then presents no major problems, although there are a few difficulties associated with multiple energy levels, degeneracy factors, and amphoteric dopants. This chapter avoids discussion of very heavily doped conditions that may result in impurity band conductivity and hopping, since these are examined in Chapter 14.

The problem we are interested in solving is that of being able to calculate, for any temperature, the carrier densities in a semiconductor that contains both deep and shallow impurities of known energy levels and densities. If this problem is solved, one also has the information required to calculate the charge states of the impurities. Conversely one would like, from measurements of the majority carrier density as a function of temperature, to be able to infer the densities (and energy levels) of all the impurities present in the semiconductor. This is not a practical approach, except in very simple instances. Generally, to evaluate a complex doping situation, one must supplement carrier density versus temperature measurements by all the ancillary experiments that are appropriate. These may include optical probing for levels, controlled addition of known dopants to provide Fermi level shifts, and direct analysis by chemical means, neutron activation, or spark emission spectroscopy.

This chapter begins with a study of the statistics of a semiconductor in which a donor, which may be deep, is partially compensated by a shallow acceptor. This leads to a brief discussion of degeneracy factors for dopants.

The problem of a semiconductor containing two independent donor levels and a partially compensating acceptor is then considered. Amphoteric dopants are briefly discussed and the chapter concludes with the statistics of multilevel impurities.

4.1 The Statistics of a Partially Compensated Semiconductor, $N_D > N_A$

Before considering the statistics of multiple-level impurities let us examine, by way of review, the statistics of a semiconductor that contains a donor of density N_D at $E_c - E_D$ and acceptors of density N_A (where $N_A < N_D$) at some level, or levels, near the valence band edge. The donor will drop electrons onto the acceptor levels and only $N_D - N_A$ electrons will be available for distribution between the donor level at $E_c - E_D$ and the conduction band. As usual in nondegenerate statistics with the Fermi level, E_F, remote from the band edge, the conduction band will be represented by an effective density of states N_c $[=2(2\pi m^* kT/h^2)^{3/2}]$ assumed located at the band edge E_c. The electron density in the conduction band is then given by

$$n = \frac{N_c}{1 + \exp((E_c - E_F)/kT)} \tag{4.1}$$

$$\simeq N_c \exp\left(-\frac{E_c - E_F}{kT}\right) \tag{4.2}$$

The density of electrons remaining in the donor levels is

$$N_D{}^0 = \frac{N_D}{1 + g\exp((E_D - E_F)/kT)} \tag{4.3}$$

where g is the degeneracy factor for the donor level. (Degeneracy factors are discussed briefly in Section 4.2.) The equation obtained by electron summation is

$$n + N_D{}^0 + N_A = N_D \tag{4.4}$$

The density of ionized donors $N_D{}^+$ is

$$N_D{}^+ = n + N_A = N_D - N_D{}^0 \tag{4.5}$$

and from (4.3) this becomes

$$N_D{}^+ = \frac{N_D}{1 + g^{-1}\exp(-(E_D - E_F)/kT)} \tag{4.6}$$

This may be rearranged as

$$\frac{nN_D^{\,+}}{N_D - N_D^{\,+}} = \frac{n(n + N_A)}{N_D - n - N_A} = gN_c \exp\left(-\frac{E_c - E_D}{kT}\right) \tag{4.7}$$

which is a form sometimes useful for curve matching.

Provided the Fermi level is several kT below the conduction band edge, the equations given may be manipulated to obtain an expression for the free-carrier density primarily in terms of N_D, N_A, and T.

$$n = \frac{2(N_D - N_A)}{\left(1 + \dfrac{N_A}{gN_c}\exp\left(\dfrac{E_c - E_D}{kT}\right)\right) + \left[\left(1 + \dfrac{N_A}{gN_c}\exp\left(\dfrac{E_c - E_D}{kT}\right)\right)^2 + \dfrac{4(N_D - N_A)}{gN_c}\exp\left(\dfrac{E_c - E_D}{kT}\right)\right]^{1/2}} \tag{4.8}$$

If $N_A = 0$, (4.8) reduces to

$$n = \frac{2N_D}{1 + \left[1 + (4N_D/gN_c)\exp\left(\dfrac{E_c - E_D}{kT}\right)\right]^{1/2}} \tag{4.9}$$

which is the readily derived result of solving the simpler quadratic obtained for n in the absence of compensation. From (4.9) the variation of electron density as a function of temperature would be as $(E_c - E_D)/2k$ if the $T^{3/2}$ variation of N_c with temperature could be neglected.

Numerical solutions of (4.8) provide the kinds of curves shown in Fig. 4.1 for the temperature dependence of the free-electron density. The curves correspond to 0, 1, 9, and 50% compensation with $N_D - N_A$ held at 10^{16} cm^{-3} and the donor level taken as 0.01 eV. The variation of the slope of the curves, with the compensation fraction, is seen to be quite pronounced. Although in this illustration the donor level considered is quite shallow, the principles of the analysis apply also to a deep donor. The analysis of course also represents the condition of a shallow donor with deep acceptors present.

If the temperature is low so that $n \ll N_A$, (4.8) tends to become

$$n \simeq \left(\frac{N_D - N_A}{N_A/g}\right) N_c \exp\left(-\frac{E_c - E_D}{kT}\right) \tag{4.10}$$

The quantity $N_D - N_A$ is readily determinable from the high-temperature (fully ionized) part of the curve. In the event that one is trying to fit an experimental n versus $1/T$ curve, (4.10) is useful because it provides some

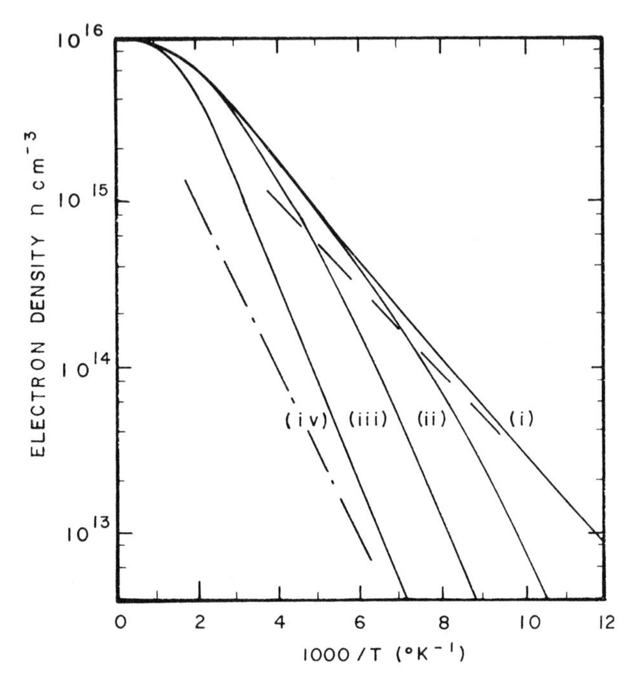

FIG. 4.1 Semilogarithmic plot of n_0 versus $1/T$ for a set of partially compensated donors under nondegenerate conditions.

For simplicity it is assumed that $N_D - N_A = 10^{16}$ cm^{-3} in each case, with $E_D - E_F = 0.01$ eV, $m_c = 0.25m_0$, $g = \frac{1}{2}$. (i) For $N_A = 0$, that is, zero compensation. (ii) For $N_A = 10^{14}$ cm^{-3}. (iii) $N_A = 10^{15}$ cm^{-3}. (iv) $N_A = 10^{16}$ cm^{-3}. The dashed line is of slope $(E_D - E_F)/2k$, and the chain dot line of slope $(E_D - E_F)/k$. (After Blakemore, 1962.)

indication of $E_c - E_D$ and of the density N_A. Then a least-squares-error fit can be attempted.

Majority carrier measurements as a function of temperature are, of course, made by Hall experiments (Putley, 1960). Alternatively, they may be made by the van der Pauw technique (1958), which allows a more arbitrary geometry for the specimen.

The Hall measurement technique for determining carrier densities is less useful as the doping becomes higher because the degree of dopant deionization with decrease of temperature becomes less. For n-type GaAs, which has a low effective mass and therefore a low effective density of states N_c (about 4.7×10^{17} cm^{-3} at 300°K), the problem arises for donor densities over about 5×10^{15} cm^{-3}. A useful technique that enables the measurements to be extended to 3×10^{17} cm^{-3} is to study the mobility in a suitable temperature range and infer the doping density from ionized impurity scattering theory (Wolfe et al., 1970).

Impurity atoms may be expected to have excited states. For instance, a donor atom has electrical neutrality whether its "donatable" electron is bound in the ground state $E_c - E_D$ or in an excited state closer to the conduction band edge. From infrared absorption measurements (Blakemore, 1962) the small amount of evidence that is available suggests that excited states of deep impurities tend to be far from the ground state and near to the semiconductor band edges. Therefore they do not appear to be of much practical importance, except in the contribution they may make in capture processes via the cascade model of Lax (1960).

4.2 Degeneracy Factors

The numerical values to assign to the degeneracy factors for dopants in semiconductors are somewhat uncertain.

Consider a group III acceptor in a conceptual group IV semiconductor in which the problem is not complicated by band degeneracy. By Fermi-Dirac statistics it may be shown that the probability $P(E)$ of the level being occupied by two electrons with paired spins (i.e., the ionized condition) is

$$P(E) = \frac{1}{1 + 2 \exp((E - E_F)/kT)} \tag{4.11}$$

and hence the degeneracy factor expected is 2. If the density of acceptors is N_A and the density of those nonionized is $N_A{}^0$, then

$$N_A{}^0 = N_A(1 - P(E))$$

whence

$$N_A{}^0 = \frac{N_A}{1 + \frac{1}{2} \exp((E_F - E_A)/kT)} \tag{4.12}$$

Similarly for a group V donor in a group IV semiconductor, the fifth electron may have either value of spin. Then Fermi-Dirac statistics show that the density of nonionized donors (i.e., those with electron retained) is

$$N_D{}^0 = \frac{N_D}{1 + \frac{1}{2} \exp((E_D - E_F)/kT)} \tag{4.13}$$

where the degeneracy factor is therefore $\frac{1}{2}$.

The problem becomes more complicated if the impurity level is regarded as created by splitting off states from a conduction or valence band with multiple or degenerate extrema. For acceptor levels of group III impurities in a semiconductor such as germanium, which has two valence bands degenerate at $k = 0$, Kohn (1957) has predicted that the degeneracy factor should be 4 instead of 2. Evidence supporting this has been obtained. For instance, the curve of hole density versus $1/T$ for germanium doped with indium (and partially compensated with antimony) tends to show a better fit to theory for $g = 4$ than for $g = 2$ (Blakemore, 1962). However, such measurements are not easy to make with the precision that is required for complete confidence on these matters.

For gold in silicon it has sometimes been considered that the acceptor and donor levels are characterized by $g_A = 2$ and $g_D = \frac{1}{2}$. These values result if it is assumed that neutral gold in silicon has three valence electrons, that an electron of either spin may leave to form the positive gold ion, and that the fourth electron joining a neutral atom to make a negative gold ion must have the proper spin to fit into the remaining bond (Bullis, 1966).

It is perhaps more reasonable to assume that gold in silicon has only a single valence electron by analogy with gold in germanium. In gold-doped germanium all of the four levels associated with a substitutional impurity with a single valence electron have been observed. In this case there are four possible states in which the electron in a neutral gold atom could reside. This electron would be the only one available to leave in forming the positive gold ion, but a second electron joining a neutral atom to make a negative gold ion could occupy any of the three remaining sites. If this were the situation, then the degeneracy factors g_A and g_D would be $\frac{2}{3}$ and $\frac{1}{4}$, respectively.

However, as discussed in Chapter 2, the resistivity calculations for gold in silicon seem more representative of experimental results if g_A and g_D are each taken as 0.125. The failure of simple reasoning to suggest suitable degeneracy factors is perhaps related to the still unresolved complexity of the gold state in silicon and its possible relation to divacancies.

Ashley et al. (1970, 1971a) have studied the statistics of gold in germanium by correlating Hall effect and resistivity data for crystals with and without partial compensation of the gold by a shallow donor (antimony). For the partially compensated condition the free-hole density as a function of temperature is given by

$$\frac{p(N_D + p)}{N_{\mathrm{Au}} - N_D - p} = \frac{N_v}{\gamma^-} \exp\left(-\frac{E_{\mathrm{Au}} - E_v}{kT}\right) \tag{4.14}$$

or if p is small

$$p = \left(\frac{N_{\mathrm{Au}} - N_D}{\gamma^- N_D}\right) N_v \exp\left(-\frac{E_{\mathrm{Au}} - E_v}{kT}\right) \tag{4.14a}$$

where γ^- is the ratio of the state degeneracies of neutral and single ionized gold centers, g^0/g^-.

Measurement of the hole density at a high temperature gives $N_{Au} - N_D$, and from (4.14a) measurement of hole density at a low temperature gives $\gamma^- N_D$. Since $N_{Au} - N_D$ is known and N_{Au} can be obtained from measurements on uncompensated crystals grown under identical conditions, the compensation density N_D can be found and hence a value of γ^- inferred. For the lowest acceptor level (0.15 ± 0.002 eV above E_v) of gold in germanium the value of γ^- obtained is 4.

From a study of germanium containing gold and also a shallow acceptor impurity, information can be obtained about the gold donor level (at 0.043 eV above E_v). At low temperatures the shallow impurities accept electrons from the gold atoms, which become positively charged. Hall measurements in the temperature range 65 to 100°K lead to a degeneracy ratio γ^+ of $\frac{1}{4}$ where γ^+ is the ratio of the state degeneracies of the donor and neutral gold centers, g^+/g^0. Since γ^- $(=g_0/g^-)$ is 4, these measurements imply that $g^+ = g^-$. If g^0 is taken as 4 from tetrahedral bonding considerations, then $g^+ = g^- = 1$. This is satisfactory for g^+, but for g^- simple counting of available states for tetrahedral bonding would suggest a larger value such as 6. Ashley et al. (1970) conclude that actual behavior is more complicated than simple counting would imply.

It should be noted that the observations of Ashley and his coworkers do not support those of Ostroborodova (1965) who studied degeneracy factors in *p*-type germanium and concluded that the degeneracy factors for all the levels were identical and equal to 2. In *n*-type gold-doped germanium Ostroborodova concluded that the degeneracy factors of the deep acceptor levels were equal to 8.

For most engineering applications, however, the problem of assigning degeneracy factors is not a particularly critical one since the exponential factors in equations such as (4.13) and (4.14) tend to be more important than the values of g, and the densities of the dopants are usually not known very accurately.

4.3 Semiconductor with Two Donor Levels That Are Independent

The statistics of a semiconductor containing two or more independent donor impurities (each capable of contributing one electron to the conduction band) and a partially compensating acceptor have been considered by several investigators (Brooks, 1955; Putley, 1960; Blakemore, 1962).

Assume the impurities are as in Fig. 4.2a with densities N_{D_1}, N_{D_2}, and N_A and energy levels E_{D_1} and E_{D_2}. If $N_A > N_{D_1}$, the semiconductor behaves in the steady state primarily as though only the donor N_{D_2} were present with an effective compensation density $N_A - N_{D_1}$.

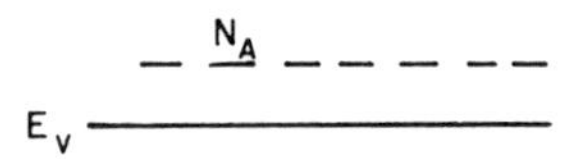

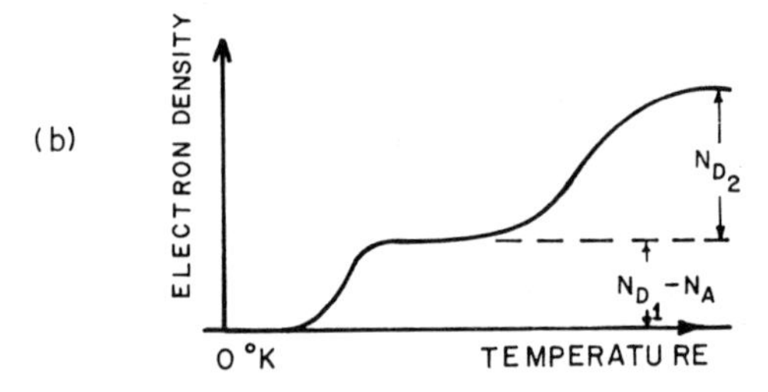

FIG. 4.2 Semiconductor with two independent monovalent donors and partial compensation of the upper donor.
(a) Energy model. (b) Variation of electron density with temperature.

The more interesting condition therefore is if $N_A < N_{D_1}$. At 0°K the lower donors will all be nonionized and the upper donors will contain $(N_{D_1} - N_A)$ electrons cm^{-3}. There will be no electrons in the conduction band until the temperature is increased, when up to $(N_{D_1} - N_A)$ electrons cm^{-3} are excited to the band from the upper donors. Then for a range of temperature very little electron density change occurs, as shown in Fig. 4.2b, until a temperature range is reached over which the deeper donors give their electrons to the conduction band. This of course assumes that the donor levels E_{D_1} and E_{D_2} are many kT apart in energy.

The problem becomes different and more complicated (Champness, 1956) if the energy levels E_{D_1} and E_{D_2} represent the two electron activation energies of a doping center that is capable of being doubly ionized. For this situation Brooks (1955) has shown that the average number of electrons occupying the double center is

$$f = \frac{2}{1 + \left[\dfrac{1 + 4\exp((E_{D_1} - E_F)/kT)}{1 + 4\exp((E_F - E_{D_2})/kT)}\right]} \tag{4.15}$$

This assumes that the first electron to be trapped can have only one direction of spin, while the second can have either. If the first electron can have either spin and the second has to pair it, then the factors 4 disappear from (4.15).

4.4 Amphoteric Dopants

Some impurities may dope a semiconductor either *p*- or *n*-type. An impurity which may be either substitutional or interstitial, depending on its density and on the heat treatment during its introduction, may exhibit this behavior. For example, indium and thallium when substitutional in silicon are deep acceptors (0.16 and 0.26 eV), but when these elements are in interstitial lattice sites there is some evidence that they act as donors. Another example of change of doping behavior is when an impurity such as germanium may occupy either gallium sites or arsenic sites in gallium arsenide depending on the doping density and the processing steps.

The action of gold in silicon is yet a third example of amphoteric doping. A gold atom may either lose or gain an electron depending on the other doping in the silicon. Three charge states (Au^+, Au^0, and Au^-) have therefore to be considered. The donor level is at $E_v + 0.35$ eV and the acceptor level is at $E_c - 0.55$ eV. The charge state of gold in silicon versus background doping is illustrated in Fig. 4.3. In strongly *p*-type material, the gold donor level is completely ionized and minority carrier (electron) capture is most efficient. As the shallow acceptor density N_A decreases, the Fermi level E_F rises and locks on to the gold donor level for $N_A \approx \frac{1}{2}N_{Au}$. As the net shallow donor density N_D increases, the Fermi level locks on to the gold acceptor level. For $N_D \gg N_{Au}$ all gold acceptor levels are ionized, again making the capture of minority carriers (holes) most effective. The temperature has

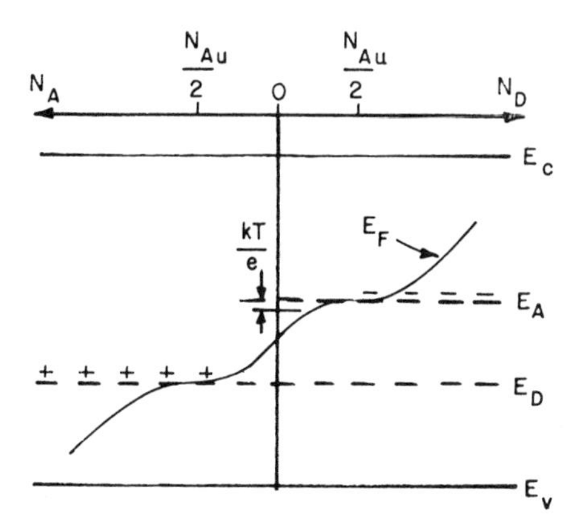

FIG. 4.3 Charge state of gold and Fermi level E_F in silicon versus shallow background doping.
The degeneracy factors are assumed as unity.

relatively little effect on the gold level occupancies unless the gold density N_{Au} is comparable with either N_A or N_D.

The behavior of gold in silicon has been commented on in Chapter 2 and is discussed further in Chapter 6.

4.5 Generalized Statistics for Multilevel Impurities

The statistics of multilevel impurities have been considered by Shockley and Last (1957). The statement of the mathematics that follows is taken from a review of the problem by Wagener (1964).

The model considered for the calculation of hole-electron equilibrium statistics is the energy level diagram shown in Fig. 4.4. This diagram shows the multiple localized energy levels introduced into the forbidden band gap of a semiconductor by a general impurity. Here E_v is of course the top of the valence band, and E_c the bottom of the conduction band. The objective of the calculation is the determination of the degree of electron occupancy of all of the local energy levels and then, by taking the proper sum, determination of how many free holes and electrons there are in the material.

Referring to Fig. 4.4, the level E_{D_1} represents the removal of one electron from the neutral impurity center into the conduction band, $E_c - E_{D_1}$ being the amount of energy needed to effect this removal. Likewise E_{D_2} represents the removal of a second electron from the impurity, requiring an

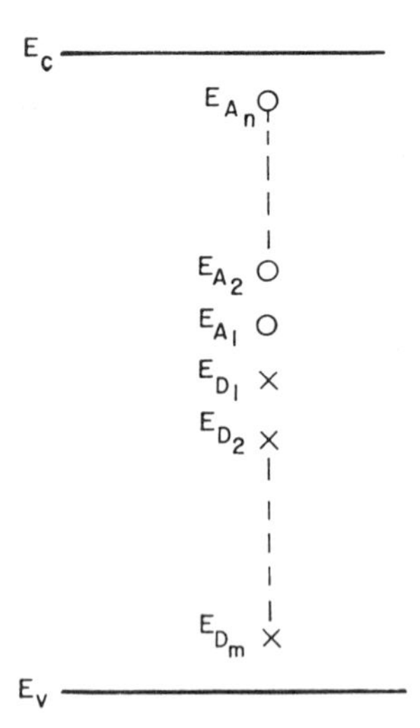

FIG. 4.4 Model illustrating the ionization states of a multilevel impurity.

energy of $E_c - E_{D_2}$. The level E_{D_m} represents the removal of the mth electron, requiring an energy of $E_c - E_{D_m}$. The removal of the $(m + 1)$st electron from the impurity center requires more energy than $E_c - E_v$, and thus will never normally occur since electron excitation from E_v to E_c will take place first. This essentially means that the level $E_{D_{(m+1)}}$ (if it exists) lies below E_v and is never observed. In a similar manner to the E_D levels, the E_A levels refer to addition of electrons to the impurity center; for instance, E_{A_n} represents the addition of n electrons to the impurity center. These electrons are taken from the valence band and thus leave free holes for electrical conduction in the valence band. The energy required to take the nth electron from the valence band and deposit it on the impurity center is $E_{A_n} - E_v$. Again, the energy required to take the $(n + 1)$st electron from the valence band and put it on the impurity is greater than $E_c - E_v$, so that this level (if it exists) is shifted above E_c and is also never observed.

The order of the various energy levels in Fig. 4.4 is not arbitrary, but must be as shown. In order to see this, one must realize that the farther below the conduction band the Fermi level lies the more electrons will be removed from the impurity centers, and the farther above the valence band the Fermi level lies the more electrons will be added to the impurity centers. In addition it is obvious that an impurity center can have only one value of charge at any one time; that is, for instance, it cannot be lacking an electron and holding an excess electron at the same time.

The occupancies of the various localized energy levels are of course not independent. Each level will be occupied by a certain fraction of the total number of impurity centers; and the sum of all of these fractions, plus the fraction of impurity centers that remain un-ionized, must equal unity. A mistake commonly made is to use the seemingly appropriate Fermi function and the density of impurity centers to calculate the occupancy of a given level. Although the occupancy of any given level depends, at all times, on the location of the Fermi level in the band gap, this procedure is only an approximation, and is valid only under certain circumstances. In general the occupancy of a given level cannot be obtained directly with a simple Fermi function, and one must resort to the statistical relation between the thermal equilibrium occupancies of two adjacent states. This key relation (Shockley and Last, 1957) can be summarized by

$$\frac{N_s}{N_{s-1}} = \frac{g_{s-1}}{g_s} \exp\left(\frac{E_F - E_{s'}}{kT}\right) \tag{4.16}$$

where N_s is the density of impurity centers which contain s excess electrons (s can be positive, zero, or negative), and N_{s-1} is the density of impurity centers which contain $s - 1$ excess electrons. The g factors are the appropriate spin level degeneracies, and for this case are probably either 1 or 2. The appropriate values are determined by considering that in changing states the electron goes from one state into one of a possible two states. Thus if the impurity gains an electron, then $g_s = 2$ and $g_{s-1} = 1$; whereas if the impurity loses an electron, then $g_s = 1$ and $g_{s-1} = 2$. (In the development below, the g factors will be replaced by 1 or 2, whichever is appropriate, as this is by far the most common situation. However, the g factors could be carried through the calculation to give even greater generality.) In the exponent of (4.16), s' is either s or $s - 1$, depending on whether acceptor or donor action is involved. The level $E_{s'}$ is always the energy of the state in which the impurity center has the greatest charge (either positive or negative). If we consider acceptor levels, then $s' = s$; if we consider donor levels, then $s' = s - 1$. Therefore if s is an acceptor level

$$N_s = \frac{1}{2N_{s-1}} \exp\left(\frac{E_F - E_s}{kT}\right) \tag{4.17}$$

Similarly, if $s - 1$ is a donor level

$$N_{s-1} = \frac{1}{2N_s} \exp\left(\frac{E_{s-1} - E_F}{kT}\right) \tag{4.18}$$

The calculation of equilibrium statistics is, then, as follows. If N is the density of impurity centers in the semiconductor material,

$$N = N_0 + \sum_{i=1}^{n} N_{A_i} + \sum_{j=1}^{m} N_{D_j} \tag{4.19}$$

where, N_0 is the density of uncharged impurity centers, N_{A_i} is the density of impurity centers that contain i excess electrons (this corresponds to level E_{A_i}), and N_{D_j} is the density of impurity centers that have lost j electrons (this corresponds to level E_{D_j}). Application of (4.17) results in

$$N_{A_1} = \frac{1}{2N_0} \exp\left(\frac{E_F - E_{A_1}}{kT}\right)$$

$$N_{A_2} = \frac{1}{2N_{A_1}} \exp\left(\frac{E_F - E_{A_2}}{kT}\right) = \left(\frac{1}{2}\right)^2 N_0 \exp\left(\frac{E_F - E_{A_1}}{kT}\right) \exp\left(\frac{E_F - E_{A_2}}{kT}\right)$$

$$\vdots$$

$$N_{A_n} = \left(\frac{1}{2}\right)^n N_0 \left[\prod_{i=1}^{n} \exp\left(\frac{E_F - E_{A_i}}{kT}\right)\right] \tag{4.20}$$

and (4.18) gives

$$N_{D_1} = \frac{1}{2N_0} \exp\left(\frac{E_{D_1} - E_F}{kT}\right)$$

$$N_{D_2} = \frac{1}{2N_{D_1}} \exp\left(\frac{E_{D_2} - E_F}{kT}\right) = \left(\frac{1}{2}\right)^2 N_0 \exp\left(\frac{E_{D_1} - E_F}{kT}\right) \exp\left(\frac{E_{D_2} - E_F}{kT}\right)$$

$$\vdots$$

$$N_{D_m} = \left(\frac{1}{2}\right)^m N_0 \left[\prod_{j=1}^{m} \exp\left(\frac{E_{D_j} - E_F}{kT}\right)\right] \tag{4.21}$$

where the $\prod$ symbol represents the product of the successive terms. The combination of (4.19), (4.20), and (4.21) gives

$$N = N_0 + N_0 \sum_{k=1}^{n} \left(\frac{1}{2}\right)^k \left(\prod_{i=1}^{k} \exp\left(\frac{E_F - E_{A_i}}{kT}\right)\right) + N_0 \sum_{l=1}^{m} \left(\frac{1}{2}\right)^l \left[\prod_{j=1}^{l} \exp\left(\frac{E_{D_j} - E_F}{kT}\right)\right]$$

Now define

$$D \equiv \frac{N}{N_0} = 1 + \sum_{k=1}^{n} \left(\frac{1}{2}\right)^k \left[\prod_{i=1}^{k} \exp\left(\frac{E_F - E_{A_i}}{kT}\right)\right] + \sum_{l=1}^{m} \left(\frac{1}{2}\right)^l \left[\prod_{j=1}^{l} \exp\left(\frac{E_{D_j} - E_F}{kT}\right)\right] \tag{4.22}$$

Thus

$$\frac{N_{A_r}}{N} = \frac{(1/2)^r \left[\prod_{i=1}^{r} \exp\left((E_F - E_{A_i})/kT\right)\right]}{D} \equiv f(E_{A_r}), \qquad r = 1, 2, 3, \ldots, n \tag{4.23}$$

and

$$\frac{N_{D_q}}{N} = \frac{(1/2)^q \left[\prod_{j=1}^{q} \exp\left((E_{D_j} - E_F)/kT\right)\right]}{D} \equiv f(E_{D_q}), \qquad q = 1, 2, 3, \ldots, m \tag{4.24}$$

In (4.23) the function $f(E_{A_r})$ gives the fraction of the N impurity centers which contain *exactly* r excess electrons, and $f(E_{D_q})$ in (4.24) gives the fraction of the N impurity centers which have lost *exactly* q electrons. It is not necessary, but it is instructive, to derive similar functions which give the fractions of the N impurity centers which contain *at least* r excess electrons and which have lost *at least* q electrons. Since N_{A_r} was defined as the density of impurity centers which contain exactly r excess electrons, then let N'_{A_r} be the density of impurity centers which contain at least r excess electrons. Thus

$$N'_{A_r} = \sum_{\alpha=r}^{n} N_{A_\alpha} \tag{4.25}$$

Similarly, if N'_{D_q} is the density of impurity centers that have lost at least q electrons, then

$$N'_{D_q} = \sum_{\beta=q}^{m} N_{D_\beta} \tag{4.26}$$

The desired functions are then

$$\frac{N'_{A_r}}{N} \equiv F(E_{A_r}) \tag{4.27}$$

and

$$\frac{N'_{D_q}}{N} \equiv F(E_{D_q}) \tag{4.28}$$

By simple algebraic manipulations, it can be shown that

$$F(E_{A_r}) = \left[1 + 2\exp\left(\frac{E_{A_r} - E_{\mathrm{F}}}{kT}\right)[\gamma]\right]^{-1} \tag{4.29}$$

$$\gamma = \frac{D - \left(\frac{1}{2}\right)^{r}\left[\prod_{i=1}^{r}\exp\left(\frac{E_{\mathrm{F}} - E_{A_i}}{kT}\right)\right]}{\sum_{\alpha=r}^{n}\left(\frac{1}{2}\right)^{\alpha-1}\left[\prod_{\substack{i=1\\ i\neq r}}^{\alpha}\exp\left(\frac{E_{\mathrm{F}} - E_{A_i}}{kT}\right)\right]} \tag{4.30}$$

and

$$F(E_{D_q}) = \left[1 + 2\exp\left(\frac{E_{\mathrm{F}} - E_{D_q}}{kT}\right)[\delta]\right]^{-1} \tag{4.31}$$

where

$$\delta = \frac{D - \left(\frac{1}{2}\right)^{q}\left[\prod_{j=1}^{q}\exp\left(\frac{E_{D_j} - E_{\mathrm{F}}}{kT}\right)\right]}{\sum_{\beta=q}^{m}\left(\frac{1}{2}\right)^{\beta-1}\left[\prod_{\substack{j=1\\ j\neq q}}^{\beta}\exp\left(\frac{E_{D_j} - E_{\mathrm{F}}}{kT}\right)\right]} \tag{4.32}$$

Equations 4.29 and 4.31 look very much like the familiar simple Fermi-Dirac distribution functions, and this similarity is discussed later.

Now that the occupancy factors for all of the levels in Fig. 4.4 have been found, hole-electron statistics are computed by counting up all of the positive and negative charges in the material and invoking charge neutrality.

Thus

$$p + N_D^{+} = n + N_A^{-} \tag{4.33}$$

where p is the density of holes in the valence band, n is the density of electrons in the conduction band, N_D^{+} is the density of bound holes on impurity centers, and N_A^{-} is the density of bound electrons on impurity centers. The value for n is obtained in the standard way, and is given by

$$n = 4\pi(2m_e^*)^{3/2}h^{-3} \int_{E_c}^{E_{max}} \frac{(E - E_c)^{1/2}\, dE}{1 + \exp((E - E_F)/kT)} \tag{4.34}$$

and in most cases can be approximated by

$$n = N_c \exp\left(-\frac{E_c - E_F}{kT}\right) \tag{4.35}$$

where N_c is the effective density of states in the conduction band. Similarly, p is given by

$$p = 4\pi(2m_v^*)^{3/2}h^{-3} \int_{E_{min}}^{E_v} \frac{(E_v - E)^{1/2}\, dE}{1 + \exp((E_F - E)/kT)} \tag{4.36}$$

and can usually be approximated by

$$p = N_v \exp\left(-\frac{E_F - E_v}{kT}\right) \tag{4.37}$$

where N_v is the effective density of states in the valence band. The term N_A^{-} is the total density of excess electrons on impurity centers and is given by

$$N_A^{-} = \sum_{r=1}^{u}\left(N_r \sum_{i=1}^{n_r} if(E_{A_i})\right) \tag{4.38}$$

where u is the number of different impurities contained in the material giving rise to acceptor levels; N_r is the density of the rth impurity of the u impurities; n_r is the total number of acceptor levels contributed by the rth impurity; and $f(E_{A_i})$ is the appropriate function, computed from (4.23), for the ith acceptor level of the n_r acceptor levels of the rth impurity. The term N_D^{+} is the total density of electrons being removed from impurity centers and is given by

$$N_D^{+} = \sum_{q=1}^{v}\left(N_q \sum_{j=1}^{m_q} jf(E_{D_j})\right) \tag{4.39}$$

where v is the number of different impurities contained in the material giving rise to donor levels; N_q is the density of the qth impurity of the v

impurities; m_q is the total number of donor levels contributed by the qth impurity; and $f(E_{D_j})$ is the appropriate function, computed by (4.24), for the jth donor level of the m_q donor levels of the qth impurity.

If the functions given in (4.29) and (4.31) are used in place of those of (4.23) and (4.24), then

$$N_A^- = \sum_{r=1}^{u} \left(N_r \sum_{i=1}^{n_r} F(E_{A_i}) \right) \tag{4.40}$$

where $F(E_{A_i})$ is given by (4.29), and

$$N_D^+ = \sum_{q=1}^{v} \left(N_q \sum_{j=1}^{m_q} F(E_{D_j}) \right) \tag{4.41}$$

where $F(E_{D_j})$ is given by (4.31).

So that, finally,

$$\begin{aligned} p - n = N_A^- - N_D^+ &= \sum_{r=1}^{u} \left(N_r \sum_{i=1}^{n_r} if(E_{A_i}) \right) - \sum_{q=1}^{v} \left(N_q \sum_{j=1}^{m_q} jf(E_{D_j}) \right) \\ &= \sum_{r=1}^{u} \left(N_r \sum_{i=1}^{n_r} F(E_{A_i}) \right) - \sum_{q=1}^{v} \left(N_q \sum_{j=1}^{m_q} F(E_{D_j}) \right) \end{aligned} \tag{4.42}$$

where all of the quantities in (4.42) have been defined above. The only unknown in (4.42) is E_F, so that when this equation is solved for E_F, then the values for n and p are known. This equation cannot, of course, be solved analytically, but using modern numerical methods the solution is readily obtained.

CHAPTER FIVE

Trap Concepts: Giant-Trap Model and Field-Dependent Effects

A deep impurity in a semiconductor may act either as a trap or as a recombination center, depending on the impurity, on the temperature, and on the other doping conditions. Consider a minority carrier captured at an impurity center. If the carrier lives a mean lifetime in the captured state and is ejected thermally to the band from which it came, we may regard the center as a trap. If, however, before thermal ejection can occur, a majority carrier is trapped, recombination will have taken place, and the impurity is acting as a recombination center. Which role a center will play depends then on the concentration of majority carriers and on the relative cross section for capture of minority and majority carriers.

To explain large capture cross sections a model has been proposed by Lax for capture into excited states and energy loss by a cascade process. This is reviewed and some of the limitations made clear. The effect of electric field on the emission of a carrier from a neutral trap by lowering of the energy barrier around the trap (the Poole-Frenkel effect) is then discussed. This is followed by a broader discussion of electric field effects on capture and generation processes at traps. The chapter concludes with a discussion of Auger recombination or trapping processes, in which energy lost by one carrier is carried away, in part or in whole, by a concurrent event involving a second carrier.

Centers that are singly charged and attractive to minority carriers are likely to act as recombination centers since the cross section for the subsequent neutral capture of a majority carrier may only be one order of magnitude lower than the minority carrier cross section, and the number of majority carriers may be sufficiently large that recombination will occur before ejection. Centers that are doubly charged and attractive to minority carriers are likely to act as traps. They will possess a large cross section for the minority carrier, but after capture will be repulsive to the majority carrier. The repulsion can reduce the cross section by many orders of magnitude so that ejection of the trapped minority carrier will be much more likely than recombination.

Capture cross sections may be determined by many different experimental techniques, as described in later chapters. However, the quantity that is determined in a typical experiment is the capture rate c $\mathrm{cm}^3\ \mathrm{sec}^{-1}$. For electrons captured from the conduction band

$$\frac{dn}{dt} = -cnN \tag{5.1}$$

where n is the density of conduction electrons and N is the density of empty traps that can accept an electron. Assume that all the electrons have the same energy E_0 and velocity v_0 $[=(2E_0/m^*)^{1/2}]$. Then in a time Δt the volume of space "swept" through by the electrons is $n\sigma v_0\ \Delta t$ cm^3 where σ cm^2 is the capture cross section. Since the density of traps is N cm^{-3}, the number of captures is

$$\Delta n = n\sigma v_0\ \Delta t N \tag{5.2}$$

Hence for this simple model

$$c = \sigma v_0 \tag{5.3}$$

The experimental constant c is an average over $c(E)$ taking into account the actual electron energy distribution. For the usual case in which a thermal equilibrium distribution is present, we may write

$$c = \langle v \rangle \sigma = \left(\frac{4kT}{m^* \pi}\right)^{1/2} \sigma \tag{5.4}$$

where

$$\sigma = \frac{\langle v\sigma(E) \rangle}{\langle v \rangle} \tag{5.5}$$

or

$$\sigma = \frac{\int E\sigma(E) \exp(-E/kT)\, dE}{\int E \exp(-E/kT)\, dE} \tag{5.6}$$

In (5.6) two factors of $E^{1/2}$, one from v and one from the Boltzmann distribution, give rise to the factor E.

The cross section most frequently quoted in experimental papers is obtained by dividing the observed rate by the root mean square velocity (Lax, 1960):

$$\sigma_{\text{rms}} = \frac{c}{(3kT/m)^{1/2}} \tag{5.7}$$

More usually this is written as

$$\sigma = \frac{c}{v_{\text{th}}} \tag{5.8}$$

where v_{th} is the thermal velocity of free electrons and at temperatures between 77 and 300°K typically has values in the range 10^6 to 10^7 cm sec^{-1}.

The cross sections reported in the literature (and reviewed in Chapter 10) cover an enormous range, 10^{-12} to 10^{-22} cm^2. The giant-trapping cross sections, 10^{-15} to 10^{-12} cm^2, are associated with attractive centers. The cross sections 10^{-17} to 10^{-15} cm^2 are usually associated with neutral cross sections, and the tiny cross sections, 10^{-22} cm^2, are associated with repulsive centers.

Although a cross section of 10^{-15} cm^2 might seem reasonable in view of the geometrical size of an impurity state, the electron not only must come to the vicinity of the center but must on arrival perform the difficult task of disposing of perhaps 0.5 eV. It may take 10 or 20 phonons to carry away this energy. Thus the geometrical cross section would have to be multiplied by the (small) probability of such an unlikely energy losing collision. A direct transition to the ground state of a trap involving such a multiphonon process is likely to have a very small cross section, say 10^{-22} cm^2, in a semiconductor where the electron–phonon interaction is weak. For repulsive centers, which possess no excited states, this is the only process available (aside from optical and Auger processes which are also weak) and we therefore expect exceedingly small cross sections. The problem therefore is to explain (1) how giant cross sections are possible for attractive centers;

(2) how large cross sections are possible for neutral centers; and (3) why the cross section is so sensitive to the charge on the center. A cascade theory of capture developed by Lax (1959, 1960) has been successful in providing partially acceptable explanations to these questions and is now a classic paper in the field of trapping.

5.1 Lax's Cascade Capture Model

The cascade model proposes that large cross sections are made possible by capture into excited states possessing large orbits. The large energy loss is accomplished by a cascade process involving the successive emission of single phonons as the electron plummets down the excited states. Such excited states of large orbit are clearly available to a coulomb attractive center, and as we shall see permit (classically calculated) enormous cross sections that depend on the charge of the center but not on details of its potential. For neutral centers a moderately long-range inverse fourth-power potential is available because of the interaction between the charge carrier and the polarizability of the center. Since the polarizability of an atom varies inversely as the square of the ionization energy, and ionization energies of centers in solids are an order of magnitude lower than free-atom ionization energies, large polarizabilities result. These polarizabilities will sometimes, but not always, be large enough to permit a ladder of excited states sufficient to allow a cascade process.

Centers with a long-range coulomb attractive potential possess a series of excited states whose radius increases without bound. The states of large radius explain the large observed capture cross section. Indeed, the total cross section for capture into all excited states could diverge. Roughly speaking, however, one must include only states whose binding energy is greater than kT. Lowering the temperature then permits contributions from states of increasing radius. Thus a rapid increase of observed cross sections with decreasing temperature might be expected for impurities at low temperatures. [This is observed, for example, in antimony-doped germanium in the range 10 to 4°K (Koenig, 1959).]

Lax assumes that

> . . . in the vicinity of binding energy kT there are states closely enough spaced for one-phonon transitions to be possible. As a result of these transitions the electron escapes, or moves closer to the ground state. The last step, from a first excited state to the ground state, may require a multiphonon transition, or the emission of radiation. This last step is undoubtedly the step limiting the rate at which electrons enter the ground state of the trap. But the cross section, as measured by the rate at which electrons leave the conduction band, is not limited nor even affected by the rate of

this last step (providing the ionization energy of the first excited state is large compared to kT): electrons caught in excited states are unavailable for conduction.

For neutral centers . . . a quasi-long range interaction (an inverse fourth-power potential) is provided by the polarizability of the center. (Traps in solids are significantly more polarizable than free atoms because of their lower ionization energies.) Since the radius at which the potential energy is kT now varies as $T^{1/2}$, we expect a much less temperature dependent cross section. Since the contributions come from smaller radii than for the Coulomb attractive case, the resulting cross section is much more sensitive to the details of the potential at shorter ranges, i.e., the chemical nature of the trapping center.

5.1.1 Classical Approximations

A calculation of the sum of the cross sections for capture into all of the excited states of a center would, indeed, be a laborious task. When highly excited states are important, however, we anticipate from the Bohr correspondence principle that a classical calculation will be valid, i.e., that the motion of the electron between collisions can be described by a classical orbit. The probability per unit time that a collision will take place can be computed at each point of the orbit by treating the electron as if it were a plane wave with the energy and momentum appropriate to that point of the orbit. This procedure is equivalent to neglecting the accelerating effect of the attractive field during the collision and is valid if the fractional energy change along the orbit is small over a time of the order of the "duration of the collision."

5.1.2 Mechanisms for Energy Loss

In order to be captured the electron must lose enough energy in a collision to go into a bound orbit. The energy can be carried away by: (1) a photon, (2) another electron or hole, (3) optical phonons, (4) acoustical phonons: (1) cross sections for radiative capture can be readily calculated from first principles, or by scaling similar calculations for proton–electron recombination. The resulting cross sections are a few orders of magnitude too low to explain observed results. (2) Auger recombination with the help of another carrier will in general lead to nonexponential decay, or in steady-state experiments to a concentration dependence of the lifetime that is not usually observed. Of course Auger recombination should occur when high carrier densities are present. (3) Optical phonons are generally not very effective in nonionic materials in producing the momentum transfers required to produce resistivity. However, each optical phonon collision transfers so much more energy to the lattice than an acoustical phonon per collision that the Joulean heat loss to the latter is often predominantly via optical phonons. We shall see later that optical phonons make a significant contribution to the room temperature capture cross section in silicon and germanium. (4) At low temperatures, at large distances from the trap, electrons will only have enough energy to emit acoustical phonons. The enormous cross sections reported in the helium temperature range must then be associated with acoustical phonons.

5.1.3 Assumptions Made by Lax

Although Lax's treatment is too lengthy to be included here in its entirety, an idea of its features can be obtained by reviewing some of the assumptions that he lists.

(1) Between collisions the motion of the electron may be described by its classical orbit in the presence of an attractive center. For a coulomb attractive center this assumption is valid at distances greater than the solid state Bohr radius.

(2) The probability per unit time of a collision with a given energy loss, at any point of the orbit, may be computed (quantum-mechanically) as if the electron were moving uniformly (in a plane wave) with the momentum appropriate to that point on the orbit. (This implies that the change in momentum on the orbit over a time interval of the order of the duration of a collision may be neglected.)

(3) The ionization energy of the ground state of the center is large compared to kT (otherwise the center would not act as a trap).

(4) For binding energies in the vicinity of kT there are enough excited states to permit the use of classical mechanics for calculating transitions into such states or between them.

(5) Motion up and down among the excited states is so rapid that electrons which escape in effect come out instantaneously (as compared with the rate at which electrons disappear from the conduction band during the capture process) so that one need not be concerned with time delays but only with the fraction $P(U)$ of electrons that are captured with binding energy U that stick (i.e., enter the ground state before emerging). The overall cross section for capture of electrons with kinetic energy E_0 can therefore be written in the form

$$\sigma(E_0) = \int \sigma(E_0, U)\, dU\, P(U) \tag{5.9}$$

where $\sigma(E_0, U)\, dU$ is the differential cross section for a collision with energy loss $E_0 + U$ into a state of binding energy U.

(6) Equation (5.9) already neglects the possible dependence of the sticking probability $P(U)$ on the angular momentum of the state as well as its binding energy.

(7) The sticking probability $P(U)$ does not depend on the history of arrival into state U. Thus if $K(U, \hbar\omega)\, d(\hbar\omega)$ is the probability (not per unit time) that an electron in state U will emit a phonon $\hbar\omega$ in $d(\hbar\omega)$ then

$$P(U) = \int_{U+\hbar\omega>0} K(U, \hbar\omega)\, d(\hbar\omega) P(U + \hbar\omega) \tag{5.10}$$

represents the probability of entering the ground state as the probability of any first step times the probability that the first step leads to eventual capture.

(8) The total probability of all possible steps including emission: $\hbar\omega > 0$ and absorption: $\hbar\omega < 0$ is by definition unity.

$$\int K(U, \hbar\omega)\, d(\hbar\omega) = 1 \tag{5.11}$$

(9) If an electron in making transitions among the states U reaches a state of positive energy ($U < 0$) it has definitely escaped. Hence $P(U) = 0$ for $U < 0$, and this explains the lower limit $U + \hbar\omega > 0$ in (5.10).

(10) In addition to [the] major assumptions concerning the use of classical mechanics and a sticking probability, depending only on binding energy, [there are] some nonessential but convenient assumptions: (a) the electrons motion is describable in terms of an isotropic effective mass (denoted simply by m). (b) The effects of transverse and longitudinal acoustic modes can be lumped together by using a single average velocity of sound and a single constant for the interaction of these modes with the electron. (c) Optical modes can be characterized by a single energy—the Einstein approximation.

The uncertainties introduced by these additional approximations, e.g., by lack of knowledge of what to choose for the appropriate effective mass and interaction constant may easily cause an uncertainty of a factor of 2 or 3 but is unlikely to change the cross section by more than one order of magnitude.

5.1.4 Results of the Analysis

The cross section obtained by Lax for coulombic attraction capture

. . . via optical phonon emission is

$$\sigma_{\text{opt}}(T) = \frac{\pi^2}{8}\frac{w}{l_c}\frac{Ze^2}{\varepsilon kT}\left(\frac{Ze^2}{\varepsilon\hbar\omega}\right)^2 \frac{\lambda}{1 - \exp(-\lambda)}\, C(\lambda) \tag{5.12}$$

where

$$C(\lambda) = \int_0^\lambda P(\lambda - x)e^{-x}\left(1 - \frac{x}{\lambda}\right)^{-5/2} dx \times \left(1 + \bar{y} - \bar{y}\frac{x}{\lambda}\right)^{1/2} \tag{5.13}$$

and

$$P(y) \approx 1 - (1 + y + \tfrac{1}{2}y^2)\exp(-\lambda) \tag{5.14}$$

is the sticking probability expressed in dimensionless units $y = U/kT$, and $\lambda = \hbar\omega/kT$. Z is the charge on the trap, ε is the dielectric constant, l_c is the mean free path for an acoustical phonon collision, w is the squared ratio of optical to acoustical matrix elements, and $\hbar\omega$ is the optical phonon energy. Equation (5.12) may be rewritten as

$$\sigma_{\text{opt}}(T) = \sigma_0\lambda[1 - \exp(-\lambda)]^{-1}D(\lambda) \tag{5.15}$$

where

$$\sigma_0 = \frac{8\pi}{15}\frac{wZ^3}{l_c}\frac{e^2}{\varepsilon kT}\left(\frac{e^2}{\varepsilon\hbar\omega}\right)^2 \tag{5.16}$$

is a unit of cross section independent of temperature.

A rough estimate of the size of the cross section in silicon can be obtained by evaluating the temperature independent unit of cross section σ_0 given by (5.16). Using $l_c \approx 320$ Å for electrons, we find that for electrons in Si

$$\sigma_0 \approx 3.5 \times 10^{-15}\, wZ^3(0.1\ \text{eV}/\hbar\omega)^2 \tag{5.17}$$

If we use as $\hbar\omega \approx 0.06$ eV for the optical phonons, we get $\sigma_0 \approx 10^{-14}w$ cm^2 for a singly charged center. At room temperature, however, $\lambda = \hbar\omega/kT \approx 2.4$ and, the temperature dependent correction factor $\lambda[1 - e^{-\lambda}]^{-1}D(\lambda)$ is about 1.5, so that the final estimated cross section $\sigma_{opt} \approx wZ^3 1.5 \times 10^{-14}$ cm^2 is larger than the 3.5×10^{-15} cm^2 observed by Bemski (1958b) for electrons in Au^+ if we assume that the coupling constant ratio has a reasonable value $w \geq 1$ from the point of view of mobility.

For holes in silicon, whose mean free path for acoustic scattering is perhaps five times smaller than for electrons, we find $\sigma_{opt} \approx wZ^3 7.5 \times 10^{-14}$ cm^2 which appears five times larger than the just quoted cross section for electrons. However w for holes is undoubtedly smaller than for electrons since the hole mobility varies at $T^{-2.3}$ whereas the electron mobility varies as $T^{-2.8}$.

The fact that the calculated cross sections are larger than the observed values is perhaps accounted for by the difficulty in evaluating the sticking probability and other factors. This sticking probability is the least satisfactory assumption of the theory. Ascarelli and Rodriguez (1962) compute it for hydrogen-type shallow donors in germanium at 4°K. The observed electron capture cross section of ionized donors such as arsenic or antimony in germanium increases about as T^{-1} to T^{-2} below 10°K, which is expected when the actual structure of the conduction band is considered but the absolute value is about five times larger than predicted. The experimental results in silicon, however, are not conclusive because impurity band conduction may become significant at these low temperatures.

The scattering into lower orbits, calculated by Lax in terms of a reciprocal mean free path, is obtained by Smith and Landsberg (1966) from quantum-mechanical considerations. These last authors obtain capture cross-section values smaller than those found by Lax by about 25% for transitions involving optical phonons in silicon, but a factor of 10 different for germanium. The observed capture cross sections of shallow ionized donors are less than expected from the giant-trap model; Antonov-Romanovskii (1963) distinguishes between gas kinetic (σ') and diffusion (σ) cross sections. Lax's model gives σ', while experimentally either one may be observed, depending on conditions. To a first approximation, neither one of these theoretical cross sections changes in the presence of small electric fields. However, in experiments, electric field strengths as low as 1000 V cm^{-1} may influence trapping parameters significantly (Dussel and Bube, 1966c).

For neutral centers, the cross section for capture with phonon emission (optical and acoustical) is reasoned by Lax as roughly proportional to the polarizability, $\alpha \sim (\Delta E)^{-2}$ where ΔE is the ionization energy of the impurity atom in the semiconductor lattice. It has been argued, for very deep levels, that the polarizability may be insufficient to account (with the phonon cascade theory) for observed neutral capture cross sections of 10^{-16} to 10^{-15} cm^2 at room temperature (Bonch-Bruevich and Glasko, 1962). Nonradiative transitions involving such neutral centers can be explained by an Auger-type process (Sheinkman, 1964, 1965) where a center changes the charge state by two or more electronic charges. As an example, the energy released upon capture of an electron by the $E_v + 0.31$ eV zinc level is sufficient to transfer a second electron from the valence band to the upper ($E_c - 0.55$ eV) zinc level. However, experimental evidence for such processes in silicon is scant and has been further weakened by recent measurements on zinc-doped silicon which indicate capture cross sections unfavorable to the Auger process (Kornilov, 1965c).

It may be concluded that the theoretical models for capture processes are only marginally satisfactory on a quantitative level. Further development and matching with experimental results are needed. This process is not helped by the scatter of experimental cross-section values reported in the literature. The spread reflects the difficulty in obtaining good material for study, and also involves in some instances the limitations of the experimental techniques used for determining the cross sections.

5.2 The Effect of Electric Field on the Emission of a Carrier from a Neutral Trap

The Poole-Frenkel effect (Frenkel, 1938) results from the lowering of a coulombic potential barrier by the electric field applied to a semiconductor. For a trap to experience the effect it must be neutral when filled and positive when empty. A trap that is neutral when empty and charged when filled will not experience the effect because of the absence of the coulomb potential.

The effect is rather similar to the Schottky barrier lowering effect in a metal-semiconductor junction (Fig. 5.1). Although the restoring force in both effects is due to coulomb interaction between the escaping electron and a positive charge, they differ in that the positive image charge is fixed for Poole-Frenkel barriers but mobile with Schottky emission. This results in a barrier lowering twice as great for the Poole-Frenkel effect, that is,

$$\Delta\phi_{\mathrm{PF}} = \left(\frac{e^3\mathscr{E}}{\pi\varepsilon}\right)^{1/2} = \beta_{\mathrm{PF}}\mathscr{E}^{1/2} \tag{5.18}$$

$$\Delta\phi_{\mathrm{S}} = \left(\frac{e^3\mathscr{E}}{4\pi\varepsilon}\right)^{1/2} = \beta_{\mathrm{S}}\mathscr{E}^{1/2} \tag{5.19}$$

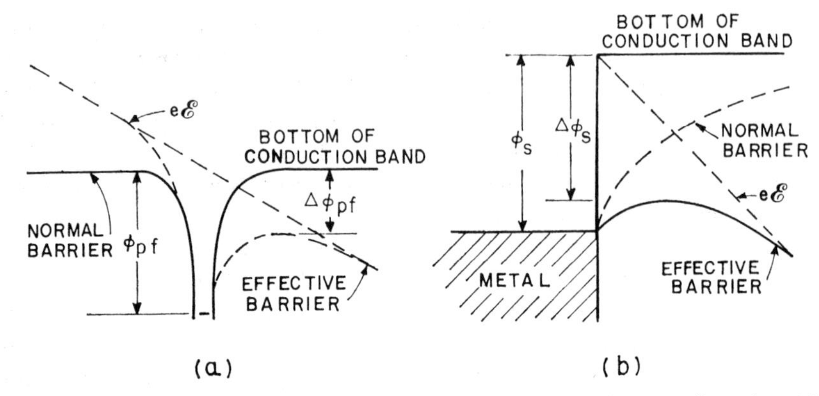

FIG. 5.1 The Poole-Frenkel effect (*a*), **compared with the Schottky effect** (*b*). **(After Yeargan and Taylor, 1968.)**

where $\Delta\phi_{PF}$ is the barrier lowering for Poole-Frenkel emission; $\Delta\phi_S$ is the barrier lowering for Schottky emission; e is the charge on an electron, $\mathscr{E}$ is the applied electric field; and ε is the high-frequency dielectric constant (Yeargan and Taylor, 1968).

The effects are governed by equations of the form

$$\text{(Poole-Frenkel)} \qquad J_{PF} = e\mu n_0 \mathscr{E} \exp\left(-\frac{\phi_{PF} - \Delta\phi_{PF}}{\tau kT}\right) \tag{5.20}$$

$$\text{(Schottky)} \qquad J_S = AT^2 \exp\left(-\frac{\phi_S - \Delta\phi_S}{kT}\right) \tag{5.21}$$

where τ is a parameter ranging from 1 to 2, depending on the position of the Fermi level, μ is the electron mobility, and n_0 is the electron density.

The concentration (n) of free carriers for an intrinsic material is

$$n = (N_c N_v)^{1/2} \exp\left(-\frac{E_c - E_v}{2kT}\right) \tag{5.22}$$

where N_c is the effective density of states in the conduction band; N_v is the effective density of states in the valence band; E_c is the lowest energy in the conduction band; and E_v is the highest energy in the valence band. For an n-type material with no acceptor sites,

$$n = \left(\frac{N_c N_D}{2}\right)^{1/2} \exp\left(-\frac{E_c - E_D}{2kT}\right) \tag{5.23}$$

where N_D is the density of donor sites and E_D is their energy level. A similar expression applies for p-type materials with no donor sites. These equations are of the form

$$n = n_0 \exp\left(-\frac{\phi}{2kT}\right) \tag{5.24}$$

where ϕ is the energy difference between the emission site and the bottom of the conduction band.

When traps or acceptor sites are present the situation may be different; since these levels are lower in energy than the bottom of the conduction band, they tend to fill first. When the number of electrons in the conduction band is small compared to either the donor or acceptor density, the concentration of free electrons is of the form

$$n = \frac{N_c(N_D - N_A)}{2N_A} \exp\left(-\frac{E_c - E_D}{kT}\right) = n_0 \exp\left(-\frac{\phi}{kT}\right) \tag{5.25}$$

We then have two limiting cases: (1) when the concentration of acceptor levels is small compared to the density of donor levels and free electrons (equation 5.24) and (2) when the concentration of excited electrons is small compared to the donor and acceptor densities (equation 5.25). In these limits, (5.20) becomes

CASE 1. $N_A \ll N_D$, $N_A \ll n$.

$$\begin{aligned} J &= e\mu n_0 \mathscr{E} \exp\left(-\frac{\phi_{\mathrm{PF}}}{2kT}\right) \exp\left(\frac{\Delta\phi_{\mathrm{PF}}}{2kT}\right) \\ &= \sigma_0 \mathscr{E} \exp\left(\frac{\Delta\phi_{\mathrm{S}}}{kT}\right) = \sigma_0 \mathscr{E} \exp\left[\left(\frac{e\mathscr{E}}{4\pi\varepsilon}\right)^{1/2} \Big/ kT\right] \end{aligned} \tag{5.26}$$

CASE 2. $n \ll N_D$, $n \ll N_A$.

$$\begin{aligned} J &= e\mu n_0 \mathscr{E} \exp\left(-\frac{\phi_{\mathrm{PF}}}{kT}\right) \exp\left(\frac{\Delta\phi_{\mathrm{PF}}}{kT}\right) \\ &= \sigma_0 \mathscr{E} \exp\left(\frac{2\Delta\phi_{\mathrm{S}}}{kT}\right) \end{aligned} \tag{5.27}$$

From (5.26), the logarithm of conductivity varies linearly with the square root of electric field for Poole-Frenkel emission.

This model is a one-dimensional planar model. The treatment for a three-dimensional case has been developed by Hartke (1968):

> The electrostatic energy of an electron which is attracted to a singly charged positive ion located at $\tau = 0$ and is under the influence of a uniform applied field $\mathscr{E}$ in the $-z$ direction may be written as
>
> $$E = -\frac{e^2}{4\pi\varepsilon\tau} - e\mathscr{E}\tau\cos\theta \tag{5.28}$$

where rationalized mks units and spherical coordinates are used and the arbitrary zero of energy is taken to be the bottom of the conduction band at $\tau = 0$. The thermal energy which a trapped electron must gain in order to escape is reduced by an amount δE which is obtained by setting $\partial E/\partial\tau = 0$ at $\tau = \tau_0$. The result is

$$\tau_0 = \left(\frac{e}{4\pi\varepsilon\mathscr{E}\cos\theta}\right)^{1/2} \tag{5.29}$$

$$\delta E = \left(\frac{e^3\mathscr{E}\cos\theta}{\pi e}\right)^{1/2} \tag{5.30}$$

where the barrier is lowered only for $0 \leq \theta \leq \frac{1}{2}\pi$. The release rate, or reciprocal lifetime, of a trapped electron may be obtained by assuming a spherically symmetric, field independent, attempt-to-escape frequency of $\nu/4\pi$ per unit solid angle and by assuming that the release rate is field independent for $\frac{1}{2}\pi \leq \theta \leq \pi$.

$$(\tau_r)^{-1} = \int_0^{2\pi} d\phi \int_0^{\pi/2} d\theta \sin\theta \left(\frac{\nu}{4\pi}\right) \exp\left(-\frac{E_T - \delta E}{kT}\right) + \int_0^{2\pi} d\phi \int_{\pi/2}^{\pi} d\theta \sin\theta \left(\frac{\nu}{4\pi}\right) \exp\left(-\frac{E_T}{kT}\right) \tag{5.31}$$

The integrals may be evaluated by substituting $t = \cos\theta$, the result being

$$\frac{\tau_{r0}}{\tau_r} = \left(\frac{kT}{\beta\sqrt{\mathscr{E}}}\right)^2 \left[1 + \left(\frac{\beta\sqrt{\mathscr{E}}}{kT} - 1\right) \exp\left(\frac{\beta\sqrt{\mathscr{E}}}{kT}\right)\right] + \frac{1}{2} \tag{5.32}$$

where the zero-field release rate, $(\tau_{r0})^{-1}$, is $\nu \exp(-E_T/kT)$ and β is defined by $(e^3/\pi\varepsilon)^{1/2}$.

The field-enhanced conductivity may be obtained by assuming that the mobility and the capture rate, or free lifetime, of conduction electrons is field independent, and by assuming that the density of occupied electron traps is reduced a negligible amount by the field.

$$\sigma = \sigma_0 \frac{\tau_{r0}}{\tau_r} \tag{5.33}$$

Thus the field- and temperature-dependent conductivity is described by (5.32) which approaches

$$\frac{\sigma}{\sigma_0} \approx \frac{kT}{\beta\sqrt{\mathscr{E}}} \left(1 - \frac{kT}{\beta\sqrt{\mathscr{E}}}\right) \exp\left(\frac{\beta\sqrt{\mathscr{E}}}{kT}\right) \tag{5.34}$$

at high fields. The conductivity predicted by (5.32) and (5.33) is shown in Fig. 5.2 from which it is apparent that the force-fitting of this curve to an exponential would produce a result which depended strongly upon the maximum field attained. In contrast to the slowly varying potential of the coulomb center, a spherically symmetric square well potential of radius b leads to a more rapid increase of conductivity with increasing electric field.

$$\frac{\sigma}{\sigma_0} = \frac{kT}{2e\mathscr{E}b} \left[\exp\left(\frac{e\mathscr{E}b}{kT}\right) - 1\right] + \frac{1}{2} \tag{5.35}$$

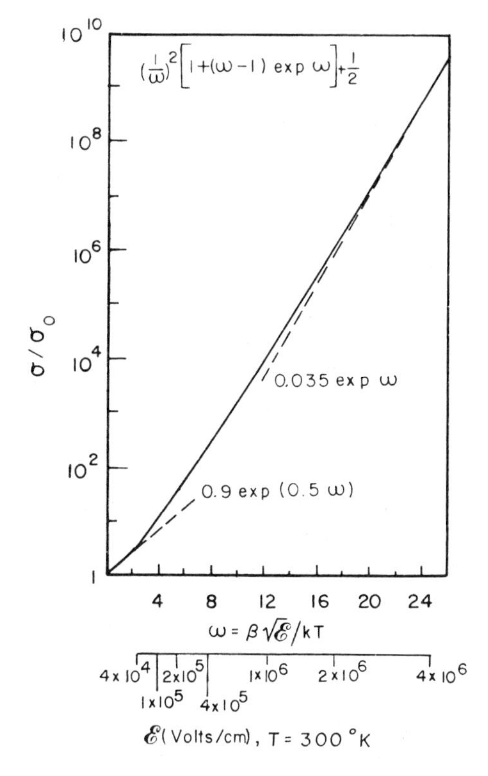

FIG. 5.2 Enhanced electrical conductivity expected for field-assisted thermal ionization of electrons trapped by a positive point charge. *Calculated for a relative dielectric constant of 6. (After Hartke, 1968.)*

The assumptions inherent to the above calculations may not be fully justified at high fields, particularly in regards to the field independent mobility. If the conduction electron drift velocity reaches a field independent saturation value, then (5.33) should be multiplied by a factor proportional to $(\mathscr{E})^{-1}$ at high fields. This would cause the conductivity shown in Fig. 5.2 to approach more closely a simple exponential behavior. The mean free time before capture of the conduction electrons could also change at high fields, particularly if capture was initially into shallow bound states.

5.3 Other Field Effects

5.3.1 Capture and Generation Processes at a Trap

In considering trapping actions the sign of the charge carrier relative to the charged state of trap during the process is important since this determines whether the process is coulombically attractive, neutral, or repulsive. It is convenient to consider the trap as an entity with a certain charge ($N_T{}^+$, $N_T{}^0$, or $N_T{}^-$) when empty. For electron capture by an $N_T{}^+$ trap the process is $N_T^+ + e \rightarrow N_T^0$ and the action is attractive. The generation process for the same trap is $N_T^0 \rightarrow N_T^+ + e$, and again is attractive. On the other hand, for a trap of intrinsic charge N_T^- when empty, the capture process is $N_T^- + e \rightarrow N_T^{2-}$ and, presumably, is repulsive because of the coulombic barrier around the N_T^- charged trap.

The possible actions for intrinsic N_T^+, N_T^0, and N_T^- states interacting with either electrons or holes are shown in Table 5.1 [This table does not include the N_T^{2-} intrinsic (empty) trap state that is involved in triple acceptor action such as $N_T^{2-} + e \rightarrow N_T^{3-}$ since this is only rarely of interest.] For multiple-energy level traps such as $N_T^0 \rightarrow N_T^- \rightarrow N_T^{2-}$, the processes are usually sufficiently well separated in energy that they may be considered one at a time. The notation A, N, R in the table is for attractive, neutral, and repulsive processes, respectively. Hence $A(e, c)$ stands for a coulombically attractive process involving electron capture and $A(e, g)$ is the inverse generation process.

The effect of electric field on the attractive-generation process is to enhance the emission by the Poole-Frenkel effect, as already discussed (Section 5.2). The magnitude of the effect is significant but not exceptionally large. For instance, for indium-doped silicon at 65°K, the hole concentration increases by about a factor of 3 for an increase of field strength from 10^2 to 4×10^3 V cm^{-1} (McCombs, 1971).

The process of electron capture by an attractive trap is generally considered to involve a capture cross section that decreases, according to $\mathscr{E}^{-3/2}$, as the field increases. This has been discussed by Dussel and Bube (1966c) who comment as follows:

The capture cross section is not determined only by the mechanisms by which a free electron loses all the energy between the free state and its final bound state, but there is the possibility that the electron may first be captured in an excited state of the center with a small energy loss requiring the emission of only a few phonons. From the excited state the electron may then proceed to the ground state by further phonon cascade processes or by other mechanisms. In this picture the capture cross section depends on the ability of the shallow excited states of the trap to hold the electron long enough to allow the transition to the ground state.

The total capture cross section can therefore be considered to be made up of two factors. One of these is a "geometrical" factor, determined by the radius such that the probability of escaping from the center is equal to that of being finally captured. The second takes account of the probability of an electron being found at that radius with the appropriate energy. In the case of a Coulomb-attractive center, the second factor is related to the probability that the free electron will emit the necessary phonons to permit it to be captured in the excited state of the center while it traverses the "geometrical diameter" of the center.

For a Coulomb-attractive center, the capture cross section would vary roughly as r_0^3, where r_0 is the "geometrical radius" of the center, i.e., the distance at which the

TABLE 5.1

Capture and Generation Processes for a Single-Process Trap and Field-Dependent Effects[a]

Process	Carrier	Intrinsic Charge of the Trap (Empty)	Initial Trap State before the Process Considered	Capture Process		Generation Process		Trap State at End of Process	Comments, and Physical Examples of Such Traps
				Equation	Field Dependence	Equation	Field Dependence		
$A(e, c)$	Electron	N_T^+	N_T^+	$N_T^+ + e \rightarrow N_T^0$	Small, or $\mathscr{E}^{-3/2}$			N_T^0	Deep donors (S, Se, Te) in Ge or Si with partial compensation.
$A(e, g)$			N_T^0			$N_T^0 \rightarrow N_T^+ + e$	$e_n\uparrow$, $\mathscr{E}\uparrow$	N_T^+	Increase of the generation rate with field in this coulombically attractive process is known as the Poole-Frenkel effect for electrons
$N(e, c)$	Electron	N_T^0	N_T^0	$N_T^0 + e \rightarrow N_T^-$	$\sigma_n^0\downarrow$, $\mathscr{E}\uparrow$ (see Fig. 13.4)			N_T^-	Au^0, Au^- in ν-type silicon. (See Chapter 13.) Coulombically neutral processes
$N(e, g)$			N_T^-			$N_T^- \rightarrow N_T^0 + e$	Small[b]	N_T^0	
$R(e, c)$	Electron	N_T^-	N_T^-	$N_T^- + e \rightarrow N_T^{2-}$	$\sigma_n^-\uparrow$, $\mathscr{E}\uparrow$			N_T^{2-}	Ge (Cu, Ag, Au, Pt, Fe, Co, Ni). Also seen for double acceptors in Si and GaAs.[c]
$R(e, g)$			N_T^{2-}			$N_T^{2-} \rightarrow N_T^- + e$		N_T^-	
$R(h, c)$	Hole	N_T^+	N_T^+	$N_T^+ + h \rightarrow N_T^{2+}$	$\sigma_p^+\uparrow$, $\mathscr{E}\uparrow$			N_T^{2+}	Double donors (with first level compensated) should give these coulombically repulsive trapping actions in Ge and Si.
$R(h, g)$			N_T^{2+}			$N_T^{2+} \rightarrow N_T^+ + h$		N_T^+	
$N(h, c)$	Hole	N_T^0	N_T^0	$N_T^0 + h \rightarrow N_T^+$	$\sigma_p^0\downarrow$, $\mathscr{E}\uparrow$			N_T^+	Au^0, Au^+ in π-type silicon. (See Chapter 13.) Coulombically neutral processes
$N(h, g)$			N_T^+			$N_T^+ \rightarrow N_T^0 + h$	Small	N_T^0	
$A(h, c)$	Hole	N_T^-	N_T^-	$N_T^- + h \rightarrow N_T^0$	Small, or $\mathscr{E}^{-3/2}$			N_T^0	Deep acceptors, such as In^-, In^0 in π-type silicon. The dependence of the generation rate on field is the Poole-Frenkel effect for holes.
$A(h, g)$			N_T^0			$N_T^0 \rightarrow N_T^- + h$	$e_p\uparrow$, $\mathscr{E}\uparrow$	N_T^-	

[a] Multiple energy level centers can be treated in this way by considering one energy-trapping level at a time.
[b] The small dependence of e_n is suggested by the calculated and experimental results of Fig. 13.4. However, see Panouis et al. (1969) for experimental evidence that e_n/e_p increases with increase of the field strength.
[c] These coulombically repulsive trapping actions cause slow-moving trap-controlled domains. (See Chapter 12)

potential energy is about $2kT$ below the maximum. For small applied fields, neither an appreciable decrease in barrier height nor a change in the critical distance will occur. At voltages such that the barrier is reduced by kT, the critical radius is decreased by a factor of $\frac{2}{3}$, and the capture cross section is decreased to about $\frac{1}{3}$ of its initial value. For still higher applied fields, the difference between the capture radius and the distance of the top of the barrier will decrease, so that for sufficiently high fields the capture radius would be practically coincident with the top of the barrier distance. As long as the center is not deformed by the field, the distance of the barrier depends on electric field as $\mathscr{E}^{-1/2}$, so that the capture cross section should decrease as $\mathscr{E}^{-3/2}$.

In an interesting study of the capture cross section σ_n^{+} for the gold donor level (E_v + 0.35 eV) in silicon, Veinger et al. (1969a) found a power dependence close to $(-\frac{3}{2})$ for the capture cross-section dependence on the applied microwave field.

In addition to the $\mathscr{E}^{-3/2}$ reasoning discussed by Dussel and Bube (1966c), the increased energy of the carriers at increased field strengths should contribute to the decrease of the effective capture cross section. However, in calculations for indium-doped silicon (65°K) over the range 10^3 to 10^4 V cm^{-1}, McCombs (1971) found the overall effect of the field on the capture rate to be smaller than $\mathscr{E}^{-3/2}$.

5.3.2 Neutral Trap Processes

The neutral capture and emission processes for the $N_T{}^0$ intrinsic trap state in gold-doped ν-type silicon (200°K) have been studied by McCombs (1971). The effect of increased field strength on the capture cross section is calculated to be a small decrease. The computed and measured carrier concentrations increase on this account by a factor of about 3 for a field increase between 10^3 and 10^4 V cm^{-1}, as shown in Fig. 13.4. The effect of the field on the thermal emission rate is calculated to be quite small for the range 10^3 to 10^5 V cm^{-1}.

This small field dependence of e_n is in reasonable agreement with the experimental observations of Tasch and Sah (1970) by photovoltaic studies of the depletion region of a reverse-biased *pn* junction. Tasch and Sah found the electric field dependences to be considerably smaller than predicted by the three-dimensional Poole-Frenkel effect in a coulombic potential, and closer to that in a polarization potential. The ratio of the thermal emission rates e_n/e_p was 16 at 300°K and 50 at 190°K under the high-field conditions of the experiment. The low-field thermal capture cross section for gold, $\sigma_n{}^0$, was found to be temperature insensitive over the range 200 to 300°K and could be matched to the Fairfield-Gokhale (1965) value of 1.7×10^{-16} cm^2 by the assumption of a degeneracy factor g_{An} of 15. This is not too inconsistent with a g_{An} of 12 expected from six conduction band valleys and two spin degeneracies. For σ_p^-, Tasch and Sah find a temperature dependence of T^{-4} consistent with Lax's impurity model. Their 300°K value is matchable with the Fairfield-Gokhale value of 1.1×10^{-14} cm^2 if a degeneracy factor g_{Ap} of 7.5 is assumed. Again this is not too unreasonable, for a system of three valence bands and two spin degeneracies for which g_{Ap} should be 6. (The experiment also provides radiative recombination cross-section values, and these are 6.3×10^{-21} cm^2 for holes and 3×10^{-21} cm^2 for electrons.)

The ratio of the emission rates for electrons and holes, e_n/e_p, is an important parameter in the kinetics of gold-doped silicon. The Tasch and Sah experiment that we have just discussed gave values of between 50 (190°K) and 16 (300°K) for the $E_c - 0.54$ eV acceptor level of ν-type silicon. Senechal and Basinski (1968) from transient capacitance (high-field) studies obtained somewhat lower results. On the other hand, from low-field photodecay studies Fairfield and Gokhale infer values as low as 0.2. These disparities led Panousis et al. (1969) to conclude that the emission ratio increases with increase in electric field. This effect can lead to charge distributions in $n^+\nu n^+$ structures that would not be expected otherwise.

In the space-charge region of a junction involving gold-doped ν-type silicon the charge balance of the gold acceptor level (if we neglect capture events) depends only on the emission ratio e_n/e_p where (Sah and Tasch, 1967)

$$N_{\mathrm{Au}^-} = \frac{N_{\mathrm{Au_{total}}}}{1 + e_n/e_p} \tag{5.36}$$

For ν-type silicon the material also contains some partially compensating donors, all of which are in the ionized ($N_D = N_D^+$) state. The charge density ρ in the depletion region is therefore

$$\rho = e(N_D^+ - N_{\mathrm{Au}^-}) = eN_D\left(1 - \frac{N_{\mathrm{Au_{total}}}}{N_D(1 + e_n/e_p)}\right) \tag{5.37}$$

Thus the sign of the net space charge depends on the ratio of $N_{\mathrm{Au_{total}}}$ to N_D and on the magnitude of e_n/e_p. If we postulate that e_n/e_p increases with field strength, we have the possibility that the net charge density is zero for a particular field strength. A consequence of this is the development of a constant field zero-space-charge region adjacent to a reverse-biased n^+ contact on gold-doped silicon. The constant field is that at which the Au^- density (controlled by the field-dependent emission ratio) just equals the density of ionized donors.

The variation of net charge density, field strength, and potential across an $n^+\nu$ junction (reverse biased) is illustrated in Fig. 5.3. The constant field

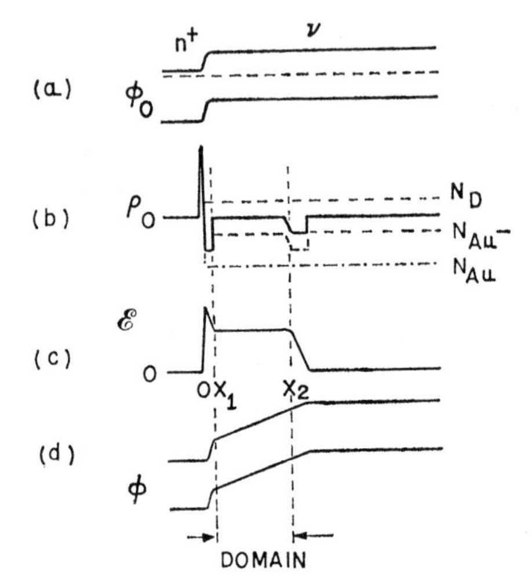

FIG. 5.3 Illustration of charge density ρ, field $\mathscr{E}$, and potential ϕ in an $n^+ \nu n^+$ structure which forms a neutral high-field domain.

The equilibrium potential is illustrated in (a), whereas the potential under bias is shown in (d). The net charge density $[e(N_{Au}-/N_D)]$ is shown as the solid line in (b) and the resulting field is shown in (c). (After Panousis et al., 1969.)

region between x_1 and x_2 in Fig. 5.3c increases in length as the reverse bias is increased. The space charge is negative in the region immediately adjacent to the n^+ contact because of the capture of electrons that diffuse from the n^+ contact. Figure 5.4 shows the results of probing such a structure. The domain region is seen to have a constant field strength of 2.9×10^{-4} V cm^{-1} for the specimen examined ($N_{Au_{total}}/N_D = 6$) and therefore from (5.37) the value of e_n/e_p at this field strength is 5. The experimental observation of the constant field region confirms that e_n/e_p is field dependent and so tends to resolve the lack of agreement of previous studies of this ratio. Since e_n/e_p increases significantly with field and e_n itself is relatively independent of field strength, it follows that e_p (and σ_p^-) must decrease with increase of field strength for the gold acceptor level.

5.3.3 "Repulsive" Capture Cross Sections

For the process $N_T^- + e \rightarrow N_T^{2-}$ the trapping action is repulsive and the trap in the N_T^- state is surrounded by a coulombic energy barrier that the electron must surmount thermally or tunnel through. Therefore one may expect the capture cross section to be very small (perhaps 10^{-21} cm^2 or less) and to be very temperature dependent. The effect of increasing the electric field should be to increase the number of high energy electrons and so increase the capture rate. Therefore the differential conductivity may become negative and a high-field region may form that moves as a domain from the cathode to the anode contact. Such action is discussed in Chapter 12.

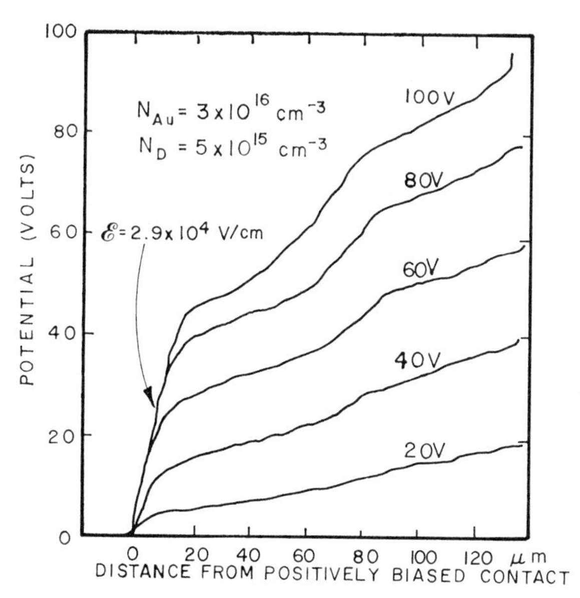

FIG. 5.4 Probe measurement of the potential variation across an $n^+ \ \nu \ n^+$ device. *For this device $N_{Au}/N_D = 6$ and the positively biased contact is on the left side of the figure. The change in slope in the ohmic region is due to a variation in the gold density near the center of the structure. (After Panousis et al., 1969.)*

Many investigators have studied trap-controlled domain action associated with repulsive capture since it was originally proposed by Ridley (1963, 1965). Such action is readily seen in studies of germanium (at less than 100°K) containing double-acceptor impurities with partial *n*-type compensation to expose the N_T^- level. Figure 5.5 shows the decrease of electron lifetime (increase of effective capture cross section) with increase of the field strength for germanium doped with copper and antimony. The differential negative-resistance effect is seen from the J-$\mathscr{E}$ curve in the figure. In related work, Zhdanova et al. (1966a) have studied the capture of electrons by the doubly charged (Cu^{2-}) copper level at $E_v + 0.32$ eV. Veinger et al. (1969b) also have studied, by microwave field pulsing, the trapping of electrons at this level.

In general, it is not at all unusual for such experiments to show the repulsive capture cross sections (such as σ_n^- or σ_n^{2-}) to have values in the range 10^{-18} to 10^{-20} cm^2 (see Tables 10.4 and 10.5) instead of the much smaller values (less than 10^{-21} cm^2) that might be expected. Bonch-Bruevich and Sokolova (1964) comment that the electron capture cross sections are often surprisingly large, even sometimes greater than the electron capture cross

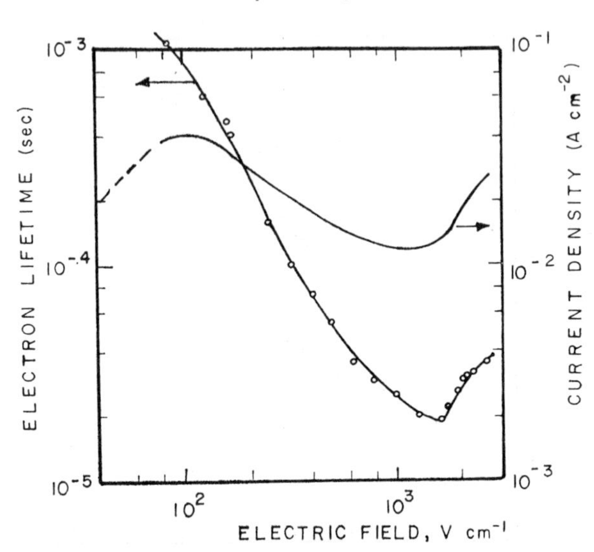

FIG. 5.5 The electron lifetime dependence on the field strength (○) and the expected $J-\mathscr{E}$ curve for germanium doped with copper and antimony (90°K). (After Kagan and Kalashnikov, 1966.)

sections of neutral centers. The weak temperature dependence of "repulsive" capture cross sections also is not consistent with the exponential temperature dependence expected from the quasiclassical concept of electrons surmounting a coulombic repulsive barrier. This weak temperature dependence, however, can be taken care of by proposing some kind of tunneling through the energy barrier, although the details of this are rather uncertain (Rogachev and Ryvkin, 1965).

Another approach to the problem, however, is to postulate that it is not an electron that is captured in the presence of coulombic repulsion, but instead a neutral particle, an exciton (Bonch-Bruevich and Sokolova, 1964). Then we have the following picture. During the first stage, an electron and hole form an exciton of the normal type. The probability of this process for a shallow exciton in germanium may be very high. It is known, of course, that a direct recombination of an electron and hole comprising an exciton is scarcely probable in comparison with thermal decay. For this reason, in the absence of impurity atoms or other structural defects, the production of excitons does not contribute appreciably to the resultant rate of vanishing of nonequilibrium charge carriers. However, in the presence of impurities, a second state is perhaps possible, namely exciton capture with its transition to a local state associated with the impurity. The concept of such local states has been introduced a number of times by various authors. Under certain conditions local states are markedly deeper than the states of moving free excitons. Consequently, the process of recombination of localized excitons could perhaps proceed at a rate that is sufficiently high to explain the experimentally observed lifetimes.

There is little direct evidence in support of this model at present. However, excitons are found to be involved in recombination in GaP.

5.3.4 Possible Auger Processes

In an Auger-type recombination or trapping process, energy lost by one carrier is carried away, in part or in whole, by a concurrent event involving a second carrier. In its simplest form this may be the interaction of two electrons in which one recombines across the band gap and the released energy is carried away by the second electron. This is not a very probable recombination process in practical semiconductors such as germanium and silicon, which usually contain many impurity recombination centers that permit energy loss to the lattice by phonon processes.

However, Sheinkman (1965) has suggested that Auger processes need reconsideration as energy loss mechanisms possibly involved when carriers are captured by multiply charged recombination centers in which a few electrons or holes may be localized. The energy released on recombination is transferred to one of the localized carriers which is then ejected into the appropriate band (Fig. 5.6a and b). The charge of the center changes by $\pm 2e$. A necessary condition for the process is the fulfillment of the energy relation

$$E_g - E_1 \geq E_2 \tag{5.38}$$

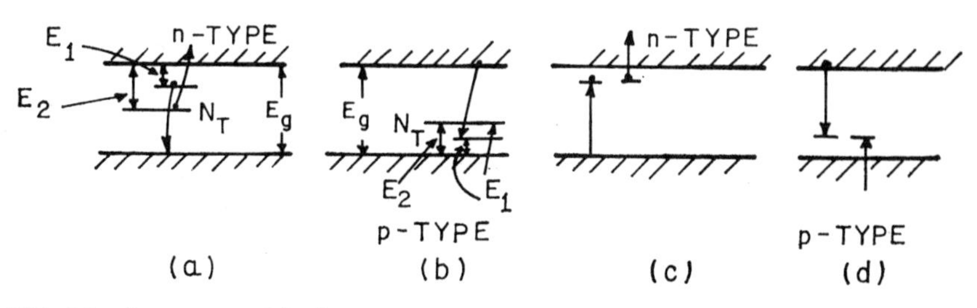

FIG. 5.6 Auger recombination processes.
(a) Auger recombination of a hole in a doubly charged filled center. (b) Auger recombination of an electron in a doubly charged empty center. (c) Auger recombination in a singly charged center with an energy transfer to an electron in a neighboring center. (d) Conduction band electron captured by a center, with energy transfer to a valence band electron. (After Sheinkman, 1965.)

From the point of view of fulfillment of the condition (5.38) the process described may take place in n-type Ge when a hole is captured by the impurities Au^{3-}, Cu^{3-}, Fe^{2-}, and Te^0 (see Fig. 5.6a), and in p-type Ge when an electron is captured by the impurities Zn^0, Cd^0, Cu^0, Co^0, Mn^0, Ni^0, Au^0, and Ag^0 (See Fig. 5.6b). In p-type Si the Auger recombination may occur when an electron is captured by the Zn^0 and Au^+ impurities.

The measured electron capture cross sections $\sigma_n{}^0$ for such impurities tend to be in the range 10^{-15} to 10^{-16} cm^2. However, such large values of cross sections for the capture by neutral centers cannot be due to the multiphonon processes of Lax because, as has been shown by Bonch-Bruevich and Glasko (1962), a center does not have a sufficient number of excited levels for reasonable values of the polarizability. The temperature dependence of $\sigma_n{}^0$ is very weak, as could be expected for the Auger recombination in contrast to the multiphonon process.

On the other hand, the Auger recombination cannot take place when an electron is captured by neutral (but deeper) levels of Ga and Al in Ge, and Ga and B in Si, or a hole is captured by the neutral level of Bi in Si. Experiment shows that here the recombination is not multiphonon but radiative (at least in Si), with a small cross section of 10^{-17} to 10^{-19} cm^2.

Sheinkman (1965) notes the large difference between the values of the cross section $\sigma_n{}^0$ for the electron capture by neutral levels of singly charged acceptors Al and Ga in Ge: $\sigma_n{}^0 < 10^{-18}$ to 10^{-19} cm^2, where the Auger process cannot take place, and of the cross section $\sigma_n{}^0$ for the electron capture by the neutral level of Cu, which lies at about the same energy distance from the bottom of the conduction band, but where the Auger process can take place. In the latter case $\sigma_n{}^0 \sim 10^{-16}$ cm^2, which is 10^2 to 10^3 times higher than $\sigma_n{}^0$ (Ga, Al). Auger effects were not observed by Norton and Levinstein (1972) in a study of the recombination of holes at singly ionized Cu in germanium.

Auger recombination lifetimes in InAs, GaSb, InP, and GaAs have been calculated by Takeshima (1972). Galkin et al. (1971) have observed band-to-band Auger recombination processes in InAs. Auger processes have also been considered as possibly significant in the recombination of bound excitons at deep-neutral-donor and deep-neutral-acceptor sites in GaP (Jayson, 1972).

CHAPTER SIX

Deep Impurities as Recombination Centers and Traps in Bulk Material

Excess carriers can be introduced into a semiconductor by photoexcitation, x-rays, γ-rays, electron and other particle irradiation, and by electrical injection at *pn* junctions. The excess carriers are created as electron-hole pairs. During and after the injection process they recombine so that the semiconductor returns to its thermal equilibrium condition. In most semiconductors such recombination is aided by the presence of recombination centers in the band gap which may be impurities, vacancies, interstitials, or dislocations.

In this chapter we consider first the simple processes of recombination and generation at a single energy level associated with a deep impurity in a bulk semiconductor. The photoconductive decay transients are then studied and shown to have two components. The application of this to indium-doped silicon is examined to illustrate some of the complexities revealed by experimental studies. Surface recombination and the interaction of traps and recombination centers are other matters considered. The function of an impurity level as a trap or as a recombination center is discussed in terms of the Sah–Shockley model which provides a graphic way of determining which processes are dominant. Finally the lifetime control problem is discussed with reference to Sah's transient lifetime maps for gold-doped silicon.

6.1 Recombination and Generation at a Single Energy Level

Recombination involves energy dissipation, and impurity or defect centers make energy dissipation by phonon emission much easier than in a perfect crystal. In semiconductors such as silicon or germanium, the band structure is indirect gap and for carrier recombination phonon emission is required for momentum considerations. Therefore the probability of recombination of electron-hole pairs across the band gap is low for these materials and would represent lifetimes at 300°K of the order of seconds (Hall, 1959). The lifetimes actually observed for these materials depend on the care taken in the crystal growth. They do not usually exceed a few hundred microseconds. If a deep impurity such as gold is allowed to enter the semiconductor, the lifetime may be reduced to tens of nanoseconds. Low lifetimes are obviously of considerable interest and practical importance to the designer of high-speed switching devices.

Carrier lifetimes may be defined as the mean times spent by excess electrons and holes in the conduction and valence bands. Consider a disturbance of electron density δn added to a thermal equilibrium level of n_0 electrons cm^{-3}. When the source of the disturbance is cut off, the semiconductor returns to equilibrium under the influence of a recombination process r and a thermal generation process g. Therefore if τ_n is the mean electron lifetime, $r = (\delta n + n_0)/\tau_n$ and $g = r_e = n_0/\tau_n$. Hence

$$\frac{dn}{dt} = g - r = -\frac{\delta n}{\tau_n} \tag{6.1}$$

and therefore the electron density as a function of time is given by

$$n = \delta n \exp\left(-\frac{t}{\tau_n}\right) + n_0 \tag{6.2}$$

Thus the mean lifetime τ_n is the time for the disturbance to be reduced to $1/e$ of its original value.

However, this is only a very simple introduction to the concept of lifetime and recombination, and a more detailed study in terms of capture cross sections and occupancy of the center(s) responsible is needed.

The kinetics of recombination, generation, and trapping at a single energy level in the band gap have been considered by Shockley and Read (1952), Hall (1957), Wertheim (1958), and others. Shockley (1958), Bemski (1958c), and Sah (1967b, c) have also considered such a center in their review papers on carrier lifetime.

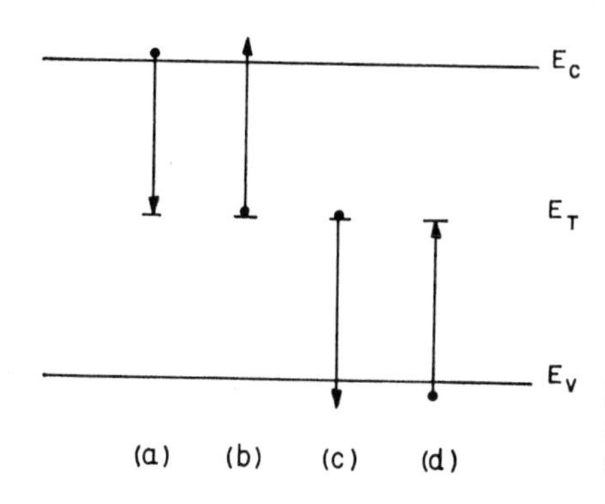

FIG. 6.1 The four processes of electron and hole capture and emission by a single-energy-level defect center.

The deep level is assumed to be capable of four processes as shown in Fig. 6.1:

(*a*) Electron capture from the conduction band to a neutral unoccupied center. This is described by $N_T{}^0 + e \rightarrow N_T{}^-$ with a capture cross section σ_n and a capture rate c_n $(=\sigma_n{}^0 v_{\text{th}})$ $\text{cm}^3\ \text{sec}^{-1}$.

(*b*) Electron emission from an occupied center to the conduction band with an emission rate e_n sec^{-1}.

(*c*) Hole capture by an occupied center, corresponding to the process $N_T{}^- + h \rightarrow N_T{}^0$ with a capture cross section σ_p and a capture rate c_p $\text{cm}^3\ \text{sec}^{-1}$.

(*d*) Hole emission by an unoccupied center with an emission rate e_p sec^{-1}.

The electron capture rate is proportional to the concentration of electrons in the conduction band n, and to the concentration of centers unoccupied by electrons, $N_T(1 - f_T)$ where f_T is the Fermi–Dirac distribution function. This function can be assumed to have its thermal equilibrium value if the carrier disturbance is small. Then

$$f_T = \left[1 + \exp\left(\frac{E_T - E_F}{kT}\right)\right]^{-1} \tag{6.3}$$

If $E_T - E_F \gg kT$, this may be simplified to the Maxwell–Boltzmann form

$$f_{T\text{MB}} = \left[\exp\left(\frac{E_T - E_F}{kT}\right)\right]^{-1} \tag{6.4}$$

The electron emission term, involved in process b, is of course dependent on the rate term e_n and the density of filled centers $N_T f_T$. Hence the electron concentration n in the conduction band varies as

$$\frac{dn}{dt} = -c_n n N_T(1 - f_T) + e_n N_T f_T \tag{6.5}$$

Similarly for the hole density change

$$\frac{dp}{dt} = -c_p p N_T f_T + e_p N_T (1 - f_T) \tag{6.6}$$

The rate coefficients may be related to each other by using the conditions $dn/dt = 0$ and $dp/dt = 0$ at equilibrium. Hence from (6.5)

$$\frac{e_{ne}}{c_{ne}} = n_0 \exp\left(\frac{E_T - E_F}{kT}\right) = n_i \exp\left(\frac{E_T - E_i}{kT}\right) \equiv n_1 \tag{6.7}$$

where e_{ne} and c_{ne} are emission and capture rates at equilibrium. Introduction of the term n_1 is merely a convenient way of simplifying our subsequent expressions. Physically n_1 is $N_c \exp(-(E_c - E_T)/kT))$ and equal to the electron density that would be in the conduction band if the Fermi level is at the impurity level E_T. Similarly from (6.6)

$$\frac{e_{pe}}{c_{pe}} = n_i \exp\left(\frac{E_i - E_T}{kT}\right) \equiv p_1 \tag{6.8}$$

It is customary to assume that the rate coefficients used in the equations may be rewritten as

$$\frac{dn}{dt} = -c_n(nN_T{}^0 - n_1 N_T{}^-) \tag{6.9}$$

and

$$\frac{dp}{dt} = -c_p(pN_T{}^- - p_1 N_T{}^0) \tag{6.10}$$

where an acceptor trapping center is assumed to have one of two states, $N_T{}^-$ or $N_T{}^0$. Then

$$N_T{}^0 = N_v(1 - f_T), \qquad N_T{}^- = N_T f_T, \qquad \text{and} \qquad N_T{}^0 + N_T{}^- = N_T$$

where N_T is the total density of traps present. The rate of change of electrons on the traps is of course

$$\frac{dN_T{}^-}{dt} = -\frac{dn}{dt} + \frac{dp}{dt} = c_n(nN_T{}^0 - n_1 N_T{}^-) - c_p(pN_T{}^- - p_1 N_T{}^0) \tag{6.11}$$

The minimum possible electron lifetime associated with the center will be obtained when all the centers are neutral, and may be written as

$$\tau_{n0} = \frac{n}{dn/dt|_{N_T{}^0 \to N_T}} = \frac{1}{c_n N_T} \tag{6.12}$$

Similarly the minimum possible hole lifetime will be when $N_T{}^- \to N_T$ and is

$$\tau_{p0} = \frac{1}{c_p N_T} \tag{6.13}$$

From (6.11) we may solve for $N_T{}^-$ in the steady-state condition where $dN_T{}^-/dt$ is zero to obtain

$$\frac{N_T{}^-}{N_T} = \frac{c_n n + c_p p_1}{c_n(n + n_1) + c_p(p + p_1)} \tag{6.14}$$

and hence

$$\frac{N_T{}^0}{N_T} = \frac{c_n n_1 + c_p p}{c_n(n + n_1) + c_p(p + p_1)} \tag{6.15}$$

Substitution for $N_T{}^-$ and $N_T{}^0$ in (6.9) and (6.10) then gives

$$\frac{dn}{dt} = \frac{dp}{dt} = -\frac{np - n_i^2}{\tau_{p0}(n + n_1) + \tau_{n0}(p + p_1)} \tag{6.16}$$

Many deep impurity centers act as deep acceptors—therefore let us consider an n-type semiconductor (of equilibrium electron density n_0) that contains an acceptor recombination center. If a disturbance such as band-gap light or other means of injection has raised the electron density to $n_0 + \delta n$ and the hole density to $p_0 + \delta p$, the return to equilibrium is controlled by

$$\frac{dn}{dt} = \frac{dp}{dt} = -\frac{(n_0 + \delta n)(p_0 + \delta p) - n_i^2}{\tau_{p0}(n_0 + \delta n + n_1) + \tau_{n0}(p_0 + \delta p + p_1)} \tag{6.17}$$

Hence for a small disturbance, bearing in mind that $n_0 \gg p_0$ and $n_0 p_0 = n_i^2$, we have

$$\frac{dp}{dt} = -\frac{n_0 \delta p}{\tau_{p0}(n_0 + n_1) + \tau_{n0} p_1} \tag{6.18}$$

The hole lifetime, $\delta p/(dp/dt)$, is therefore

$$\tau_p = \frac{\tau_{p0}(n_0 + n_1) + \tau_{n0} p_1}{n_0} \tag{6.19}$$

$$\tau_p \simeq \tau_{p0} \tag{6.20}$$

Consider, for example, n-type silicon with 10^{16} electrons cm^{-3} in the conduction band and 5×10^{15} gold atoms cm^{-3} present in the material.

The silicon will be of resistivity 1 Ω-cm at 300°K (from Fig. 2.4) and virtually all of the gold atoms will be in the form Au^- corresponding to the acceptor level at $E_c - 0.54$ eV. The capture cross sections observed for the processes $Au^- + h \rightarrow Au$ and $Au + e \rightarrow Au^-$ are of the order of $\sigma_p^- = 10^{-15}$ cm^2 and $\sigma_n^0 = 5 \times 10^{-16}$ cm^2, respectively. From these values we obtain

$$\tau_{p0} = (\sigma_p^- v_{\text{th}} N_T)^{-1} = (10^{-15} \times 10^7 \times 5 \times 10^{15})^{-1} = 2 \times 10^{-8} \text{ sec}$$

$$\tau_{n0} = (\sigma_n^0 v_{\text{th}} N_T)^{-1} = (5 \times 10^{-16} \times 10^7 \times 5 \times 10^{15})^{-1} = 4 \times 10^{-8} \text{ sec}$$

Now $n_1 \simeq p_1 \simeq 2 \times 10^{10}$ cm^{-3} and is negligible compared with n_0, which is 10^{16} cm^{-3}. Hence from (6.20) the minority carrier lifetime τ_p for this crude example is seen to be about 2×10^{-8} sec. Thus the gold is effective in reducing minority carrier lifetime for fast response devices. In p-type silicon the gold donor level at $E_v + 0.35$ eV performs a similar role. This matter is taken up again later in this chapter (Section 6.7) and also in Chapter 8 on the effects of deep impurities in junction devices.

6.2 Further Lifetime Analysis

The treatment given in the preceding section is based on a number of simplifications such as $\delta n = \delta p$; $c_n, c_p = c_{ne}, c_{pe}$; low injection levels; small densities of recombination centers; and a single-level center. As these simplifications are relaxed the problem becomes more complicated and the expressions obtained are increasingly cumbersome.

For the condition where δn is not equal to δp, although the injection level is low, Shockley (1958) has obtained

$$\tau_p = \left[\frac{\tau_{n0}(p_0 + p_1) + \tau_{p0}[n_0 + n_1 + N_T(1 + n_0/n_1)^{-1}]}{n_0 + p_0 + N_T(1 + n_0/n_1)^{-1}(1 + n_1/n_0)^{-1}}\right]_{\delta n \rightarrow 0} \tag{6.21}$$

and

$$\tau_n = \left[\frac{\tau_{p0}(n_0 + n_1) + \tau_{n0}[p_0 + p_1 + N_T(1 + p_0/p_1)^{-1}]}{n_0 + p_0 + N_T(1 + p_0/p_1)^{-1}(1 + p_1/p_0)^{-1}}\right]_{\delta p \rightarrow 0} \tag{6.22}$$

If N_T is small, the two lifetime expressions reduce to the single form

$$\tau_0 = \frac{\tau_{p0}(n_0 + n_1) + \tau_{n0}(p_0 + p_1)}{n_0 + p_0} \tag{6.23}$$

For large values of injected carrier density Shockley and Read (1952) derive the expression

$$\tau = \tau_0 \left(\frac{1 + \delta n(\tau_{n0} + \tau_{p0})/[\tau_{p0}(n_0 + n_1) + \tau_{n0}(p_0 + p_1)]}{1 + \delta n/(n_0 + p_0)}\right) \tag{6.24}$$

where τ_0 is given by (6.23). This shows that as δn is increased the lifetime may be expected to change from τ_0. For a strongly n-type sample the expression may be rewritten in terms of the conductivity change $\delta\sigma$ and the ratio of electron to hole mobility b as shown below

$$\tau \simeq \tau_0 \left(\frac{1 + \Delta[(\tau_{po} + \tau_{no})/\tau_{po}]}{1 + \Delta} \right) \tag{6.25}$$

where Δ is $\delta\sigma b/(\sigma_0(1 + b))$.

Sandiford (1957) has shown that a more complete treatment of the Shockley–Read–Hall-type center for a photoconductive decay transient gives, for both δp and δn, expressions of the form:

$$\delta p = Ae^{-t/\tau_i} + Be^{-t/\tau_t} \tag{6.26}$$

where A and B are determined by the initial conditions. For

$$[c_p(N_T^- + p_0 + p_1) + c_n(N_T + n_0 + n_1)]^2 \gg 4c_pc_n[N_T^-N_T + N_T^-(n_0 + n_1) + N_T(p_0 + p_1)] \tag{6.27}$$

which is nearly always the case,

$$\tau_i = \left(c_p \left[p_0 + p_1 + N_T \left(1 + \frac{p_0}{p_1} \right)^{-1} \right] + c_n \left[n_0 + n_1 + N_T \left(1 + \frac{n_0}{n_1} \right)^{-1} \right] \right)^{-1} \tag{6.28}$$

$$\tau_t = \frac{\dfrac{p_0 + p_1 + N_T(1 + p_0/p_1)^{-1}}{c_nN_T} + \dfrac{n_0 + n_1 + N_T(1 + n_0/n_1)^{-1}}{c_pN_T}}{n_0 + p_0 + N_T(1 + n_0/n_1)^{-1}(1 + n_1/n_0)^{-1}} \tag{6.29}$$

Thus the decay of δp and δn is not expressed exactly by a simple exponential decay but shows double exponential behavior for times of the order of τ_i from the initial conditions. However, for a particular example where $c_p = 10^{-9}$ cm^3 sec^{-1}, $c_n = 10^{-7}$ cm^3 sec^{-1}, $N_T = 10^{13}$ cm^{-3}, $n_0 = 10^{14}$ cm^{-3}, and $p_1 = 3 \times 10^{16}$ cm^{-3}, other values being small, we find that $\tau_t = 400$ μsec and $\tau_i = 0.025$ μsec. Hence it is seen that τ_i tends to be negligible compared with τ_t. Since δp and δn have the same form of solution it follows that τ_t, the lifetime under transient conditions, is equal for both holes and electrons, even for N_T large.

For N_T small, (6.23) and (6.29) have the same value. When N_T is large, this is not so and (6.29) must be used for the interpretation of photoconductive decay measurements.

The transient recombination problem has been studied by Wertheim (1958) who arrives at expressions identical with (6.28) and (6.29) except that they are expressed in the form

$$\tau_i = [c_p(p_0 + p_1 + N_T^-) + c_n(n_0 + n_1 + N_T^0)]^{-1} \tag{6.30}$$

$$\tau_t = \frac{\tau_{n0}(p_0 + p_1 + N_T^-) + \tau_{p0}(n_0 + n_1 + N_T^0)}{n_0 + p_0 + (N_T^- N_T^0 / N_T)} \tag{6.31}$$

In electron-bombarded n-type silicon, the behavior of the lifetime as a function of bombardment and temperature is shown in Fig. 6.2. Bombardment produces an energy level 0.27 eV above the edge of the valence band. The imperfections associated with this level act as recombination centers which can be introduced in controlled density. Wertheim specializes (6.31) for this case, and obtains

$$\tau_t = \frac{1}{c_p N_T} + \frac{1 + p_1/N_T}{c_n n_0} \tag{6.32}$$

This equation indicates that a lower limit in lifetime is reached as the bombardment continues and N_T is increased. In practice a minimum is observed because n_0 decreases as bombardment progresses. This is observed in Fig. 6.2 at 10^{17} electrons cm^{-2}. The equation adequately describes the behavior of the lifetime as a function of temperature for various values of bombardment.

Other problems considered by Wertheim include (a) direct recombination in conjunction with recombination through centers and (b) recombination in crystals containing two centers. For the latter if $N_1 + N_2 < n_0 + p_0$, it may be shown that

$$\frac{1}{\tau} = \frac{((1 + \mu_1)/\tau_1) + ((1 + \mu_2)/\tau_2)}{1 + \mu_1(1 + \nu_1) + \mu_2(1 + \nu_2)} \tag{6.33}$$

where, in n-type material,

$$\begin{aligned} \mu_1 &= N_1^-/[p_0 + p_{11} + (n_0 + n_{11})c_{n1}/c_{p1}] \\ \mu_2 &= N_2^-/[p_0 + p_{12} + (n_0 + n_{12})c_{n2}/c_{p2}] \\ \nu_1 &= c_{n1}(n_0 + n_{11})/[c_{n2}(n_0 + n_{12}) + c_{p2}(p_0 + p_{12})] \\ \nu_2 &= c_{n2}(n_0 + n_{12})/[c_{n1}(n_0 + n_{11}) + c_{p1}(p_0 + p_{11})] \end{aligned} \tag{6.34}$$

where n_{11}, p_{11} are the electron and hole densities with the Fermi level at the first impurity level, and n_{12}, p_{12} are similarly defined for the second impurity level.

It may be seen from (6.33) and (6.34) that the customary addition of reciprocal time constants is justified only when $\mu_i \ll 1$ and $\mu_i \nu_i < 1$. Both conditions are met when the recombination center density is sufficiently small.

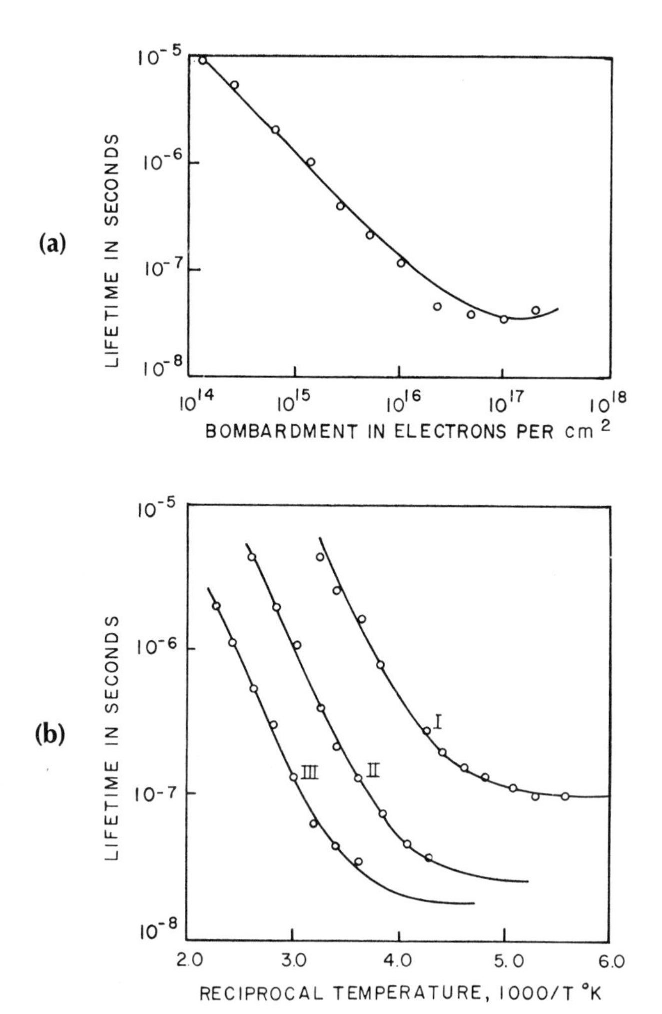

FIG. 6.2 Lifetime in *n*-type silicon (7 Ω-cm) containing an acceptor level at E_v + 0.27 eV induced by electron bombardment.

(a) Lifetime versus bombardment density, n_e. Curve computed for $c_n = 9.5 \times 10^{-8}$, $c_p = 1.6 \times 10^{-5}$, $p_1 = \frac{1}{2}N_v \exp(-0.27/kT)$, $n_a = 6.2 \times 10^{14}$, and $N = 5 \times 10^{-3} n_e$. (b) Lifetime as a function of temperature for three amounts of bombardment: (I) 1.4×10^{14}, (II) 1.4×10^{15}, and (III) 1.1×10^{16} electrons cm^{-2}. (After Wertheim, 1958.)

For recombination centers near the middle of the gap, this requires that

$$\frac{c_{pi}N_i^-}{c_{ni}n_0} < 1 \qquad \text{where} \quad i = 1, 2$$

This suggests that deviations may be found when one or both of the centers are negatively charged, since then c_p tends to be larger than c_n.

The carrier lifetime in semiconductors with two interacting or two independent recombination levels has also been studied by Choo (1970). Baicker and Fang (1972) have made theoretical studies of silicon containing two sets of monovalent traps or one set of divalent traps. Srour and Curtis (1972) consider similar problems in radiation induced defect studies of silicon.

6.3 Lifetime in Indium-Doped Silicon

A recombination problem of particular interest is that of donor- or acceptor-limited lifetime where a deep impurity determines the carrier concentration and also limits the recombination. The usual approximations must be reexamined because the density of centers may be much larger than the majority carrier concentration.

Consider p-type material with a deep acceptor (such as indium in silicon) and assume $N_T^0 \gg n_0 + n_1$. If the sample contains N_D donors cm^{-3}, and N_T acceptors cm^{-3} with larger ionization energy and in great concentration, so that the latter will have a dominant effect on lifetime, then

$$N_T^0 = N_T - p_0 - N_D, \qquad N_T^- = p_0 + N_D \tag{6.35}$$

The lifetime then becomes, according to Wertheim,

$$\tau_t = [c_n(N_T - p_0 - N_D)]^{-1} + [c_p(2p_0 + p_1 + N_D)]^{-1} \tag{6.36}$$

At low temperature, provided N_D is sufficiently small, the two time constants become

$$\tau_t = [c_p(2p_0 + p_1 + N_D)]^{-1} \tag{6.36a}$$

$$\tau_i = [c_n(N_T - p_0 - N_D)]^{-1} \tag{6.36b}$$

Behavior of this type has been observed in heavily indium-doped silicon with $N_T = 1.4 \times 10^{17}\ \text{cm}^{-3}$ and $N_D < 10^{12}\ \text{cm}^{-3}$ (Fig. 6.3). The solid line consists of (6.36) at high temperature and continues as (6.36b) at low temperature. The rising curve at low temperature is (6.36a). The temperature dependence of the capture cross sections of indium was derived from a comparison of the data with these equations.

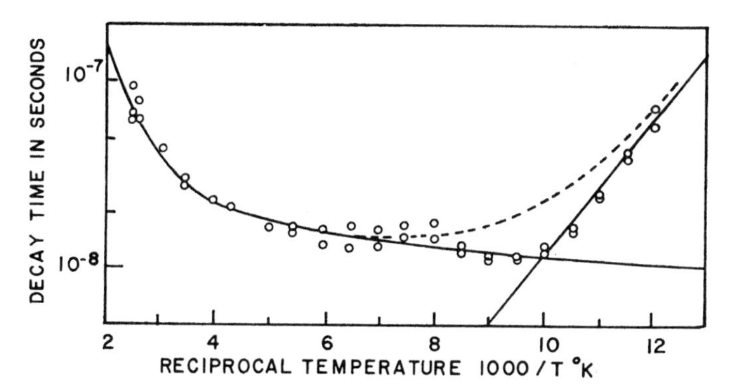

FIG. 6.3 Lifetime (mean life) in indium-doped silicon.
Curve computed with $c_n = 1.1 \times 10^{-14}T^{-1}v_n$, $c_p = 1.5 \times 10^{-9}T^{-2}v_p$; v_n, v_p *are average thermal velocities of electrons and holes;* $p_1 = \frac{1}{2}N_v \exp(-0.16/kT)$. *(After Wertheim, 1958.)*

In the case of an acceptor, the cross section for hole capture, σ_p, is expected to be larger than the cross section for electron capture, σ_n, because the former is aided by coulomb attraction, whereas the latter is not. At room temperature the lifetime is consequently dominated by the first term of (6.36). Using the relation

$$\frac{N_T^{\,0}}{N_T^{\,-}} = \frac{p_0}{p_1}$$

one obtains

$$\tau_t = \frac{p_1}{c_n p_0^{\,2}(1 + N_D/p_0)} \tag{6.37}$$

which indicates that the lifetime will depend on the inverse square of the carrier concentration when the donor concentration is small.

The same problem, indium in p-type silicon, however, has been given further study by Pokrovskii and Svistunova (1969). Their results for crystals containing 4×10^{15} and 1.2×10^{17} indium atoms cm^{-3} show the lifetime to be independent of the acceptor concentration (Fig. 6.4). This independence means that the recombination rate is governed by electron transitions, not to indium atoms, but to some other centers N_R present in the sample at nearly constant concentration. Then for small deviations from equilibrium one obtains

$$\tau = (c_n N_R + c_{\text{In}} N_{\text{In}})^{-1} \tag{6.38}$$

where c_n is the capture rate term for the unknown recombination center(s) and N_R is the density of the center(s). Since τ is found to be independent of N_{In} it is inferred that $c_n N_R \gg c_{\text{In}} N_{\text{In}}$. The fact that the Wertheim lifetimes

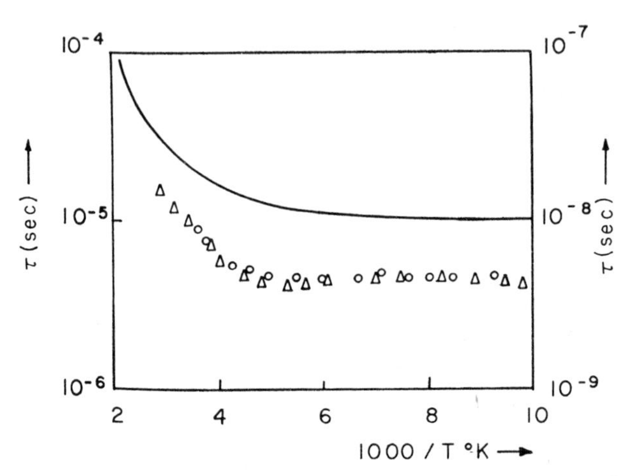

FIG. 6.4 Temperature dependence of the time constant τ describing the relaxation of the nonequilibrium conductivity in indium-doped silicon.
(O) $N_{\text{In}} = 4 \times 10^{15}\ cm^{-3}$; (Δ) $N_{\text{In}} = 1.2 \times 10^{17}\ cm^{-3}$. *The solid curve (scale on the right-hand side) represents the data of Wertheim (1958),* $N_{\text{In}} = 1.4 \times 10^{17}\ cm^{-3}$. *(After Pokrovskii and Svistunova, 1969.)*

are much lower than those of Pokrovskii and Svistunova is attributed to a greater density of recombination centers. Since Wertheim's studies were early in the development of silicon crystal technology it is not unlikely that his material did indeed contain a high density of other recombination centers.

The point to be stressed is that in deep impurity studies there is always the possibility (some would say, probability) that an unexpected center may be contributing to the effect that is under examination. Therefore wherever possible, one must check that the effect seen is correctly related to the added impurity density as well as varying with temperature as expected.

The capture cross sections that are of interest in indium-doped silicon are $\sigma_n{}^0$ for the process $N_{\text{In}}{}^0 + e \rightarrow N_{\text{In}}{}^-$, and $\sigma_p{}^-$ for the process $N_{\text{In}}{}^- + h \rightarrow N_{\text{In}}{}^0$. Since the latter process is one involving coulombic attraction, we expect $\sigma_p{}^-$ to be large, perhaps 10^{-13} to 10^{-14} cm^2, corresponding to $c_p{}^-$ of 10^{-6} to 10^{-7} cm^3 sec^{-1}. It has been measured by several methods, as described in Chapter 10, with general agreement among the laboratories involved.

The electron capture cross section $\sigma_n{}^0$ for indium in silicon is still uncertain in value. The Pokrovskii and Svistunova approach to the problem has been to study the lifetime of n-type silicon containing antimony as the shallow donor dopant and indium as the deep acceptor. As shown in Fig. 6.5(a), at high temperatures τ is independent of the indium concentration and

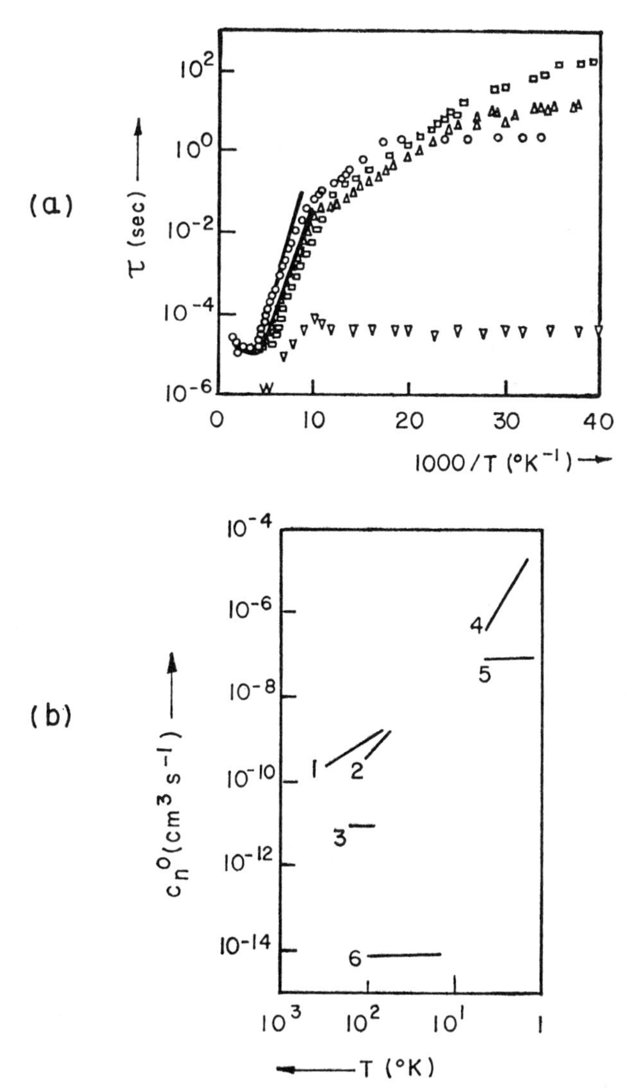

FIG. 6.5 (a) Temperature dependence of τ in indium- and antimony-doped n-type silicon.

(□) $N_A = 1.4 \times 10^{15}$ cm^{-3}, $N_D = 6.8 \times 10^{15}$ cm^{-3}; (△) $N_A = 4.5 \times 10^{15}$ cm^{-3}, $N_D = 2.7 \times 10^{16}$ cm^{-3}; (○) $N_A = 6 \times 10^{16}$ cm^{-3}, $N_D = 8.7 \times 10^{16}$ cm^{-3}; (▽) $N_A = 1 \times 10^{17}$ cm^{-3}, $N_D = 3 \times 10^{18}$ cm^{-3}. (After Pokrovskii and Svistunova, 1969.)

(b) Temperature dependence of the electron capture coefficients $c_n^{\,o}$ of neutral group III acceptors in silicon.

(1) indium from conductivity decay; (2) indium from double-injection features; (3) indium from generation-recombination noise; (4) boron from broadening of the electron cyclotron resonance line in the presence of high-intensity intrinsic illumination; (5) indium from combined optical and paramagnetic resonance techniques; (6) from the intrinsic photoconductivity of 3×10^{18}cm^{-3} antimony-doped silicon containing some indium.

is equal to 10 μsec. Thus it is approximately equal to the value of τ observed in n-type silicon doped with antimony only. In this temperature range the recombination rate is determined by electron and hole capture by residual recombination centers. On lowering the temperature the hole capture by negatively charged indium atoms becomes effective. At the same time τ is determined by the rate of the thermal excitation of holes from indium atoms to the valence band [i.e., the process $\mathrm{In}^0 + e(\text{valence}) \rightarrow \mathrm{In}^- + h$] followed by their recombination with electrons on the residual centers. This leads to an exponential increase of τ with temperature, the activation energy being equal to the thermal ionization energy of In, E_{In}. In the case of relatively high temperatures and small deviations from equilibrium ($\delta n, \delta p \ll n_0$) one has

$$\tau = \tau_{n0}\left(1 + \frac{N_{\mathrm{In}}}{p}\right) \tag{6.39}$$

Here $p = N_v \exp(-E_{\mathrm{In}}/kT)$; N_v is the effective density of states of the valence band; and $\tau_{n0} \approx 10$ μsec is the time constant of intrinsic photoconductivity relaxation in n-type silicon doped with antimony only. This theoretical dependence is shown in Fig. 6.5 by the solid curves. On further lowering the temperature the thermal reemission rate of holes from indium atoms decreases exponentially and becomes insignificant at about 100°K. On the other hand, the ionization energy of antimony ($E_D = 0.039$ eV) is essentially smaller than E_{In} (0.16 eV). Therefore, the equilibrium concentration of free electrons, n_0, remains high enough in this temperature range that the recombination rate is determined by electron transitions to indium atoms occupied with holes (i.e., to In^0). For this condition Pokrovskii and Svistunova (1969) have shown that

$$\tau = (c_n{}^0 n_0)^{-1} \tag{6.40}$$

On lowering the temperature n_0 decreases exponentially with an activation energy E_D, hence τ increases in this range with the same activation energy. At low temperatures the equilibrium concentration of electrons in the conduction band becomes so small that τ is determined by direct donor-acceptor transitions. In this temperature range τ is only weakly temperature dependent and is given by

$$\tau = (c_A{}^D N_D{}^0)^{-1} \tag{6.41}$$

Here $c_A{}^D$ is the interimpurity recombination coefficient and $N_D{}^0$ is the concentration of neutral antimony atoms.

Thus in n-type silicon the electron capture coefficient of a neutral acceptor, $c_n{}^0$, may be determined in a certain temperature range only. This is the range where the thermal reemission of holes from acceptors is insignificant while the thermal free-electron concentration, n_0, is still high enough and interimpurity recombination plays no serious role. In this temperature range the electron capture coefficient of neutral indium in silicon is given by (6.40) and in the Pokrovskii and Svistunova studies turns out to be equal to 5×10^{-15} cm^{-3} sec^{-1}. This represents a capture cross section of the order of 5×10^{-22} cm^2, which is unexpectedly small for a nonrepulsive capture event (see Chapter 10). Such capture cross sections are seen for radiative recombination processes in silicon.

6.4 Surface Recombination Effects

In lifetime measurements, particularly where the carrier disturbance, or injection, is produced by light, one must be careful to distinguish the relative contributions of surface and bulk recombinations to the lifetime observed. Unless the illumination is chosen to be nearly monochromatic with a small absorption coefficient, the light may create a very nonuniform carrier distribution with high electron-hole pair concentrations near the illuminated surface. There is then considerable recombination at the surface and it is given by $es\delta p$, where δp is the excess minority carrier density near the surface and s is a constant known as the surface recombination velocity since it has the dimensions of centimeters per second.

Under appropriate conditions the observed lifetime may be expressed as

$$(\tau_{\text{obs}})^{-1} = (\tau_B)^{-1} + (\tau_s)^{-1} \tag{6.42}$$

where τ_s is the lifetime due to surface recombination and τ_B is the bulk lifetime. The two contributory time constants can in principle be separated by experiments which involve changing the cross-sectional dimensions of the sample. If a bar of rectangular cross section $2A \times 2B$ is involved and the carriers are considered to be uniformly generated, Shockley (1950) has shown that τ_s is related to s by the expression

$$(\tau_s)^{-1} = D\left(\frac{\eta^2}{A^2} + \frac{\xi^2}{B^2}\right) \tag{6.43}$$

where D is the minority carrier diffusion constant and $sA/D = \eta \tan \eta$ and $sB/D = \xi \tan \xi$. The smallest pair of roots, η_0 and ξ_0, correspond to the longest lifetime and are of principal importance.

With careful cleaning and etching, the surface recombination velocities observed for germanium and silicon may be less than 10^3 cm sec^{-1}. On the other hand, the surface recombination velocities of many compound semiconductors tend to be greater than 10^5 cm sec^{-1}.

6.4.1 Recombination at a Single Deep Surface Level

Consider an n-type semiconductor with a surface containing a uniform density N_r of centers per unit area, all at one energy level $e\phi_s$ below the center of the band gap E_i. Such states tend to accept electrons and cause a depletion layer near the surface as shown in Fig. 6.6.

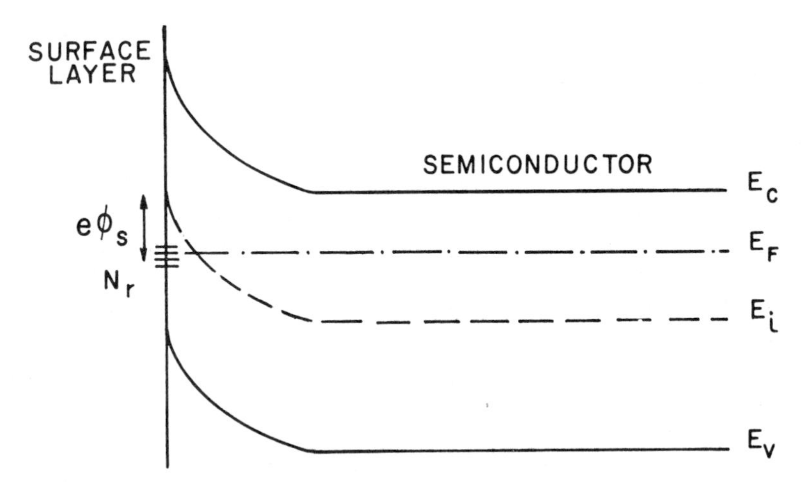

FIG. 6.6 Hypothetical energy band diagram for an n semiconductor with acceptor states at the surface at an energy $e\phi_s$ below the center of the band gap.

The rate of recombination through the center is determined by its capture cross sections and by the availability of holes and electrons at the surface. An analysis of the statistics of the recombination processes for the case in which the hole and electron distributions are nondegenerate yields for the rate of recombination per unit area

$$r_s = \frac{N_r c_p c_n (p_b + n_b)\,\delta p}{c_p(p_s + p_{s1}) + c_n(n_s + n_{s1})} \tag{6.44}$$

where c_p and c_n are the capture probabilities per center per unit time for holes and electrons when the centers are filled and empty, respectively, and are given by the product of the corresponding capture cross section with the thermal velocity; p_s and n_s are the free-carrier densities at the surface; p_{s1} and n_{s1} are the surface densities if the Fermi level is at the energy level of the centers; p_b and n_b are the bulk equilibrium carrier densities;

and δp is the excess minority carrier density at the surface just inside the space charge layer. If it is further assumed that the densities p_s and n_s at the surface are in equilibrium with the densities in the interior; then, from (6.44), the surface recombination velocity is given by

$$s = \frac{r_s}{\delta p} = \frac{N_r c_p c_n (p_b + n_b)}{n_i c_n \exp\left((E_F - E_i + e\phi_s)/kT\right) + n_i c_p \exp\left(-(E_F - E_i + e\phi_s)/kT\right)} \tag{6.45}$$

where $e\phi_s$ is negative in value if it lies below the center of the band gap.

The expression given is basically similar to that for single-energy-level recombination centers in a bulk semiconductor, following Shockley–Read–Hall treatments. The validity of this kind of surface recombination expression has been studied by Gräfe (1971) and others. It is only moderately acceptable since it is based on a grossly simplified model of a much more complicated situation.

Several excellent reviews of surface state work are available. Intrinsic surface state effects have been studied by Davison and Levine (1970), Henzler (1971), Fairbairn (1971), and others. Experimentally for Si and Ge the intrinsic surface state distributions appear as pairs of donor–acceptor bands with the donor band lying below the acceptor band. The pair bands are not symmetrical with respect to the center of the band gap but tend to be displaced somewhat toward the valence band edge. The surface states do not appear to be conductive. Theory does not predict pairing of the bands: the reason for this, presumably, is that surface reconstruction effects are not taken into account. For some semiconductors with unreconstructed surfaces the surface state bands tend to appear much closer to the bulk bands, and the agreement between theory and experiment is satisfactory.

Extrinsic surface states may be caused by adsorbates of unknown or known species, or by interfaces between two solid materials, as for semiconductor–insulator, semiconductor–semiconductor and semiconductor–metal interfaces.

The field of metal-insulator-semiconductor physics has been reviewed by Goetzberger and Sze (1969). Surface state densities at SiO_2–Si interfaces tend to be in the range 10^{11} to 10^{13} cm^{-2} and lie mainly in two bands. Such bands, for instance, may be 0.15 eV wide with peaks at $E_c - 0.12$ eV and $E_v + 0.08$ eV. The results are very dependent on the processing technology. Useful measurement techniques have been discussed recently by Castagné and Vapaille (1971), Fogels and Salama (1971), Schroder and Guldberg (1971), Amelio (1972), Zuev et al. (1972), and others.

Heine (1965) has considered surface states at metal-semiconductor interfaces, and Kar and Dahlke (1972) have looked at states caused by metal diffusion through thin (20–40 Å) SiO_2 films on silicon. Semiconductor–semiconductor interface states have been discussed by Milnes and Feucht (1972).

Clean cleaved (110) gallium arsenide surfaces have been studied by Fischer and coworkers by photoelectric emission, contact potential, and surface photovoltage measurements. The results are interpreted in terms of a band of surface acceptors with a density of at least 2×10^{13} cm^{-2} eV^{-1} at a lower band edge which lies 0.85 eV below the conduction band edge (Dinan et al., 1971). The density of states at SiO_2–GaAs interfaces has been studied by Kern and White (1970). The range found is from 10^{13} down to the low 10^{11} cm^{-2} eV^{-1} depending on the processing. The Si_3N_4–GaAs interface has been found to have a state density in the 10^{12} cm^{-2} range (Foster and Swartz, 1970). Pulsed field effect experiments on "real" gallium arsenide surfaces have been made by Valahas et al. (1971). In oxygen-doped gallium arsenide a surface level was found 0.45 eV below the conduction band edge, and this appeared to be related to the bulk oxygen doping. In CdS, Gatos and his coworkers have applied photovoltage spectroscopy and other techniques to the study of surface states with considerable success (Lagowski et al., 1972).

6.5 Recombination in a Semiconductor Containing an Electron Trap and a Recombination Center

An impurity center acts as a trap if the probability of reemission of a trapped carrier to its energy band of origin, as discussed earlier, is substantially greater than its transition to the other energy band for recombination. This is illustrated in Fig. 6.7 where an electron is shown falling into a trap E_T and being reemitted to the conduction band (step 2) rather than recombining with a hole (step 2*a*). After some time in the conduction band the electron is retrapped (3) for a period and reemitted (4), and finally recombines through a recombination center E_R.

A trapping center differs primarily from a recombination center, then, in the fact that the capture rates for the two kinds of carriers are very different so that direct communication from the conduction band to the valence band through the trap is improbable. Figure 6.7 therefore implies that the hole capture cross section of the E_T level is very low. This could be true, for instance, if the level E_T represents a doubly ionized donor (with two positive charges). Such a level would be singly positively ionized after step 1 and would have a very small attraction for a hole, namely a low-capture cross section for step 2*a*.

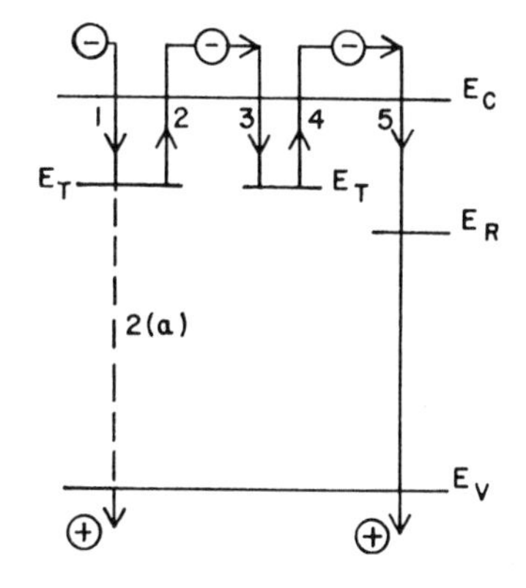

FIG. 6.7 Electron trapping at the level E_T followed by reemission to the conduction band and subsequent recombination through another center, E_R.

The time spent by the carrier in the conduction band is the lifetime (mean life) of the electron, but trapping affects the time spent in the traps which can exceed the lifetime of the carrier by orders of magnitude. This effectively reduces the mobility of the carrier and gives an apparent long decay constant.

Although we have considered an electron trap in Fig. 6.7, trapping of holes is also very common in wide-gap semiconductors. Wertheim (1958) has analyzed the behavior of a hole trap in n-type material with the assumptions that the electron-capture cross section of the trap is zero and that the semiconductor contains a recombination center. The result obtained for the hole lifetime, assuming a small density of recombination centers and small fractional filling of the trap, is

$$\tau = \frac{[1 + c_T(N_T^- + p_T)/c_pN_R^-](c_Tp_T)^{-1} + [1 + p_1(N_T^- + p_T)/p_TN_R^-][c_n(n_0 + n_1)]^{-1}}{1 + p_1N_R^0N_T^-/p_TN_R^-(n_0 + n_1)} \tag{6.46}$$

where c_T and N_T^- relate to the trap and c_n, c_p, N_R^- and N_R^0 relate to the recombination center. The denominator will approximate unity, provided the recombination centers lie below the Fermi level and the trapping centers below the recombination centers.

At low temperature, where $p_T < N_T^-$, the equation may be written

$$\tau = \frac{c_pN_R^- + c_TN_T^-}{(c_Tp_T)(c_pN_R^-)} + \frac{N_R^- + p_1}{(c_nN_R)n_0} + \frac{p_1N_T^-}{(c_nN_R)p_Tn_0} \tag{6.47}$$

The first term is the reciprocal of the net rate at which holes are captured by the recombination center. This follows since c_Tp_T is the rate of regeneration from the traps and $c_pN_R^-/(c_pN_R^- + c_TN_T^-)$ is the fraction which is captured by the recombination center. The rest are retrapped. The second term is the electron capture time of the recombination centers

$$\tau_n = (c_nN_R)^{-1}(N_R^- + p_1)n_0^{-1} \tag{6.48}$$

We now define the regeneration time $\tau_g = (c_T p_T)^{-1}$, the hole-trapping time $\tau_T = (c_T N_T^{-})^{-1}$, and the hole capture time of the recombination center $\tau_p = (c_p N_R^{-})^{-1}$, and rewrite (6.47) in the form

$$\tau = \tau_g \tau_p \tau_T^{-1} + \tau_g + \tau_n + p_1 N_T^{-} (c_n n_0 p_T N_R)^{-1} \tag{6.49}$$

This is similar to the result obtained by Hornbeck and Haynes (1955) in a treatment that did not consider the individual processes of the recombination center.

6.6 General Consideration of Trapping and Recombination Kinetics

Sah and Shockley (1958) in considering the problems of trapping and recombination have shown that it is worthwhile to introduce concepts that help define which processes are dominant for various doping conditions. The first concept is that carrier concentrations n^* and p^* exist at which the four capture and emission processes of Fig. 6.1 proceed at equal rates. These quantities are known as the "equality carrier concentrations."

If the capture of electrons from the conduction band by a center is equal to the capture of holes from the valence band and to the emission of electrons and emission of holes, we see from (6.9) and (6.10) that

$$n^* = \frac{c_p}{c_n} p_1 \tag{6.50a}$$

and

$$p^* = \frac{c_n}{c_p} n_1 \tag{6.50b}$$

It follows that $n^* p^* = p_1 n_1 = n_i^2$, which is the thermal equilibrium condition, as expected. The Fermi level corresponding to n^* and p^* is defined by

$$n^* = n_i \exp\left(\frac{E_T^* - E_i}{kT}\right) \tag{6.51}$$

and is known as the "equality Fermi level." In a semiconductor containing a given deep center level E_T, the actual Fermi level can be varied by the shallow doping provided. However, E_T^* depends on the ratio of the capture cross sections. For instance, Fig. 6.8a represents a material for which $c_n > c_p$ (or $e_n > e_p$) and the trap level is assumed to be above the center of the band gap. Under these conditions E_T^* is well below the center of the

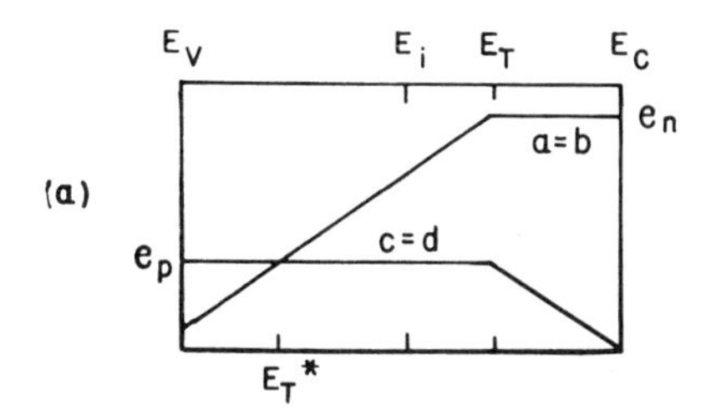

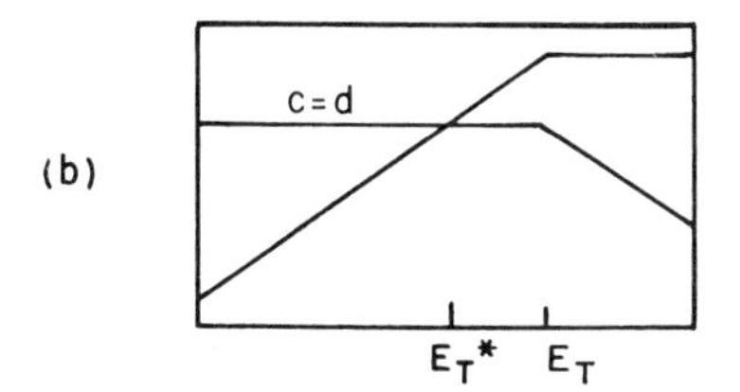

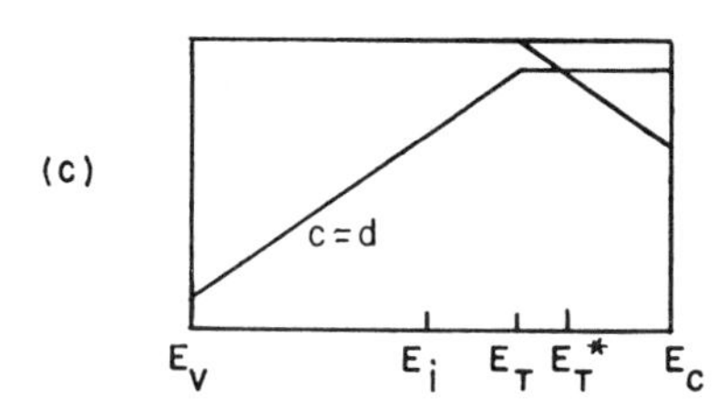

FIG. 6.8 The equilibrium rate of electron or hole capture or emission as a function of Fermi level with E_T fixed and $E_T{}^*$ varied. *(a) $e_n > e_p$, $c_n > c_p$, or $e_n/e_p > 1 > c_p/c_n$, electron trap; (b) $e_n/e_p = c_n/c_p = 1$; (c) $e_n/e_p < 1 < c_p/c_n$, hole trap. Processes a and b are electron capture and emission, respectively, and processes c and d are hole capture and emission (see Fig. 6.1). (After Sah, 1967b.)*

band gap. The capture and emission processes of Fig. 6.1 are identified by the symbols a, b, c, and d in Fig. 6.8.

It is then seen that if the material in Fig. 6.8a is n-type doped or lightly p-type doped ($E_F > E_T^*$), the deep center behaves predominantly as an electron trap since electron capture (process a) is greater than hole capture (process d). On the other hand, if the material is heavily p doped, hole trapping dominates since process d is greater than process b. The other two diagrams, Fig. 6.8b and c, are for $c_n = c_p$ and $c_p > c_n$ and are interpreted in a similar way.

The diagrams of Fig. 6.8 are for the thermal equilibrium condition where p and n are related by $n_i{}^2$. However, the concepts involved can be applied to the nonequilibrium condition by the use of a diagram with $\log_{10} p$ and $\log_{10} n$ as the coordinates (Sah and Shockley, 1958). For a single-energy-level center the diagram is as shown in Fig. 6.9. The carrier concentration space is divided into four quadrants with the equality point (p^*, n^*) as the

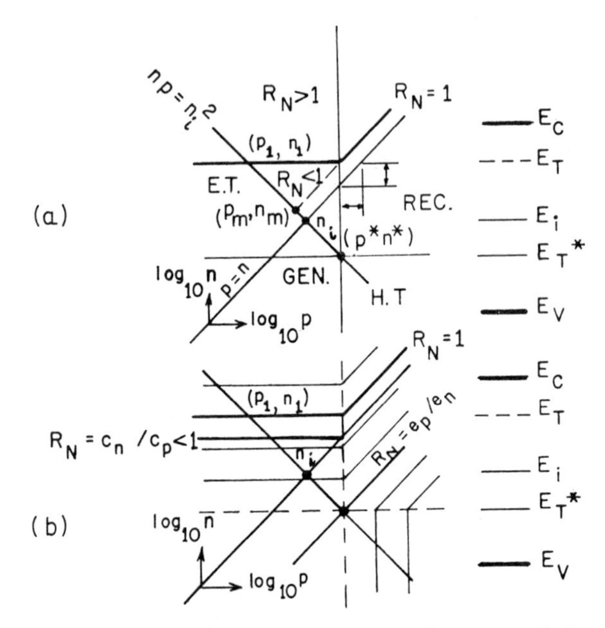

FIG. 6.9 The criteria of recombination, generation, electron- or hole-trapping dominance for nonequilibrium conditions shown in the carrier steady-state concentration space.

The steady-state charge distribution on the defect centers is illustrated in (a) and (b) by the R_N = constant lines. E.T. indicates electron trapping; REC indicates recombination; GEN indicates generation; and H.T. indicates hole trapping. (After Sah, 1967b.)

center. The lines for $pn = n_i^2$ and for $p = n$ appear on the diagram as straight lines intersecting at the value n_i. The energy band diagram is shown to the right of Fig. 6.9*a* and *b* since it can be related to the equilibrium majority carrier concentration by the Maxwell–Boltzmann expression normally used for nondegenerate semiconductor statistics.

Under steady-state conditions, there is another convenient family of lines which can be added to this diagram which represents the conditions of constant charge on the impurity center. This family of lines, denoted by R_N = constant, is the ratio of N_T^- to N_T^0 for the steady-state condition. The equation of these lines can be obtained from the steady-state condition $dN_T^-/dt = 0$. This gives (6.14)

$$\frac{N_T^-}{N_T} = \frac{c_n n + c_p p_1}{c_n(n + n_1) + c_p(p + p_1)}$$

and

$$R_N = \frac{N_T^-}{N_T^0} = \frac{N_T^-}{N_T - N_T^-}$$

$$= \frac{c_n n + c_p p_1}{c_p p + c_n n_1} \tag{6.52}$$

$$= \frac{c_n(n + n^*)}{c_p(p + p^*)} \tag{6.53}$$

$$= \left(\frac{p^* p_1}{n^* n_1}\right)^{1/2} \frac{n + n^*}{p + p^*} \tag{6.54}$$

Now, following Sah's discussion closely for a few paragraphs, these reference lines for a set of assigned constant values of R_N correspond to the lines of constant charge distribution on the center for a particular pair of steady-state concentrations of electrons and holes.

The construction of the line $R_N = 1$ is illustrated in Fig. 6.9*a*. From (6.54) for $n \gg n^*$ and $p \gg p^*$ in the recombination-dominated region, we have

$$R_N = 1 = \left(\frac{p^* p_1}{n^* n_1}\right)^{1/2} \frac{n}{p} \tag{6.55}$$

which shows that the $R_N = 1$ line is a $45°$ line in the $\log_{10} n - \log_{10} p$ diagram passing through the point $[(p^* p_1)^{1/2}, (n^* n_1)^{1/2}]$ if it is extended to the equilibrium line $np = n_i^2$. For small p and p_1, (6.52) gives

$$R_N = 1 \approx \frac{c_n n}{c_n n_1}$$

from which

$$n = n_1 \tag{6.56}$$

since $p^* > n^*$ and $n_1 \gg n^*$ for the particular case shown in this figure. This shows that $R_N = 1$ is a horizontal line passing through the point (p_1, n_1). These two asymptotic limiting lines are used to construct the $R_N = 1$ line in Fig. 6.9*a* where the rounded corner is not shown for simplicity. The fact that the $R_N = 1$ line must pass through the point (p_1, n_1) can be readily observed from (6.52) using $np = n_i^2 = p_1 n_1$.

The two regions of the carrier concentration space divided by the $R_N = 1$ line give the following charge conditions on the centers. Above the $R_N = 1$ line we have $R_N = N_T^-/N_T^0 > 1$ so that most of the centers are occupied by electrons, while below the $R_N = 1$ line we have $R_N < 1$ so that most of the centers are empty. A family of $R_N =$ constant lines is shown in Fig. 6.9*b*, which illustrates this effect. The $R_N = 1$ line is the half-occupancy line.

Another interesting feature of the particular example illustrated is that there is a considerable range of carrier concentration in the p-type region where most of the centers are empty but still act as electron traps. This region is bounded by $n_1 > n > n^*$ and $p < p^*$, as evident in Fig. 6.9*a*. Such an unusual situation comes from the relative position of E_T and E_T^*, which for this case satisfies $E_T > E_i > E_T^*$. In terms of the kinetic coefficients, this result comes from the extremely large electron emission rate e_n, or small electron capture coefficient c_n.

Sah (1967b) discusses the physical processes which govern the steady-state charge distribution in the four quadrants shown in Fig. 6.9*a*.

Depleted region ($n \ll n^*$ *and* $p \ll p^*$, *generation dominant*). In this region, the emission processes dominate over the capture processes, so electron-hole pair generation dominates over the recombination and trapping events. The steady-state charge distribution is governed by the two emission constants as can be seen from (6.52), which gives

$$R_N = \frac{N_T^-}{N_T^0} = \frac{c_p p_1}{c_n n_1} = \frac{e_p}{e_n} \tag{6.57}$$

Flooded region ($n \gg n^*$ *and* $p \gg p^*$, *recombination dominant*). In this region, the capture processes dominate and the recombination events predominate over the other processes so that the steady-state charge distribution is governed by the two capture rate constants. From (6.52), we have

$$R_N = \frac{N_T^-}{N_T^0} = \frac{c_n n}{c_p p} \tag{6.58}$$

Electron-dominated region ($n \gg n^*$ *and* $p \ll p^*$, *electron trap*). In this region electron-trapping events dominate and the steady-state distribution is determined mainly from the balance of the electron capture and emission processes with holes playing no important role. Thus the charge distribution ratio from (6.52) is approximately

$$R_N = \frac{N_T^-}{N_T^0} \approx \frac{n}{n^*} = \exp\left(\frac{E_T - E_T^*}{kT}\right) \tag{6.59}$$

Hole-dominated region ($n \ll n^*$ *and* $p^* \ll p^*$, *hole trap*). The property in this region is similar to that of the electron-dominated region with the hole capture and emission processes predominantly establishing the charge condition of the centers.

For a more detailed study of the use of this kind of diagram the reader is referred to Sah (1967b, c). An equivalent circuit approach to the calculation of lifetimes is also developed in these papers.

To summarize, we see that a single-level impurity Shockley–Read–Hall center can be characterized by three parameters chosen from a combination of the six parameters E_T, E_T^*, c_n, c_p, e_n, and e_p. Whether the center acts as a recombination site, a generation site, or an electron- or hole-trapping site depends not only on the numerical values of these three parameters but also on the concentration of electrons and holes at a given condition of the sample.

6.7 Transient Lifetime Maps for Gold-Doped Silicon

The transient for a single-energy-level recombination center has been shown by Sandiford (1957), Wertheim (1958), and others (equations 6.28 through 6.31) to be represented by two time constants τ_i and τ_t, where for many typical conditions, $\tau_i < \tau_t$. In the notation of equality carrier densities, n^*, p^*, and for the condition $N_T \to 0$, the two expressions obtained are

$$\tau_i \equiv \tau_- \simeq [c_n(n + n_1) + c_p(p + p_1)]^{-1} \tag{6.60}$$

and

$$\tau_t \equiv \tau_+ \simeq \frac{[\tau_{p0}(n + n_1) + \tau_{n0}(p + p_1)][\tau_{p0}(n + n^*) + \tau_{n0}(p + p^*)]}{\tau_{p0}(n + n_1)(n + n^*) + \tau_{n0}(p + p_1)(p + p^*)} \tag{6.61}$$

For a given impurity center τ_- and τ_+ contour lines can be plotted on a chart that represents the $\log_{10} n$, $\log_{10} p$ plane. This has been done by Sah (1967b), for gold in silicon.

Gold in silicon has two energy levels in the band gap: an acceptor at $E_T - E_i = 0.020$ eV with $c_p = 1.15 \times 10^{-7}$ cm^3 sec^{-1} and $c_n = 1.65 \times 10^{-9}$ cm^3 sec^{-1}; and a donor at $E_i - E_T = 0.208$ eV with $c_p = 2.40 \times 10^{-8}$ cm^3 sec^{-1}. [These capture rates are from the cross section data of Fairfield and Gokhale (1965) with the thermal velocity taken as 1.15×10^7 cm sec^{-1}.] The energy gap of silicon may be taken as 1.12 eV and n_i as 1.4×10^{10} cm^{-3} at 300°K, and contour maps of τ_- and τ_+ have been constructed for each of the two gold levels on the assumption that the other does not exist. Figure 6.10 is for the acceptor level and Figure 6.11 is for the donor level. The principal conclusions to be drawn from Figs. 6.10 and 6.11 are (a) that τ_+ is indeed usually the largest time constant and (b) that gold doping to the 10^{16} cm^{-3} density can produce lifetimes as low as a few tens of nanoseconds (depending on the carrier density conditions) but that much lower lifetimes, say 10^{-9} sec^{-1} or below, are not readily achievable in this way. This is confirmed in Section 8.2 where we review the observed lifetimes for injected minority carriers in silicon diodes that have been diffused with gold to various densities.

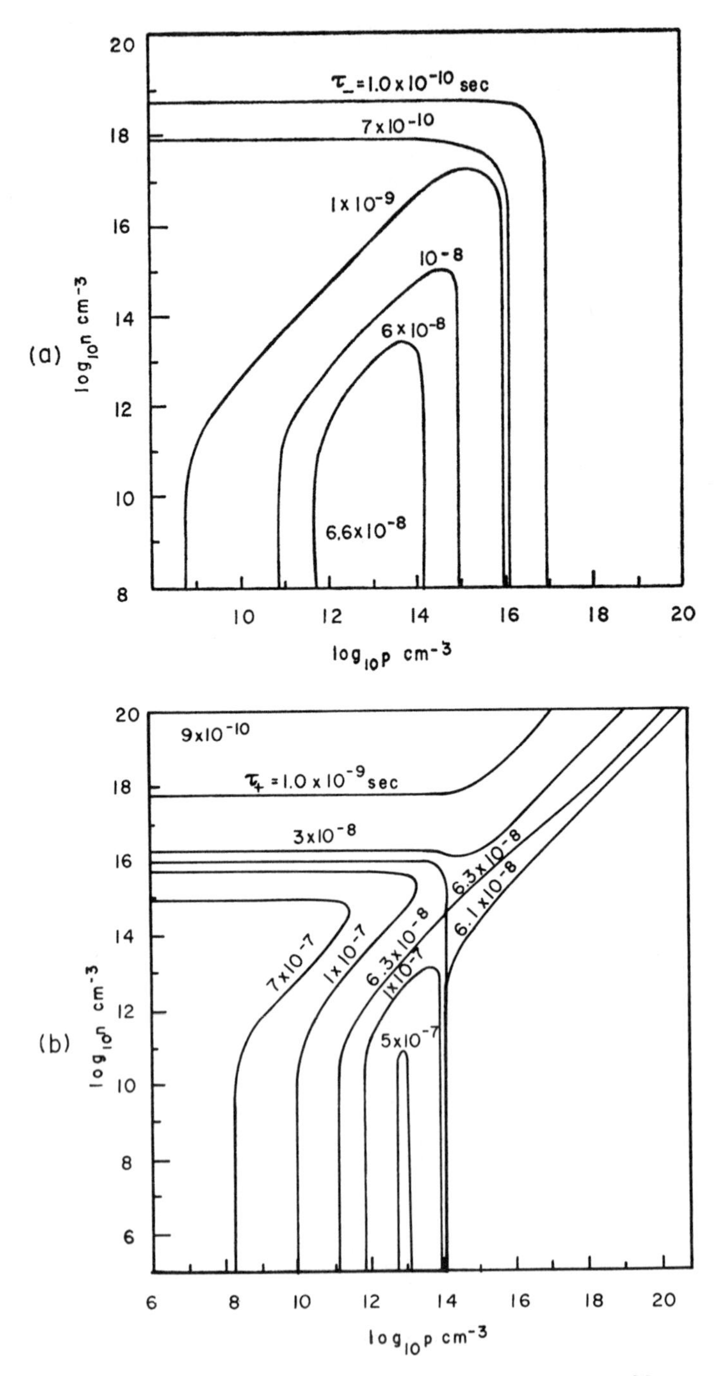

FIG. 6.10 The transient decay time constant contour maps of the gold acceptor level in silicon with $N_T = 10^{16}$ cm^{-3} and at 300°K.
(a) τ_- and (b) τ_+. (After Sah, 1967b.)

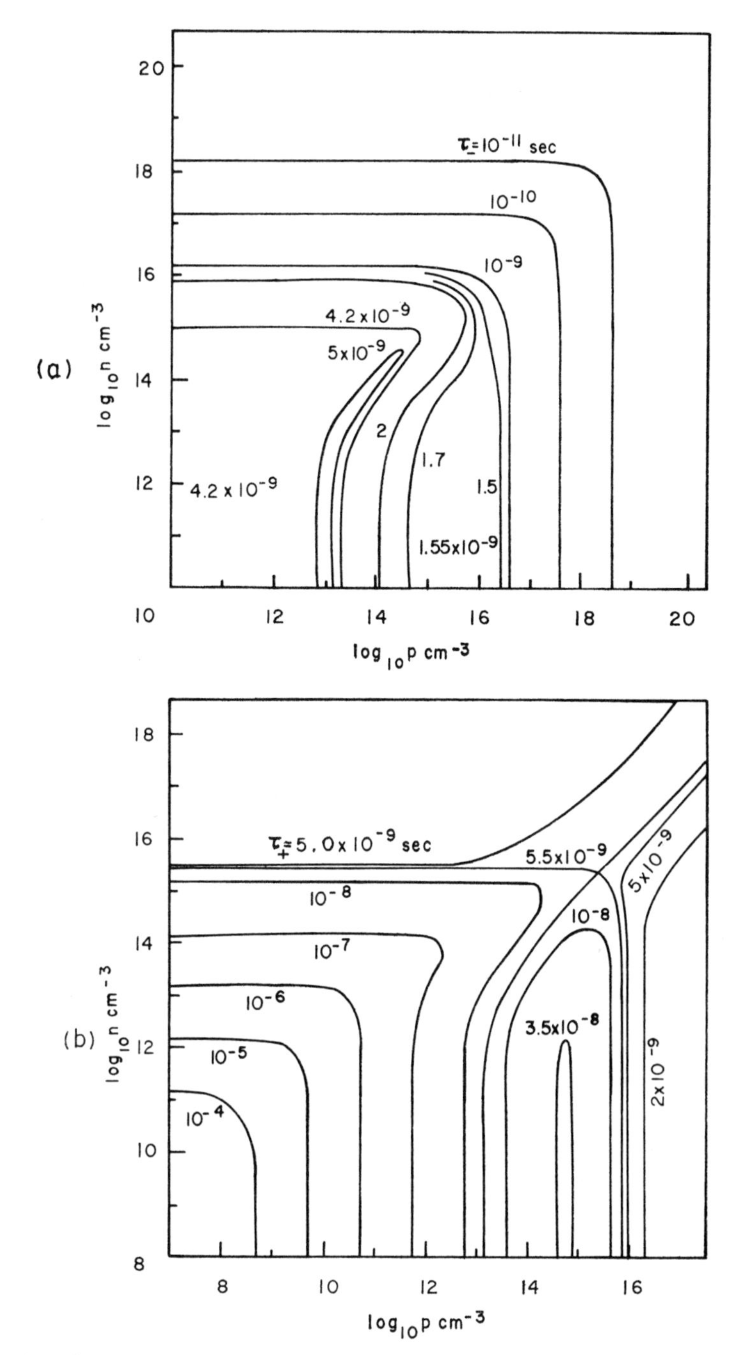

FIG. 6.11 The transient decay time constant contour maps of the gold donor level in silicon with $N_T = 10^{16}$ cm^{-3} and at 300°K.
(a) τ_- and (b) τ_+. (After Sah, 1967b.)

Rigorous solution of the problem of transients in gold-doped silicon is of course rather more complicated because the gold can assume three energy states, Au^0, Au^+, and Au^-, and these cannot be treated as independent of each other. This has been the subject of detailed study by Sah (1967b, c) as part of a general treatment of multiple-energy-level impurity centers.

CHAPTER SEVEN

Photoconductivity and Phototransient Effects in Bulk Semiconductors Containing Deep Impurities

Photoconductivity in semiconductors tends to be dominated by deep impurity effects. The discussion in this chapter is confined to bulk semiconductors, not containing junctions, and begins with impurity photoconductivity associated with a single deep acceptor level which communicates with the valence band. This is made specific by a discussion of the photoconductivity behavior of indium-doped silicon.

Transient effects in impurity photoconductivity are considered and equations are given for rise and fall times. A model is then presented for photoconductivity transients in a semiconductor that contains a deep acceptor in the upper half of the band gap. This is illustrated with reference to the behavior of silicon doped with silver.

Intrinsic photoconductivity, induced by band-gap light, is then examined with a deep center present acting either as a recombination center or as a trap to enhance the quantum efficiency. Photoconductivity effects involving light of two wavelengths are next considered, where quenching, superlinearity effects, and charge transfer between deep impurities may occur. The chapter concludes with a brief discussion of photoconductive detector performance.

7.1 Extrinsic Photoconductivity, Associated with a Single Deep Acceptor Level

Impurity photoconductivity, in a simple form, involves a deep impurity that is photoionized by light of energy, $h\nu$, that is less than the semiconductor band-gap energy but greater than the impurity activation energy. If the deep impurity is a donor at some energy below the conduction band, the photoinduced carriers are electrons. These are assumed to remain in the conduction band until recaptured by the deep donor, and so the conduction of the semiconductor is increased by the applied light. The concept of the impurity photoconductivity effect is therefore at first statement very simple. However, interesting and less obvious behavior emerges when the effect is studied in practice. To illustrate this we consider the example of indium in silicon.

Indium is a substitutional impurity in silicon with a deep acceptor level 0.155 to 0.160 eV from the valence band edge as shown in Fig. 7.1*a*. At room temperature most of the indium is ionized since the level is not very

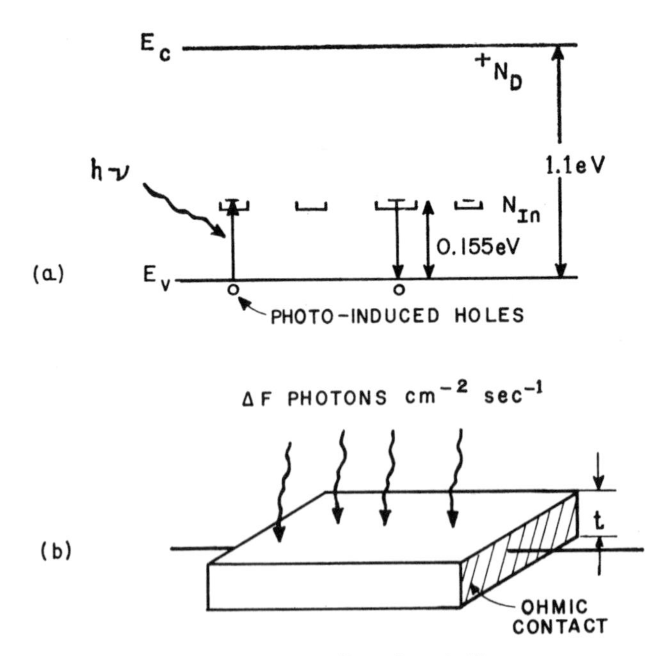

FIG. 7.1 Impurity photoconductivity in indium-doped silicon.
(a) Energy diagram for indium-doped silicon showing photoexcitation of holes by the process $In^0 + h\nu \rightarrow In^- +$ *hole. (b) Silicon bar, of thickness t, with a photon flux* ΔF cm^{-2} sec^{-1} *applied to the front face.*

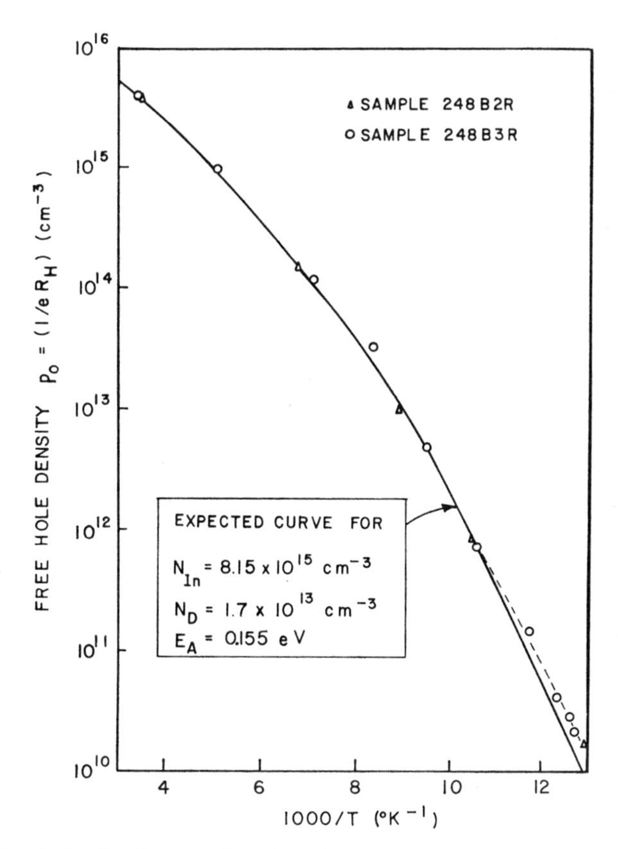

FIG. 7.2 Free-hole density as a function of inverse temperature for a specimen of indium-doped silicon. (After Blakemore and Sarver, 1968.)

deep in relation to the thermal energy kT. However, at 100°K only a small fraction (say 10^{-3}) of the indium will be thermally ionized, as seen from Fig. 7.2, and the resistivity of the semiconductor may be several thousand ohm-centimeters. In such material a small density of shallow donors has been provided to overcompensate any shallow acceptors that might otherwise be active. The densities of the indium, N_{In}, and the donor, N_D, can be inferred by careful fitting of the Hall data of Fig. 7.2 with the equation

$$\frac{p_0(N_D + p_0)}{(N_{\text{In}} - N_D - p_0)} = \frac{N_v}{\beta} \exp\left(-\frac{E_{\text{In}}}{kT}\right) \tag{7.1}$$

where p_0 is the thermal hole density, E_{In} is 0.155 eV, β is the acceptor occupancy spin factor (taken as 6 for the indium ground state), and $N_v = 2.18 \times 10^{15} T^{3/2}$ states cm^{-3}.

To determine the effect of light on the conductivity it is necessary of course to work at low temperatures, say in the range 70 to 180°K, and to apply chopped monochromatic radiation so that lock-in amplifier detection techniques may be used to combat noise and drift effects. One is concerned therefore with the rms volume-averaged density of excess holes Δp which results from a known rms photon flux ΔF (photons per square centimeter per second). If there are $N_{\mathrm{In}} - N_D - p$ neutral indium acceptors per cubic centimeter and the photoionization cross section is σ_i cm^2 for a particular photon energy, then the optical absorption coefficient is

$$\alpha = \sigma_i(N_A - N_D - p) \tag{7.2}$$

in the absence of competing absorption processes. Consider a photon flux ΔF cm^{-2} sec^{-1} applied to the front face of an indium-doped silicon bar of thickness t cm, as shown in Fig. 7.1*b*. The reflectivity R is dependent on the refractive indices of the silicon and air (n_2, n_1), and for normal incidence is given by

$$R = \frac{(n_2 - n_1)^2}{(n_2 + n_1)^2} \tag{7.3}$$

If the specimen is thick, $t \gg 1/\alpha$, all the photons that enter the bar will be absorbed. Although it may be expected that each photon creates one carrier pair, this is not something that can be accepted without verification, and therefore a quantum efficiency Q (carrier pairs per photon) is introduced. For a thick specimen and complete absorption, therefore, the rate of optical generation of holes is

$$G_0 = \frac{Q\Delta F(1 - R)}{t} \ \mathrm{cm}^{-3}\ \mathrm{sec}^{-1} \tag{7.4}$$

where the rate has been averaged over the thickness of the specimen (although of course it is greatest at the front face). However, for impurity light the absorption coefficient is usually such that complete absorption is not obtained for specimens of reasonable thickness ($t < 1$ cm). For incomplete absorption the expression becomes

$$G_0 = \frac{Q\Delta F(1 - R)[1 - \exp(-\alpha t)]}{t[1 - R\exp(-\alpha t)]} \ \mathrm{cm}^{-3}\ \mathrm{sec}^{-1} \tag{7.5}$$

The product of the generation rate and the hole lifetime gives the sustained excess hole density Δp. Hence

$$\tau_p Q = \frac{\Delta p}{G_0/Q} \tag{7.6}$$

or if $Q = 1$, then

$$\tau_p = \frac{\Delta p}{G_0} \tag{7.7}$$

The photoconductive response as a function of photon energy for a specimen of indium-doped silicon is shown in Fig. 7.3 for a temperature of 93°K. The fact that the response is not completely flat across the energy range from 0.155 to 1.1 eV reflects the dependence of the capture cross section on the photon energy. The excess hole density Δp can be determined from the change of conductivity on illumination, provided the hole mobility is known from Hall and dark resistivity measurements at the appropriate temperature. Hence if the photon flux is known, the hole lifetime can be determined from (7.7) and (7.5). The reciprocal of the hole lifetime obtained for the 8.15×10^{15} cm^{-3} indium-doped silicon is shown as a function of temperature in

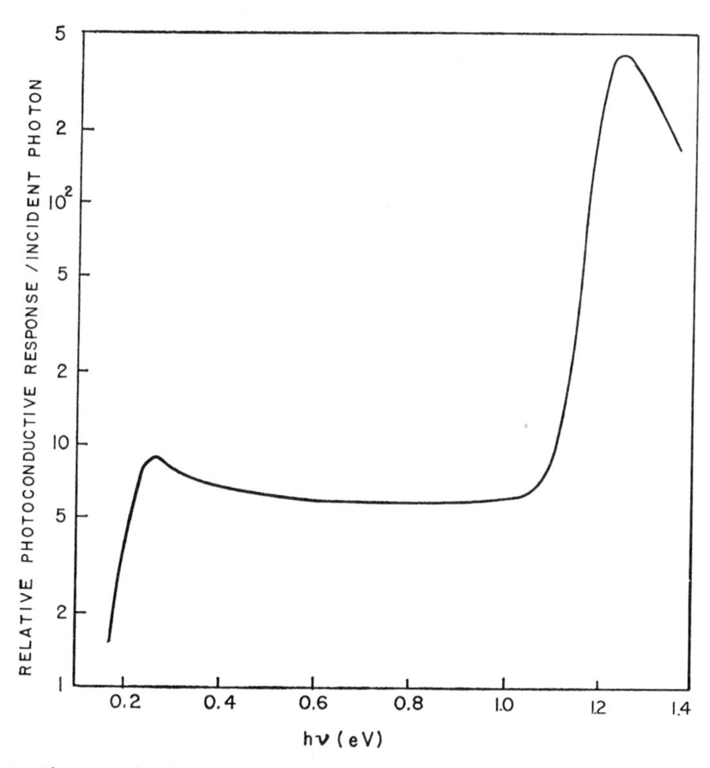

FIG. 7.3 Photoconductive response versus photon energy for indium-doped silicon (sample 214-B1, $N_{\text{In}} = 4.55 \times 10^{17}$ cm^{-3}, $N_D = 1.75 \times 10^{14}$ cm^{-3}) at 93°K. (After Blakemore and Sarver, 1968.)

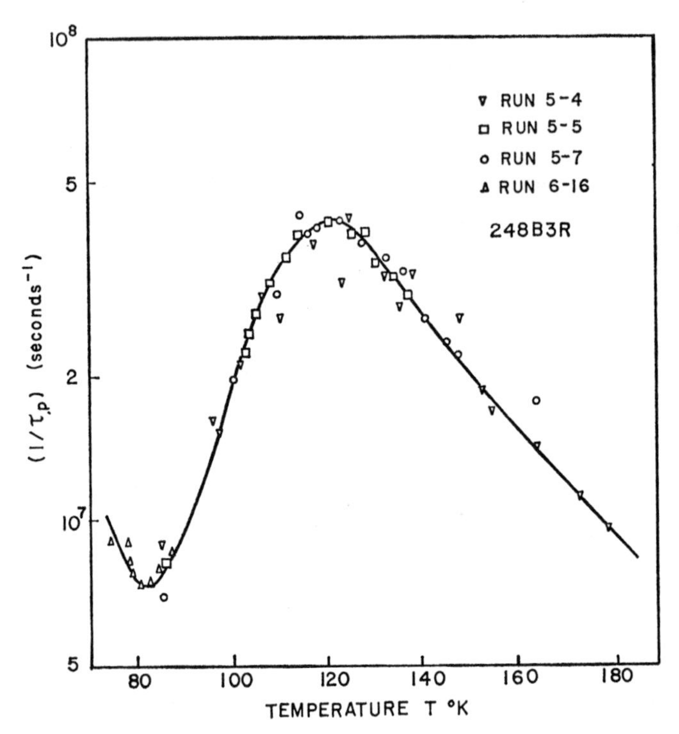

FIG. 7.4 Reciprocal of hole lifetime as a function of temperature for indium-doped silicon. (After Blakemore and Sarver, 1968.)

Fig. 7.4. The variation with temperature reflects the fact that the density of In^- atoms, which are the recombination centers, and the recombination cross section σ are both variables with temperature. The thermal generation and recombination equations are

$$g = v_{th}\sigma_p p_1(N_{In} - N_D - p) \tag{7.8}$$

and

$$r = v_{th}\sigma_p p(N_D + p) \tag{7.9}$$

where

$$p_1 = \frac{N_v}{\beta} \exp\left(-\frac{E_{In}}{kT}\right) \tag{7.10}$$

and v_{th}, the mean thermal velocity for the Boltzmann distribution of free holes, is

$$v_{th} = \left(\frac{8kT}{\pi m_v^*}\right)^{1/2} \tag{7.11}$$

If the optical generation is small, the lifetime for excess carriers expected from (7.8) and (7.9) is

$$\tau = [v_{th}\sigma_p(N_D + 2p + p_1)]^{-1} \tag{7.12}$$

Since the temperature variations of v_{th}, p, and p_1 are known, and τ is given by Fig. 7.4, the capture cross section variation with temperature may be derived. Blakemore and Sarver (1968), after making some allowance for size of the mobile hole mean-free-path relative to the average spacing of ionized capture centers, obtain the results shown by circles in Fig. 7.5. There is good agreement with previous studies at about 80°K, and there is seen to be a decrease of capture cross section with increase of temperature above about 100°K. This decrease is probably related to a decline in the sticking probability for recombination through excited states. The observed results are not unreasonable in terms of what is known of indium excited states and

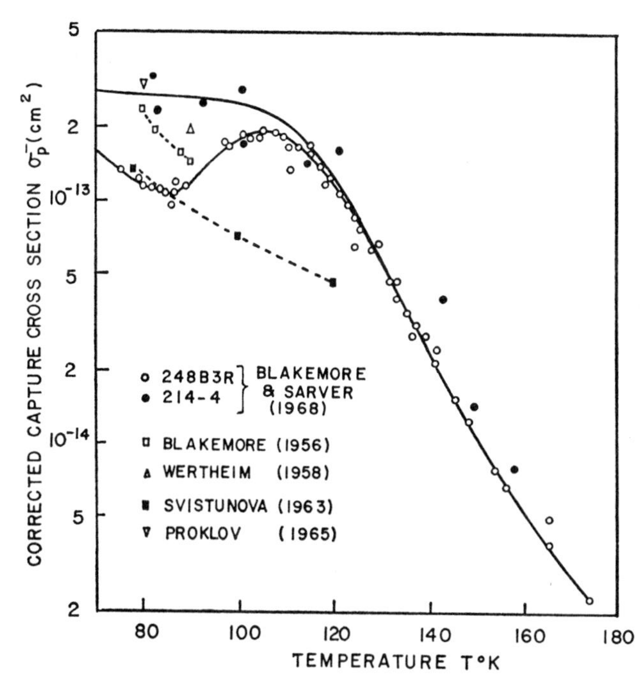

FIG. 7.5 Capture cross section for capture of free holes by ionized indium acceptors. *The Blakemore and Sarver (1968) and Blakemore (1956) results are from extrinsic photoconductivity. The Wertheim (1958) results are from transient electron-hole recombination; the Svistunova (1963) results are from electron-hole recombination and trapping; and the Proklov (1965) data are from generation-recombination noise studies. (After Blakemore and Sarver, 1968.)*

multiphonon processes in silicon but a complete model for the decline is still awaited.

Another feature of extrinsic photoconductivity that is not fully understood is the complete spectral dependence of the photoionization cross section. This has been modeled by Lucovsky (1965), Bebb and Chapman (1967). and others. However, the observed photoionization cross section for indium in silicon appears to be anomalously small for photon energies below about 0.23 eV, as shown by curve (*d*) in Fig. 7.6. More recently Messenger and Blakemore (1971) have reported 3.3×10^{-17} cm^2 at 0.3 eV.

The relation between radiative electron-capture cross section and the optical-absorption cross section for impurities in semiconductors has been derived by Bebb (1972) without recourse to detailed balance arguments.

Steady-state photoconductivity in the presence of traps has been considered by Curtis (1968) and illustrated with respect to irradiated germanium. He finds that a situation can exist in which the lifetime versus excess density curve shows no indication of trapping and yet incorrect recombination parameters are obtained if trapping is ignored.

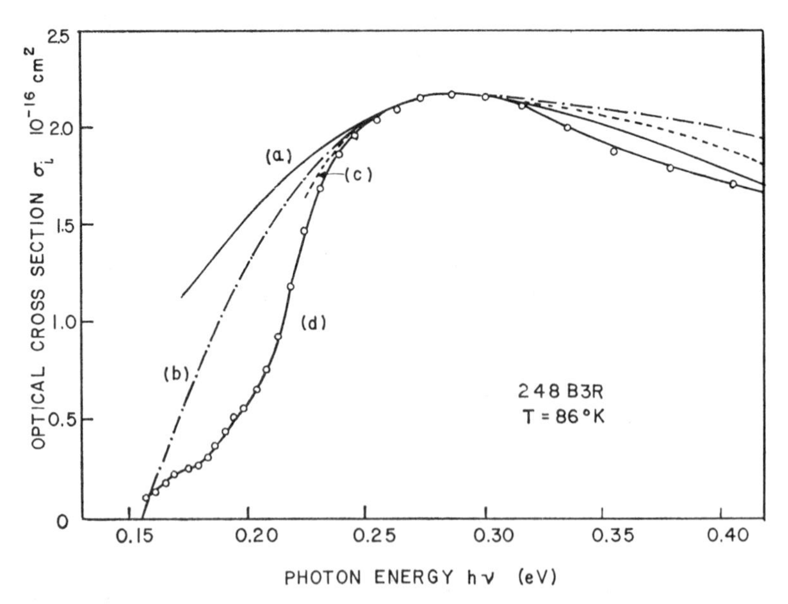

FIG. 7.6 Photoionization cross section for indium in silicon as function of photon energy.

(a) From optical transmission, Burstein et al. (1956). (b) Lucovsky model (1965) using square-well potential approximation. (c) Quantum defect model of Bebb and Chapman (1967). (d) Blakemore and Sarver photoconductivity measurements (1968). Curves b, c, and d have been normalized to match the maximum ordinate of curve a. (After Blakemore and Sarver, 1968.)

The effect of light intensity on extrinsic photoconductivity is a matter that we have not yet discussed. It is obvious that the number of carriers that can be achieved in the conduction or valence band (whichever is appropriate) under high light intensities is limited by the impurity density. Therefore the relationship of photoconductivity to photon flux may be expected to depart from linearity. Depending on the parameters, the photoconductivity for moderate to high light levels may be roughly dependent on the square root of the photon flux and then approach saturation at high light intensities. The saturation level should be related to the impurity density. The problem has been considered by Ryvkin and others (1961) and Fig. 7.7 shows computed results for the steady-state density of nonequilibrium carriers Δn_{ss} versus light intensity for two temperatures. The density of deep centers was assumed to be 10^{12} cm^{-3} and the model was for

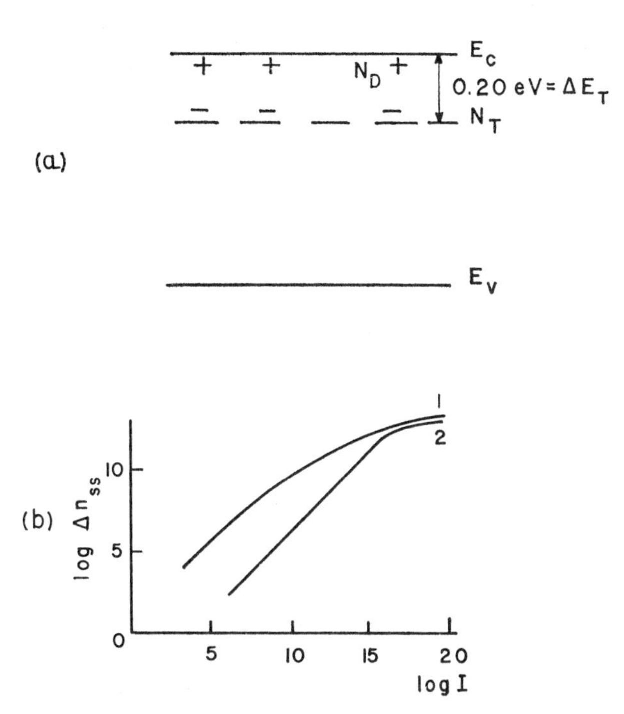

FIG. 7.7 Effect of high illumination intensity on the steady-state photoinduced carrier density.

(a) Model showing a deep acceptor 0.2 eV below the conduction band and $N_D < N_T$. *(b) Calculated dependence on light intensity for (1)* $T = 77°K$ *and (2)* $T = 190°K$. [*After S. M. Ryvkin, L. G. Paritskii, R. Yu. Khansevarov, and I. D. Yaroshetskii, "Investigation of the Kinetics of Impurity Photoconduction as a Method of Determination of Parameters of Local Levels," Sov. Phys. Solid State, 3, 185 (1961).*]

excitation of electrons to the conduction band from a deep acceptor above the center of the band gap that has received electrons from a shallow donor, as in Fig. 7.7*a*.

7.2 Transient Effects in Impurity Photoconductivity

The transient effects in impurity photoconductivity may become quite complex if anything other than a very simple model is considered or if high light intensities are assumed. However, for the model of Fig. 7.7*a*, assuming small light intensities that do not significantly complicate matters, the turn-on and turn-off transients for a square wave of light may be fairly simply characterized.

The deep acceptor impurity will be in one of two states, N_T^- or N_T^0, where N_T^0 is equal to $N_T - N_T^-$ and N_T^- is equal to $N_D - n$ where n is the density of electrons in the conduction band if the shallow donor density N_D is assumed completely ionized at all times. The thermal recombination rate at the deep center depends on the electron density n and the density of neutral centers N_T^0 and therefore is given by

$$r = v_{\text{th}} \sigma_n^0 n (N_T - N_D + n) \tag{7.13}$$

where σ_n^0 is the thermal capture cross section. In equilibrium, in the dark, this is matched by a thermal generation rate which depends on N_T^- and is

$$g = v_{\text{th}} \sigma_n^0 n_1 (N_D - n) \tag{7.14}$$

If the temperature is low, so that n is small and the value of N_T^- is very closely that of N_D, the condition $r = g$ gives

$$n_1 = \frac{N_c N_T}{N_D} \exp\left(-\frac{\Delta E_T}{kT}\right) \tag{7.15}$$

and

$$n_0 \simeq \frac{n_1 N_D}{N_T - N_D} \tag{7.16}$$

Consider a photon flux passing through the material of I photons cm^{-2} sec^{-1}. If σ_i is the photoionization cross section for capture of a photon by an electron at the deep acceptor level with transfer to the conduction band with unity quantum efficiency, the generation rate produced by photons is

$$g_{\text{ph}} = \sigma_i (N_D - n) I \tag{7.17}$$

From (7.13) to (7.17)

$$\frac{dn}{dt} = \sigma_i(N_D - n)I + v_{\text{th}}\sigma_n{}^0 n_1(N_D - n) - v_{\text{th}}\sigma_n{}^0 n(N_T - N_D + n) \quad (7.18)$$

Considering that $n = n_0 + \Delta n$, $\Delta(N_T) = -\Delta n$, and $N_{TO}{}^- + n_0 = N_T{}^- + n$, we obtain

$$\frac{d(\Delta n)}{dt} = \sigma_i N_{TO}{}^- I - v_{\text{th}}\,\sigma_n{}^0\left[\frac{\sigma_i I}{v_{\text{th}}\sigma_n{}^0} + n_0 + \Delta n + n_1 + (N_T - N_{TO}{}^-)\right] \quad (7.19)$$

Equations of this form have been discussed by Ryvkin (1964b). If the illumination level is low so that $\Delta n \ll N_{TO}{}^-$, then (7.19), for the application of a square wave of light intensity I, has a simple exponential rise

$$\Delta n_i = \Delta n_{ss}\left(1 - \exp\left(-\frac{t}{\tau_{\text{on}}}\right)\right) \quad (7.20)$$

where

$$\tau_{\text{on}} = [v_{\text{th}}\sigma_n{}^0(n_1 + N_T{}^0 + n_0) + \sigma_i I]^{-1} \quad (7.20a)$$

For the turn-off transient the decay is

$$\Delta n_d = \Delta n_{ss}\exp\left(-\frac{t}{\tau_d}\right) \quad (7.21)$$

where

$$\tau_d = [v_{\text{th}}\sigma_n{}^0(n_1 + N_T{}^0 + n_0)]^{-1} \quad (7.21a)$$

From (7.20a) and (7.21a) it is seen that the turn-on transient is faster than the turn-off transient, and depends on the illumination level as illustrated in Fig. 7.8. Also from the equations

$$\frac{1}{\tau_{\text{on}}} - \frac{1}{\tau_d} = \sigma_i I \quad (7.22)$$

Therefore if $1/\tau_{\text{on}}$ and $1/\tau_d$ are plotted as a function of light intensity, subtraction should give a straight line of slope σ_i, as sketched in Fig. 7.8*b*. This is therefore a method of determining the photoionization cross section. Ryvkin states that this allows the determination of σ_i at very low impurity densities (about 10^{12} cm^{-3}) for which the usual optical absorption method would not be applicable.

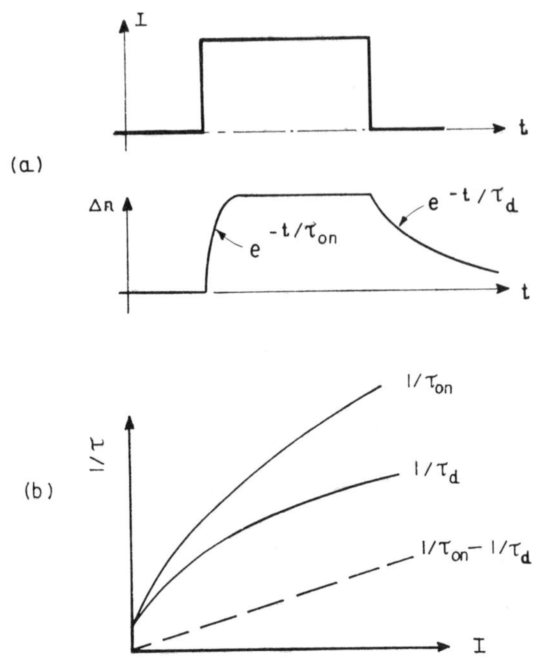

FIG. 7.8 Extrinsic photoconductivity transients produced by a square pulse of illumination.
(a) The turn-on transient is shorter than the turn-off transient. (b) The slope of the broken line gives the photoionization capture cross section.

7.3 The Photoconductivity Decay of Silver-Doped Silicon

As a specific example of a photoconductivity decay transient it is interesting to consider silver in high-resistivity silicon. The silver provides an acceptor level at $E_c - 0.29$ eV that acts as an electron trap during and after band-gap illumination. (Silver in silicon under suitable doping conditions may also exhibit a donor level at 0.26 to 0.32 eV above the valence band edge, but this is not involved here.) A model has been developed, therefore, for the decay of photoconductivity after a pulse of band-gap light is applied to such a semiconductor containing a deep acceptor with an energy level in the upper half of the band gap, partially compensated by a shallow donor. A closed-form solution is obtainable for the return to the equilibrium conductivity if appropriate simplifications are made (Smith and Milnes, 1971).

7.3.1 Theory

Consider a high-resistivity semiconductor that contains a deep acceptor level N_T in the upper half of the band gap and a shallow donor, $N_D < N_T$, providing partial compensation. The deep impurity may have the charge states N_T^- or N_T^0 as shown in Fig. 7.9*a*. If the deep impurity also has a donor level, it is assumed that this is in the lower half of the band gap and plays no role in the phototransient that is considered.

In equilibrium in the absence of illumination the electron density in the conduction band is given by

$$n_{\mathrm{eq}} = N_D^{\ +} - N_{T\mathrm{eq}}^{-} + p \tag{7.23}$$

If we assume that the Fermi level E_{F} is in the upper half of the band gap, we can neglect p and obtain

$$n_{\mathrm{eq}} = N_D^{\ +} - N_{T\mathrm{eq}}^{-} \tag{7.24}$$

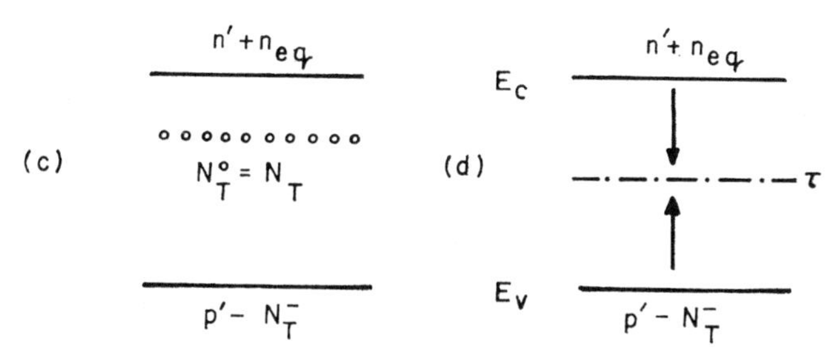

FIG. 7.9 Energy band diagrams showing the proposed sequence of events before the relatively long decay of electrons combining with silver centers begins.

(a) Dark equilibrium conditions. (b) Excess electrons and holes $n' = p'$ *created by the pulse of band-gap illumination. (c) Loss of electrons from the deep acceptor caused by recombination with holes. (d) Recombination of remaining holes with electrons through a recombination center. (After Smith and Milnes, 1971.)*

The electron density may also be expressed in terms of the Fermi level by

$$n_{eq} = N_c \exp\left(-\frac{E_c - E_F}{kT}\right) \tag{7.25a}$$

$$= n_i \exp\left(\frac{E_F - E_i}{kT}\right) \tag{7.25b}$$

where the symbols have their usual meanings. Yet another equation involving n_{eq} may be obtained by considering the balance of recombination and emission of electrons, namely,

$$\frac{dn}{dt} = 0 = -n_{eq} N_T{}^0 \sigma_n{}^0 v_{th} + e_n N_T{}^- \tag{7.26}$$

where e_n is the emission constant. From (7.25) and (7.26) the emission constant may be shown, following Shockley and others, to be

$$e_n = n_1 \sigma_n{}^0 v_{th} \tag{7.27}$$

where

$$n_1 = N_c \exp\left(-\frac{E_c - E_T}{kT}\right) \tag{7.28a}$$

$$= n_i \exp\left(\frac{E_T - E_i}{kT}\right) \tag{7.28b}$$

Consider now the sequence of events after a pulse of band-gap light has been applied to the semiconductor and produces electron-hole pairs $n' = p'$ as shown in Fig. 7.9*b*. The deep acceptor center is assumed relatively undisturbed by the pulse of light. Two processes now begin acting to restore equilibrium. The first process is taken to be that the holes depopulate the deep acceptors of their electrons. This is assumed to be fast and to leave all the deep centers N_T in their neutral state. The approximate time constant for this can be estimated from

$$\tau_p = (N_T{}^- \sigma_p{}^- v_{th})^{-1} \tag{7.29}$$

where $\sigma_p{}^-$ is the capture cross section for the coulombically attractive process $h + N_T{}^- \rightarrow N_T{}^0$. If $N_T{}^-$ is on the order of 10^{14} cm^{-3}, $\sigma_p{}^-$ is 10^{-14} cm^2, and v_{th} is 10^7 cm sec^{-1}, then τ_p is 10^{-7} sec. The second process is that the remaining holes recombine with electrons (Fig. 7.9*d*) through a recombination center which is assumed to exist in the material. For simplicity we have assumed that the recombination lifetime is of the order of a microsecond and that the process shown in Fig. 7.9*c* is completed before 7.9*d* begins. This may not be true and in this event the deep acceptor center would not be completely depopulated. This would change the boundary conditions for the next event that we particularly wish to consider, namely the relatively slow return of nonequilibrium electrons to the deep acceptor level.

After the events of Fig. 7.9*a* to 7.9*d* we are left with a density of n electrons cm^{-3} in the conduction band, where

$$n = (n' + n_{\text{eq}}) - (p' - N_T^-) \tag{7.30a}$$

$$= n_{\text{eq}} + N_T^- \tag{7.30b}$$

since $n' = p'$. Using (7.24) this leaves $n = N_D^+$ as the initial condition after the assumed rapid transients of hole capture and recombination have passed. In summary the net result of an intense pulse of light may be to remove all the electrons from the deep acceptor. Such a condition may not be readily achievable in practice. However, this boundary condition does not affect the basic equations for the decay which now follow.

Consider the conduction electrons returning to the equilibrium trapped state on the deep level centers. The rate equations describing the recovery are

$$\frac{dn}{dt} = -nN_T{}^0\sigma_n{}^0v_{\text{th}} + (N_T - N_T{}^0)n_1\sigma_n{}^0v_{\text{th}} \tag{7.31}$$

and

$$\frac{dN_T{}^0}{dt} = \frac{dn}{dt} \tag{7.32}$$

Solution of these two equations begins by integration of (7.32). This gives

$$N_T{}^0(t) = n(t) + c \tag{7.33}$$

where c is a constant of integration. Using the equilibrium value

$$n(\infty) = n_{\text{eq}} = N_D^+ = N_{T\text{eq}}^-$$

and from

$$N_T = N_T^- + N_T{}^0$$

we have

$$c = N_T - N_D^+ \tag{7.34}$$

and therefore

$$N_T{}^0(t) = n(t) + N_T - N_D^+ \tag{7.35}$$

Substitution of (7.35) into (7.31) gives

$$\frac{1}{\sigma_n{}^0v_{\text{th}}}\frac{dn}{dt} = -n(t)(n(t) + N_T - N_D^+) + (N_T - (n(t) + N_T - N_D^+))n_1 \tag{7.36}$$

Collecting terms in (7.36) we have

$$\frac{1}{\sigma_n{}^0 v_{\text{th}}} \frac{dn}{dt} = -(n(t))^2 - n(t)(N_T - N_D{}^+ + n_1) + n_1 N_D{}^+ \tag{7.37}$$

This equation can be integrated in closed form using

$$\int \frac{dx}{A + 2Bx + Cx^2} = \frac{1}{2(B^2 - AC)^{1/2}} \ln \left(\frac{Cx + B - (B^2 - AC)^{1/2}}{Cx + B + (B^2 - AC)^{1/2}} \right) \tag{7.38}$$

when $AC < B^2$. In our equation

$$A \equiv n_1 N_D{}^+, \qquad B \equiv -\tfrac{1}{2}(n_1 + N_T - N_D{}^+), \qquad C \equiv -1 \tag{7.39}$$

so the inequality is always satisfied.

If we now define $B^+ = B + (B^2 - AC)^{1/2}$ and $B^- = B - (B^2 - AC)^{1/2}$ and at $t = 0$ let $n = n_0$, the expression becomes

$$\frac{Cn + B^+}{Cn + B^-} \frac{Cn_0 + B^-}{Cn_0 + B^+} = \exp\left[-2(B^2 - AC)^{1/2} \sigma_n{}^0 v_{\text{th}} t\right] \tag{7.40}$$

which is easily solved for n. Substituting $C = -1$ and defining

$$K = \frac{n_0 - B^+}{n_0 - B^-}$$

we obtain

$$n = \frac{B^+ - B^- K \exp\left[-2(B^2 - AC)^{1/2} \sigma_n{}^0 v_{\text{th}} t\right]}{1 - K \exp\left[-2(B^2 - AC)^{1/2} \sigma_n{}^0 v_{\text{th}} t\right]} \tag{7.41}$$

As a check on the algebra, limiting values can be substituted. If $t = 0$, (7.41) reduces to $n(0) - n_0$. For the limit $t \to \infty$, we obtain $n = B^+$. This may be confirmed by returning to (7.37) in which $t \to \infty$ implies $dn/dt \to 0$ or $n^2 - 2B_n - n_1 N_D{}^+ = 0$. Direct solution of this gives

$$n = B + (B^2 - AC)^{1/2}$$

which is B^+.

Another check is to assume that N_D, n_0, and n_1 are small compared to N_T. Then B^+ is zero and B^- is $-N_T$. Substitution of these values in (7.41) gives

$$n = \frac{N_T(n_0/(N_T + n_0)) \exp(-\sigma_n{}^0 v_{\text{th}} N_T t)}{1 - n_0/(N_T + n_0) \exp(-\sigma_n{}^0 v_{\text{th}} N_T t)} \tag{7.42}$$

$$\simeq n_0 \exp(-\sigma_n{}^0 v_{\text{th}} N_T t) \tag{7.43}$$

So in this limit the expression becomes an exponential decay with the time constant that might be expected by simple physical reasoning.

Returning to an inspection of the more general expression, (7.41), we see that the same exponential appears in both the numerator and the denominator. Whether the exponential in the denominator is significant or not depends on the size of the term K relative to unity. If K is small, then the expression reduces to

$$n = B^{+} - B^{-}K \exp\left[-2(B^{2} - AC)^{1/2}\sigma_{n}{}^{0}v_{\text{th}}t\right] \tag{7.44}$$

If t is zero, $n = B^{+} - B^{-}K$ where B^{-} is a negative number; and if t is infinite, then $n = B^{+}$. The expression for K is

$$K = \frac{n_0 - B^{+}}{n_0 - B^{-}}$$

$$= \frac{n_0 + \frac{1}{2}(n_k + N_T - N_D{}^{+}) - [\frac{1}{4}(n_1 + N_T - N_D{}^{+})^2 + n_1 N_D{}^{+}]^{1/2}}{n_0 + \frac{1}{2}(n_1 + N_T - N_D{}^{+}) - [\frac{1}{4}(n_1 + N_T - N_D{}^{+})^2 + n_1 N_D{}^{+}]^{1/2}} \tag{7.45}$$

If the term $n_1 N_D{}^{+}$ can be neglected compared with the other term in the radical (a condition which exists if N_T is substantially larger than $N_D{}^{+}$ and n_1), then (7.45) reduces to

$$K = \frac{n_0}{N_T + n_0 + n_1 - N_D{}^{+}} \tag{7.46}$$

For a trap depth $E_T = 0.29$ eV, n_1 is about 10^{14} cm^{-3}, and if we assume $N_D = 10^{14} = n_0$ and $N_T = 10^{15}$ cm^{-3}, then K is 0.09. If n_0 is less than $N_D{}^{+}$, which happens if all the $N_T{}^{-}$ centers cannot capture holes, K will be further reduced. We conclude that for a wide range of conditions that are likely to be of physical interest K is small compared to unity, and therefore the transient is adequately characterized by the simple exponential form (7.44) rather than by the more general form (7.41).

Consider now the exponential in (7.44) which may be rewritten as

$$\exp\left(-2(B^{2} - AC)^{1/2}\sigma_{n}{}^{0}v_{\text{th}}t\right)$$
$$= \exp\left(-2[\tfrac{1}{4}(n_1 + N_T - N_D)^2 + n_1 N_D]^{1/2}\sigma_{n}{}^{0}v_{\text{th}}t\right) \tag{7.47}$$

If the quantities N_T, N_D, and $n_1(E_T)$ are known, then the slope of the decay gives $\sigma_n{}^0$. If the temperature is raised so that $n_1 > N_T$ and N_D, and $\sigma_n{}^0$ is known, then the trap depth can be determined independently of sample parameters. Alternatively, if $\sigma_n{}^0$ is not known, E_T and $\sigma_n{}^0$ can be found by measuring the decay slope at two temperatures, provided $\sigma_n{}^0$ is not a strong function of temperature. At low temperatures, n_1 becomes unimportant and N_T can be determined from the slope if the other parameters are known.

As N_T is increased, the decay rate increases (an effect that would be expected) and increasing N_D^+ decreases the rate. If the values of N_T and N_D^+ are accurately known, the variation in decay rate at low temperature can be related to the temperature dependence of σ_n^0.

7.3.2 Experimental Studies of the Photoconductivity Decay in Silicon Containing Silver

To verify the theoretical predictions presented above, several silicon samples doped with silver were prepared and their photoconductivity decay transients examined. The silicon slices were initially *n* type of about 200 and 50 Ω-cm, corresponding to shallow donor densities of 2.3×10^{13} and 9.3×10^{13} cm^{-3}. Silver was diffused into the samples at 1100 and 1200°C for times sufficient to achieve uniform doping and then quenched. Neutron activation studies (Smith and Milnes, 1970) gave silver concentrations of 5.6×10^{14} and 2×10^{15} cm^{-3}, respectively. The percentage of silver which is electrically active is not known for certain but appears to be very high.

The photoconductivity measurements were made with the sample connected in series with a 15-V battery and a 2-kΩ resistor. A spark source (30 kV, 0.02 μF) was used to excite the sample at a repetition rate of a few sparks per second. The fall time of this excitation was fast (about 0.2 μsec) relative to the portion of the photoconductivity decay that was to be studied.

Each photoconductive transient was found to have a fast-falling initial section, of duration less than 1 μsec, before the long final decay began. This fast transient may represent sweepout effects in the presence of the decreasing tail of the spark illumination or the capture of holes by the silver acceptor centers ($N_{\text{Ag}}^- + h \rightarrow N_{\text{Ag}}^0$) and subsequent recombination of the remaining holes by the processes suggested in Fig. 7.9*c* and *d*. Two decay transients are shown in Fig. 7.10. The top trace shows a fast-falling initial section before a rather long decay begins. The lower curve was made after the sample was moved away from the light to reduce the intensity. Note the relative decrease in the initial spike. This can perhaps be explained by saying that the silver centers have captured holes and the spike represents recombination of the relatively smaller number of holes remaining through a recombination center. These events take place during the first microsecond.

Figure 7.11 shows the photocurrent decay transient for the 200-Ω-cm (2.3×10^{13} cm^{-3} donors) sample diffused at 1100°C (5.6×10^{14} cm^{-3} silver centers) tested at three temperatures. Consider first the 425°K curve and return to (7.47). At this temperature n_1 is 5.4×10^{15} cm^{-3}, assuming

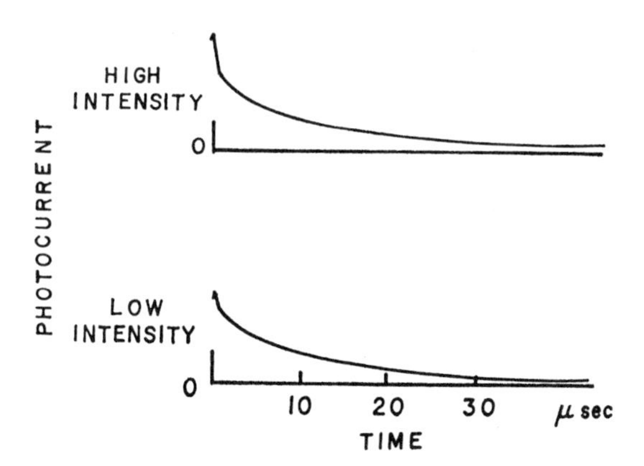

FIG. 7.10 Transients showing the effect of reduced light intensity on the initial spike photoconductivity decay for silicon with silver. (After Smith and Milnes, 1971.)

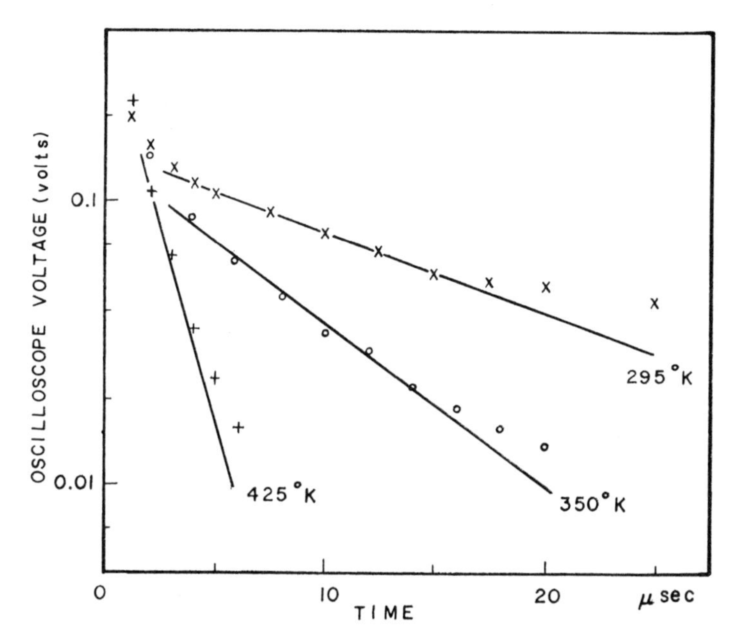

FIG. 7.11 Photodecay curves for silicon with silver concentration 5.6×10^{14} cm^{-3} and a shallow donor concentration 2.3×10^{13} cm^{-3} (sample 200N11).

The slopes of the straight lines are obtained from theory, and the temperature dependence is seen to be correct. (After Smith and Milnes, 1971.)

$E_T = 0.29$ eV. Since n_1 is much greater than N_T or N_D, the only unknowns are E_T and $\sigma_n{}^0$. Solving for $\sigma_n{}^0$ gives 10^{-17} cm^2, which is in good agreement with the value reported by Thiel and Ghandhi (1970). Assuming that $\sigma_n{}^0$ does not change significantly between 425 and 350°K we can test the assumed value of E_T. The degree of fit confirms the E_T value used. For the 295°K curve the N_T may be expected to be dominant in the slope expression since n_1 is about 20% of the silver density if all 5.6×10^{14} cm^{-3} atoms are active. The fit of the theoretical line to the experimental points at 295°C is therefore an indication that all the silver is indeed active. It should be noted that only the slope of the line is being fitted and the lines are arbitrarily positioned.

Figure 7.12 shows the photocurrent decay transients for the 50-Ω-cm material diffused at 1200°C and tested at 77, 295, and 375°K. Again, the

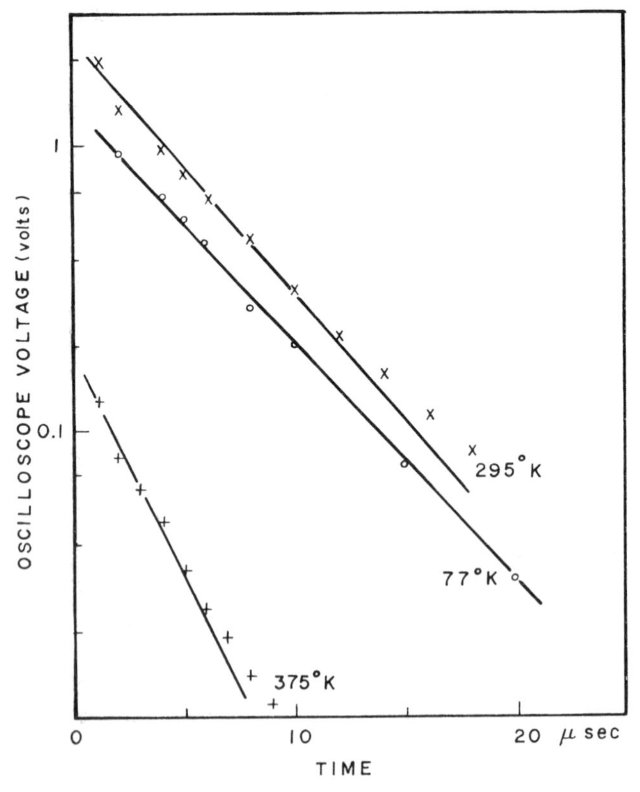

FIG. 7.12 Photodecay curves for silicon with $N_{Ag} = 2 \times 10^{15}$ cm^{-3} and $N_D = 9.3 \times 10^{13}$ cm^{-3} (sample 50N12). (After Smith and Milnes, 1971.)

high-temperature data confirm the E_T and $\sigma_n{}^0$ values used. The 77 and 295°K lines imply that $\sigma_n{}^0$ is a very weak function of temperature. At room temperature and below, the n_1 term is small so N_T and $\sigma_n{}^0$ are the only unknowns. The slopes of the two lines are nearly equal. Therefore if the active silver concentration is not a function of temperature, then neither is $\sigma_n{}^0$.

Figure 7.13 shows the dependence of the photocurrent decay on N_{Ag}. The theoretical slopes are seen to be in good agreement with the experimental curves. From the agreement found in Figs. 7.11 to 7.13, it may be concluded that the theory represents the experimental decays quite well.

Several points, however, should be noted. The data plotted are voltages read from oscilloscope traces, without modification. A scale factor is necessary to convert the oscilloscope reading to carrier density, which the theory produces. To make the adjustment it is necessary to know the mobility of electrons as a function of temperature and the actual initial electron density. If a mobility is assumed, n_0 can be chosen to produce correspondence between the numerical values from theory and experiment. However, the values n_0 so obtained tend to be less than the N_D values. This probably indicates that the illumination was not intense enough in the experiments to

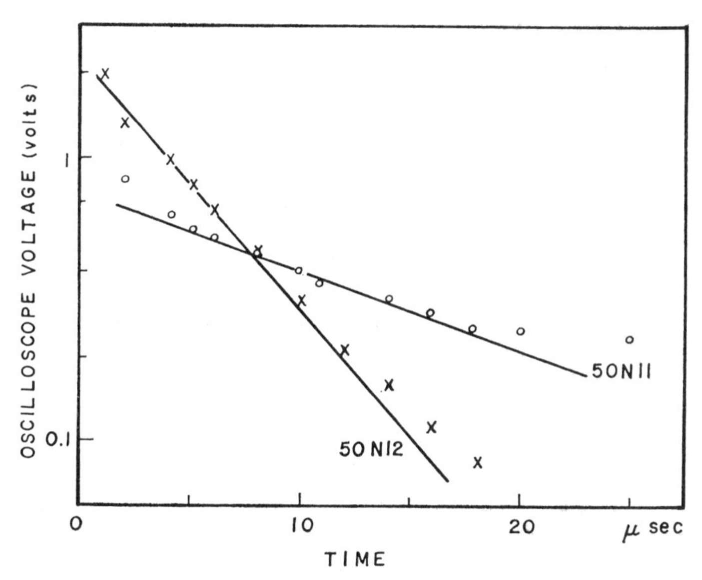

FIG. 7.3 Photodecay curves to show the change with silver concentration: 5.6 × 10^{14} cm^{-3} (for specimen 50N11) and 2 × 10^{15} cm^{-3} (specimen 50N12), at room temperature. (After Smith and Milnes, 1971.)

achieve $n_0 = N_D$, which would result from complete depopulation of the silver acceptors. As discussed in the theoretical section, n_0 is not a factor in determining the decay exponential, so the matter is not of critical importance.

The model, however, is inadequate to explain the tendency of the experimental data to curve from the straight-line slopes of the theory. The Fermi level in the materials studied lies between 0.3 and 0.4 eV from the conduction band. This means that the assumption that holes and the donor centers of silver can be neglected is not strictly valid. Interactions between holes, donors, acceptors, and electrons are probably responsible for the increased carrier densities observed. At low temperatures these effects can be expected to be much less since the Fermi function becomes quite sharp. It is therefore of interest to note that the 77°K data points (Fig. 7.12) in fact show no signs of the departure from linearity seen at higher temperatures.

7.4 Intrinsic Photoconductivity, Induced by Band-Gap Light

7.4.1 Deep Center Acting as a Recombination Center

If the deep impurity has capture cross sections ($\sigma_n{}^0$, $\sigma_p{}^-$) for electrons and holes that are comparable in magnitude, the impurity acts primarily as a recombination center. Band-gap light is applied, and we will assume it creates electron-hole pairs at the rate g_L cm^{-3} sec^{-1}. Also assume that the deep impurity has only two permissible states, $N_T{}^0$ and $N_T{}^-$. The kinetic equations are then

$$\frac{dn}{dt} = g_L + \nu N_T{}^- \exp\left(-\frac{E_c - E_T}{kT}\right) - v_{\text{th}}\sigma_n{}^0 N_T{}^0 n \tag{7.48}$$

and

$$\frac{dp}{dt} = g_L + \nu N_T{}^0 \exp\left(-\frac{E_T - E_v}{kT}\right) - v_{\text{th}}\sigma_p{}^- N_T{}^- p \tag{7.49}$$

where ν in the thermal generation terms is a vibration frequency of the order of 10^9 sec^{-1}. If the thermal generation terms can be neglected, the steady-state solutions are

$$\Delta n = \frac{g_L}{v_{\text{th}}\sigma_n{}^0 N_T{}^0} \tag{7.50a}$$

and

$$\Delta p = \frac{g_L}{v_{\text{th}}\sigma_p{}^- N_T{}^-} \tag{7.50b}$$

The change of conductivity of the material is therefore

$$\Delta\sigma = e(\Delta n \mu_n + \Delta p \mu_p) \tag{7.51}$$

For a bar of length L, cross-sectional area A, and negligible dark conductivity, the change of conductance caused by the light is

$$\Delta G = \Delta\sigma \frac{A}{L} = e\left(\frac{\mu_n}{v_{\text{th}}\sigma_n{}^0 N_T{}^0} + \frac{\mu_p}{v_{\text{th}}\sigma_p{}^- N_T{}^-}\right)\frac{g_L A}{L} \tag{7.52}$$

which may be rewritten

$$\Delta G = e(\mu_n \tau_n + \mu_p \tau_p)\frac{g_L A}{L} \tag{7.53}$$

If the occupancy of the recombination centers is small, which will be true at low light levels, then (7.53) predicts a linear relationship between the photoconductance and the light intensity.

In (7.53) the photoconductance is inversely proportional to the length of the bar. However, if the matter is reexamined on the assumption that there is a certain illuminating power available, W, which can be applied to the bar irrespective of its dimensions, then

$$g_L = C\frac{W}{AL} \tag{7.54}$$

where C is a constant involving the optical injection efficiency. Therefore (7.53) shows that ΔG becomes proportional to L^2. For a voltage V applied, the current in the photoconductor caused by the applied light is

$$\Delta I = \Delta G V = \frac{Ce(\mu_n \tau_n + \mu_p \tau_p)W}{L^2} \tag{7.55}$$

Returning to the basic carrier equations for the model under consideration, (7.48) and (7.49), it is seen that except for greatly simplified conditions, it is necessary to solve them numerically. When this is done for conditions of high light intensities the photoconductivity response tends to become a sublinear function of the light flux.

One problem in intrinsic photoconductivity is that the absorption depth of the light is normally quite small (say 10^{-3} cm or less) and therefore modeling on the basis of uniform carrier generation throughout the photoconductive bar is inadequate under some conditions.

7.4.2 Deep Center Acting as a Trap and Enhancing the Quantum Efficiency

Consider a semiconductor containing a deep acceptor level below the middle of the band gap so that communication with the valence band is easier than with the conduction band. This acceptor level of density N_T may be partially filled with electrons from a shallow donor, where $N_D < N_T$. The semiconductor therefore may be of fairly high resistance under dark conditions provided that the temperature is not such that the deep acceptor is thermally ionized. When band-gap light is applied, the processes that may be expected to happen are as illustrated in Fig. 7.14, (1) to (4) being the important ones.

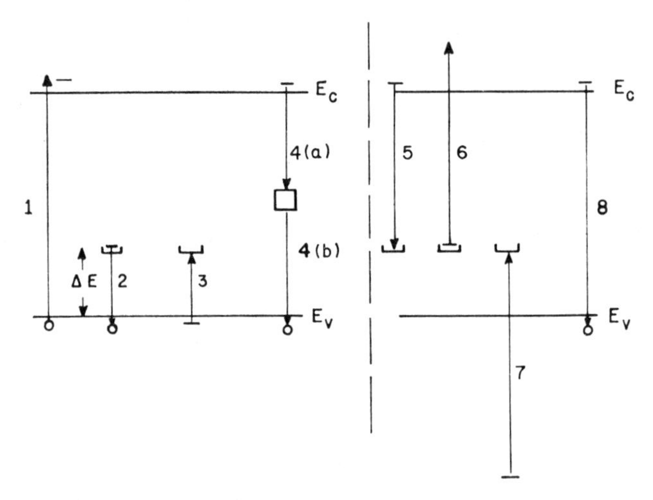

FIG. 7.14 Some processes in an intrinsic photoconductor with a hole trap ΔE above the valence band.
(1) Optical or thermal generation of an electron-hole pair. (2) Hole is trapped by a charged deep acceptor center. (3) Hole is untrapped (thermally) by a neutral deep acceptor center. (4) Recombination of free electrons with free holes through a recombination center. Processes (5) to (8) are regarded as not significant in this model.

In process (2) it is seen that the optically generated holes may be captured by the process $N_T^- + h \rightarrow N_T^0$. Reemission to the valence band depends on the thermal process (3) and the rate term involved is $\nu N_T^0 \exp(-\Delta E/kT)$. If ν is 10^9 sec^{-1}, N_T^0 is 10^{15} cm^{-3}, ΔE is 0.3 eV, and T is 250°K; the numerical value of the term is about 10^{18} reemissions cm^{-3} sec^{-1}. The lifetime of a hole τ in a trap may therefore be of the order of 10^{-3} sec. This time is likely to be much longer than the transit time for the electrons between the ohmic contacts of the photoconductor of length L. The transit time in a drift field of V/L V cm^{-1} is

$$\tau_d = \frac{L}{\mu_n V/L} = \frac{L^2}{\mu_n V} \tag{7.56}$$

For example, if L is 0.1 cm, V is 1 V, and μ_n is 10^3 cm^2 V^{-1} sec^{-1}, the transit time is 10^{-5} sec. If a photon creates an electron-hole pair and the hole is immediately trapped by the deep acceptor, the electron remains available for conduction until the hole is untrapped and recombines with it. Therefore in effect the electron travels 100 times around the circuit (i.e., $10^{-3}/10^{-5}$) during the lifetime of the trapped hole. The electron is not lost to the photoconductor when it first drifts to the ohmic contact on the anode side of the bar since a new electron is injected at the cathode ohmic contact to maintain space-charge neutrality in the band. The ratio of the trapped hole lifetime to the transit time for electrons therefore increases the sensitivity of the photoconductor and is termed the quantum gain factor. Although highly desirable from the viewpoint of steady-state sensitivity, a long hole-trapping lifetime limits the transient response or frequency response of the photoconductor. Since the sensitivity is proportional to the trap lifetime one is working within the constraint of a roughly constant gain-bandwidth product.

In very high resistance semiconductors the dielectric relaxation time may exceed the carrier drift time, and space-charge-limited current effects may affect the sensitivity (R. L. Williams, 1969b; Kanazawa and Batra, 1972).

The photoconductive model of Fig. 7.14 involves only one hole trap and a recombination center. This is not adequate to explain many of the effects observed in actual photoconductors. For example, Shirafuji (1966) in a study of n-type gallium arsenide presents analysis for two hole trap levels and a recombination center. However, he concludes from the results of photoconductivity studies over a wide temperature range that six species of acceptor hole traps were active at depths of 0.33, 0.2, 0.13, 0.085, 0.06, and 0.02 eV. On the other hand, for gold-doped germanium Jayakumar and Ashley (1972) find good agreement between experimental results and the predictions of their model.

Other investigations of photoconductivity have shown that both electron and hole traps must be taken into account to explain the experimental results. It is beyond the scope of this chapter to cover all these variations and interested students of these matters are referred to the index of the bibliography for further access to the subject. In this connection it must be observed that many photoconductivity studies relate to the II–VI compound semiconductors and materials such as lead sulfide, lead selenide, and lead telluride. Papers on these are not really within the scope of the bibliography. However, the references to Aven and Prenner (1967) and to the proceedings of the international conferences on II–VI semiconducting compounds (1967) and on photoconductivity (1969) (Pell, 1971) should provide the desired leads.

7.5 Photoconductivity Effects Involving Light of Two Wavelengths

From the concepts that have been developed in this chapter it is obvious that if the occupancy of traps is changed, say by the application of light of less than band-gap energy, some effect may be expected on the photoconductive response to band-gap light. The addition of other impurities is of course another way, a permanent one, of changing the occupancy of traps that are active in the photoconductive process.

Consider, for example, the band-gap photoconductivity of a semiconductor containing a deep acceptor level, E_T. If there is also a shallow acceptor present so that the material is p type, one expects limited photoconductivity sensitivity (assuming that the deep acceptor is too far from the conduction band to act as an electron trap). However, if a shallow donor is added that overcompensates the shallow acceptor so that the material becomes lightly n type, the deep acceptor is then able to act as a hole trap. The band-gap photoconductivity sensitivity then becomes high, although the transient response is slowed by the trapping time constant. If light of energy $E_T - E_v$ is applied, in addition to the band-gap illumination, it excites electrons from the valence band to the deep acceptor atoms that are neutral and so prevents them from acting as hole traps. The photosensitivity to the band-gap light is therefore decreased. This is known as "quenching" of photoconductivity by long-wave length secondary illumination. The effect is found in a wide range of semiconductors, when the doping conditions are favorable.

Thermal quenching of photoconductivity is also possible, and the modeling of this has been studied recently by Kalashnikov et al. (1971). Thermal and field quenching of the photoconductivity of nickel-doped silicon have been considered by Lebedev and Mamadalimov (1972).

Extrinsic illumination may also cause charge exchange between two deep level impurities, or between two charge states of the same impurity, via either the conduction band or the valence band and so modify the photoconductivity. This has been studied by Ivanov and Ryvkin (1962), Kogan et al. (1965), Kornilov (1965a), and others. Mutual enhancement of photoconductivity in GaAs:Cr under the excitation of light of two wavelengths has been studied by Gutkin et al. (1972).

Another aspect of this subject concerns the dependence of photoconductive effects on the light flux intensity. In general the photoconductivity increases linearly with the light flux until sublinear effects tend to cause bending over and near saturation. This, however, is not always so, and semiconductors are found in which the photoconductive response increases more rapidly than the light flux, for a limited range. The superlinearity,

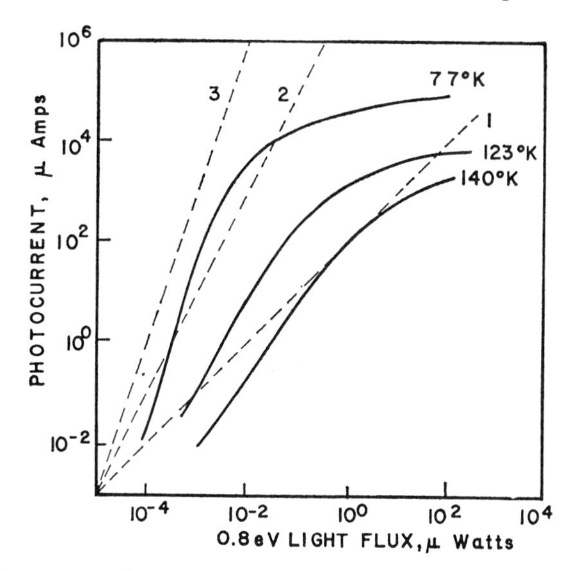

FIG. 7.15 Photocurrent for manganese-doped germanium as a function of light intensity, showing a variation with a power of light intensity greater than unity (dashed lines indicate powers of 1, 2, and 3). (After Newman and Tyler, 1959.)

as it is usually termed, is illustrated in Fig. 7.15 for manganese-doped *n*-type germanium. In general superlinearity can be envisaged as caused by increased gain due to increased trapping action, as the light flux in a certain intensity range changes the occupancy of the traps in a complex situation, involving perhaps several traps or multiply charged traps. Other aspects of high light injection have been studied by Ashkinadze et al. (1972) for silicon and by Glinchuk et al. (1972) for germanium.

Some of the quenching and charge-exchange ideas may be made clearer in the next section which considers particular phototransient curves for semiconductors with various deep impurity doping conditions.

7.5.1 Some Interesting Phototransient Curves under Complex Trapping and Illumination Conditions

Light applied to some semiconductors may cause the conductivity to decrease instead of increase. This is known as negative photoconductivity, or as quenching. This effect is shown in Fig. 7.16*a* for 2 μm radiation applied to *n*-type gold-doped germanium at 77°K. The conductivity at first rises sharply under the illumination, then decreases slowly ($\tau \sim 1$ sec) to a value lower than before the radiation was turned on. When the 2-μm light is

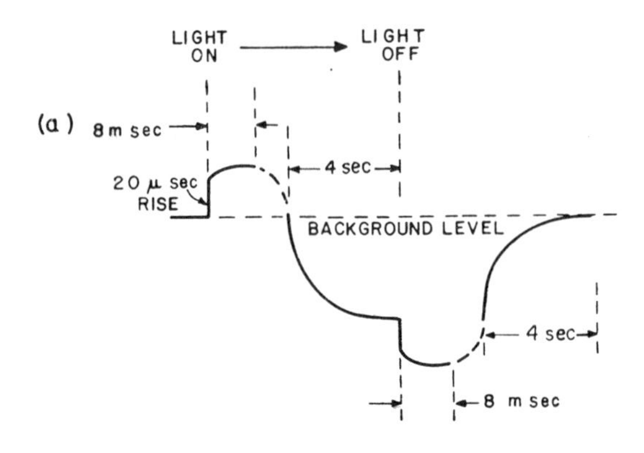

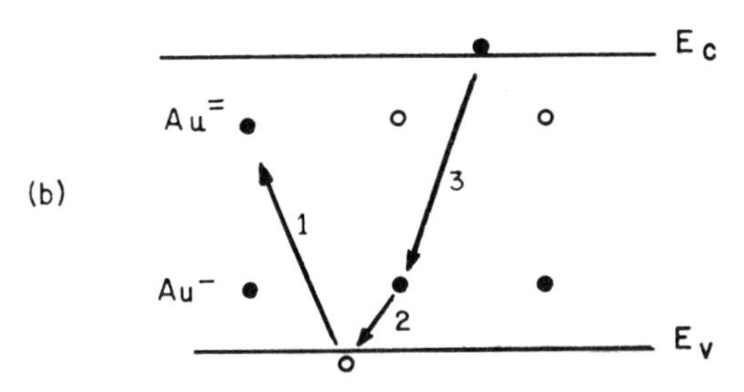

FIG. 7.16 Negative photoconductivity in gold-doped germanium when exposed to extrinsic illumination (2 μm wavelength).
(a) Transient of conductivity. (b) Model postulated. (After Johnson and Levinstein, 1960.)

turned off the conductivity initially decreases sharply, then increases slowly to the level set by 300°K background radiation. The model proposed by Johnson and Levinstein (1960) for the quenching consists (Fig. 7.16*b*) of a three-step process of (1) exciting an electron from the valence band to a vacant upper state, (2) capture of a freed hole by an electron in the lower gold state Au^-, and (3) recombination of a conduction band electron with the hole in the lower state. Excitation is also taking place from the lower and upper gold levels to the conduction band.

The initial spike when the light is turned on is due to the rapid rate of capture of holes in the lower gold state (step 2) for which the time constant is approximately 10^{-6} sec. While electron excitation from and recombination to the lower and upper levels are processes which, by themselves, can come to equilibrium with the radiation field, the process from the valence band to the upper level followed by hole capture in the lower level is one which can never come to equilibrium with the radiation field. Hence the background level of conductivity must shift to achieve a balance; that is, conduction band electrons will be required to fill the vacancies created by hole capture in the lower gold states. The net result will be a reduction or a "quenching" of the "dark carrier" electron concentration in the conduction band during the "light on" cycle.

Negative photoconductivity has been observed in many semiconductors, including germanium doped with cobalt, nickel, iron, and copper (Paritskii et al., 1971).

In cobalt-compensated n-type silicon, at room temperature, similar negative photoconductivity effects are observed (Fig. 7.17a), although the time scale is faster. The cobalt acceptor levels are Co^- at 0.35 eV above the valence band and Co^{2-} at 0.53 eV from the conduction band. The model is

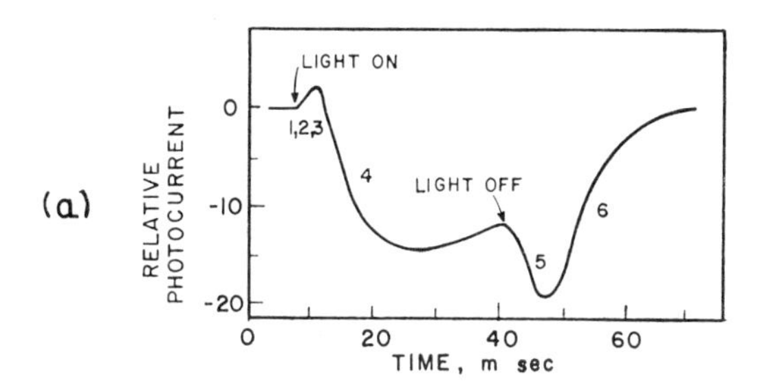

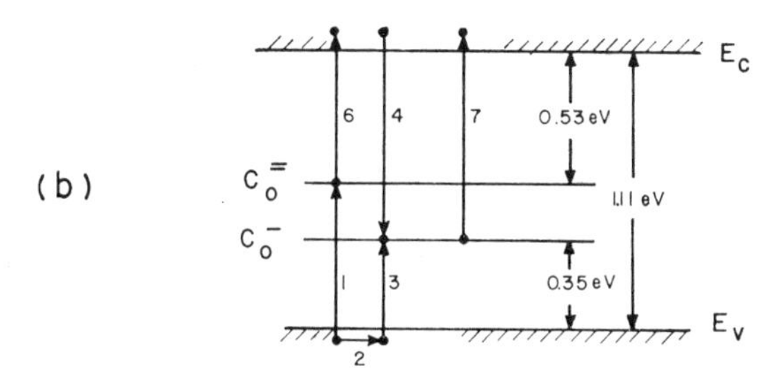

FIG. 7.17 Negative photoconductivity with chopped extrinsic light in cobalt-compensated n-type silicon at room temperature.

(a) Photocurrent transient produced by extrinsic light ($h\nu < E_g$). (b) Energy transitions involved. (After Penchina et al., 1966.)

illustrated in Fig. 7.17*b*. When the light is turned on, electrons are excited from the valence band to an upper impurity level which is not completely filled with electrons (1) thus leaving a hole to give some positive photoconductivity (2). The hole rapidly recombines at a lower impurity level which is normally filled with electrons (3) (the slow rise in positive photoconductivity is due mainly to the shape of the light pulse). An electron from the conduction band is then trapped at the lower level (4) thus decreasing the conductivity below the dark level (time constant about 4 msec). When the light is turned off, holes are no longer created in the valence band, thus further decreasing the conductivity (5), but electrons are then slowly (time constant, about 6 msec) ionized thermally from the upper level to the conduction band (6), thus increasing the conductivity back to the dark level. At increased photon energy (but still extrinsic) one expects to find an additional positive photoconductivity due to direct excitation of electrons from the lower impurity level to the conduction band (7).

Turning now to another type of phototransient, let us consider enhancement of photoconductivity in semiinsulating GaAs. Optical enhancement is obtained by light of photon energy 0.9 eV. The enhancement is observed even after the removal of the excitation causing it, and at 77°K it persists for a long time (more than 1 h) even in the presence of the intrinsic light. Enhancement is also obtained if an ac electric field of 5×10^3 V cm^{-1} is applied. The peak value of the field-enhanced photoconductivity may be as much as ten times larger than the steady-state photoconductivity.

The energy level model, derived by standard activation energy and photoconductivity procedures, for the semiinsulating gallium arsenide that exhibits enhancement is given in Fig. 7.18*a*. The enhancement effect is seen if a sequence of light pulses intrinsic-extrinsic-intrinsic is applied (Fig. 7.18*b*), when it is found that the extrinsic pulse results in an increase in the amplitude of the next intrinsic pulse. The effect is a maximum for extrinsic radiation in the energy range 1.1 to 0.9 eV, and becomes very small for photon energies below 0.85 eV.

When an attempt is made to provide an explanation of the enhancement in terms of the four energy levels shown in Fig. 7.18*a* it is found that they are not sufficient. Let us follow through such an attempt as given by Shah and Yacoby (1968). The condition of our sample on gradual removal of the intrinsic light will be as follows: Levels 1 and 2, being trapped levels which come into rapid equilibrium with the conduction band, will be practically empty. Level 3 will be nearly filled because the dark conductivity measured as a function of temperature revealed that the carriers responsible for dark conductivity at low temperature come from level 3. Level 4 contributes electrons to the impurity photoconductivity and it will therefore be nearly filled after the removal of the intrinsic light.

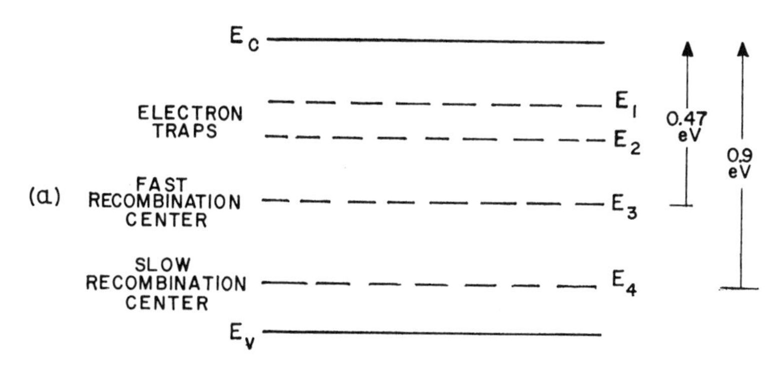

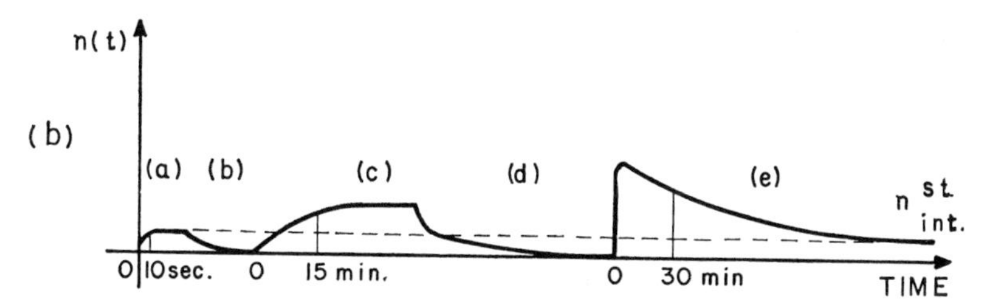

FIG. 7.18 Photoconductivity studies of semiinsulating gallium arsenide. *(a) Energy diagram showing two electron-trapping levels and fast and slow recombination centers. (b) Illustration of the sequence of operations carried out for investigating the effect of the impurity light excitation of the sample on its intrinsic photoconductivity: (a) rise of the intrinsic photoconductivity; (b) decay of the intrinsic photoconductivity; (c) and (d) rise and decay, respectively, of the impurity light-enhanced intrinsic photoconductivity. (After Shah and Yacoby, 1968.)*

On application of the impurity light, electrons from level 4 are excited to the conduction band and the trapping levels. After its removal, the conduction band and therefore the trapping levels become empty and because of charge neutrality levels 3 and 4 must be nearly filled again. Thus in this model, containing only the four levels with properties as described above, an impurity photoconductivity cycle leaves the sample unaffected; therefore this model cannot explain the enhancement.

However, the experimental results are satisfactorily explained by considering an additional impurity level, level 5, which has a high capture coefficient for the holes, σ_{p5}.

Because of large σ_{p5}, level 5 would be very nearly empty of electrons after an intrinsic photoconductivity cycle. On the other hand, the impurity light creates free electrons but no free holes. Hence an impurity cycle will leave level 5 partially filled with electrons and level 4 partially emptied; that is, the impurity light brings about a charge transfer from level 4 to level 5. Hence if the sample is exposed to the intrinsic light after an impurity photoconductivity (PC) cycle, both electrons and holes will be created. Since level 5 is partially filled with electrons and since σ_{p5} is large, the free holes will be captured by level 5. This will increase the electron lifetime in the conduction band. Hence the density of free electrons will be larger, giving rise to the observed enhancement of intrinsic photoconductivity. However, it is clear that this is not the steady state. The partially filled level 5 will be gradually emptied and level 4 will be gradually filled. Thus the intrinsic photoconductivity will approach its steady-state value with a time constant comparable to that of level 4. This is precisely what is observed experimentally.

The enhancement due to an applied electric field is similarly attributed to a reduction of electrons on level 4, where this is assumed to be caused by the electric field effect on the capture cross-section coulomb barrier (after Dussel and Bube, 1966c).

As a further example consider phototransients in GaP of high resistivity. Resistivities of 10^4 to 10^{14} Ω-cm (300°K) may be obtained by diffusing copper into either *p*- or *n*-type GaP (Goldstein and Perlman, 1966). The moderate-resistivity material obtained by diffusion of copper into *n*-type GaP was found to have a large photoconductive response for band-gap radiation ($h\nu > 2.25$ eV). Photoconductive gains (ratio of electron lifetime to electron transit time) in excess of 10^4 were measured in some samples at fields of 10^3 V cm^{-1} (0.1 cm long, with 100 V applied). The GaP appeared to be characterized by electron traps about 0.60 eV below the conduction band edge, recombination levels near the center of the band gap, and hole traps about 0.7 eV above the valence band edge (Fig. 7.19*a*). The effect of extrinsic 1-eV radiation, as shown in Fig. 7.19*b*, is to slow the transient response to a subsequent pulse of band-gap light. This could be caused by the 1-eV radiation emptying the electron traps, which then take electrons from the conduction band during the next pulse of band-gap light.

Simultaneous application of infrared radiation (about 1 eV) reduces (quenches) the photoconductivity with time constants of the order of seconds. When the band-gap excitation is sufficiently small, however, the infrared radiation produces an increase (stimulation) in conductivity. Thermal probe measurements show that under these conditions the stimulated conductivity is *p* type, which indicates that the quenching transition produces an increase in the number of holes in the valence band. These basic effects are illustrated in Fig. 7.20 for zero, low, and high levels of band-gap excitation. In the presence of band-gap radiation, simultaneous infrared radiation sufficiently close to the band-gap energy can produce

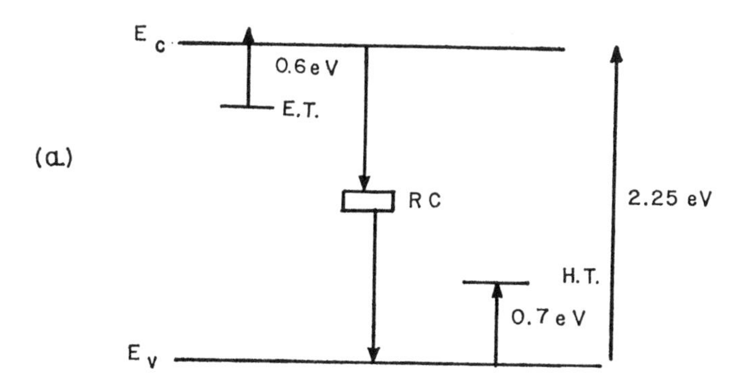

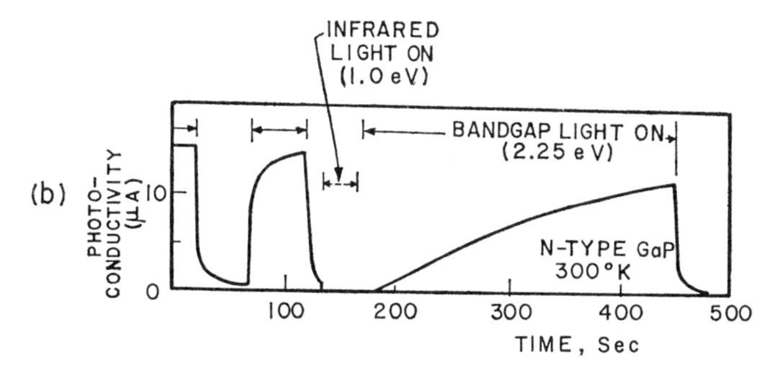

FIG. 7.19 Photoconductivity of high-resistivity GaP.
(a) Energy diagram showing electron traps (E_c − 0.6 eV) and hole traps (E_v + 0.7 eV). (b) Phototransients showing the effect of extrinsic infrared light in slowing the subsequent band-gap light transient. (After Goldstein and Perlman, 1966.)

various combinations of transient increases and decreases in the photo-conductivity depending on the respective light intensities and the specific infrared energy.

One difficulty associated with the interpretation of photoconductivity kinetic data is that model selection and interpretation are often uncertain and ambiguous. The problem is that photoconductivity gives information only on the free carriers in the band and does not directly provide information on the occupancy of the traps. Arkad'eva and coworkers (1962) have shown that some information on trap occupancy can be obtained by short-pulse long-wavelength probing of trap levels during an intrinsic wavelength phototransient. This is illustrated in Fig. 7.21. The technique is well discussed in Ryvkin's book (1964b).

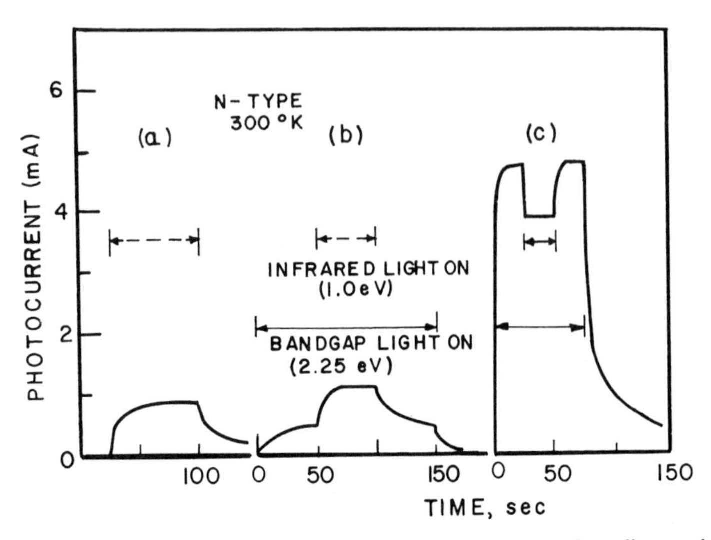

FIG. 7.20 Photoconductivity of high-resistivity GaP showing the effects of simultaneous infrared and band-gap illumination. (After Goldstein and Perlman, 1966.)

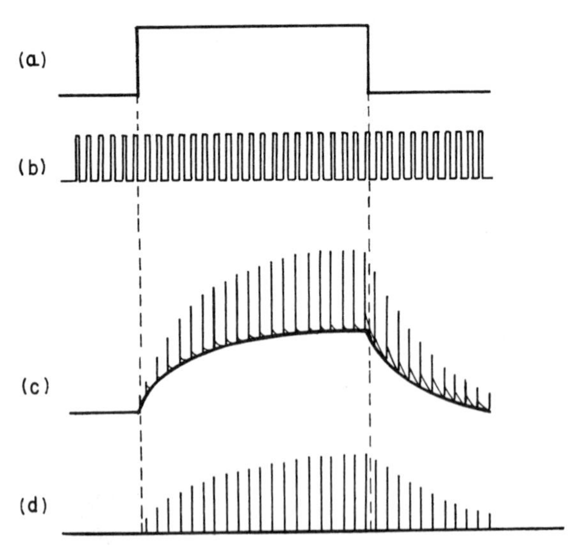

FIG. 7.21 Continuous probing of a photoconductivity transient.

(a) Light pulse versus time which produces the process being studied. (b) Long-wavelength probing pulses. (c) Oscillogram of the complete photoresponse. The continuous curve in (c) represents the variation with time of the free-carrier density, while the envelope (d) of the peaks due to the probing pulses represents the relaxation of the carrier density at the trap level being investigated. (After Arkad'eva et al., 1962.)

7.6 Photoconductive Detector Performance

The extrinsic photoconductivity of semiconductors containing deep impurities is of considerable practical importance as the basis of a class of infrared photodetectors. Excellent reviews of the fundamental concepts have been given by Petritz (1959) and Kruse et al. (1962). International conferences on photoconductivity were held at Atlantic City in 1954, at Cornell University, Ithaca, New York in 1961 (*Proceedings*, H. Levinstein, Ed., 1962), and at Stanford University, Palo Alto, California in 1969 (*Proceedings*, E. M. Pell, Ed., 1971). General discussions of photoelectric device effects have been given by Ambroziak (1968) and by Pankove (1971).

Important factors in characterizing photodetector performance include the detectivity D^*, the spectral response, the response time, and the temperature of operation. The detectivity versus wavelength for germanium detectors containing various dopants is shown in Fig. 7.22. The detectivity cannot exceed the fundamental limit set by the photon noise of the background radiation and the field of view, which is given by one of the dotted lines shown in the figure.

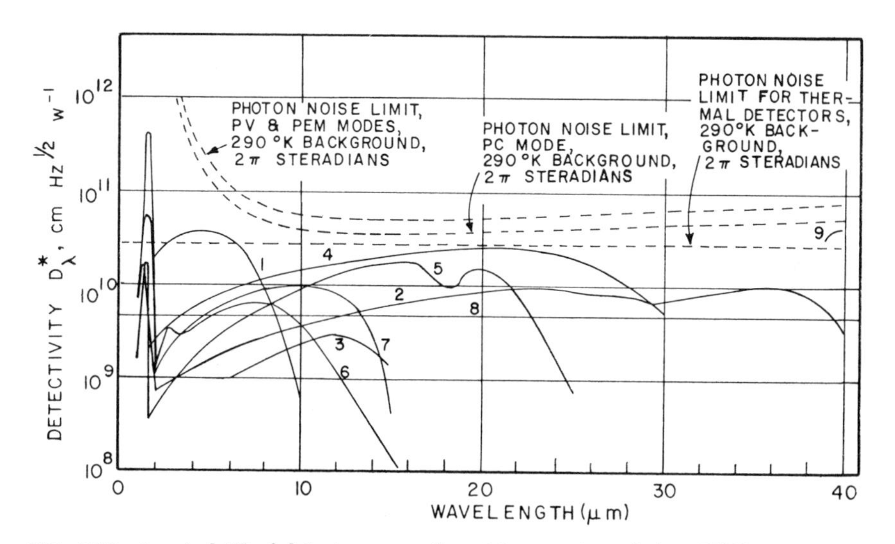

FIG. 7.22 Spectral D^*_λ of detectors operating at temperatures below 77°K.
(1) Ge:Au, 65°K, PC (900 Hz). (2) Ge:Zn, 4.2°K, PC (800 Hz). (3) Ge:Zn, Sb, 50°K, PC (900 Hz). (4) Ge:Cu, 4.2°K, PC (900 Hz, 60° field of view). (5) Ge:Cd, 4.2°K, PC (500 Hz, 60° field of view). (6) Ge–Si:Au, 50°K, PC (100 Hz). (8) NbN superconducting bolometer, 15°K (360 Hz). (9) Carbon bolometer, 2.1°K (13 Hz). (After Kruse et al., 1962.)

TABLE 7.1

Photoconductive Detector Performance

Detector Material and Dopant	Relevant Energy Level of the Dopant (eV)	Long-Wavelength Limit[a] (μm)	Peak Detectivity $D^*_{\lambda p}$(cm (Hz)$^{1/2}$ W^{-1})	Response Time (μsec)	Temperature of Operation Typical (°K)	Ref.[b]
Si:Au	$E_c - 0.54$	2.3	9.5×10^9	<1	300	Ing and Gerhard, 1965
Si:In	$E_v + 0.16$	7.6		0.1	77	Blakemore and Sarver, 1968
Si:Zn	$E_v + 0.55$	2.2			77–300	Loebner et al., 1966
						Kornilov, 1965a
Ge:Au	$E_v + 0.15$	6.9[a]	$1.7\text{–}4 \times 10^{10}$	<1	77–65	Kruse et al., 1962
Ge:Cd	$E_v + 0.05$	21.5[a]	1.8×10^{10}		<25	Kruse et al., 1962
Ge:Co	$E_v + 0.08$	15	10^{10}		5.5	Schultz et al., 1962
Ge:Cu	$E_v + 0.04$	27[a]	2.5×10^{10}		<20	Kruse et al., 1962
Ge:Ga	$E_v + 0.011$	~11			4.2	Jeffers and Johnson, 1968
Ge:Hg	$E_v + 0.083$	15	3×10^{10}		40	Whelan and Distelhorst, 1962
Ge:Zn	$E_v + 0.03$	40[a]	1×10^{10}	<0.01	4.2	Kruse et al., 1962
GaAs	$E_c - 0.055$	~250		<1	4.2	Stillman et al., 1968
InAs	$E_g = 0.35$	3.8[a]	1.4×10^8	0.2	295	Kruse et al., 1962
InSb	$E_g = 0.35$	6.9	$2.5 \times 10^9\text{–}6 \times 10^{10}$	<2	195–77	Kruse et al., 1962
InSb:Ag	$E_v + 0.039$	32				Engeler et al., 1961

[a] At 50% of peak response, otherwise from $\lambda = 1.24/\Delta E_{(\text{impurity})}$.
[b] Also consult index to the bibliography under photoconductivity.

Table 7.1 presents a summary of the performance of some photoconductive detectors. The shallower the impurity that is providing the photoconductivity effect, the greater is the infrared wavelength limit, and the lower must be the temperature of operation. A low temperature of operation ensures that the impurity is not thermally ionized and can respond to the light that is to be detected. In small-band-gap intrinsic photoconductors the low temperature ensures that the dark conductivity is suitably low. The speed of response of most of the detectors is seen from Table 7.1 to be usually in the range 10^{-8} to 10^{-6} sec.

Laser and maser developments in the last ten years or so have resulted in the concept of communications systems with the signal information modulated at microwave frequencies onto the very high carrier frequencies. This has led to an examination of photoconductors as mixers and demodulators of microwave frequency signal information on optical carrier beams. Sommers and Teutsch (1964) have shown that if an intrinsic photoconducting semiconductor is driven by an ac bias of microwave frequency (above the dielectric relaxation frequency), it can be used as a broad-band envelope demodulator for optical signals with high current gain. The current gain results from the fact that the photocarriers at the microwave bias frequency can traverse the sample many times in the lifetime of the photoinduced carrier. They conclude that for bandwidths up to 1 GHz the photoconductor should outperform junction diode demodulators by a large margin. Coleman et al. (1964) in considering a similar mode of operation suggest that impurity photoconductors, such as gold-doped germanium, might be used although the low optical absorption coefficients and phase-matching considerations are recognized as possible problems.

CHAPTER EIGHT

Recombination and Trapping Effects of Deep Impurities in Semiconductor Junctions

Generation and recombination centers in *pn* junctions play an importa role in determining the current-voltage and capacitance characteristics of junctions. In general these centers are most effective when located in energy somewhere near the center of the band gap because then communication between the conduction and valence bands through the centers tends to be greatest. The generation-recombination centers may be impurity atoms located interstitially or substitutionally in the lattice or crystal lattice defects or surface effects. In this study of deep impurity effects in junctions we begin with a brief review of the Sah–Noyce–Shockley model. This has become the basic treatment for generation-recombination effects in junctions.

As discussed in Chapter 6, deep impurities may be used for control of the minority carrier lifetime. This can result in junctions with reverse recovery times in the nanosecond range, but also affects the forward and reverse characteristics of diodes. These matters are examined in Section 8.2.

The effects of deep impurities on the frequency dependence of junction capacitance, and capacitance and reverse current transients in depletion regions are studied in Sections 8.3 and 8.4. These transients provide fundamental information on capture cross sections as well as affecting junction performance.

Such studies help to explain phenomena seen in wide-band-gap semiconductor junctions where deep impurity effects are sometimes unavoidable.

The chapter concludes with comments on the effects of deep impurities in bipolar and MOS field effect transistors.

8.1 Generation and Recombination Current Caused by Deep Generation-Recombination Centers in *pn* Junctions (Sah–Noyce–Shockley Model)

The Sah–Noyce–Shockley model assumes that there are single-level generation-recombination centers that are uniformly distributed and have an energy E_T that is somewhere near the intrinsic energy level E_i for the band-gap. Figure 8.1 then shows the recombination and generation processes expected in the *p*, *n*, and depletion regions of a junction. In the *p* region (Fig. 8.1*a*), the centers are mostly empty since the Fermi level for

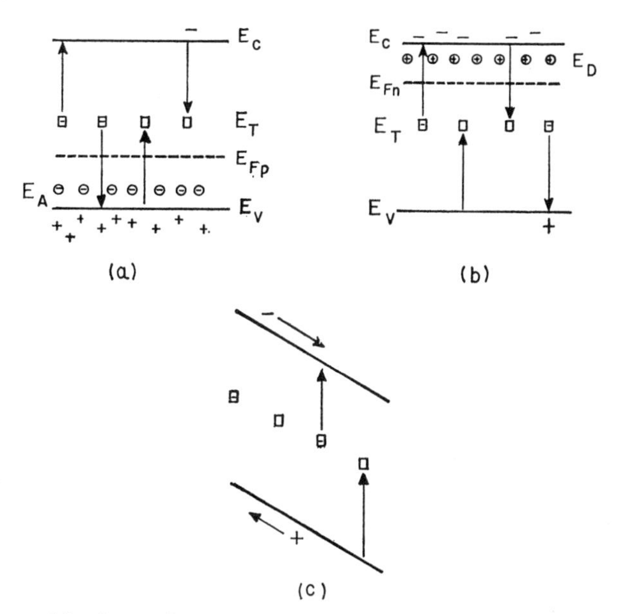

FIG. 8.1 Recombination and generation processes in a semiconductor. *(a) p region; (b) n region; (c) intrinsic with swept field. (After Sah et al., 1957.)*

holes is below the trap level E_T. Hence the centers are ready to capture injected electrons and pass them on to the valence band, thus accomplishing a recombination process. Similarly in the *n*-type side of the junction the center facilitates the recombination of the injected holes with electrons (Fig. 8.1*b*).

In the space-charge region of the junction under reverse bias voltage conditions (Fig. 8.1*c*), the centers generate holes and electrons which are separated by the electric field. For each carrier pair generated one electron flows in the external circuit and the current is therefore

$$J_g A = -\frac{eWn_i A}{2\tau_0} \tag{8.1}$$

where A is the diode cross-sectional area, W is the width of the space-charge layer, and $n_i/(2\tau_0)$ is the generation rate where n_i is the intrinsic carrier density. The lifetime, τ_0, is defined from the generation rate and of course depends on the generating centers present. Also contributing to the leakage current of the junction are the minority carriers generated within one diffusion length distance of the space-charge region. The generation rate for the p region is n_{p0}/τ_{n0} and for the n region is p_{n0}/τ_{p0}. The currents therefore are

$$J_1 A = \frac{en_{p0}L_n A}{\tau_{n0}} \tag{8.2a}$$

$$J_2 A = \frac{ep_{n0}L_p A}{\tau_{p0}} \tag{8.2b}$$

If we assume for simplicity that $n_{p0} = p_{n0}$, that $\tau_{n0} = \tau_{p0} = \tau_0$, and that $D_n = D_p$, the ratio of the space-charge $J_g A$ current to the diffusion region current is

$$\frac{J_g}{J_1 + J_2} = \frac{n_i}{n_{p0}}\left(\frac{W}{4L_n}\right) = \frac{n_{n0}}{n_i}\left(\frac{W}{4L_n}\right) \tag{8.3}$$

It follows that the component of current from the depletion region may be large for semiconductors of large band gap corresponding to small n_i. For example, the current ratio for a silicon junction may be 3000 compared with about 0.1 for a germanium junction of comparable parameters (2 Ω-cm, 1 μm depletion width, 1 μsec lifetime).

Under forward bias conditions the carrier concentrations in the space-charge region are large and substantial recombination occurs. The recombination where $p = n$ in the space-charge region is

$$r = \frac{n_i \exp(eV/2kT)}{2\tau_0} \tag{8.4}$$

since $n = p = n_i \exp(eV/2kT)$ from $np = n_i^2 \exp(eV/kT)$. This recombination rate falls exponentially with distance on either side of the pn interface with a characteristic length $kT/e\mathscr{E}$ where $\mathscr{E}$ is the electric field at the junction. The recombination current density is therefore

$$J_r = \frac{(kT/e\mathscr{E})en_i}{\tau_0} \exp\left(\frac{eV}{2kT}\right) \tag{8.5}$$

On the other hand, the simple Shockley treatment for a junction in which the currents are diffusion limited gives

$$J_d = \frac{2en_pL_0}{\tau_0} \exp\left(\frac{eV}{kT}\right) \tag{8.6}$$

assuming again that $n_{p0} = p_{n0}$ and $L_n = L_p = L_0$ and $\tau_n = \tau_p = \tau_0$. The relative magnitudes of J_r and J_D in these simplified expressions may then be studied. For a space-charge layer width of 0.1 μm under forward bias conditions and τ_0 of 1 μsec in a typical diode it is found that the recombination current is significant in relation to the diffusion current at applied biases of the order of 5 to 10 kT/e. Since the total current density is the sum of the recombination and diffusion components, the total current varies slower than the ideal formula exp (eV/kT).

The model has been developed in detail by Sah–Noyce–Shockley (1957) with the results shown in Fig. 8.2. When the reverse current is due primarily to generation in the depletion region the current depends on the depletion width which varies as $V^{1/3}$ for a graded junction. This explains the slope of 3 shown in Fig. 8.2*a*. The forward characteristic shows a roughly exponential dependence exp $(eV/\eta kT)$ where η varies between 1 and 2 for different sections of the characteristic. By matching theoretical and experimental curves for silicon junctions Sah et al. infer that the recombination centers involved are within a few kT of the center of the band gap.

For high current gain in a transistor it is important that as much of the current as possible be carried by the injection component rather than by the emitter-base depletion region recombination component. The computed current gain $[I_C/I_E = (J_1 + J_2)/(J_1 + J_2 + J_r)]$ from the model is shown in Fig. 8.3 as a function of current density for various values of W/L. The gain is seen to be low at low current densities because of the significant recombination component. However, most transistors are operated at much higher current densities (amperes per square centimeter) than the densities considered in Fig. 8.3 and there are other factors that contribute to an α dependence on injection level. The recombination variation of α with current density is significant in the turn-on of *pnpn* transistor switches. This problem has been considered by various workers since 1957. One of the most recent treatments is that of Osipov (1968) who considers four-layer structures with deep impurity levels.

Although the Sah–Noyce–Shockley treatment has gained general acceptance the assumptions made are not really representative of conditions in a *pn* junction over a wide current range. The problem is complicated and the factors contributing to the term η in the $I = I_0$ exp $(eV/\eta kT^{-1})$ type of representation are many and varied. Many commercial diodes show characteristics with η in the range from 1.2 to 1.8, depending on the voltage,

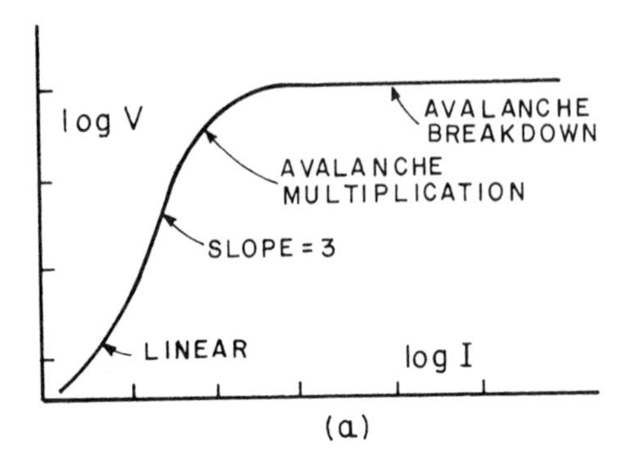

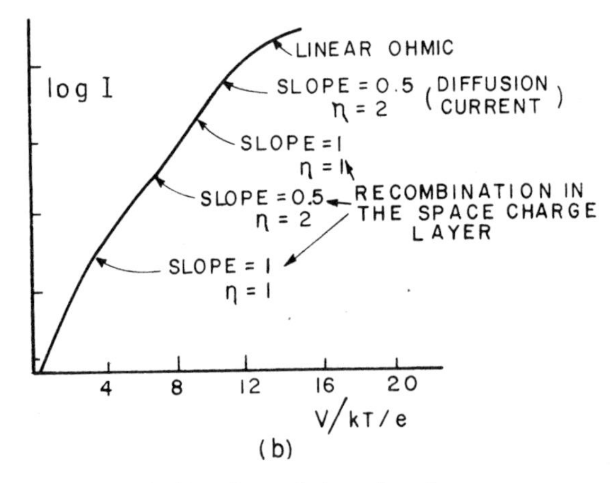

FIG. 8.2 Characteristics of a linearly graded *p-n* junction.
(a) reverse bias curve; (b) forward bias curve. (After Sah et al., 1957.)

the temperature, and the method of fabrication. Buckingham and Faulkner (1969) have shown that this effect might be expected from a theory in which Shockley–Read–Hall traps are localized in one side of the depletion region. An ion-pairing mechanism could result in traps forming on the p side and not the n side of the junction. There are some commonly occurring impurities, for instance iron, which in p-type material are ionized and therefore will readily form ion pairs with the acceptor impurity ions at diffusion temperatures. The same impurity in n-type material is not ionized, and the formation of ion pairs with acceptor impurities in n-type material is therefore not a likely process. Consequently, a diffused junction may be expected to contain ion pairs (assumed to be the predominant trapping centers) restricted to the p region. Hence there is some physical basis for modeling the junction in this way.

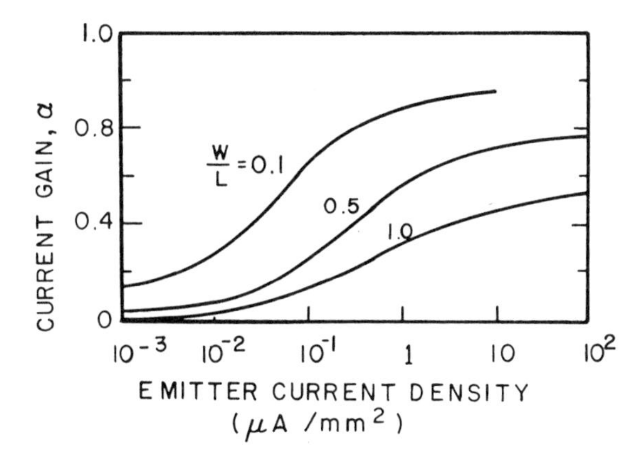

FIG. 8.3 Variation of *p-n-p* silicon transistor dc alpha with emitter current density. (After Sah et al., 1957.)

The high current transition from $\eta = 1$ to 2 for diodes is discussed by McKelvey (1966). For transistors, Ebner and Gray (1966) have shown that η may also involve lateral bias effects, and be different for *pnp* and *npn* structures.

In p^+nn^+ structures, Venkateswaren and Roulston (1972) conclude that recombination mechanisms at the high low junction interface must be considered.

8.2 Observed Effects of Deep Impurities on Diode Characteristics

8.2.1 Reverse Recovery Time of Diodes

An important feature of deep impurities in semiconductors is the reduced lifetime that can be obtained. This can result in diodes with reverse recovery times in the nanosecond range.

The transient of current that results when the voltage is switched from forward to reverse on a diode-resistor circuit is shown in Fig. 8.4*b*. The time t_{I} for the simple model of an abrupt diode of uniform bulk lifetime τ is given by

$$\operatorname{erf}\left(\frac{t_{\mathrm{I}}}{\tau}\right)^{1/2} = \frac{1}{1 + |I_r/I_f|} \tag{8.7}$$

This expression is therefore one way of estimating the bulk lifetime of a diode from the storage time t_{I}. Alternatively, it may be shown that the total time $t_r = t_{\mathrm{I}} + t_{\mathrm{II}}$ is approximately equal to 0.9τ. This simplified approach

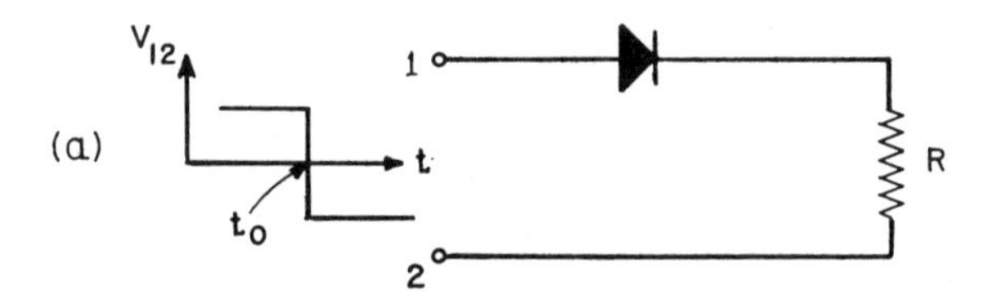

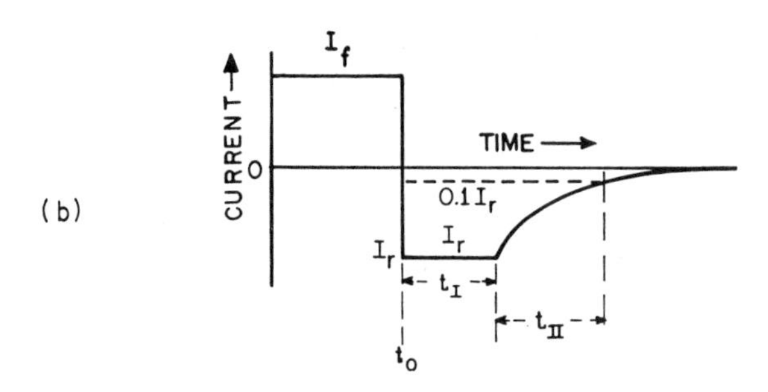

FIG. 8.4 Diode reverse recovery theoretical behavior.
(a) Circuit voltage reversed at t_0. (b) Diode current transient.

was used by Bakanowski and Forster (1960) in studies of gold-doped diffused silicon computer diodes.

In the *n* side of a *pn* junction with gold present the active recombination level is the acceptor level at $E_c - 0.54$ eV. In the *p* side the active level is the gold donor level at $E_v + 0.35$ eV. The general lifetime expression (Chapter 6)

$$\tau = \left(\frac{n_0 + n_1}{n_0 + p_0}\right) \tau_{p0} + \left(\frac{p_0 + p_1}{n_0 + p_0}\right) \tau_{n0} \tag{8.8}$$

therefore reduces to

$$\tau_1 \cong \tau_{p0} = \frac{1}{\sigma_{p0} v_p N_{\mathrm{Au}}} \tag{8.9}$$

in the *n*-type side, and to

$$\tau_2 \cong \tau_{n0} = \frac{1}{\sigma_{n0} v_n N_{\mathrm{Au}}} \tag{8.10}$$

in the *p*-type side.

From Bemski (1958b), the values of $\sigma_{p0}v_p$ and $\sigma_{n0}v_n$ are in the range 1.4 to 3.5×10^{-8}, and therefore Bakanowski and Forster infer that the observed recovery time should be, to a fair approximation,

$$t_r \cong \frac{2.5 \times 10^7}{N_{\text{Au}}} \tag{8.11}$$

Since gold densities of several times 10^{16} cm^{-3} are achievable, diode recovery times of nanoseconds may be expected. The observed recovery times for various gold diffusion temperatures are given in Fig. 8.5. Good agreement is seen for the line which represents (8.11) with N_{Au} taken from

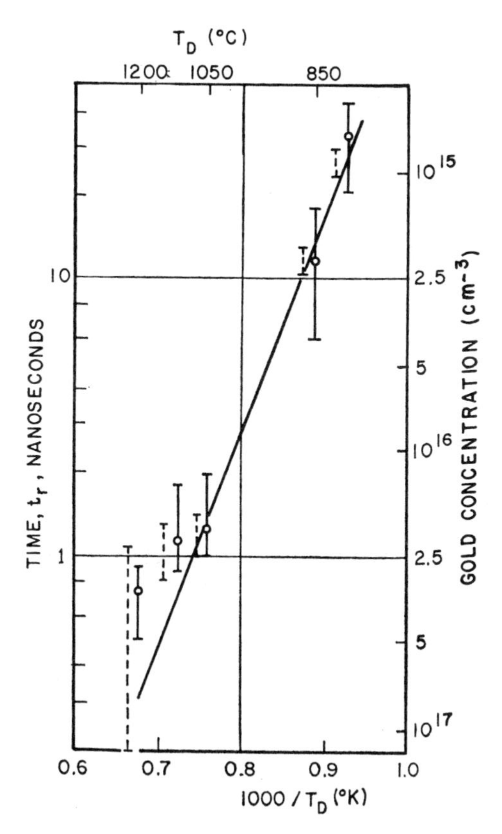

FIG. 8.5 Silicon diode reverse recovery time as a function of the inverse of the gold diffusion temperature.

Gold concentrations shown on the right-hand side were calculated from solubility data. The solid line was computed from (8.11). (After Bakanowski and Forster, 1960.)

gold solubility-temperature data. However, there is some uncertainty about the σv_{th} products used, as noted by Fairfield and Gokhale (1965) and Bullis (1966).

Other deep impurities have been studied as lifetime controllers in diodes. Figure 8.6 shows the results observed by Ghandhi and Thiel (1969) for nickel in silicon. In this figure it is the hole lifetime in *n*-type silicon that is being plotted rather than the diode recovery time. The expression that gave the calculated line was

$$\tau_p \cong (\sigma_p v_p N_r)^{-1} \tag{8.12}$$

where $\sigma_p v_p$ was 6.7×10^{-7} cm^3 sec^{-1} and N_r is the active nickel concentration, assumed to be nearly all doubly ionized.

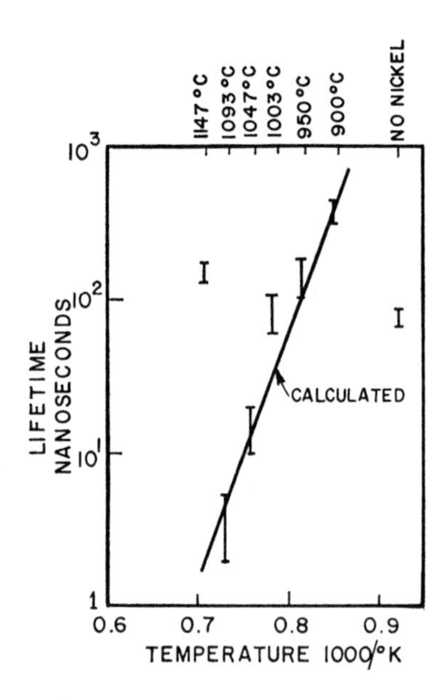

FIG. 8.6 Lifetime in nickel doped silicon versus nickel diffusion temperature. (After Ghandhi and Thiel, 1969.)

Nickel is seen to be less effective than gold for producing short lifetimes. This is because the active density of nickel that can be achieved (10^{14} to 10^{15} cm^{-3}) is less than that of gold. Other impurities, such as silver, have been considered as lifetime controllers in silicon, but the evidence available so far suggests that none is as effective as gold, although platinum is now under examination.

In work with gold-doped structures care must be taken that the processing does not result in loss of active gold. The relative effectiveness of boron and phosphorus glasses as gold-gettering agents is shown in Fig. 8.7. Shaklee

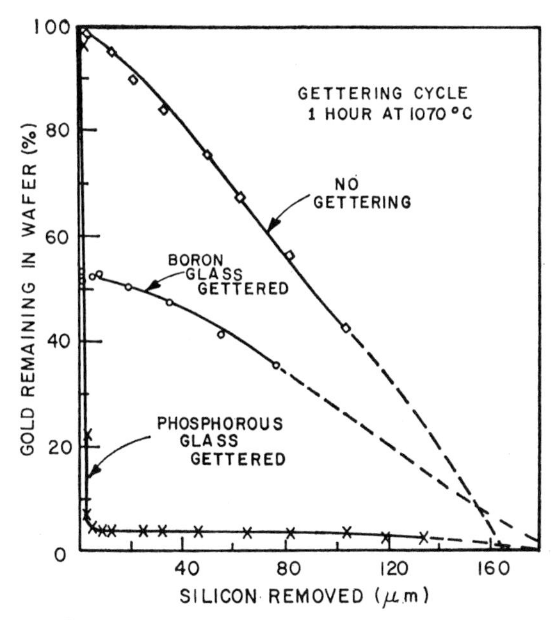

FIG. 8.7 Relative effectiveness of boron and phosphorus glasses on silicon as gold-gettering agents. (After Adamic and McNamara, 1964.)

et al. (1969) have recently studied the effects of heat treatments in the 700 to 850°C range on gold-doped p^+n diodes. The recovery times of diodes are typically found to increase by a factor of 30 or so within the first few hours of heat treatment as seen in Fig. 8.8. The effect is attributed to gold becoming interstitial rather than substitutional, and ceasing to act as a recombination center. Another adverse effect to be guarded against is that if too much gold is present, there is the possibility that some region of the diode may be converted to high-resistance material. Excess voltage drop or negative-resistance characteristics may then be observed. Peters and Shipley (1968) have studied such action in the forward transient characteristics of gold-doped silicon p^+nn^+ diodes. Oscillations and inductive effects are seen in certain current ranges.

8.2.2 Effect of Gold on the Forward and Reverse Characteristics of Diodes

The presence of gold in a diode tends to increase the current flowing at a given voltage. This, presumably, is because of an added recombination component in the space-charge region as well as the changes inherent in the bulk lifetime reduction. The results in Fig. 8.9 show that the effect is more

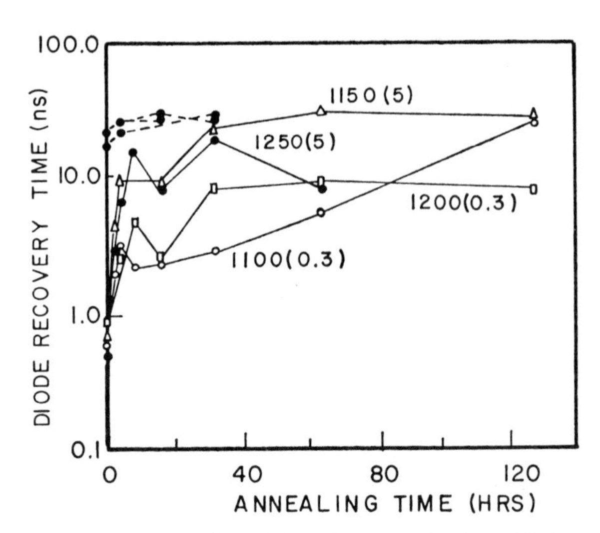

FIG. 8.8 Diode recovery time of gold-doped silicon diodes with heat treatment at 750°C.

The numbers on the curves denote initial boron diffusion temperatures and expitaxial resistivity (Ω cm). The dashed curves are for control diodes which had no gold doping. The control diodes were heat treated at 850°C. (After Shaklee et al., 1969.)

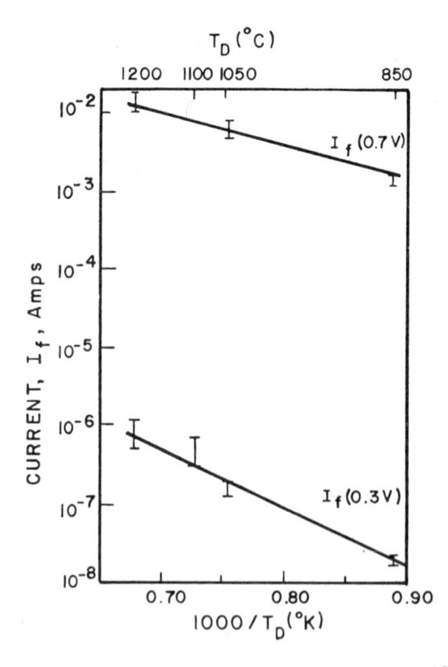

FIG. 8.9 Forward current as a function of inverse gold diffusion temperature for silicon diodes. (After Bakanowski and Forster, 1960.)

pronounced at low currents ($V_a = 0.3$ V), where the straight line represents a variation roughly proportional to N_{Au}, than at higher currents ($V_a = 0.7$ V) where the line represents a $(N_{Au})^{1/2}$ dependence. The complexity of the problem precludes a closed-form mathematical solution.

For the reverse current, Bakanowski and Forster (1960) derive an expression of the form

$$I_r = \frac{e n_i N_{Au} W_{1/2} A}{(\sigma_p v_p)^{-1} + (\sigma_n v_n)^{-1}} \tag{8.13}$$

where $W_{1/2}$ is one-half the total space-charge width. The observed results are shown in Fig. 8.10 where the straight line represents the prediction of (8.13). The agreement is seen to be acceptable.

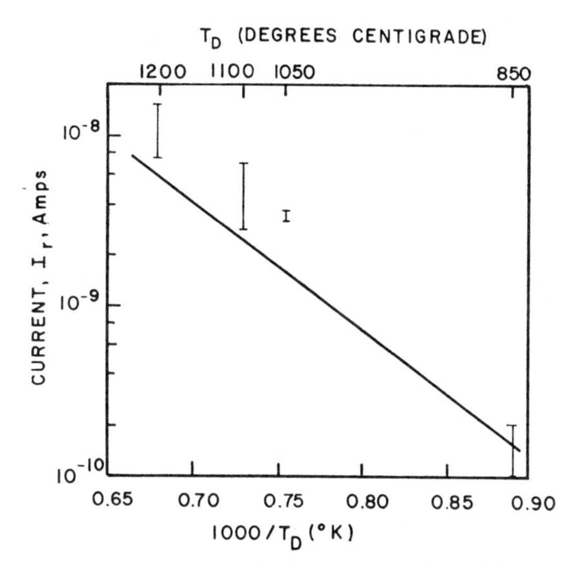

FIG. 8.10 Reverse current of silicon diodes as a function of inverse gold diffusion temperature for a reverse bias of 10 V. (After Bakanowski and Forster, 1960.)

The presence of a deep impurity such as gold may also be expected to affect the breakdown voltage of diodes. For a p^+n diode, the effective doping in the n region is $N_D - N_{Au}$ since the gold is in the form of $N_{Au}{}^-$ centers. The voltage for breakdown V_B is therefore given by

$$V_B = \frac{\varepsilon \mathscr{E}^2_{max}}{2e(N_D - N_{Au})} \tag{8.14}$$

where $\mathscr{E}_{max}$, the field strength at the p^+n junction interface, is 3.1×10^5 V cm^{-1} for avalanche in silicon. The effect of the gold therefore is to widen the depletion region and increase the voltage for breakdown. If the reverse voltage is maintained for a long period, the N_{Au}^- state slowly reverts to N_{Au}^0, by ionization and sweepout of the electrons, and the depletion region narrows. The breakdown voltage then decreases to the value corresponding to N_D. (Similar effects have been observed by Antell and White (1968) in gallium arsenide junctions containing oxygen.)

For gold-doped n^+p silicon junctions the problem is more complex and Sah (1967b) has shown that because of different electron and hole emission rates the gold in the p region can be in the N_{Au}^- state. Therefore the effective space-charge density is $N_A + N_{Au}$ and the breakdown voltage for pulsed conditions is given by

$$V_B = \frac{\varepsilon \mathscr{E}_{max}^2}{2e(N_A + N_{Au})} \tag{8.15}$$

Hence the breakdown voltage is decreased by the presence of gold in the n^+p diodes. This has been confirmed experimentally and the results are shown in Fig. 8.11. A detailed study of the spatial variation of the charged gold concentration in silicon p-n step junctions has been made by Schroder et al. (1968).

Another reason why the breakdown voltage of a junction may be reduced by the introduction of a deep impurity is related to the tendency at high

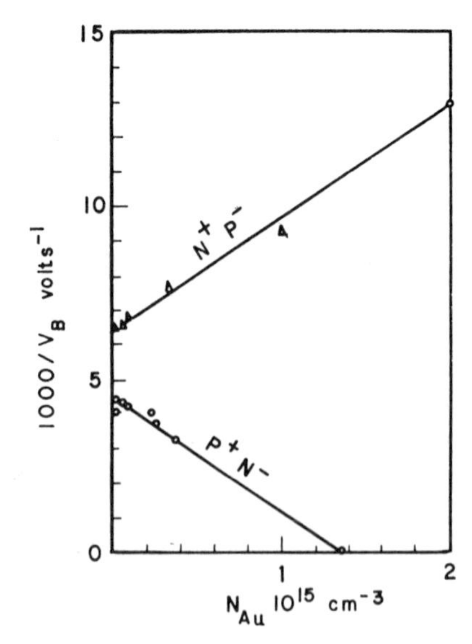

FIG. 8.11 The reciprocal breakdown voltage $1000/V_B$ as a function of the gold concentration (atoms per cubic centimeter) determined by pulse recovery lifetime method for gold-doped silicon n^+p^- and p^+n^- mesa junctions at room temperature.

Gold diffusions were made between 800 and 1000°C, and for most cases, the solid solubility limit was not reached. (After Sah and Schroder, 1968.)

concentrations of most impurities to introduce precipitates. These cause field-strength inhomogeneities, and microplasma form; the usual effect is a softening of the voltage breakdown knee. This has been studied for gold and copper in silicon n^+p junctions (Ogawa, 1966).

8.3 Variation of Junction Capacitance with Frequency when Deep Impurities Are Present

The usual depletion approximation for the capacitance of a reverse-biased *pn* junction involves several simplifications. The measuring frequency must be low, so that dispersion at the space-charge edge is negligible. If either one of the bulk sections is degenerate, the free carriers diffuse into the depletion region and the capacitance is greater than expected from the depletion approximation at reverse biases of less than a few volts (Chang, 1966; Carter et al., 1972). Traps or deep centers in a *pn* junction will cause effects similar to free-carrier dispersion, but related to the capture and emission probabilities of these deep centers for carriers.

The reverse-biased capacitance of a p^+n junction depends on the depletion width W which exists in the lightly doped side of the junction. If this is determined solely by a shallow donor of density N_D, the expression for the capacitance is

$$C = \frac{\varepsilon}{W} = \left[\frac{e\varepsilon N_D}{2(V_D - V_a)}\right]^{1/2} \tag{8.16}$$

where the voltage V_a is substituted as a negative value for reverse bias conditions and V_D is the diffusion voltage of the diode.* An expression of this form is also valid for Schottky barriers on n-type semiconductors.

The capacitance is measured, of course, by superimposing a small alternating voltage on the reverse bias battery level of voltage and making a bridge measurement. Equation 8.16 is valid only if the edge of the depletion region follows the frequency of the applied ac voltage. If the junction contains deep impurities whose ionization states must change with change of the applied bias voltage, the response time for the system may not allow the capacitance to follow the applied frequency unless this is very low. The capacitance at a low frequency such as 10^2 to 10^3 Hz is then greater than that measured at a high frequency such as 10^5 to 10^7 Hz.

This effect has been studied by Sah and Reddi (1964) for gold-doped silicon p^+n step junctions over the frequency range 10 Hz to 30 MHz at

* In this and subsequent expressions the capacitace has been given for a unit area.

room temperature. The deep impurity (gold) density was less than the density of shallow donors in these studies. This is typical of deep impurity doping in practical devices. It is not usually desirable that the material become high in resistance because of overcompensation of the shallow doping density. However, some overcompensation may be unavoidable in graded junction regions.

The problem of there being deep donors present with a density that exceeds the concentration of shallow donors has been considered by Perel' and Éfros (1968). Berman (1971) has considered the problem of the capacitance of an alloyed p^+n junction when the diode base contains not only shallow donors (the main impurity) but also deep donors and deep acceptors that exchange electrons with both the conduction and the valence band. The influence of incomplete ionization of impurities has been considered by Nuyts and Van Overstraeten (1971).

The behavior of an n^+p junction where the deep impurity is an acceptor, and there is no shallow acceptor, has been studied by Schibli and Milnes (1968). This is treated in detail in Section 8.3.2. In these studies the depletion edge is usually considered to be abrupt and the deep impurity level is usually represented by a single time constant. Roberts and Crowell (1970) and Crowell and Nakano (1972) in considering Schottky barrier junctions have relaxed these constraints. When the high- and low-frequency capacitances are obtained with and without the abrupt space-charge approximation it appears that the approximate approach is in reasonable agreement with the more exact treatment. The single-time-constant treatment, where the response, say, of a high-level acceptor is taken as depending on a characteristic time $1/c_n(n + n_1)$ and the interaction with the valence band is neglected, is only an approximation because the trap behavior influences the electron density n in the region considered.

Other more exact treatments include the numerical calculations of Forbes and Sah (1969) and EerNisse (1971). However, Oldham and Naik (1972) point out that there are some advantages in approximate treatments. Using a truncated space-charge approximation, which is tantamount to a single time constant treatment for a given trap level, they are able to obtain reasonably accurate estimates of frequency dependence and to include effects that otherwise would be intractable such as the interaction of two defect levels.

In the sections that follow we consider first the Sah–Reddi model and then discuss briefly the Schibli–Milnes and Crowell–Nakano treatments. Examination of the Forbes–Sah equations and of the Oldham–Naik model with its two-level theory applied to neutron-irradiated junction field effect transistors, is left to individual study.

8.3.1 The Sah–Reddi Model (p^+n)

The energy diagram for a p^+n junction with gold as a deep impurity, N_T, which provides an acceptor level $E_{Au}{}^-$ and a donor level $E_{Au}{}^+$, is shown in Fig. 8.12 for the condition of reverse bias. The net charge state as a function of position (Fig. 8.12*a*) has a fairly sharp step from $N_D - N_T$ to N_D at the position y_t where the Fermi level in the n material crosses the deep acceptor level. If the bias voltage is changed, this step must change in position as suggested in Fig. 8.12*d*. This involves relatively slow emission or capture processes and is the reason why the capacitance is not independent of the frequency of measurement.

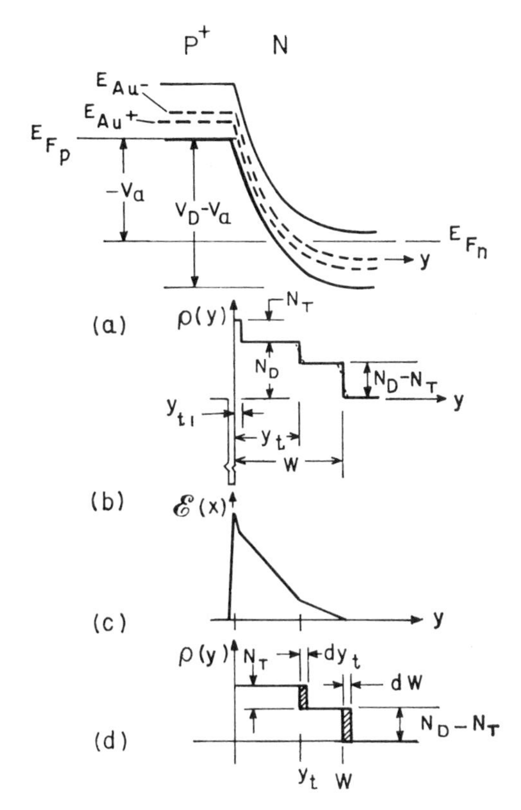

FIG. 8.12 Energy, charge, and field diagrams for a reverse-biased gold-doped silicon p^+n **diode. (After Sah and Reddi, 1964.)**

For very low frequencies the change in charge for a voltage change dV_{DC} is

$$dQ_{\mathrm{DC}} = dQ(y_t) + dQ(W) = eN_T dy_t + e(N_D - N_T)\, dW \tag{8.17}$$

and

$$dV_{\mathrm{DC}} = \frac{e}{\varepsilon}\,(N_T y_t\, dy_t + (N_D - N_T) W\, dW) \tag{8.18}$$

Hence the low frequency or dc junction capacitance is

$$C_{\mathrm{DC}} = \frac{dQ_{\mathrm{DC}}}{dV_{\mathrm{DC}}} = \frac{\varepsilon[N_T\, dy_t + (N_D - N_T)\, dW]}{[N_T y_t\, dy_t + (N_D - N_T) W\, dW]} \tag{8.19}$$

This neglects the effect of the gold donors, which is permissible since y_{t_1} (Fig. 8.12*a*) is very small and close to the p^+ region.

From Poisson's equation

$$W - y_t = \left(\frac{2\varepsilon\phi_t}{e(N_D - N_T)}\right)^{1/2} \tag{8.20}$$

where $\phi_t = (E_{\mathrm{F}n} - E_{\mathrm{Au}}{}^-)/e$. Also

$$V_D - V = \left(\frac{N_T}{N_D - N_T}\right)\phi_t + \frac{e}{2\varepsilon}\left(N_D W^2 - 2N_T(W - y_t)W\right) \tag{8.21}$$

From (8.19), (8.20), and (8.21) the dc or low-frequency capacitance is found to be

$$C_{\mathrm{DC}} = \frac{C_0}{[1 - (N_T/N_D)\phi_t/(V_D - V)]^{1/2}} \tag{8.22}$$

where $C_0 = [e\varepsilon N_D/2(V_D - V)]^{1/2}$ and is the capacitance without the gold present.

The other frequency range of particular interest is that for which the gold acceptor centers in the region near y_t cannot respond because the time constant is long. For example, the time constant given by the emission rates of electrons $(c_n n_1)$ and of holes $(c_p p_1)$ may be of the order of 0.005 sec. At the depletion edge region $(y = W)$, however, the free carriers present allow charging and discharging time constants of the order of 10^{-6} to 10^{-9} sec. Hence there is a high-frequency range for which the gold does not charge and discharge at y_t but does follow at the depletion edge W. The capacitance in this range is given by

$$C_{\mathrm{AC}} = \frac{dQ(W)}{dV(W)} = \frac{\varepsilon}{W} \tag{8.23}$$

The expression for the depletion width is

$$W = \frac{N_T}{N_D}\left(\frac{2\varepsilon\phi_t}{e(N_D - N_T)}\right)^{1/2} + \frac{2\varepsilon(V_D - V - \phi_t(N_T/N_D))^{1/2}}{eN_D} \tag{8.24}$$

Then from (8.23) and (8.24) it may be shown that

$$C_{\text{AC}} = \frac{C_0}{(N_T/N_D)([N_D/(N_D - N_T)][\phi_t/(V_D - V)])^{1/2} + (1 - (N_T/N_D)\phi_t/(V_D - V))^{1/2}} \tag{8.25}$$

Although expressions for the capacitance at low and high frequencies are readily obtained, the treatment for the intermediate-frequency range is difficult, and numerical solutions are the only approach.

The experimental results observed by Sah and Reddi (1964) for a p^+n gold-doped diode (Fig. 8.13) are in good agreement with their model. A substantial difference between the high- and low-frequency capacitance values is observed. The bias dependence expected from the theory is given by

$$\left(\frac{C_{\text{AC}}}{C_{\text{DC}} - C_{\text{AC}}}\right)^2 = \frac{N_D}{N_T}\left(\frac{N_D}{N_T} - 1\right)\left(\frac{V_D - V}{\phi_t} - \frac{N_T}{N_D}\right) \tag{8.26}$$

This predicts a linear dependence of $[C_{\text{AC}}/(C_{\text{DC}} - C_{\text{AC}})]^2$ on $V_D - V$ and

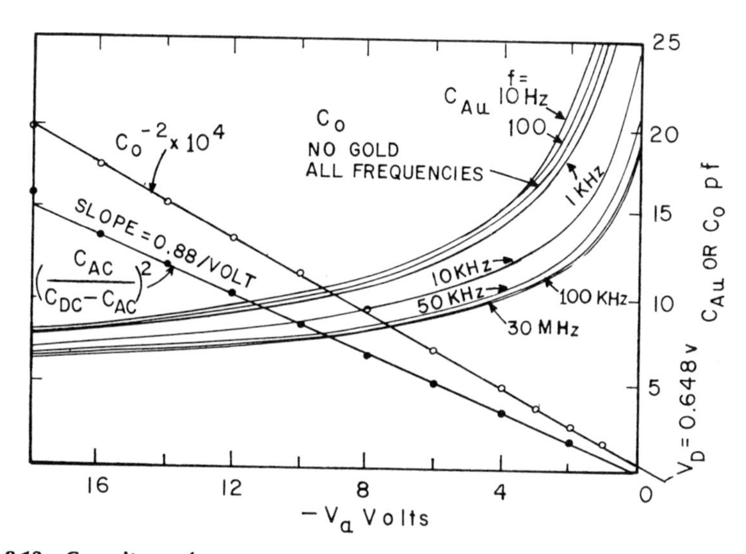

FIG. 8.13 Capacitance-frequency variation for a gold-doped p^+n silicon diode. (After Sah and Reddi, 1964.)

this is seen to be verified in Fig. 8.13, where the 10 and 10^5 Hz capacitance values produced the straight line shown. The gold doping inferred from the slope of the line was 7.3×10^{15} atoms cm^{-3}.

8.3.2 Capacitance-Frequency Dependence where the Deep Impurity Is the Dominant Dopant on One Side of the Junction

In many junctions containing deep impurities it is not unusual for a region to exist in which the deep impurity is the dominating dopant. This is particularly likely to occur in studies of large energy gap semiconductors but may also be observed in silicon if an excess amount of deep impurity is present.

The model considered in this section is that of an n^+p junction in which the p region is formed by a deep acceptor such as indium in silicon, 0.16 eV above E_v (Schibli and Milnes, 1968). At room temperature indium in silicon is nearly all ionized, and the hole density in the bulk represents the doping density of indium atoms. At a low temperature such as 77°K the indium is almost completely un-ionized and the resistivity of the silicon for typical deep doping densities (10^{15} to 10^{16} atoms cm^{-3}) may be of the order of 10^4 Ω-cm (assuming no shallow acceptor is present).

In the model, Fig. 8.14*b*, the p region may be considered to be (a) doped entirely by the deep acceptor, (b) doped by the deep acceptor and a shallow acceptor, or (c) doped by the deep acceptor and partly compensated by a shallow donor. The small-signal capacitance for all these doping conditions may be expected to depend on frequency. The low-frequency capacitance C_{LF} is determined by the sum of deep and shallow impurities, but the high-frequency capacitance C_{HF} is given by the free-carrier density. As a consequence, for a net shallow donor density on the p side in the presence of a higher concentration of deep acceptors, the free-hole density may be quite small. The lower plateau capacitance C_{HF} can then be several decades below the low-frequency capacitance.

For convenience of comparison, some diagrams from the Sah–Reddi model are also included in Fig. 8.14.

8.3.2.1 Equivalent Circuit for the Capacitance

If the junction is sufficiently reverse biased, the holes and electrons are spatially separated and only one carrier species has to be considered at either depletion layer edge. The analysis is further simplified if one side of the junction is much more heavily doped than the other, as in Fig. 8.14*a* or *b*.

The capacitance of a *pn* junction equals that of a parallel plate capacitor with the area of the plates the same as that of the *pn* junction, and with the

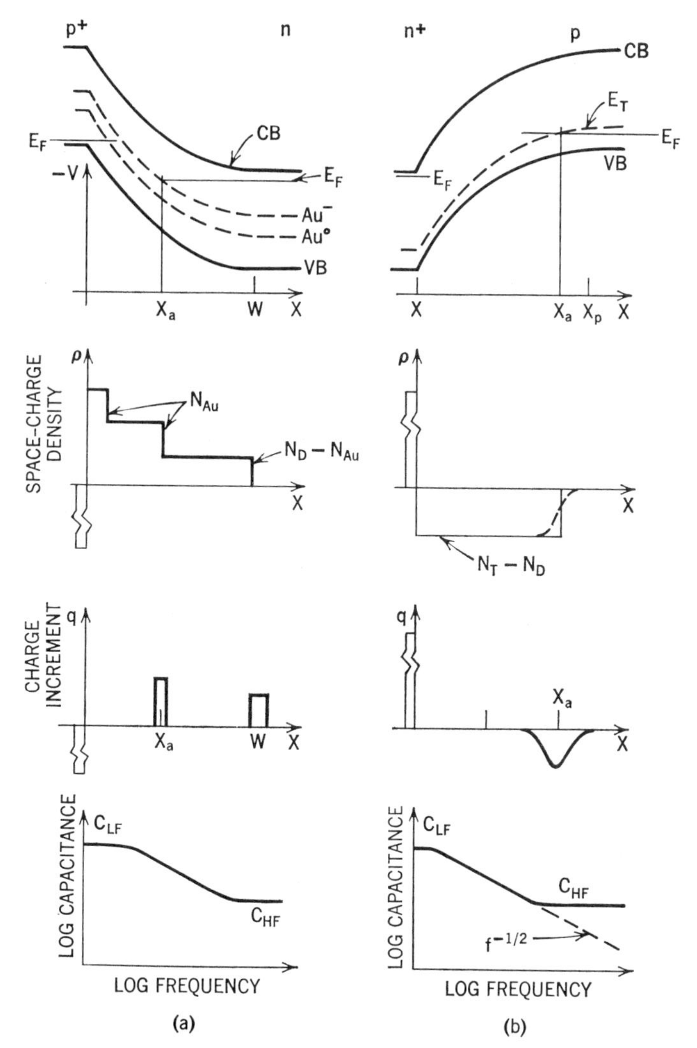

FIG. 8.14 Diagram of reverse-biased *pn* junction containing a deep acceptor, with distribution of space charge ρ, charge increment q, and frequency response of the capacitance.

(a) The Sah-Reddi model for gold in n-type silicon. The Fermi level E_F on the n side intersects the gold level at a large angle, and therefore the charge state of this level changes rather abruptly at x_a. (b) Model typical of indium-doped silicon. The indium level changes the charge state gradually near x_a. (After Schibli and Milnes, 1968.)

dielectric constant that of the semiconductor. The plates are located at the respective centers of the positive and negative incremental space-charge distributions. In the Sah–Reddi model (Fig. 8.14*a*) the incremental charge density q consists of components at x_a and W arising from the gold level and the normal depletion edge. These charge increments are shown diagrammatically as separated components with W remote from x_a, since the gold acceptor level is at about mid band gap and the shallow n-type doping level N_D exceeds N_{Au} considerably.

For the model considered in Fig. 8.14*b*, however, the primary interest is in the capacitance behavior at low temperatures for a specimen in which the deep dopant density substantially exceeds that of the shallow dopants. Therefore the material may be of high resistivity and the Fermi level quite close to the deep dopant level. Under these conditions the free-carrier density is small and the charge increment is more appropriately represented as a single hump of charge (Fig. 8.14*b*) rather than the two charge increments shown at x_a and W in Fig. 8.14*a*.

The energy band and charge diagrams are shown redrawn in Fig. 8.15 to indicate more clearly the $n^+\pi$ configuration under consideration. A

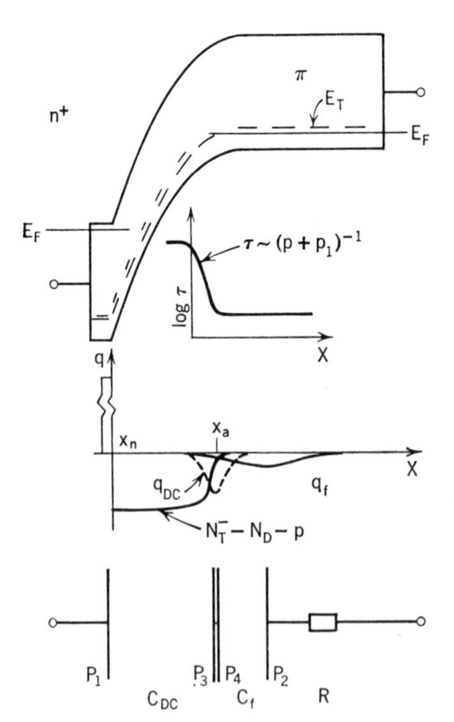

FIG. 8.15 Diagrams for capacitance equivalent circuit of indium-doped silicon $n^+\pi$ junction. *The position of the C_{DC} capacitor plate is given by the low-frequency charge increment q_{DC}. At higher frequencies the charge increment q_f may extend well into the π bulk; the spacing of the C_f capacitor plates is the distance between the centers of q_{DC} and q_f. R is the series resistance of the π bulk, and also includes the losses of C_f. (After Schibli and Milnes, 1968.)*

small-signal alternating voltage superimposed on a direct voltage bias results in a charge increment (dashed line) near x_a at low frequencies (and of course a corresponding charge increment of opposite sign in the n^+ region). At higher frequencies the charge increment moves further from the junction as the indium-level charge state is less able to follow the ac voltage, and the result is the q_f distribution shown by the solid line in Fig. 8.15. The small-signal capacitance at frequency f may then be represented by a parallel plate model with plates P_1 and P_2 located at the effective centers of the charge increment profiles.

This model may be made more useful in the analysis that follows by imagining two connected plates P_3 and P_4 inserted between P_1 and P_2 at the position x_a. Obviously this does not modify the capacitance value C since the charges on P_3 and P_4 will be equal and opposite, and these plates are infinitely closely spaced. In this fashion a series connection of two capacitors, C_{DC} and C_f, reflects the particular space-charge conditions in the device, as explained below.

In Fig. 8.15, q_{DC} (dashed line) is the incremental space-charge distribution due to an applied, small ac signal of very low frequency. Since q_{DC} spreads over much of the transition region, the hole density p and consequently also the time constant of the deep level response

$$\tau = \frac{1}{c_p^-(p + p_1)} \tag{8.27}$$

vary across q_{DC}. The actual space-charge increment q approaches the limiting low-frequency distribution q_{DC} when most of q comes from regions where $f \ll 1/\tau$. The spacing of the C_f plates is then infinitesimal, because the actual q nearly coincides with q_{DC} and the device capacitance C ($C \approx C_{LF}$ at these low frequencies) approaches C_{DC}.

At higher frequencies the response of the deep level near x_a decreases because $f > 1/\tau$ at x_a. Therefore, some of the space-charge increment q will come from regions with a faster deep level response, that is, smaller τ. Consequently the center of the space-charge distribution q will be farther to the right where the hole density is greater (Fig. 8.15), and the capacitance decreases.

The time constant τ is smallest in the π bulk region where the hole density is at its thermal value p_0. At frequencies f in excess of

$$\tau^{-1} = c_p^-(p_0 + p_1) \tag{8.28}$$

the incremental space charge extends deep into the π bulk region where the time average of the space charge is zero, as indicated in Fig. 8.15 by the spread of q_f into the π section where the energy bands are drawn horizontally.

The center of q_f is relatively far to the right, and consequently the capacitance decreases substantially at these frequencies. Of course, the distribution q always obeys Poisson's equation subject to suitable constraints.

The dispersion considered here is a consequence of the slow deep level response and is relatively independent of the dc bias, that is, of the distance $x_a - x_n$. On the other hand, the position x_a of the incremental space-charge distribution at low frequencies q_{DC} depends on the dc bias voltage. The effects of bias and frequency are thus separated. This is reflected in the equivalent circuit representation for the junction capacitance in the form of a series connection of two elements, C_{DC} and C_f, with C_{DC} depending on bias, but not on frequency, while the frequency-dependent element C_f is independent of bias.

The calculation of the frequency dependence is not simple for the general case. However, once this is done for one value of reverse bias (or C_{DC}), then the frequency dependence of the junction capacitance at another value of reverse bias (or C_{DC}) can be predicted using the model proposed here.

8.3.2.2 Analytical Treatment for the n^+p Junction Model

The first part of Fig. 8.14*b* represents the band diagram for the n^+p model under reverse bias voltage. The space-charge density ρ is shown in the second part of Fig. 8.14*b*, and the dashed curve portion indicates the gradual dispersion of space charge at the depletion edge in place of the usually assumed abrupt edge. The third part of Fig. 8.14*b* represents the incremental charge density change when a voltage increase ΔV is applied in the reverse direction. The voltage increment, from Poisson's equation, may be expressed as

$$\Delta V = e \int_{x_n{}^+}^{\infty} \frac{q}{\varepsilon} (x - x_n{}^+)\, dx \tag{8.29}$$

and the charge increment is

$$\Delta Q = e \int_{x_n{}^+}^{\infty} q\, dx \tag{8.30}$$

where $x_n{}^+$ indicates the lower limit of integration as just to the right of the junction at x_n. The small-signal capacitance is then given by

$$C = \frac{\Delta Q}{\Delta V} = \frac{\varepsilon \int_{x_n{}^+}^{\infty} q\, dx}{\int_{x_n{}^+}^{\infty} (x - x_n{}^+) q\, dx} \tag{8.31}$$

To study the small-signal capacitances it is therefore necessary to determine q, the incremental charge density, as a function of position, temperature, and frequency.

The deep impurity in this model may be either neutral or negatively charged, with densities designated as N_T^0 and N_T^-. Shallow acceptors or donors, if present, do not change their charge state with ΔV, provided the Fermi level is not close to the shallow energy levels. The incremental charge balance equation is therefore

$$q = \Delta p - \Delta n - \Delta N_T^- \tag{8.32}$$

or, as an approximation,

$$q = \Delta p - \Delta N_T^- \tag{8.33}$$

since $\Delta n \ll \Delta p$ in p-type material. The rate equation for the trap level is

$$\frac{dN_T^-}{dt} = -c_p^- p N_T^- + c_p^- p_1 N_T^0 \tag{8.34}$$

which gives for equilibrium

$$N_T^- = \frac{p_1 N_T}{p + p_1} \tag{8.35}$$

and hence for sufficiently small signals

$$\Delta N_{T\,\max}^- = \frac{-p_1 \Delta p N_T}{(p + p_1)^2} \tag{8.36}$$

If a single time constant τ is adequate, then the response of the deep acceptor level to a small sinusoidal input may be represented by

$$\Delta N_T^- = \Delta N_{T\,\max}^- \frac{1 - j\omega\tau}{1 + \omega^2\tau^2}$$

$$q = \Delta p \left[1 + \frac{p_1 N_T}{(p + p_1)^2} \times \frac{1 - j\omega\tau}{1 + \omega^2\tau^2} \right] \tag{8.37}$$

where Δp, p, τ, and N_T depend on position x; j is the imaginary unit; and $\omega = 2\pi f$ with f the signal frequency. In the p-type material considered here, $\tau = 1/[c_p^-(p + p_1)]$ since $c_n^0(n + n_1) \ll c_p^-(p + p_1)$. Capacitance and loss angle are obtained by substituting (8.37) into (8.31) and separating real and imaginary parts, with the result

$$C_p - jG_p \frac{1}{\omega} = \frac{\varepsilon(I_1 - j(\omega/\omega_1)I_2)}{I_3 - j(\omega/\omega_1)I_4} \tag{8.38}$$

where C_p and G_p are the parallel equivalent circuit capacitance and conductance.

In (8.38), $\omega_1 = c_p^- p_1$ and

$$I_1 = p_1 \int_{x_n^+}^{\infty} \frac{\Delta p}{p_1} \left(1 + \frac{N_T/p_1}{(1 + p/p_1)^2 + (\omega/\omega_1)^2}\right) dx \tag{8.39}$$

$$I_2 = p_1 \int_{x_n^+}^{\infty} \frac{\Delta p}{p_1} \left(\frac{N_T/p_1}{(1 + p/p_1)^2 + (\omega/\omega_1)^2}\right)\left(\frac{1}{(1 + p/p_1)}\right) dx \tag{8.40}$$

$$I_3 = p_1 \int_{x_n^+}^{\infty} \frac{\Delta p}{p_1} (x - x_n^+) \left(1 + \frac{N_T/p_1}{(1 + p/p_1)^2 + (\omega/\omega_1)^2}\right) dx \tag{8.41}$$

$$I_4 = p_1 \int_{x_n^+}^{\infty} \frac{\Delta p}{p_1} (x - x_n^+) \left(\frac{N_T/p_1}{(1 + p/p_1)^2 + (\omega/\omega_1)^2}\right)\left(\frac{1}{(1 + p/p_1)}\right) dx \tag{8.42}$$

Equations 8.38 to 8.42 involve p as a function of x and computer solutions are too cumbersome to be practical unless simplifying assumptions are made that provide $p(x)$.

Trap Density Less than Density of Shallow Level. For frequencies very much less than the dielectric relaxation frequency ($f \ll e\mu_p p/\varepsilon$), the holes near x_p (Fig. 8.14) are in quasiequilibrium and obey Boltzmann statistics. Since the trap density is small ($N_T \ll N_A$, N_D), their delayed response does not alter the shape of the space-charge distribution in the transition region significantly. Thus, to a first approximation, this space-charge distribution moves back and forth with amplitude Δx. Under these conditions, the change in hole density can be written as

$$\Delta p = \frac{dp}{dx} \Delta x \tag{8.43}$$

where dp/dx is computed from dc conditions and Δx is independent of position. Boltzmann statistics provide the relationship

$$p = p_0 \exp\left(-\frac{eV}{kT}\right) \tag{8.44}$$

where V, which is a function of x, is the electrostatic potential of the p-doped region, and is taken as zero at x_p. Consequently, p_0 is the equilibrium hole density at x_p where the space charge is practically zero. Introduction of the symbol v for $\exp(-eV/kT)$ at this stage allows the following expression

$$\left(\frac{dv}{dx}\right)^2 = v^2 \left(\frac{2e^2}{kT\varepsilon} \left\{p_0(v - 1) - N_A \ln(v) - N_T \left[\ln(v) - \ln\left(\frac{p_0 v + p_1}{p_0 + p_1}\right)\right]\right\}\right) \tag{8.45}$$

to be derived from Poisson's equation (Schibli and Milnes, 1968).

If the doping densities N_A and N_T are position independent, then (8.45) may be readily evaluated numerically for v, and hence for V and p as functions of x. The boundary condition needed is that $dv/dx = 0$ for $x \geq x_p$ since the semiconductor is neutral outside the depletion region. The result of the integration, p as a function of x, allows the capacitance frequency dependence to be calculated from (8.38) to (8.42).

Capacitance C_p and loss angle δ have been computed numerically as a function of frequency f (Fig. 8.16*a*) for indium-doped silicon at 77°K with

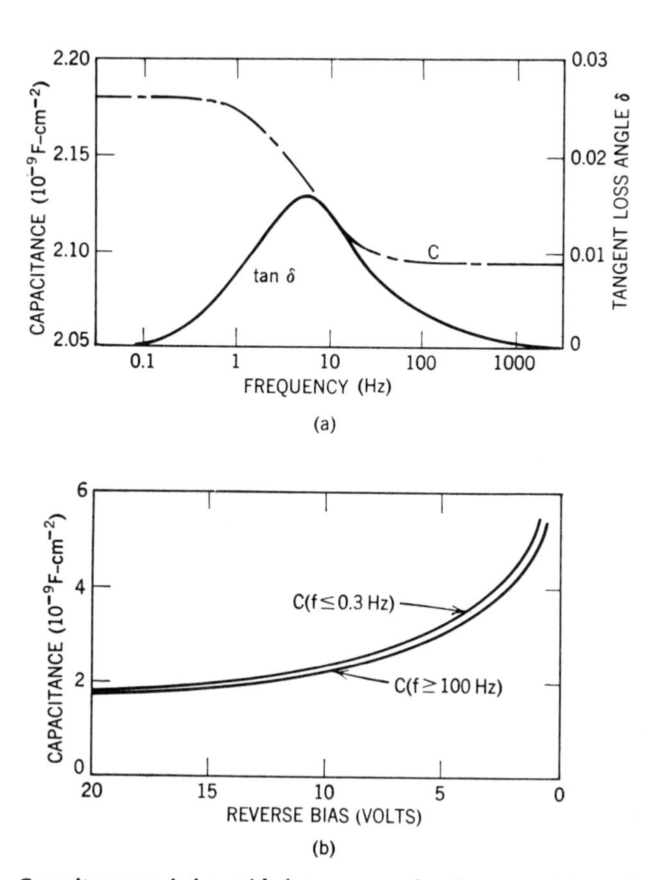

FIG. 8.16 Capacitance variation with frequency and with reverse bias voltage. *(a) Computed frequency dependence of indium-doped (2×10^{15} cm^{-3}) silicon n^+p step junction at 77°K, with a shallow acceptor density of 10^{16} cm^{-3}. (b) Theoretical capacitance versus reverse bias at low and high frequencies. The same parameters were used as in (a). The low-frequency capacitance is proportional to $V^{-1/2}$. The relative change of capacitance with frequency is only a few percent because the indium density is relatively small and its level is not very deep. The frequency effect becomes weaker at larger reverse bias. (After Schibli and Milnes, 1968.)*

a shallow acceptor density $N_A = 10^{16}$ cm^{-3}, an indium density $N_T = 2 \times 10^{15}$ cm^{-3}, and the hole capture cross section of indium $\sigma_p^- = 7 \times 10^{-14}$ cm^2. The capacitance changes with frequency by about 5% in this particular example, which is for 12.5 V reverse bias, and may be represented by the equivalent circuit elements $C_{DC} = 2180$ pF cm^{-2} at 12.5 V reverse bias, and C_f decreasing from infinity below 0.3 Hz to 55,000 pF cm^{-2} above 100 Hz. The reverse bias dependence $C(V)$ has been computed (Fig. 8.16*b*) by varying C_{DC}, which obeys a $V^{-1/2}$ relation in the case of a step junction. The frequency dependence becomes relatively weaker at larger reverse bias, an observation similar to Sah and Reddi's on gold-doped p^+n junctions.

The magnitude of the relative capacitance change is quite small for the case considered, where the shallow doping exceeds the deep level density. The effect, however, would be larger for deeper levels than the $E_v + 0.16$ eV indium level.

Two observations are of interest at this point. First, the frequency dependence computed so far is due to the delayed deep level response on the *p*-side depletion edge; similar calculations could also be done for the space-charge region. This, however, is not necessary in this particular case of indium-doped silicon at 77°K, since the indium level is fully ionized in the *n* bulk and throughout the depletion region and hence does not change the charge state there.

Secondly, this analysis assumed (equation 8.43) that the space-charge distribution at the edge of the depletion region follows the applied small ac voltage without much distortion. This condition is best satisfied when only a relatively small number of slow deep levels is present. It is also fulfilled approximately when the trap density is greater than the shallow density, provided the frequency is so low that the capacitance is close to its low-frequency value. The high-frequency approximation for the condition of a deep level density in excess of a shallow compensating one is considered in the next section.

Deep Level Density in Excess of the Shallow Compensating Density. The type of semiconductor doping considered here results in semiinsulating material at low temperatures. As an example consider *p*-type silicon containing $N_T = 2$ to 4×10^{16} cm^{-3} indium and $N_D = 4 \times 10^{13}$ cm^{-3} phosphorus. The free-hole density p is a few times 10^{10} cm^{-3} for this material at liquid-nitrogen temperatures, and the resistivity is then about 10^4 Ω-cm. At 300°K, however, much of the indium is ionized and the hole density may exceed 10^{16} cm^{-3}. A *pn* junction on such material has some unusual capacitance temperature and frequency properties.

In the depletion region of such a junction, under reverse bias conditions, the indium level becomes ionized by thermal generation and a sweepout process, as discussed earlier. Therefore the depletion width and hence the low-frequency junction capacitance are the same as would be obtained on a p region containing 2 to 4 $\times$ 10^{16} cm^{-3} shallow acceptors. The capacitance, of course, is about 1000 times that observed on a few times 10^{10} cm^{-3} net shallow acceptors. (This effect of deep level doping of course accounts for the large capacitances sometimes observed for high-resistivity semiconductor materials.)

The response of the indium level to the ac signal decreases at higher frequencies because the rates of transition of holes between the indium level and the valence band are rather slow. As a consequence, the junction capacitance decreases at frequencies greater than

$$\frac{1}{2\pi\tau} = \frac{c_p^-(p + p_1)}{2\pi}$$

where τ is the time constant of the deep level response in the p bulk.

For high-resistivity bulk conditions the decrease of capacitance with frequency is very substantial. This indicates that the distributed space-charge increment q_f produced by the alternating voltage of frequencies above $1/(2\pi\tau)$ extends deep into the bulk region as shown in Fig. 8.15. Under these conditions the contribution to the total incremental space charge from the narrow transition region is insignificant. The charge distribution q_f is of course described by the rate equation for the deep level and by Poisson's equation.

The ac impedance of this junction can be represented by a lossy capacitance $\bar{C}$, the losses being due to the delayed trap response. The bar over a quantity indicates a complex value. From Schibli and Milnes (1968)

$$\bar{C} = \frac{\varepsilon}{(x_a - x_n) + (1/\bar{a})} = \frac{1}{1/C_{\text{DC}} + 1/\bar{C}_f} \tag{8.46}$$

where $\bar{a}$ is given by

$$\bar{a} = \left\{\frac{e^2}{kT\varepsilon} p \left[1 + N_T \frac{p_1}{(p + p_1)^2} \frac{1 - j\omega\tau}{1 + \omega^2\tau^2}\right]\right\}^{1/2} \tag{8.47}$$

C_{DC} is the bias-dependent equivalent capacitance, and $\bar{C}_f$ is the frequency-dependent part (Fig. 8.15); $\bar{C}_f$ can be represented by its equivalent series elements, C_f and R_s, given by

$$\bar{C}_f^{-1} = C_f^{-1} + j\omega R_s = (\bar{a}\varepsilon)^{-1}$$

or

$$C_f^{-1} = \varepsilon^{-1} \operatorname{Re} (\bar{a})^{-1}$$

and

$$\tan \delta = \frac{\operatorname{Im} (\bar{a})}{\operatorname{Re} (\bar{a})} \tag{8.48}$$

which becomes, by substituting (8.47) for $\bar{a}$,

$$C_f = \left(\frac{p\varepsilon e^2}{kT}\right)^{1/2} \left\{ \mathrm{Re}\left[\left(1 + N_T \frac{p_1}{(p+p_1)^2} \times \frac{1 - j\omega\tau}{1 + \omega^2\tau^2}\right)^{-1/2}\right]\right\}^{-1} \tag{8.49}$$

$$\tan\delta \approx \tan\left[\tfrac{1}{2}\tan^{-1}(\omega\tau)\right] \tag{8.50}$$

where Re and Im indicate real and imaginary parts, respectively. Equations 8.49 and 8.50 are asymptotic expressions, valid only for $\tau^{-1} \ll \omega \ll \tau_{\mathrm{rel}}^{-1}$, with $\tau = [c_p^-(p + p_1)]^{-1}$. Since most of the space-charge increment comes from $x > x_p$, the appropriate value for p in (8.49) is the thermal hole density. The term $\tau_{\mathrm{rel}} = \varepsilon/(e\mu_p p)$ is the dielectric relaxation time. The loss angle δ of the frequency-dependent $\bar{C}_f$ approaches $\frac{1}{4}\pi$, but the loss angle of the actual device capacitance is somewhat less, because C_{DC} is essentially lossless.

Since (8.49) is valid only for $\omega\tau \gg 1$, it becomes, neglecting the free-carrier density,

$$C_f = \left(\frac{e^2\varepsilon}{kT}\frac{2pN_T}{(p+p_1)}\frac{\omega_1}{\omega}\right)^{1/2} \tag{8.51}$$

where $c_p^-(p + p_1)$ has been substituted for τ^{-1}, and $\omega_1 = c_p^- p_1$.

With the further assumption that $p \gg p_1$, which is almost always true, C_f becomes

$$C_f = \left(\frac{2e^2\varepsilon}{kT} N_T \frac{\omega_1}{\omega}\right)^{1/2} \tag{8.52}$$

It is seen by inspecting (8.47) that under the assumptions made above, a is independent of the free-hole density. Thus, even if a gradient in the shallow donor density is present, $\bar{a}$ remains independent of position, and therefore (8.52) is also true if the shallow compensating donor density is a function of position.

8.3.2.3 Temperature and Bias Dependence of the Break Frequency

The analyses in the previous section indicate that the capacitance versus frequency plots of *pn* junctions with deep levels contain characteristic breaks. These break points should in principle permit determination of various parameters of the deep dopant.

For small trap densities the frequency effect is small (a few percent) and no closed-form solution for the frequency dependence of the capacitance has been worked out. However, if the deep level impurity is dominant, the frequency effect is large and a convenient analytical expression is available (equation 8.52). The intersection of the low-frequency value $C_{\mathrm{LF}} \approx C_{\mathrm{DC}}$ and the straight-line high-frequency approximation C_f (equation 8.52), as

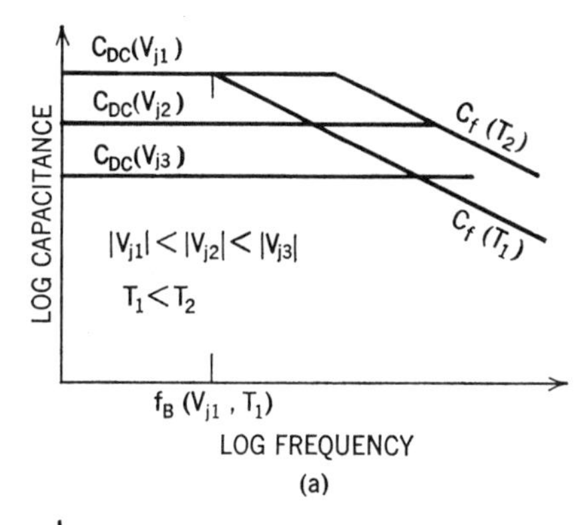

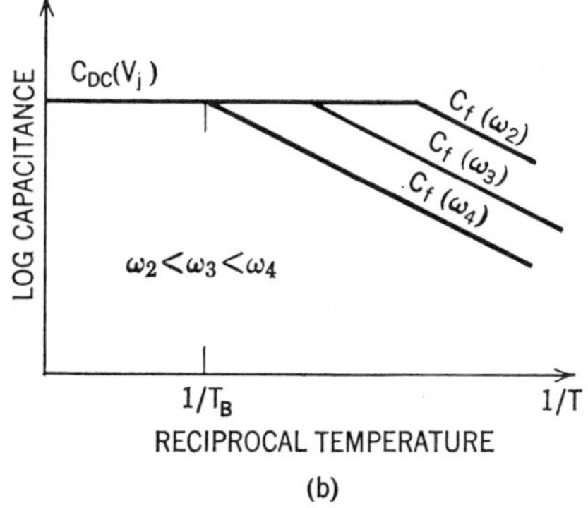

FIG. 8.17 Straight-line approximations to the equivalent elements C_f and C_{DC}. C_f is independent of dc bias.
(a) The frequency f_B at the break point increases as C_{DC} is lowered by increasing the reverse bias. (b) The break point $1/T_B$ moves toward higher temperatures as the frequency is increased. The slope is essentially half the activation energy of the deep level. (After Schibli and Milnes, 1968.)

shown in Fig. 8.17, is at

$$f_B = \frac{2e^2\varepsilon}{kT} N_T \frac{\omega_1}{2\pi C_{DC}{}^2} \tag{8.53}$$

Thus, if $C_{DC}{}^2 = \varepsilon N_T e/(2V_j)$ as for step functions, then

$$\frac{\omega_B}{\omega_1} = \frac{4\,V_j}{(kT/e)} \tag{8.54}$$

where V_j is the total dc voltage across the junction. Hence from the break frequency ω_B, it is possible to determine ω_1 $(=c_p{}^- p_1)$. If the energy level of the deep acceptor impurity is known, then p_1 can be calculated, and therefore the break frequency gives the capture cross section $\sigma_p{}^-$ (corresponding to the process $N_T{}^- + h \rightarrow N_T{}^0$).

The variation of capacitance with temperature also follows from the equations that have been derived. This variation is shown in a linearly idealized form in Fig. 8.17*b*. The break on a log C versus $1/T$ plot occurs at a temperature T_B such that

$$\frac{E_T}{kT_B} = \ln\left(\frac{2N_v N_T e\varepsilon c_p^-}{\omega C_{DC}^2 g} \frac{e}{kT_B}\right) \tag{8.55}$$

or, for step junctions,

$$\frac{E_T}{kT_B} = \ln\left(\frac{4c_p^- N_v}{\omega g} \frac{eV_j}{kT_B}\right) \tag{8.56}$$

where g is the degeneracy factor of the deep level.

The straight-line slope in the linear approximation of Fig. 8.17*b* represents the temperature dependence of the capacitance C_f at low temperatures. Substituting $c_p^- p_1$ for ω_1 in (8.52) where

$$p_1 = \frac{N_v}{g} \exp\left(-\frac{eE_T}{kT}\right)$$

gives the expression

$$C_f = \left[\frac{2e^2\varepsilon}{kT} N_T \frac{c_p^- N_v}{g\omega} \exp\left(-\frac{eE_T}{kT}\right)\right]^{1/2} \tag{8.57}$$

Hence if the capture cross section σ_p^- is relatively temperature independent, the slope of C versus $1/T$, assuming C_f dominates over C_{DC}, corresponds to half the activation energy of the deep impurity. (Schibli and Milnes, 1968).

8.3.3 Schottky Barrier Capacitance-Frequency Analysis

A closed-form expression for the low-frequency capacitance-voltage relationship of a Schottky barrier with an arbitrary distribution of deep energy levels has been given by Roberts and Crowell (1970). This has been extended by Crowell and Nakano (1972) in the form of a universal function which predicts the change in elastance between two frequencies at any bias and temperature. This leads to a method of using measurements at a single temperature to deduce the deep and shallow level impurity concentrations, the capture cross section, and the impurity energy level of the deep level.

Figure 8.18 is a schematic energy band diagram for an n-type Schottky barrier diode with one or more completely ionized shallow donor levels and one deep donor level. The deep level is completely ionized near the surface of the semiconductor junction and partially ionized in the bulk. In the diagram $e\phi$ is the barrier energy and $e\Psi(x)$ is the energy of the conduction band (measured from the Fermi energy level E_F) at x, the distance from the metal-semiconductor interface; $e\Psi_f$ and $e\Psi_s$ are values of $e\Psi(x)$ measured far in the bulk and at the metal-semiconductor interface,

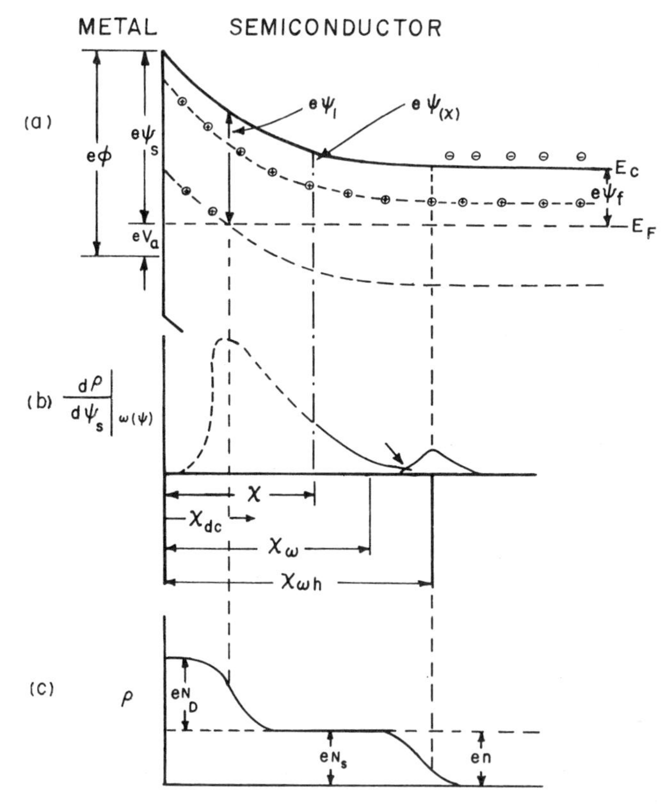

FIG. 8.18 Energy and charge diagrams for a Schottky barrier containing deep and shallow impurities.

(a) Electron energy band diagram. (b) Increment of charge density with voltage: for low-frequency voltage, solid plus dotted line; for higher frequency voltage, solid line only. (c) Total charge distribution within the transition region. (After Crowell and Nakano, 1972.)

respectively. The energy level of the deep impurity with respect to the conduction band is $e\Psi_i$; V_a is the applied bias. It is assumed that there is no gradient in the electron imref even when an ac voltage is applied. Nearly zero imref gradient over most of the depletion region is reasonable for forward and small reverse bias voltages. In most cases, even for appreciable reverse bias, the imref should be constant throughout the regions in which deep levels change their state of charge.

The free-electron concentration is given by the Boltzmann approximation

$$n = N_c \exp\left[-\frac{e\Psi(x)}{kT}\right] \tag{8.58}$$

where N_c is the effective density of states, k is the Boltzmann constant, and T is the temperature. The validity of Shockley–Read recombination statistics is assumed and only electron trapping and reemission processes are considered. At a given point x [where $\Psi = \Psi(x)$] there is a frequency $\omega(\Psi)$ such that $\omega[\Psi(x)]\tau \leq 1$ for all impurities between x and the bulk semiconductor (here τ is the response time of the deep level) and $\omega[\Psi(x)]\tau \geq 1$ for all impurities between x and the metal. This frequency is given by the relationship

$$\omega(\Psi) = \sigma_n v[n(\Psi) + n_1] \tag{8.59}$$

where σ_n is the capture cross section for electrons, v is the electron thermal velocity, $n(\Psi)$ is the free-electron concentration at the potential energy $e\Psi$, and n_1 is the free-electron concentration at $n(\Psi_1)$. When $\omega\tau \gg 1$, the ac stored charge contribution is out of phase with the ac voltage and the deep level contribution is conductive. When $\omega\tau \ll 1$, the stored charge on the deep level is essentially in phase with the ac bias and the deep level contribution is capacitive. Accordingly, an abrupt transition in this frequency response may be assumed and the conductance contributions neglected. Hence at the frequency $\omega[\Psi(x)]$ only the charges beyond x respond in phase with the ac voltage and charges between the metal-semiconductor interface and x do not respond. Figure 8.18*b* shows the derivative of the charge with respect to the surface potential versus depletion width. The first peak is due to impurity ions and the second peak is due to free electrons near the bulk. The charge distribution itself is shown in Fig. 8.18*c*.

If $\Psi_s - \Psi_1 \gg kT/e$, the elastance difference ΔS for two frequencies is independent of Ψ_s and is found to be

$$\Delta S(\omega, 0) \equiv \frac{1}{C(\omega, \Psi_s)} - \frac{1}{C(0, \Psi_s)} = \int_{\infty}^{\Psi(\omega)} \frac{\sigma(\Psi)}{\varepsilon\rho 2(\Psi)} \frac{d\rho}{d\Psi}\, d\Psi \tag{8.60}$$

The expression for the charge density $\rho(\Psi)$ is

$$\rho(\Psi) = eN_D\left[(R + f_i) - (R + f_F)\exp\left(-\frac{e(\Psi - \Psi_f)}{kT}\right)\right] \tag{8.61}$$

where

$$R \equiv \frac{N_s}{N_D} \tag{8.62}$$

N_s and N_D are the shallow and deep donor concentrations, respectively, and R is the ratio of free-electron concentration to deep level concentration when the deep level is occupied by electrons (for an n-type semiconductor).

$$f_i = f_i(\Psi) = \left[1 + \exp\left(-\frac{e(\Psi - \Psi_i)}{kT}\right)\right]^{-1} \tag{8.63}$$

where f_i is the probability of ionization of the deep level at $x(\Psi)$, and

$$f_F = f_i(\Psi_f) \tag{8.64}$$

f_F is the probability of ionization of the deep level in the bulk. It is convenient to define a new parameter F which can be expressed as

$$F \equiv \frac{e(\Psi_i - \Psi_f)}{kT} \tag{8.65}$$

The larger F is, the greater will be the probability of freezeout of carriers in the bulk.

In normalized form the elastance difference in (8.60) becomes

$$\begin{aligned} C_D \Delta S(\omega, 0) &\equiv \frac{C_D}{C(\omega, \Psi_s)} - \frac{C_D}{C(0, \Psi_s)} \\ &= 2^{1/2} \int_{\infty}^{Y(\omega/\omega_1)} \\ &\times \frac{[\ln(f_F/f_i\lambda) - R\ln\lambda + (R + f_F)(\lambda - 1)]^{1/2}(1 - f_i)f_i}{[f_i + R - (R + f_F)\lambda]^2} \, dY \end{aligned} \tag{8.66}$$

where C_D, a normalizing capacitance, is

$$C_D \equiv \left[\frac{\varepsilon e^2(N_s + N_D)}{kT}\right]^{1/2} \tag{8.67}$$

$$\lambda \equiv \frac{1 - 1/f_i}{1 - 1/f_F} \tag{8.68}$$

$$Y \equiv \frac{e(\Psi(\omega) - \Psi_i)}{kT} \tag{8.69}$$

and

$$\omega_1 = \sigma_n v_{th} n_1 \tag{8.70}$$

C_D is the flat-band capacitance when all the donors are ionized. The flat-band capacitance is that associated with the Debye length for screening in the bulk semiconductor with all donors ionized. Thus the normalized elastance difference (8.66) is a function of shallow to deep donor ratio R, the value of F associated with the degree of bulk freezeout, and the frequency in units of ω_1.

If a deep acceptor level of concentration N_A exists instead of a deep donor level, since the acceptor is neutral when not occupied by an electron, C_D

becomes

$$C_D = \left[\frac{\varepsilon e^2 N_s}{kT}\right]^{1/2} \tag{8.71}$$

and R becomes

$$R = \frac{N_s}{N_A} - 1 \tag{8.72}$$

If the shallow impurity concentration is overcompensated, the barrier must be considered as a p-type Schottky barrier.

Numerically calculated values of the normalized elastance difference are shown in Fig. 8.19 as a function of ω/ω_1 for selected values of R and F. For

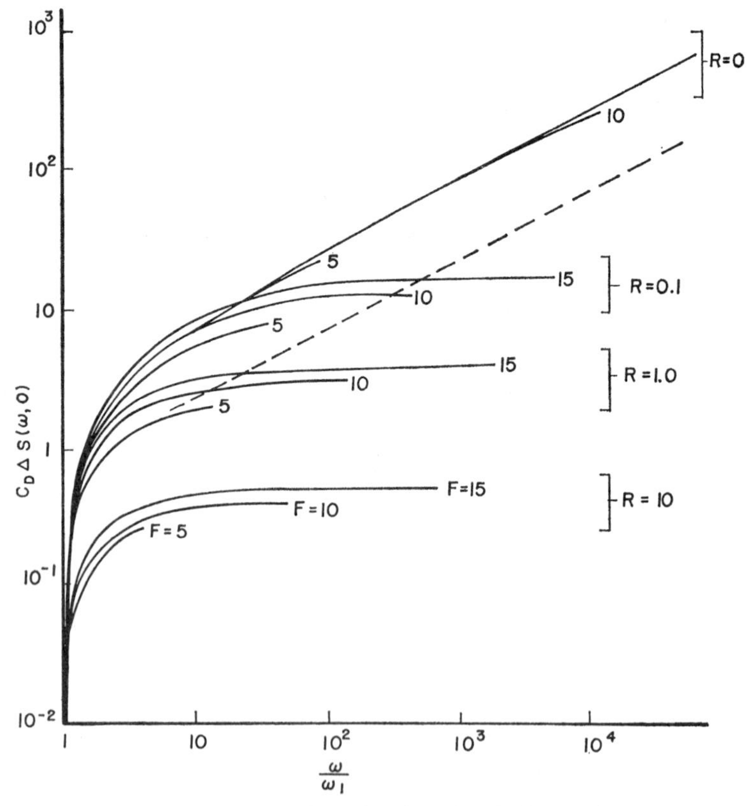

FIG. 8.19 Normalized elastance difference of a Schottky barrier between frequencies ω_1 and ω versus ω in units of ω_1.
The curves are for selected degrees of freezeout F and ratio of shallow to deep impurity concentrations R. The dashed line gives the asymptotic results of Schibli and Milnes (1968) for no shallow donor present and is seen to be a factor of 4 lower. (After Crowell and Nakano, 1972.)

frequencies below ω_1 the predicted elastance difference is zero because charges on all the impurity levels are assumed to respond in phase with the driving voltage. This introduces an abrupt low-frequency transition in barrier capacitance which stylizes the analysis.

The dashed line shown in Fig. 8.19 is an asymptotic result derived by Schibli and Milnes (1968) for $N_s = 0$. It is interesting to note that in this asymptotic region their result differs from the Crowell–Nakano elastance result. Since $\ln \Delta S \propto \ln \omega^2$ in this region, if capture cross sections are to be deduced the Schibli–Milnes result will yield capture cross sections that are smaller by a factor of 16. This scale factor exists in spite of the fact that the calculation assumes a range of time constants of a factor of 2 exp F and the Schibli and Milnes result involves a single time constant associated with an equivalent parallel capacitance of a single series RC circuit.

By comparing experimental and theoretical capacitance-frequency curves for a single temperature, the density, energy level, and capture cross section of the deep impurity can be determined, in principle, although the accuracy is problematic.

In more recent work, Crowell has concentrated on formulating a model that is sufficiently precise to be relevant yet simple enough to assist the experimentalist in developing insight into the problem. He has developed a differential equation approach for $p-n$ junctions or Schottky barriers that gives the complex capacitance as a function of depletion layer width without the necessity of following the separate development of current and voltage. This approach provides a simple intuitive result, backed up by a more exact computer analysis. The theory yields lumped constant circuit elements in which ratios of capacitances equate to ratios of concentrations of impurity species, and time constants are the actual trapping time constants of the deep levels. The treatment suggests the desirability of measuring C^{-2} versus V with precision as far into the forward bias region as is practically possible. With a phase-sensitive system and suitable calibration procedures, meaningful capacitance measurements can be made on diodes with Q factors as low as 0.01. Crowell is presently applying the study to In-doped silicon in a Hf-Si Schottky barrier system.*

The frequency dependence of GaAs Schottky barrier capacitances, with a not too deep donor, has been studied by Hesse and Strack (1972).

8.3.4 Other Aspects of Capacitance Deep Impurity Effects

From the studies developed so far, it is clear that caution must be exercised in interpreting capacitance measurements in semiconductor structures if

* M. Beguwala and C. R. Crowell, to be published, 1973.

there is the possibility that deep impurities are present. Even if there is no obvious reason to expect deep impurity effects, it is wise to make capacitance measurements over wide frequency and temperature ranges to establish that normal shallow impurity behavior is indeed being observed.

Copeland (1969) has devised a technique for directly plotting the inverse doping profile of semiconductor wafers. This involves driving a Schottky diode deposited on the surface with a small constant 5-MHz current and varying the depth of the depletion layer by changing the dc bias. The inverse doping profile $n^{-1}(x)$ is obtained by monitoring the voltage across the diode at the second harmonic frequency, which is proportional to n^{-1}, and at the fundamental frequency, which is proportional to x. If deep impurities are present, there is obviously a possibility that they will disturb such a plotter.

Low-frequency second harmonic measurements have been used by Schibli (1972) to determine impurity profiles in indium-doped silicon n^+p junctions.

Zohta (1970 and 1972) and Zohta and Ohmura (1972) have shown that in the presence of deep centers not only the capacitance C itself but also $\Delta V/\Delta(C^{-2})$ is dependent on the signal frequency of measurement. These studies have been applied to examination of gold-doped silicon p^+n junctions. Gold levels in silicon have also been studied in MOS structures (Sixou and Nuzillat, 1972).

Amsterdam (1970) has reported that when Schottky barriers are used to evaluate lightly doped n-type (10^{15} cm^{-3}) epitaxial GaAs on n^+ substrates, the diodes do not always have the expected varactor characteristics. The capacitance variation with bias is sometimes too small and the voltage intercept of the $1/C^2$ versus V plot too high. Amsterdam is able to interpret the results in terms of traps in the epilayer. Impedance effects in GaAs p-n junctions with silicon doping in the p region have been studied by Dmitriev et al. (1972).

Capacitance energy level spectroscopy of deep-lying semiconductor impurities using Schottky barriers has been studied by Roberts and Crowell (1970); Wronski, (1970), and Gol'dberg and Tsarenkov (1972).

In general deep impurity effects on capacitance are undesirable. However, Atalla and Kahng in U.S. Patent No. 3,176,151 (1965) describe a varactor diode with deep-lying impurities that exhibits a high figure of merit.

A further example of the unusual capacitance effects that can be observed are the measurements of Kroko (1966) on bulk water-clear, compensated, single-crystal silicon carbide over the frequency range 1 to 100 MHz and the temperature range 77 to 750°K. These show frequency and temperature dependences that still await interpretation.

8.4 Capacitance and Reverse Current Transients in Depletion Regions

When a voltage step is applied to a reverse-biased junction, capacitance and current transients are obtained. If the depletion region contains a deep impurity and if the temperature is low, the transients may be slow enough for convenient measurement. Information therefore can be obtained on the thermal emission rate terms (e_n, e_p) of the impurity by such experiments.

In gold-doped silicon diodes the initial capacitance corresponding to a reverse bias is governed by the forward current flowing in the diode prior to the switching to the reverse bias (L. S. Berman, 1972).

Experiments of this nature may have many variations since the voltage cycle may have several forms and the charge state of the impurities may be changed by photoinjection, current injection, or the temperature of the specimen. Measurements of this kind have been developed systematically by Sah and his coworkers for a number of years.

In an initial experiment Sah et al. (1967) applied a square wave of

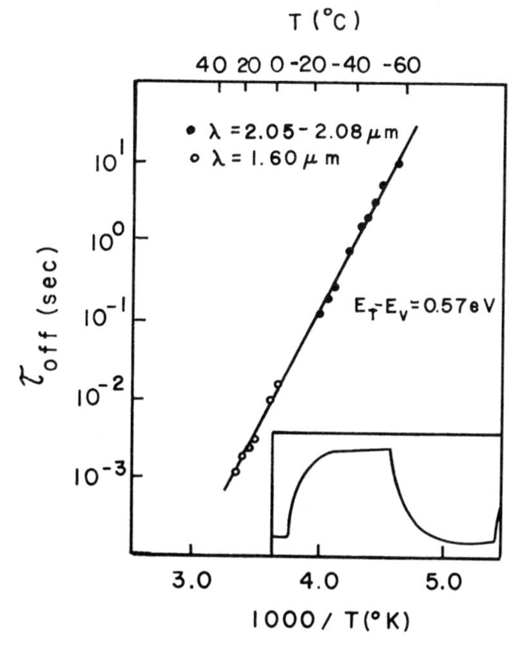

FIG. 8.20 The measured decay time constant for a gold doped silicon diode as a function of inverse temperature.

The insert shows the observed photocurrent that results when mechanically chopped monochromatic light is incident on the junction. (After Sah et al., 1967.)

monochromatic light to a reverse-biased p^+n silicon junction containing gold. The turn-off time for the photocurrent transient varied with temperature with an activation energy of 0.57 eV, as expected for the gold acceptor level (Fig. 8.20). The model for the experiment gives $t_{\text{off}} = (e_n + e_p)^{-1}$. However, Senechal (1968) has pointed out some difficulties of interpretation in such an experiment.

More recently Sah et al. (1969a, b) have resolved these problems by dark current, dark capacitance, and photocapacitance transient methods to obtain emission rates for holes and electrons at the gold acceptor level and for holes at the gold donor level. Figure 8.21 shows that the emission rate variation with temperature can be observed over many decades. Furthermore since the field strengths in the depletion region are high, the technique

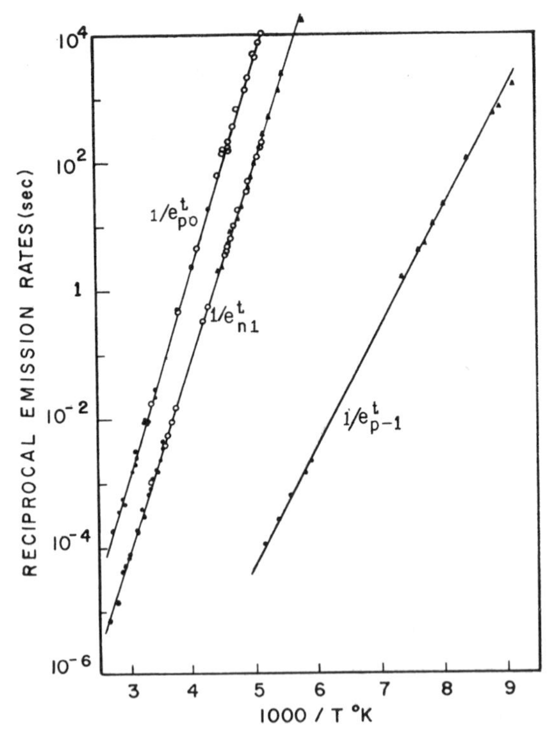

FIG. 8.21 The reciprocal thermal emission rates of holes and electrons at the gold acceptor level in silicon, $1/e_{p0}{}^t$ and $1/e_{n1}{}^t$, at 10^4 V cm^{-1} and of holes at the gold donor level, $1/e_{p-1}{}^t$, at 10^5 V cm^{-1}.

The solid dots are dark current transient data; the dark triangles are dark capacitance transient data; and the circles are photocurrent data using a square light pulse. (After Sah et al., 1969.)

provides some evidence on barrier lowering of the impurity potential by the field.

In a useful attempt to systematize such studies Sah et al. (1970b) have prepared a paper describing more than ten variations of the experiment. A few of these are given below as examples of the kinds of procedures involved.

A low-temperature, dark *C-V* and *I-V* transient study is illustrated in Fig. 8.22. Lines *a* and *c* show that no light is applied and that the temperature is low. The voltage waveform applied is given in line *b*, and *d* shows the resulting C^2 transient for a single-level acceptor center in a p^+n junction. Line *e* is the C^2 transient expected for a single-level donor center. Line *f* is the current transient expected. (The letters *A* and *D* in the illustration indicate acceptor and donor, respectively.) The difference of capacitance before and after switching to $V_J = 0$ provides a numerical result for $N_{TT}e_n/(e_n + e_p)$ at 300°K. If N_D is known from the data for a control sample, then the *C-V* data of a sample containing deep impurities give N_{TT}. Therefore the ratio e_n/e_p can be obtained from the transient capacitance study. This experiment may be varied by the addition of band-gap light.

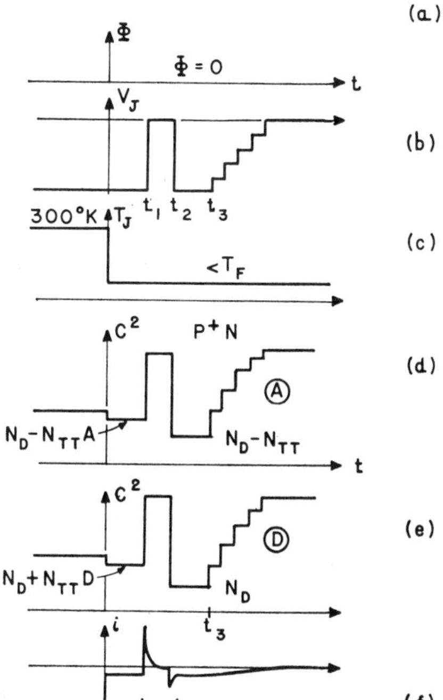

FIG. 8.22 Low-temperature dark *C-V* and *I-V* experiments. (After Sah et al., 1970.)

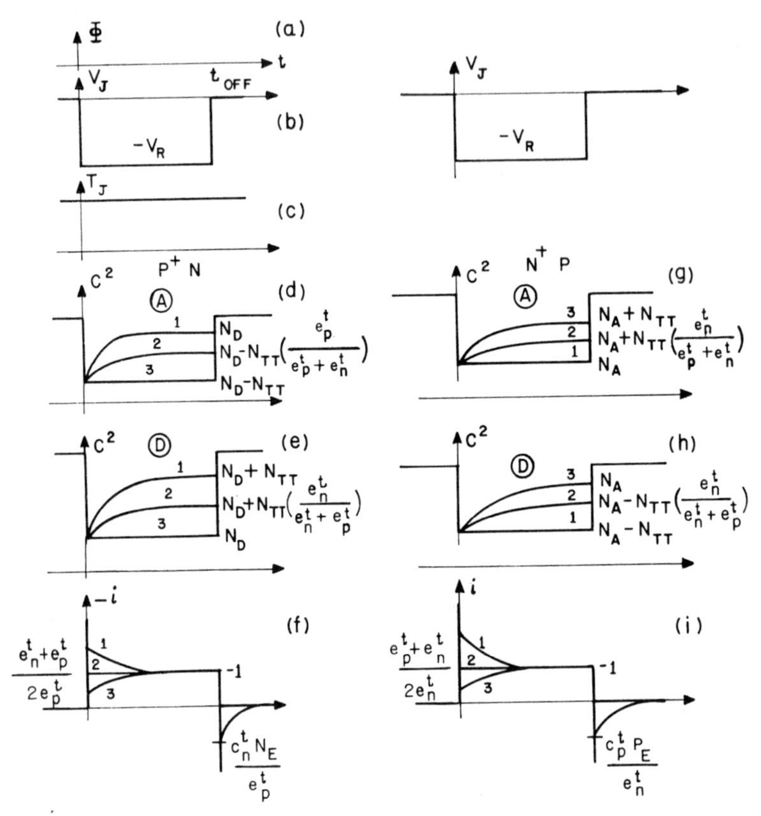

FIG. 8.23 High-temperature transient experiment without light. (After Sah et al., 1970.)

Figure 8.23 illustrates a class of high-temperature dark experiments for p^+n and n^+p junctions. The junction voltage is switched between zero and $-V_R$. By study of the C^2 and current transients the information that can be extracted from such a set of experiments includes e_n, e_p, C_n, C_p, and ΔE_T.

Another class of experiments is illustrated in Fig. 8.24 where the p^+n junction is primed by a pulse of band-gap light and then the capacitance transient is produced by light of energy between the trap level ΔE_T and the band gap. By study of both the capacitance and current transients it is possible to determine the optical emission rates $e_n{}^0$, $e_p{}^0$ and the thermal capture rates c_n, c_p from this type of experiment.

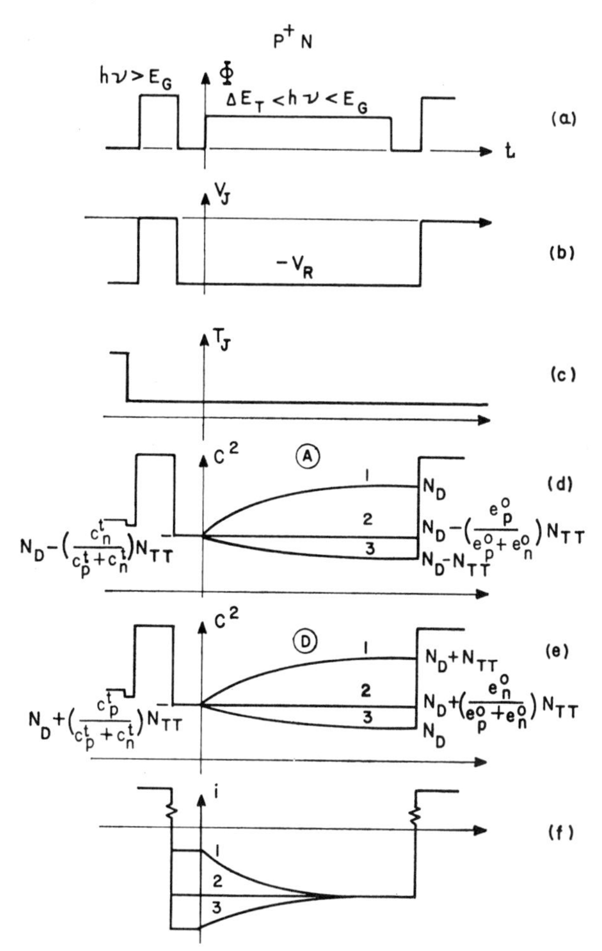

FIG. 8.24 Low-temperature photoexperiment with carrier injection by band-gap light in a p^+n diode. (After Sah et al., 1970.)

The capacitance transients of gold-doped p^+n silicon junctions and Schottky barriers on n-type silicon have been studied also by Senechal and Basinski (1968). The spatial variation of the charged gold concentration in silicon p^+n and n^+p step junctions has been examined by Schroder et al. (1968). Capacitance transients for Schottky barriers on gallium arsenide have been studied by Williams (1966) and by Furukawa (1967).

The induced photovoltaic effect and change in transition-layer capacitance in alloyed copper-doped GaP diodes have been studied by Grimmeiss and Olofsson (1969), and the results explained by hole capture at acceptor levels.

A method to evaluate energy level and density of hole and electron traps from capacitance measurements on Schottky barriers or *pn* junctions has been presented by Ikoma and Jeppsson (1972). The capacitance difference of the *pn* junction between the reverse biased steady state and the state immediately after the voltage is switched from forward to the same reverse voltage is measured. By investigating the occupancy factor for electrons at the deep level, using Shockley–Read statistics, for reverse and forward biased junctions the capacitance change is related to the deep level. Working formulas, from which both hole and electron traps can be evaluated, are derived. The analysis includes traps that are distributed in the bulk of the material as well as traps that are introduced at and near the interface where the junction is formed. The method is shown to allow detection of deep level densities down to about 2×10^{11} cm^{-3} for GaAs layers having doping densities of 1×10^{15} cm^{-3}.

8.5 Effects of Deep Impurities in Other Junction Structures

Deep impurity effects in bulk semiconductors and conventional diode junctions have been discussed, and their effects in other junction structures now need mention.

8.5.1 Deep Impurities in Bipolar Transistors

The current amplification of a transistor containing deep impurities in the base may tend to rise as the injection level is increased since the minority carrier lifetime is a function of the injection. This problem has been considered by several workers (Osipov, 1968). In early designs of *pnpn* structures, this effect was used to contribute to the turn-on process.

In phototransistors containing deep impurities the variation of minority carrier lifetime with injection level may be useful in producing a photosensitivity that increases with the illumination intensity. This has been studied by Varlamov et al. (1968) for gold-doped germanium phototransistors and by Osipov and Kholodnov (1972).

The effects of deep impurities have been found prominent in gallium arsenide bipolar transistors where they cause trapping effects (Strack, 1966; Antell and White, 1968). Strack found that the cutoff frequency (5 MHz) of sulfur–magnesium double-diffused transistors was well below the calculated value and attributed the effect to deep level imperfections. Doping with iron, a deep acceptor, during the emitter formation process was found, however, to increase the cutoff frequency to 150 MHz and extended the temperature operating range of the devices. Further studies of GaAs transistors have been reported by Doerbeck et al. (1968).

Radiation can of course produce deep defect levels in transistors and the effects are normally deleterious.

8.5.2 Deep Impurities in Field Effect or MOS Transistors

In field effect transistors of GaAs, Turner (1966) has observed pronounced looping of the characteristics. This is attributed to traps arising from the fabrication process when an epilayer, grown on a semiinsulating GaAs substrate, is used for the transistor channel. Very few studies, however, have been made of FET structures containing deep impurities and most of the interest has centered on MOSFET structures. Oldham and Naik (1972), however, have considered the effects of two levels of defects in neutron-irradiated junction field effect transistors.

Iwauchi and Tanaka (1968) have studied gold-doped and copper-doped silicon MOS transistors. As might be expected the transistor characteristics are shifted by the presence of the deep impurity and in general no particularly useful effect is achieved. Richman (1968) reports that the threshold voltages of MOS field effect *n*-channel transistors shift in the positive direction when gold is diffused into the silicon substrate from the back side. The evidence suggests that this is caused in part by a negative charge sheet of ionized gold acceptor states in the silicon just below the Si–SiO_2 interface.

Reddi (1965) has shown that tunable high-pass filter characteristics can be expected from MOS structures in which a deep generation-recombination center plays a role.

MOSFET structures containing deep impurities also provide an interesting means of studying the depletion region generation-recombination noise spectrum of the impurity over a wide range of frequency (Sah and Yau, 1969).

Deep trapping levels are also found in insulators such as SiO_2 or Si_3N_4 or at the interface between the nitride and oxide in MNOS transistors. By application of a suitable gate bias the charge occupancy of the trapping levels can be changed and the turn-on voltage of the transistor is altered. One model for the transfer of charge to the deep traps involves tunnel injection as indicated in Fig. 8.25. A memory action is obtained which can be erased by the application of a gate bias of opposite polarity. The memory can be of very long duration. However, problems of fabricating controllably and of fully understanding the action must be resolved before successful applications can be envisaged.

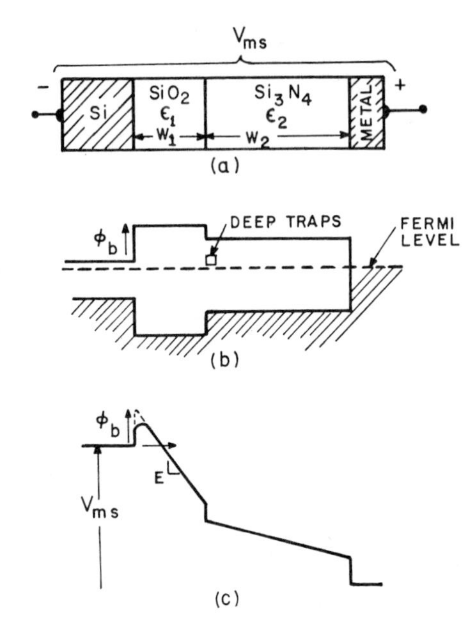

FIG. 8.25 Charging of deep traps in MNOS structures according to a tunnel-injection model.

(a) Schematic diagram of MNOS memory element. (b) Simplified energy band diagram for zero applied bias. (c) Conduction band variation of positive metal–silicon bias showing the tunnel injection. (J. G. Simmons, private communication.)

8.5.3 Counting Structures Involving Deep Impurities

Cohen (see Panousis, 1968) has reported "counting" behavior in gallium arsenide *p-i-n* diodes in which the diodes conduct once for every n applied voltage pulses. Similar action has also been reported by Panousis (1968) for gold-doped silicon *p-i-n* diodes at low temperatures. For silicon, the range of stable counting by two ($n = 2$) extends from a pulse repetition rate of less than 1 Hz to greater than 100 kHz. The device behavior is explained by a double-injection model in which the diode can be in one of two states: a high-impedance state in which the current is predominately single-carrier space-charge limited, and a low-impedance state in which a neutral plasma is set up in the i region. The transition from the high to the low state occurs after a certain fraction of the deep centers has become occupied by the injected minority carriers. Because of the finite capture time associated with this process, several voltage pulses may be required in order to inject a sufficient charge into the centers to switch the diode to the low-impedance state, each pulse leaving the diode in a "sensitized" high-impedance state. Counting behavior should be expected if the decay from the low-impedance state is much more rapid than that from the high-impedance state. Indeed this is the case since decay from the low-impedance state involves the capture of electrons and holes from the high-density plasma, whereas decay from the high state is predominately due to emission of carriers from the deep traps. A simple statistical model allows the determination of the rate at which excess electrons, holes, and trapped charges decay from the "sensitized" high-impedance state, and hence gives an indication of the life of this state.

8.5.4 Stress Effects in Junctions Containing Deep Impurities

Large stress effects are observed in *pn* junctions containing deep impurities. Figure 8.26 shows the effect on the saturation currents of reverse-biased germanium n^+p junctions doped with copper at various concentrations. Similar effects have been observed for copper in silicon by Yamashita et al. (1966). The effect is related to high concentrations of interstitial impurities and depends on the quenching cycle used when the impurities are frozen into the lattice.

The piezoresistive behavior of silicon containing nickel has been studied by Zucker et al. (1969) and of germanium containing nickel by Traum et al. (1969). Stress-dependent deionization of the deep acceptor levels is observed.

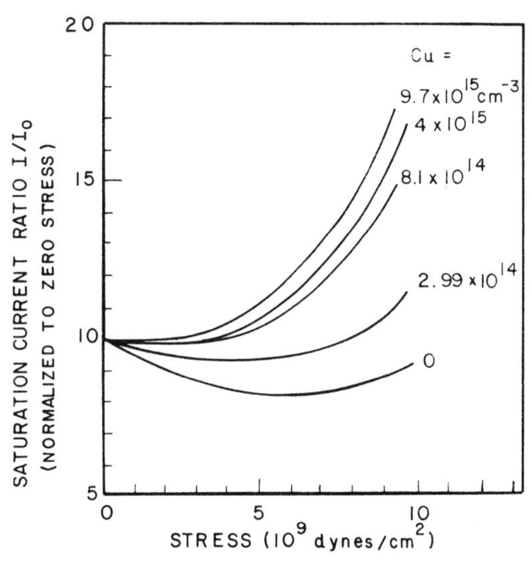

FIG. 8.26 Variations of saturation current I/I_0 with uniaxial stress on germanium n^+p junctions doped with copper. (After Oda and Matukura, 1967.)

8.5.5 Deep Impurity Effects in Tunnel Diodes

Impurities with energy levels lying deep in the band gap have been shown to contribute to the excess current of tunnel diodes of germanium, silicon, and gallium arsenide, as sketched in Fig. 8.27. This is generally an undesirable effect since a high ratio of peak to valley current is usually wanted in a tunnel diode.

The effect, however, can be used as a form of tunnel spectroscopy to determine the energy level of the impurity. For example, the copper level in GaAs has been determined by Fistul' (1965) to be at 0.44 ± 0.013 eV from the valence band edge. As another example, the Au–Ge trap state in $Al_xGa_{1-x}As$ has been studied by an examination of photosensitive impurity-assisted tunneling (Andrews et al., 1972).

Similar tunnel spectroscopy studies are possible in heavily doped Schottky barrier diodes.

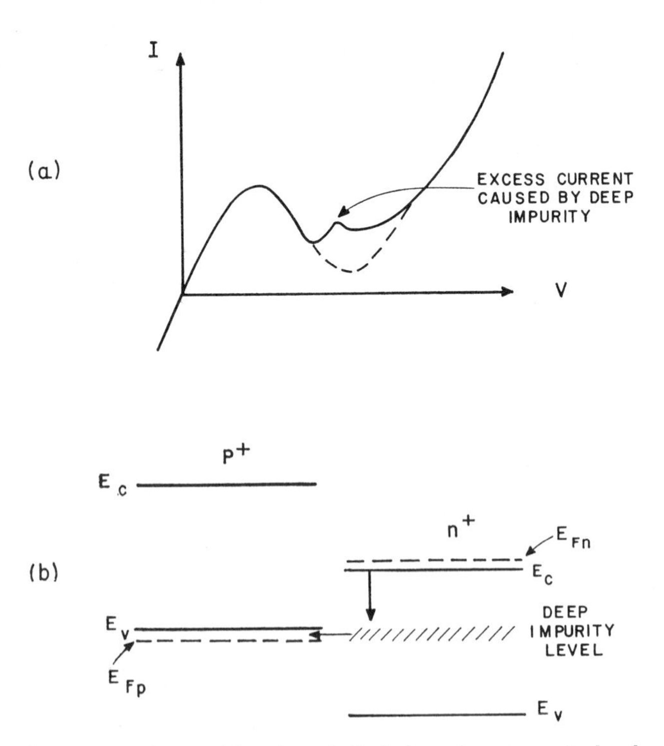

FIG. 8.27 Excess current caused in a tunnel diode by a deep impurity level.

8.5.6 Long Diodes Containing Deep Impurities

Modulation of diffusion length occurs in long diodes containing deep impurities as the injection level is increased (Stafeev, 1959). This leads to a number of effects such as increased photosensitivity and negative-resistance effects. The effects have been studied by various workers including Stafeev (1959), Gribnikov (1965), Kazarinov et al. (1967a, b), Osipov (1968), Baranenkov and Osipov (1969), Garin and Stafeev (1972), Osipov and Kholodnov (1972).

CHAPTER NINE

Thermally Stimulated Current Studies of Traps

At low temperatures, deep impurity levels in a high-resistance semiconductor can be studied by the method of thermally stimulated current.

There are a number of techniques available, but basically the experiment is conducted as follows. The sample is mounted in a vacuum cryostat and cooled. It is excited by illumination to place the traps in a nonequilibrium state. Occasionally some other method of excitation such as a current or voltage transient is used. The temperature at which this step takes place must be low enough so that equilibrium is reestablished very slowly. The sample is then heated at a linear rate and the variation in conductivity or the emitted light, resulting from return to equilibrium, is monitored versus temperature. Comparison of this measurement with one done without the initial excitation illumination shows the presence of current or glow peaks associated with the emptying of traps within the semiconductor. It is then possible to determine the position of these trap levels within the energy gap. From the experiment the density of the traps or their capture cross sections may sometimes be determined, although not usually with great accuracy. It is usually possible to infer from other observations on the material whether the energy level observed is measured with respect to the conduction or the valence band. However, if this is not absolutely clear, it may be necessary to measure the Hall coefficient while the thermally stimulated current (TSC) run is in progress (Bube and MacDonald, 1962; Kulshreshtha and Yunovich, 1966).

Thermally stimulated glow (TSG) effects were first reported on NaCl excited by x-rays. The field was then inactive until Randall and Wilkins

(1945) considered its application to electroluminescent materials. Since that time many papers have been published, and thermally stimulated glow and current measurements have proved to be useful ways to study traps in insulators and semiconductors. The temperature dependence of photoluminescence excitation spectra has been found effective in the study of short recombination time impurity states in GaP (Dishman, 1972).

The main emphasis in this chapter, however, is on a review of thermally stimulated current theory and measurements.

9.1 Quasiequilibrium Analysis

Some theoretical analysis is necessary to relate conductivity peaks and impurity levels. There are generally three approaches to the problem, namely, the quasiequilibrium treatment and fast and slow retrapping analyses.

The quasiequilibrium approach (Bube, 1960), requires the following assumptions:

1. Recombination in the material involves retrapping of the emitted carrier by the level under consideration.
2. The level is initially completely filled.
3. Equilibrium is assumed between the conduction band and the trap, so that a Fermi level can be defined throughout the temperature rise.
4. The peak current occurs when the Fermi level crosses the trap level.

Under these conditions, the energy level of the trap is

$$\Delta E_T = kT_m \ln\left[\frac{N_c(T_m)}{n^m}\right] \tag{9.1}$$

where ΔE_T is the level of the trap below the conduction band edge, k is Boltzmann's constant, T_m is the temperature of the conductivity peak, ln is the logarithm to the base e, N_c is the effective density of states in the conduction band at T_m, and n^m is the electron concentration at T_m.

There are two drawbacks to this approach. The first is that assumption 2 is unlikely to be true. In general it is not possible to know the initial occupancy of a particular level, and this will greatly influence n^m. The second problem concerns assumption 4. It will be shown later that the trap level ΔE_T can be derived from general considerations that are independent of assumption 4. It is then found that (9.1) is true only if the trap occupancy at T_m is half the total number of traps.

Assuming no retrapping, and with n^m computed from a temperature-independent capture cross section of 10^{-14} cm^2 for the traps of a typical semiconductor, (9.1) reduces to

$$\Delta E_T = 23kT_m \tag{9.2}$$

Thus a thermally stimulated current peak at 100°K represents a trap level of about 0.2 eV. In practice T_m is found to be increased by an increase in the heating rate, and this is an important way of determining the trap depth. However, for the approximations represented by (9.2), the effect of heating-rate dependence of T_m is negligible with respect to the uncertainty caused by the usual lack of precise knowledge of the capture cross section. Fortunately the logarithmic form of (9.1) means that a change in the cross section from 10^{-12} to 10^{-18} cm^{-2} changes the numerical factor in (9.2) only from 21 to 28. In general, however, (9.2) is usable only for very crude estimates of trap depths and is not generally acceptable because of the assumptions involved.

9.2 General Equations

A more fundamental approach to the thermally stimulated current problem is that given by Randall and Wilkins (1945), and later expanded by Saunders and Jewitt (1965), and is based on the equations

$$\frac{dn_i}{dt} = -n_i N_c \sigma_i v_{\text{th}} \exp\left(-\frac{\Delta E_i}{kT}\right) + n(N_i - n_i)\sigma_i v_{\text{th}} \tag{9.3}$$

$$\frac{dn}{dt} = -\frac{n}{\tau} - \sum_i \frac{dn_i}{dt} \tag{9.4}$$

where n_i is the density of electrons in the ith impurity level, n is the density of electrons in the conduction band, N_c is the effective density of states in the conduction band, v_{th} is the thermal velocity of electrons, σ_i is the capture cross section of the ith level, N_i is the density of the ith level, and τ is the recombination time of electrons (i.e., the average time it takes for an electron in the conduction band to recombine with a hole by passing through the center N_i, or through other impurity recombination centers).

The left-hand side of (9.3) is the net rate of change of electrons in the ith level. The first term on the right-hand side is the thermal release of the n_i trapped electrons. The density of states is $N_c = 2(2\pi m^* kT/h^3)^{3/2}$, the quantity $N_c \sigma_i v_{\text{th}}$ is the attempt-to-escape frequency, and the exponential gives the probability of escape. The second term on the right-hand side of (9.3) represents the retrapping of the conduction band electrons, n, by the unoccupied traps, $N_i - n_i$, which have a capture cross section for electrons of σ_i.

Equation 9.4 involves the recombination of conduction electrons. The first term on the right assumes that free electrons interact with a recombination center which can be described by an effective constant lifetime τ. The center is otherwise left unspecified. This is a rather gross simplification of the problem but usually appears to be justified experimentally. Saunders (1967) and Dussel and Bube (1966d) discuss this approximation more fully. The second term is recapture by the traps. If more than one trap is present, it is assumed that they are of significantly different energy so that they do not interact.

In general there will be $N + 1$ coupled equations where N is the number of impurity levels considered. If a general solution to these equations could be found in closed form, the exact shape of the thermally stimulated current curve would be known in terms of basic parameters. These parameters could then be determined from experiments. Unfortunately, these equations form a set of nonlinear coupled equations which cannot be solved without a considerable number of assumptions.

9.3 Fast Retrapping Analysis

One of the simplest approaches to the problem is to assume that only one impurity level acts at a given time and that the recombination time τ is long in comparison with the retrapping action. In general these two assumptions are not compatible when only one level exists. If there are several impurity levels, there is no reason to assume that retrapping will not involve all of them. In this case, the effective value of τ would depend on the occupancy and capture cross section of all of the other levels. Computed solutions have been made (Sacks, 1970) to study this interaction. However, for initial simplification the assumptions of one level and τ infinite are made and (9.4) becomes

$$\frac{dn}{dt} = -\frac{dn_i}{dt} \tag{9.5}$$

Now, to a good approximation, the peak current will occur where n is a maximum since mobility is not a strong function of temperature. Then the peak current will occur at the temperature T_m at which $dn/dt = 0$. This implies from (9.5) that $dn_i/dt = 0$. Using this in (9.3) gives

$$-n_i N_c \exp\left(-\frac{\Delta E_i}{kT}\right) + n(N_i - n_i) = 0 \tag{9.6}$$

where n_i, N_c, and n are taken to be evaluated at T_m. Solving for ΔE_i and letting $\eta = n_i/N_i$ be the relative occupancy at T_m we obtain

$$\Delta E_i = kT_m \ln\left(\frac{N_c}{n} \times \frac{\eta}{1-\eta}\right) \tag{9.7}$$

This expression was first derived by Boer et al. (1958). It is easily seen that to obtain (9.1), it is necessary to set $\eta = \frac{1}{2}$. If we now let

$$n = N_c \exp\left(-\frac{E_c - E_F}{kT}\right)$$

as in the case of thermal equilibrium, and use this in (9.6), we can solve for $E_c - E_F$ as follows:

$$E_c - E_F = E_c - E_T - kT_m \ln\left(\frac{\eta}{1-\eta}\right) \tag{9.8}$$

Again if $\eta = \frac{1}{2}$, the assumption that E_F crosses E_T at T_m is true. To use (9.7) in general, however, it is necessary to know the impurity density as well as the trap occupancy at T_m to calculate η.

9.4 General Analysis

A more complete analysis of (9.3) and (9.4) has been made by Saunders and Jewitt (1965). They assume that the current peaks occur sufficiently far apart for two conditions to prevail:

1. The contribution of electrons from all but one trap may be neglected.
2. The occupancy of all of the other traps remains sufficiently constant so that retrapping by these levels may be accounted for in the term involving τ and be temperature independent.

Under these conditions, the summation term in (9.4) may be neglected. If it is now assumed that $n \ll n_i$ and $n_i \ll N_i$, then (9.3) and (9.4) may be solved and the conductivity $\sigma(T) = e\mu_n n$ can be shown to be

$$\sigma(T) = \frac{e\tau\mu_c N_c \sigma_i v_{\text{th}}}{1 + \tau N_i \sigma_i v_{\text{th}}} \hat{n}_i \exp\left(-\frac{\Delta E_i}{kT} - \frac{1}{\beta}\int_{T_0}^{T} \frac{N_c \sigma_1 v_{\text{th}} \exp(-\Delta E_i/kT)\, dT}{1 + \tau N_i \sigma_i v_{\text{th}}}\right) \tag{9.9}$$

where β is the rate of temperature rise defined such that $T(t) = T_0 + \beta t$ and $\hat{n}_i$ is equal to the initial occupancy of the ith level at the temperature T_0. Now the following explicit temperature dependencies are assumed: $N_c \propto T^{3/2}$; $v_{\text{th}} \propto T^{1/2}$; $\sigma_i \propto T^{-b}$; and $\tau =$ constant. Then (9.9) can be shown to have a maximum at $T = T_m$ such that

$$\frac{\Delta E_i}{kT_m} = \ln\left(\frac{T_m^2}{\beta}\right) + \ln\left(\frac{N_c \sigma_i v_{\text{th}} k}{\Delta E_i}\right) - \ln\left(1 + \tau N_i \sigma_i v_{\text{th}}\right) \tag{9.10}$$

provided ΔE_i is large in relation to kT_m. Two cases can now be distinguished:

CASE 1. Slow retrapping: $\tau N_i \sigma_i v_{\text{th}} \ll 1$. In this case, (9.10) becomes

$$\frac{\Delta E_i}{kT_m} = \ln\left(\frac{T_m^2}{\beta}\right) + \ln\left(\frac{N_c \sigma_i v_{\text{th}} k}{\Delta E_i}\right) \tag{9.11}$$

and if $\sigma_i \propto T^{-2}$, as is often assumed, the second term on the right-side becomes independent of temperature since $N_c \propto T^{3/2}$ and $v_{\text{th}} \propto T^{1/2}$. Then a plot of $1/T_m$ versus $\ln(T_m^2/\beta)$ for various heating rates should give a straight line of slope $\Delta E_i/k$, and the intercept yields the capture cross section. If $b \neq 2$, then (9.11) becomes

$$\frac{\Delta E_i}{kT_m} = \ln\left(\frac{T_m^{4-b}}{\beta}\right) + C \tag{9.12}$$

where C is independent of temperature. The value of b must be found by trial and error so that a straight line is obtained in a plot of $1/T_m$ versus $\ln(T_m^{4-b}/\beta)$. We can then proceed to determine ΔE_i and σ_i as above.

CASE 2. Fast retrapping: $\tau N_i \sigma_i v_{\text{th}} \gg 1$. Equation 9.10 then becomes

$$\frac{\Delta E_i}{kT_m} = \ln\left(\frac{T_m^{3.5}}{\beta}\right) + C' \tag{9.13a}$$

where C' is independent of temperature and

$$C' = \ln\left(\frac{N_c}{\tau N_i}\right) - \frac{3}{2}\ln(T_m) - \ln\left(\frac{\Delta E_i}{k}\right) \tag{9.13b}$$

In this case, no information can be derived about σ_i, but the quantity τN_i can be computed.

The intermediate situation where $\tau N_i \sigma_i v_{\text{th}} \sim 1$ can usually be resolved to reasonable accuracy by using either case 1 or case 2.

9.5 Numerical Evaluation

The basic equation, (9.9), has been evaluated numerically to give some feel for its meaning. Conductivity versus temperature curves for energy levels 0.2 and 0.4 eV and heating rates 0.1 and 1°K sec^{-1} are shown in Fig. 9.1.

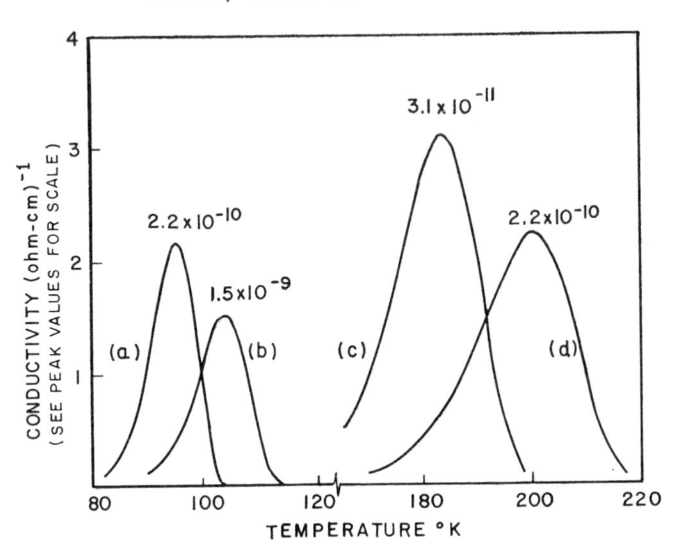

FIG. 9.1 Calculated thermally-stimulated conductivity versus temperature showing effect of heating rate on T_m.
(a) $\Delta E_i = 0.2\,eV$, $\beta = 0.1°K\ sec^{-1}$; (b) $\Delta E_i = 0.2\,eV$, $\beta = 1°K\ sec^{-1}$; (c) $\Delta E_i = 0.4\,eV$, $\beta = 0.1°K\ sec^{-1}$; (d) $\Delta E_i = 0.4\,eV$, $\beta = 1°K\ sec^{-1}$. (Courtesy P. C. Smith.)

The trap density assumed was $N_i = 10^{15}\ \mathrm{cm}^{-3}$ and the lifetime τ was taken as 10^{-6} sec and assumed independent of temperature. The capture cross section σ_i was taken as $10^{-16}\ \mathrm{cm}^2$ at 100°K and assumed to vary as T^{-2}. The initial occupancy of the trap $\hat{n}_i$ was assumed to be $10^{13}\ \mathrm{cm}^{-3}$, and the other parameters were taken as those appropriate to silicon. The increase of the temperature for the peak current with increase in the heating rate is one of the most obvious features of the curves.

The peak of the TSC curve can be determined by finding the maximum of (9.9). By setting $\partial\sigma/\partial T = 0$ at $T = T_m$ the following expression is obtained

$$\exp\left(-\frac{\Delta E_i}{kT_m}\right) = \frac{(1 + \tau N_i \sigma_i v_{\mathrm{th}})\Delta E_i \beta}{N_c \sigma_i v_{\mathrm{th}} k T_m{}^2} \tag{9.14}$$

This expression has been evaluated for various values of β under two limiting conditions: (1) slow retrapping, when $\tau N_c \sigma_i v_{\mathrm{th}} \ll 1$; and (2) fast retrapping, when $\tau N_c \sigma_i v_{\mathrm{th}} \gg 1$. Under these approximations (9.14) reduces to particularly convenient expressions since in the first case variations of $N_i \tau$ are not important, and in the second σ_i drops out. Figure 9.2 shows the results of such calculations for the fast retrapping approximation, and can be used to find the approximate temperature shift to be expected from a TSC experiment. The curves are computed for electrons in silicon with an effective

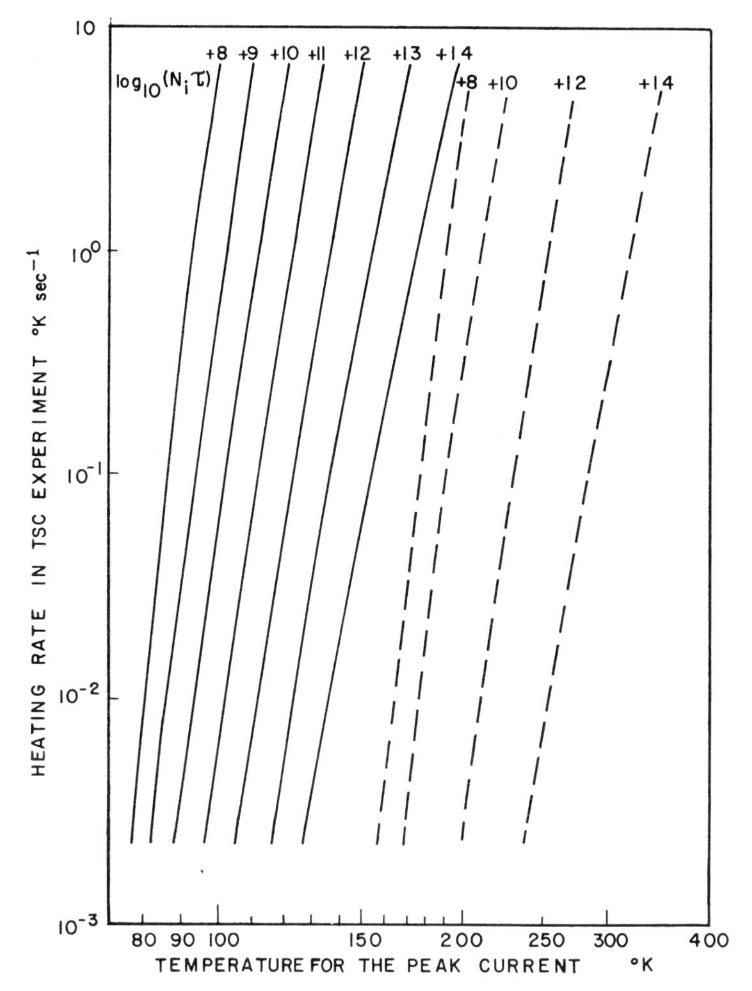

FIG. 9.2 Calculated variation of the temperature for the peak current versus the heating rate.

The solid lines are for $\Delta E_i = 0.2$ eV and the dashed lines for $\Delta E_i = 0.4$ eV. The assumption of fast retrapping was made, so that $\tau N_c \sigma_i v_{th} \gg 1$. (Courtesy P. C. Smith.)

mass of 0.5 of the free-electron mass. This factor would be different for other materials, but its effect on the general conclusions are not as important as errors in estimating the other parameters that determine which curve to use.

Experimentally, such curves are determined by measuring T_m for various values of β and plotting ln $(T_m{}^n/\beta)$ versus $1/T_m$—where n is empirically determined for a straight-line plot. For slow retrapping $n = 4 - b$ where

$\sigma_i \propto T^{-b}$, and for fast retrapping $n = 3.5$. The slope of the line is proportional to ΔE_i and the intercept is proportional to σ_i or $N_i\tau$ for slow or fast retrapping, respectively.

The thermally stimulated current curve observed for a specimen of semiinsulating GaAs (oxygen–copper doped) is shown in Fig. 9.3. The cryostat was heated to 400°K and then cooled to 80°K before illumination of the specimen with band-gap light. This preheat cycle helped to produce repeatable results. Increasing the sample light exposure time prior to heating increased the peak height but did not change the temperature for the peak. We see later in this chapter that it is not unusual for a much more complex TSC curve to be observed for semiinsulating GaAs, depending on the illumination conditions.

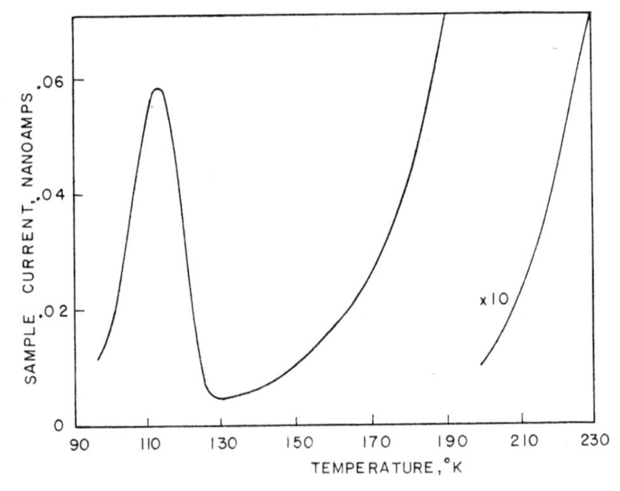

FIG. 9.3 Thermally stimulated current curve for semiinsulating GaAs (oxygen–copper doped) with light applied before heating.
The heating rate β *was* 0.1°K sec^{-1}. *(After Sacks, 1970.)*

Descriptions of the practical details of thermally stimulated current systems may be found in various places in the literature, including Saunders and Jewitt (1965) and Sacks (1970).

The experimental results of the variation of peak temperature with heating rate for the GaAs of Fig. 9.3 are shown in Fig. 9.4.

To observe the temperature shift of the peak as a function of heating rate, an expanded horizontal scale was used so that the peak was spread out over

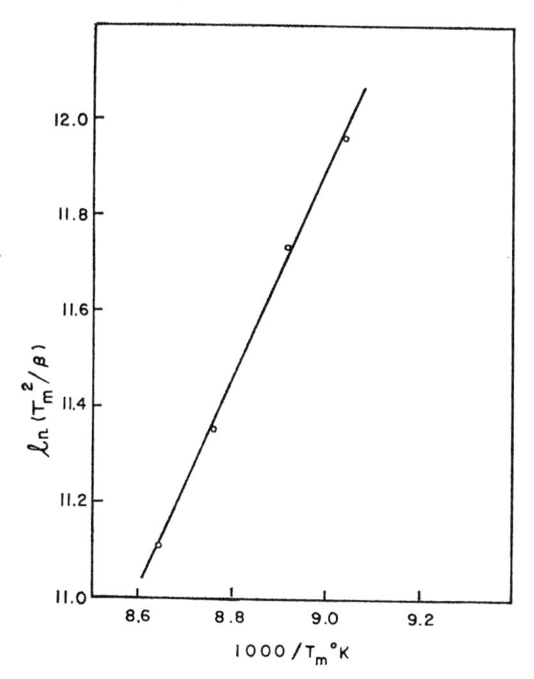

FIG. 9.4 The peak temperature shift with heating rate for a thermally stimulated current peak in GaAs.
The slope of the line suggests a trap depth of 0.19 eV, which is associated with copper in GaAs. (After Sacks, 1970.)

the entire horizontal axis. In this manner, the peak temperature could be determined to within 0.5°K. Four heating rates were chosen between 0.08 and 0.20°K sec^{-1}. A greater range would have been desirable, but because of the power input limitation of the cryostat heater, further increase in heating rate was not possible. The plot of $\ln (T_m{}^2/\beta)$ versus $1000/T_m$ results in a straight line, which would be expected for $\sigma_i \propto T^{-2}$ and slow retrapping conditions. The slope of the line corresponds to an energy level of 0.19 ± 0.02 eV. The intercept of the line, as discussed earlier with reference to (9.10), provides an estimate of the capture cross section. However, since this is proportional to exp (−intercept), an error in determining the intercept causes a large error in the capture cross section. From Fig. 9.4 the capture cross section, converted to a 300°K value, was estimated to lie in the range 2×10^{-18} to 2×10^{-20} cm^2. Fuller et al. (1967) have observed a 0.19-eV level above the valence band in GaAs and attributed it to a Cu–Te complex acting as an acceptor.

9.6 Energy Level from Initial Rise Slope

Several authors, Garlick and Gibson (1948) and Bräunlich (1967), have shown that under certain conditions, the light emitted at the beginning of a thermoluminescent experiment is given by

$$I \propto \exp\left(-\frac{\Delta E_i}{kT}\right)$$

where ΔE_i is the energy of the trapping center involved relative to the appropriate band edge. Bräunlich (1967) has also similarly shown that the initial increase of electron density dn/dt or dn/dT in a thermally stimulated current measurement is related to ΔE_i in the case of slow retrapping.

For the GaAs of Fig. 9.3, a difficulty with the initial rise technique becomes apparent. Theoretically, the current before and after a single peak should go to zero. Unfortunately, as Fig. 9.3 shows, this is not the case. If it is assumed that a correction to the current can be made so that only curtent due to the level in question is considered, then the initial rise can be plorted. A problem in accuracy can arise because at the lowest temperatures, where the initial rise slope is theoretically most accurate, the value of the current is small and so the correction to the current is greatest. For the plot of Fig. 9.5 the lowest point on the thermally stimulated current curve, $I = I_0$, was

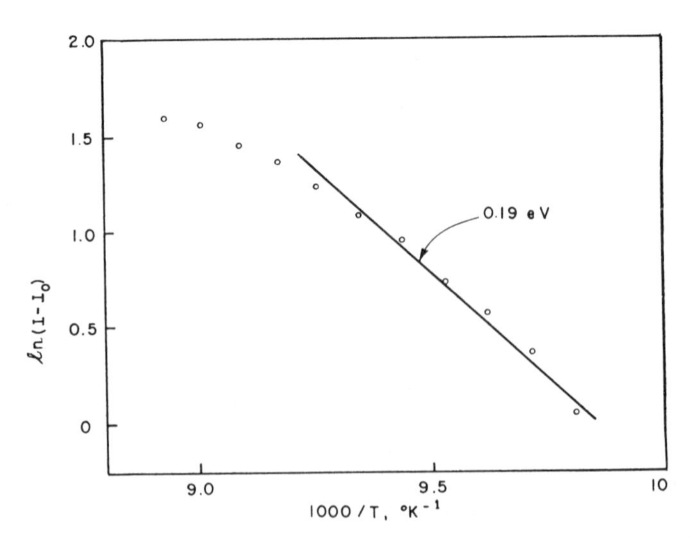

FIG. 9.5 Initial rise technique for determining the trap level for a thermally stimulated current peak in GaAs. (After Sacks, 1970.)

taken as the zero line, and the function ln $(I - I_0)$ was plotted versus $1000/T$. The line drawn on Fig. 9.5 has a slope of 0.19 eV and the limits of an acceptable fit are estimated to be ± 0.02 eV. This result is in excellent agreement with the energy level deduced from the T_m shift measurements.

The validity of the initial current rise technique for determining the trap depth has been studied by Haake (1957) and Bräunlich (1967) who find limitations in its usefulness. On the other hand, Sacks (1970) finds the theoretical accuracy in determining the trap depth good to within a few percent (Fig. 9.6) for parameter ratios in the range of his interest. However, experimentally, there are difficulties associated with allowing for background current and overlap of adjacent current peaks.

Numerous other authors have proposed methods to analyze the thermal stimulation data. Most start with (9.3) and (9.4) above and make other simplifying assumptions to obtain an expression for ΔE_i as a function of T_m. Grossweiner (1953) used a simplified form assuming slow retrapping, and expanded the integral in a series. He assumed $\sigma_T \propto T^{-2}$ and obtained

$$\Delta E_T = \frac{1.51 k T_m T_{1/2}}{T_m - T_{1/2}} \tag{9.15}$$

where $T_{1/2}$ is the temperature at half-maximum on the rising portion of the curve. Booth (1954) and Hoogenstraaten (1958) have made analyses for

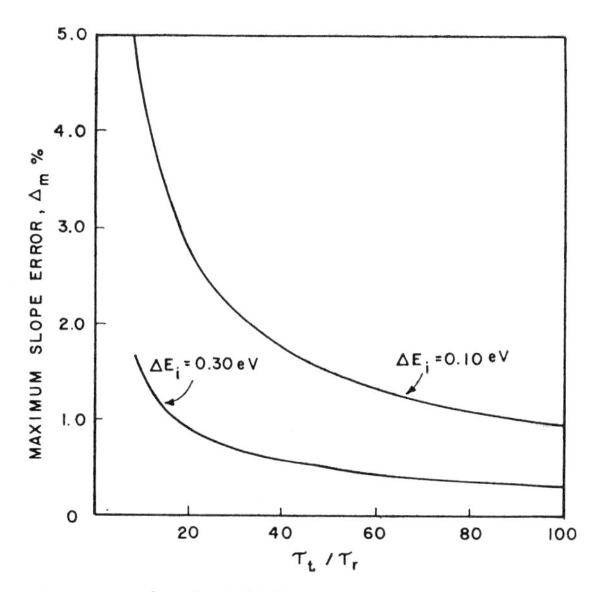

FIG. 9.6 Percentage error for the initial rise slope as a function of the ratio τ_t/τ_r (where $\tau_t = 1/N_i\sigma_i v_{th}$). (After Sacks, 1970.)

$\sigma_T \propto T^{-b}$ and slow retrapping. Their results are similar to those of Saunders and Jewitt (1965) with $\tau N_c \sigma_T v_{\text{th}} \ll 1$.

The methods above represent the main approaches to analysis of TSC data. However, some variations on these methods have been presented by other authors.

Keating (1961) uses an approach similar to Grossweiner but assumes that $\sigma_T \propto T^{-b}$. In this method ΔE_T, σ_T, and b can be computed for no retrapping using the half-maximum temperature values and T_m. Halperin and Braner (1960) approximate the curve with a triangle and obtain two equations (fast and slow retrapping) using T_m and $T_{1/2}$, the low-side half-maximum temperature. Cowell and Woods (1967) develop a numerical curve-fitting approximation using Haering and Adams' theory (1960). Their procedure involves adjusting ΔE_T to best fit theoretical and experimental curves. Boiko et al. (1960) break the problem into a number of different cases and obtain expressions for each. Dittfeld and Voigt (1963) and Nicholas and Woods (1964) present comparisons of the various theories and Broser and Broser–Warminsky (1955) consider differences in TSC and thermally stimulated luminescence experimental data. Saunders (1967) also considered this difference and concludes that if τ is a strong function of temperature, the TSC temperature maximum occurs at a significantly higher temperature. The data of Broser and Broser–Warminsky (1955) confirm this experimentally. Saunders concludes, however, that experimentally τ is not usually a strong function of temperature since a number of recombination centers are generally present.

Wright and Allen (1966) discuss the effects of various light intensity, energy, and cross section relationships between several levels on the TSC curve. They conclude that if retrapping is significant, T_m will increase with increasing initial trap occupancy. They also point out that all levels cannot be observed since the emptying of a level requires the cooperation of another level. Kulshreshtha and Goryunov (1966) have considered this point and give an expression relating temperature shift and initial occupancy. Determining the initial occupancy is a considerable problem but if this can be overcome, the temperature shift versus n_T suggests a method of determining the extent of retrapping qualitatively.

A number of other experimental techniques have been reported for obtaining TSC curves. These involve space-charge-limited currents, capacitor and diode structures, polarization charge relaxation and ac field methods (Tkach, 1972).

Devaux and Schott (1967) develop expressions for space-charge-limited peaks and trap exhaustion current peaks. In the first case

$$\frac{\Delta E_T}{kT_m} = \ln\left[\frac{2\alpha e\mu\nu k T_m{}^2}{\beta \Delta E_T}\left(1 + \frac{kT_m}{\Delta E_T}\right)\right] \tag{9.16}$$

where α is a constant related to the dielectric constant and ν is the thermal escape frequency. In the exhaustion case

$$\frac{\Delta E_T}{kT_m} = \ln\left[\frac{AkT_m^{\ 2}}{\beta \Delta E_T}\right] \tag{9.17}$$

where A is a constant proportional to the constant field in the device.

9.7 Interaction of Two Trap Levels

Solution of (9.3) and (9.4) presents some difficulties, even if only one trap level is considered, and becomes more difficult if two independent trap levels are present. However, if the number of electrons trapped on each impurity is always much less than the total number of impurities, the equations may be linearized and a computer solution becomes practical (Sacks, 1970).

Figure 9.7 shows the results obtained for a semiconductor with two impurity levels, 0.3 and 0.4 eV, below the conduction band edge. The values used were $N_1 = N_2 = 10^{17}$ cm^{-3}, $\sigma_1 = \sigma_2 = 10^{-16}$ cm^2 at 300°K with T^{-2} variation in temperature, a lifetime τ of 10^{-8} sec, and a heating rate β of 0.1°K sec^{-1}.

Curve (a) was computed with only one level at a time and the results superimposed. Curve (b) was computed by using the full set of equations, which included both levels. It is seen that the first peak has been shifted to a lower temperature by 11°K in the two-level calculation. Also the second peak has increased in size at the expense of the first. Both of these changes can be explained by retrapping at the 0.4-eV level of electrons emitted by the 0.3-eV level. Since the time constant for trapping associated with the 0.4-eV level is

$$(\sigma_2 N_2 v_{\text{th}})^{-1} \approx 2 \times 10^{-8} \text{ sec}$$

this mechanism can compete with the recombination time of 10^{-8} sec used in the calculation. Hence the effective recombination time for a given trap depends on the parameters of the other traps present. The second peak was not shifted in temperature because at this higher temperature, the emission rate of the first level had increased sufficiently to make it ineffective as a retrapping center for free carriers. The computer solution shows that as the 0.3-eV level empties, the 0.4-eV level occupancy increases, resulting in a lower current at the first peak and a higher current when the 0.4-eV level finally empties.

The effect of retrapping on the temperature peak versus heating-rate curves for the levels used above is shown in Fig. 9.8. A least-squares-error routine was used to fit straight lines for each set of five points used. The full lines were computed by considering each level independently, and the

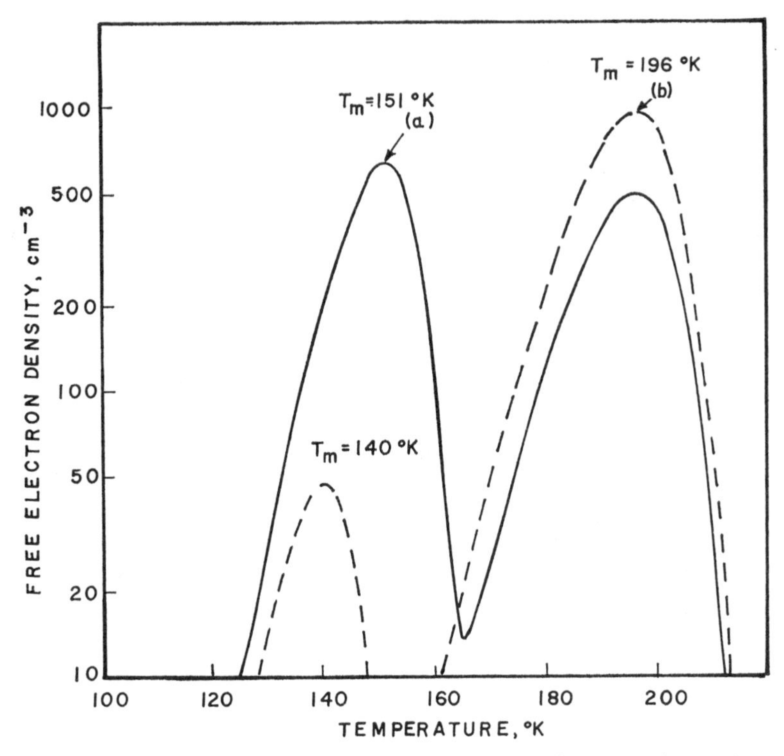

FIG. 9.7 Thermally stimulated carrier density for two levels, 0.30 and 0.40 eV. *(a) superposition of two single-level calculations; (b) calculation using two levels. (After Sacks, 1970.)*

dashed line by considering both levels. There is no dashed curve for the 0.4-eV level since there was no change when both levels were considered. However, retrapping by the 0.4-eV level has shifted the curve for the 0.3-eV level as well as changing its slope.

The error in energy for the single-level calculation occurred because the limit of the slow retrapping approximation had been reached, that is, $N_i \sigma_i v_{th} \tau \simeq 5$. For the double-level calculation, an additional error occurred because the effective lifetime of the model varied with temperature. Unexpectedly, these errors were in opposite directions and gave an answer close to the correct answer of 0.3 eV. In all three cases, the capture coefficients inferred from the line intercepts were a factor of 30 lower than the actual value used in the computer calculations.

Some caution must be used therefore in interpreting the results of thermally stimulated current measurements if more than one trap level is present at energy levels which allow overlap interactions to occur.

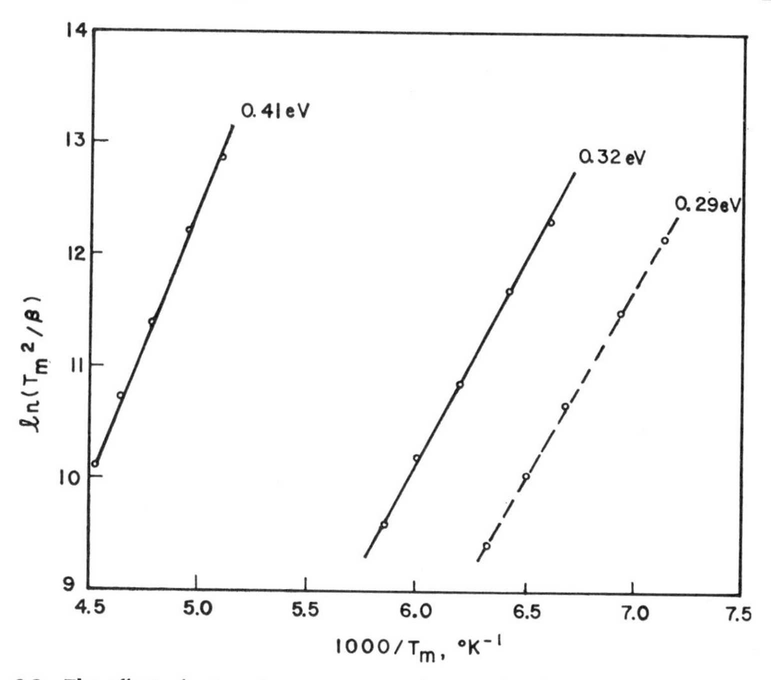

FIG. 9.8 The effect of retrapping on computed curves involving temperature peak T_m and heating rate β. (After Sacks, 1970.)

9.8 Some Experimental Studies with Silicon

Numerous investigators have used TSC and thermal glow to study a wide range of materials. A comprehensive review of such experiments is not possible here but a selection of results will illustrate the range and scope of the method. In this section we consider in particular TSC studies of silicon.

In 1965 Wysocki used the technique to study neutron-irradiated silicon but obtained disappointing results. He concluded that the thermally stimulated measurements were only marginally applicable since the signals observed tended to be obscured by the dark current. The next attempt was made by Atalla et al. (1966) in an examination of the properties of zinc-doped silicon. For high-resistivity material (10^3 to 10^4 Ω-cm at 300°K) they observed large peaks with respect to the dark current. They analyzed their data by equating the quasi-Fermi level to the trap depth at the peak as proposed by Bube (1955) and by the initial rise method of Garlick and Gibson (1948). By comparing their computed energy levels with the literature, they infer one peak to be associated with an *A* center (oxygen interstitial-silicon vacancy) and a second peak to be due to either a *C* center (a silicon divacancy) or an *E* center (phosphorus–silicon-vacancy pair). As for a third peak, they say, "There remains the possibility that the intermediate trap is associated with one of the well-known defects above the valence band, or with a donor state produced by interstitial zinc atoms."

Mirdzhalilova and Paritskii (1967) reported TSC measurements on $n^+ - p$ structures of silicon and germanium. The silicon device showed a single peak at about 135°K. Using $E_F = E_T$ at the peak they place the energy level at 0.24 eV. They do not state what impurity or defect this represents. A similar experiment was made by Schade and Herrick (1969). Their device was an off-the-shelf high-voltage p^+nn^+ silicon rectifier in a glass package. They observed four peaks and used $24kT_m = \Delta E_T$ to find the trap energy associated with each peak. By comparing their computed energy with published energy levels they produced a table of possible impurities. Their interpretations must be regarded as highly speculative. More recently Saunders and Wright (1970) used thermally stimulated charge release in an MOS capacitor to study interface states. They found too much structure in the experimental results to interpret. Their situation is very complex since charge can be released from the oxide or silicon as well as from the interface states they wish to study.

A simpler experimental situation was that of Bassett and Hogarth (1972) who examined the thermally stimulated current from the gold acceptor trapping level in silicon.

We conclude from this brief review that, although it is rather easy to see structure in TSC experiments, it is very difficult to obtain information about traps from the data. In particular, until recently there has been an almost complete absence of experiments with pure materials containing known trap levels. Atalla et al. (1966) knew that the silicon they studied contained zinc but they were unwilling to identify a peak definitely as associated with the impurity. All other reports are attempts to identify an unknown impurity by determining the energy level. In general the energy level alone is not sufficient to determine uniquely what element is causing the peak. As is shown subsequently, a wide range of energy levels and capture cross sections produce a peak at a particular temperature.

Consider now in some detail thermally stimulated current studies (Smith and Milnes, 1972) with silver-doped silicon which show some measure of success in correlating the experimental data with information on the impurity found by independent methods. The equations describing the interactions between the conduction band and the silver acceptor or recombination centers are

$$\frac{dN_{\mathrm{Ag}^-}}{dt} = -N_{\mathrm{Ag}^-}N_c\sigma_n{}^0v_{\mathrm{th}}\exp\left(-\frac{\Delta E_T}{kT}\right) + n(N_{\mathrm{Ag}} - N_{\mathrm{Ag}^-})\sigma_n{}^0v_{\mathrm{th}} \tag{9.18}$$

$$\frac{dn}{dt} = -\frac{dN_{\mathrm{Ag}^-}}{dt} - \frac{n}{\tau} \tag{9.19}$$

The first equation states that the rate at which electrons, n cm^{-3}, are added to the silver acceptor, N_{Ag^-}, is the difference between reemission and capture. The capture cross section for the reaction $Ag^0 + n \rightarrow Ag^-$ is $\sigma_n{}^0$, and v_{th} is the thermal electron velocity. Equation 9.19 accounts for conduction electron recapture and recombination, where τ is a recombination lifetime. A solution of the two equations involves assuming $dn/dt \ll dN_{Ag^-}/dt$, which permits restating (9.19) in the form $n = -\tau dN_{Ag^-}/dt$. It is also assumed that all traps will not be filled, giving $N_{Ag} \gg N_{Ag^-}$. Including these assumptions in (9.18) and using a linear heating rate β such that $T = T_0 + \beta t$, we find

$$\int_{N_{Ag^-}(0)}^{N_{Ag^-}(t)} \frac{dN_{Ag^-}}{N_{Ag^-}} = \int_0^t \frac{N_c(1°K)\,T^{3/2}\sigma_n{}^0 v_{th} \exp\left(-\Delta E_T/kT\right)}{1 + \tau N_{Ag}\sigma_n{}^0 v_{th}}\, d\left(\frac{T - T_0}{\beta}\right) \tag{9.20}$$

At this point consider the problem that is to be studied experimentally. The results from Smith and Milnes (1972) for $N_{Ag} = 2 \times 10^{15}$ cm^{-3} indicate that $\sigma_n{}^0 v_{th} = 10^{-10}$ cm^3 sec^{-1} and $\tau = 10^{-7}$ sec. For these values $\tau N_{Ag}\sigma_n{}^0 v_{th}$ may be neglected with respect to 1. Let us assume that the temperature coefficient of $\sigma_n{}^0 v_{th}$ is zero. While the assumption has some effect, it is slight compared to other possible sources of errors.

The conductivity is given by

$$\sigma(T) = e\mu_n n = -e\mu_n\tau \frac{dN_{Ag^-}}{dt} \tag{9.21}$$

Then if (9.20) is used to compute the derivative we obtain

$$(T) = e\mu_n\tau N_{Ag^-}(t = 0) N_c(1°K)\,T^{3/2}\sigma_n{}^0 v_{th} \times \exp\left(-\frac{\Delta E_T}{kT} - \frac{N_c(1°K)\sigma_n{}^0 v_{th}}{\beta}\int_{T_0}^{T} T^{3/2} \exp\left(-\frac{\Delta E_T}{kT}\right) dt\right) \tag{9.22}$$

The integral can be evaluated numerically using Simpson's rule in a straightforward computer calculation. Figure 9.9 shows the results for two values of the heating rate β. The theoretical results have been given on a current scale rather than expressed in carrier density. The current is derived from

$$i_{TSC} = \frac{\sigma(T) A V_a}{L} \tag{9.23}$$

where the conductivity is given by (9.21).

In these computations the area of the specimen was taken as 10^{-3} cm^2, the applied voltage V_a was 15 V, and the specimen length L was 0.57 cm. The mobility $\mu_n(T)$ was taken as $3500(T/200)^{-2.5}$ cm^2 V^{-1} sec^{-1}. The value used is somewhat unimportant since $N_{Ag^-}(t = 0)$ must also be set,

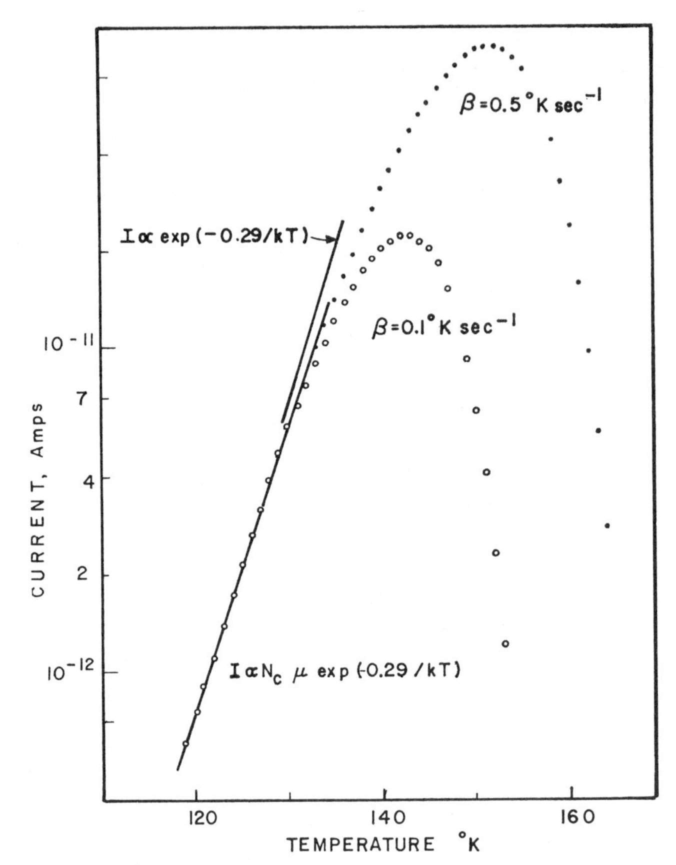

FIG. 9.9 Thermally stimulated current curves calculated from theory using values appropriate to experimental studies of silver-doped silicon where silver is an acceptor with a level 0.29 eV below the conduction band edge.
The curves are for heating rates of 0.1 and 0.5°K sec^{-1}. (After Smith and Milnes, 1972.)

which was done by comparison with experiment, and 10^{14} cm^{-3} was chosen. This probably is nonconstant between each run of the experiment but it is a scale factor which only affects the position of the curve on the current axis. The quantity $\mu_n \tau N_{Ag^-}$ $(t = 0)$ appears as a multiplying factor in determining the current. Each of these quantities can be varied within a factor of 2 or 3 and still will represent a reasonable value.

Several points should be made about Fig. 9.9. Qualitatively it can be seen that as the heating rate increases the temperature of the peak shifts to higher values. Also the amplitude increases with increase of the heating rate. The third consideration is that the slope of the initial rise is well fitted by a line $I \propto T^{-1} \exp(-\Delta E_T/kT)$. Using $I \propto \exp(-\Delta E_T/kT)$ leads to a slightly different fit as shown. The straight lines intersect at 120°K. This indicates the degree of error caused by neglecting nonexponential temperature dependences. Note also that changing the heating rate has no effect on this portion of the curve.

Although it is interesting to look at the correlation between a theoretical current and an experimental measurement, more can be learned about the problem if we consider what the known and unknown variables are. Experimentally we can determine the heating rate, the temperature T_m at which the current is a maximum, and the slope of the initial rise. From these we would like to determine the trap depth E_T and the capture coefficient $\sigma_n{}^0 v_{\text{th}}$. The current is a maximum when $d(\ln(\sigma))/dT$ is zero and $T = T_m$; so from (9.22)

$$\beta = \frac{N_c(1^\circ\text{K})\sigma_n{}^0 v_{\text{th}} \exp(-\Delta E_T/kT)}{(\Delta E_T/kT_m^{3.5} - 1/T_m^{2.5})} \simeq \frac{N_c(1^\circ\text{K})\sigma_n{}^0 v_{\text{th}} \exp(-\Delta E_T/kT)}{(\Delta E_T/kT_m^{3.5})} \tag{9.24}$$

The second term of the denominator is on the order of several percent of the first and will be dropped for convenience. This produces an error in the absolute relation, but most of the dependence is in the exponential term so the important features are not affected. The assumption of a zero temperature dependence for $\sigma_n{}^0 v_{\text{th}}$ also serves to cloud the problem, although probably not making a serious effect.

In general both ΔE_T and $\sigma_n{}^0 v_{\text{th}}$ are unknown. The procedure needed then is to find T_m for several values of β and solve simultaneously. Figure 9.10 shows computed curves for this procedure. Since the intersection of the curves represents the values that are being sought for ΔE_T and $\sigma_n{}^0 v_{\text{th}}$, the problem in the experiment now becomes clear. Namely, the intersection is at a very shallow angle, and small experimental errors may produce large shifts in the intersection point.

The obvious approach to obtain better accuracy is to make the curves intersect at a larger angle. However, this is not easy experimentally. As noted before, the amplitude of the peak decreases as the heating rate goes down. This places a lower limit on β since the signal current eventually falls into the leakage and noise of the measuring system. Increasing the heating rate also presents problems because of the measuring system design. If the cryostat is to cool to the operating temperature range in reasonable time, it must be closely coupled to its heat sink. This, however, requires large power to heat quickly and there is difficulty in coupling the heater coil to the piece to be heated. As a result the heating rate is limited by heater failure.

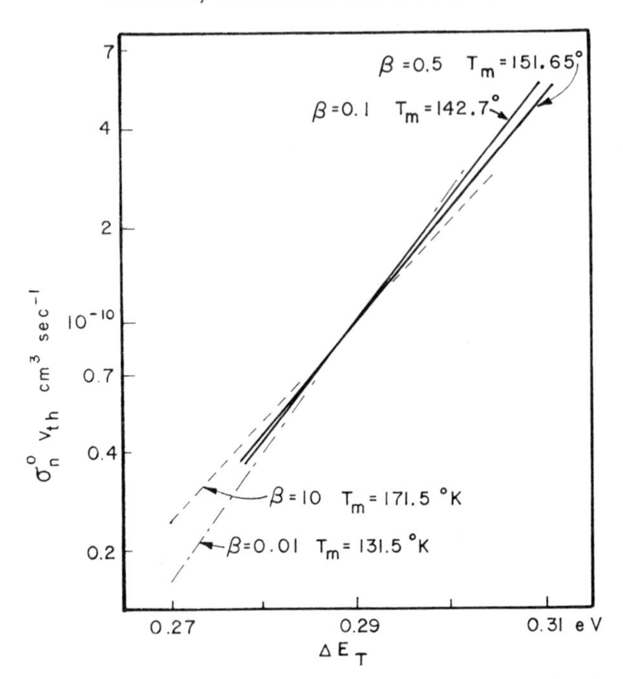

FIG. 9.10 Values of $\sigma_n{}^0 v_{th}$ and ΔE_T which give the same computed peak temperature for a given heating rate.

When two heating rates are considered, the intersection of the two lines represents the simultaneous solution for the trap depth and the capture term. (After Smith and Milnes, 1972.)

Conversely if the sample is well isolated, the power requirements become less but the cool-down time becomes impossibly long. Even by overcoming this, the problem would not be greatly reduced. Figure 9.10 includes curves for $\beta = 0.01$ and $10°K\ sec^{-1}$, a range of 1000. The temperature peaks are then separated by 40°K but the angle of intersection is still rather shallow.

In summary, it is very easy to state that the peak temperature will shift by a given number of degrees if the trap energy and capture coefficient are known, but working in the opposite direction is quite difficult and requires very accurate experimental results. The magnitude of the difficulty is illustrated in Fig. 9.11. This is a plot similar to Fig. 9.10. The $\beta = 0.1$ line is the same and is drawn for exact results of $\Delta E_T = 0.29$ eV and $\sigma_n{}^0 v_{th} = 10^{-10}\ cm^3\ sec^{-1}$. Assume that this line ($\beta = 0.1$, $T_m = 142.7°K$) is without error, but that there is an uncertainty of $\pm 1°K$ in determining T_m for $\beta = 0.5$. The correct value should be $T_m = 151.65°K$ in order to correspond to $\Delta E_T = 0.29$ eV and $\sigma_n{}^0 v_{th} = 10^{-10}\ cm^3\ sec^{-1}$, but for $T_m = 150.65$ and $152.65°K$ the lines computed are as shown in the figure.

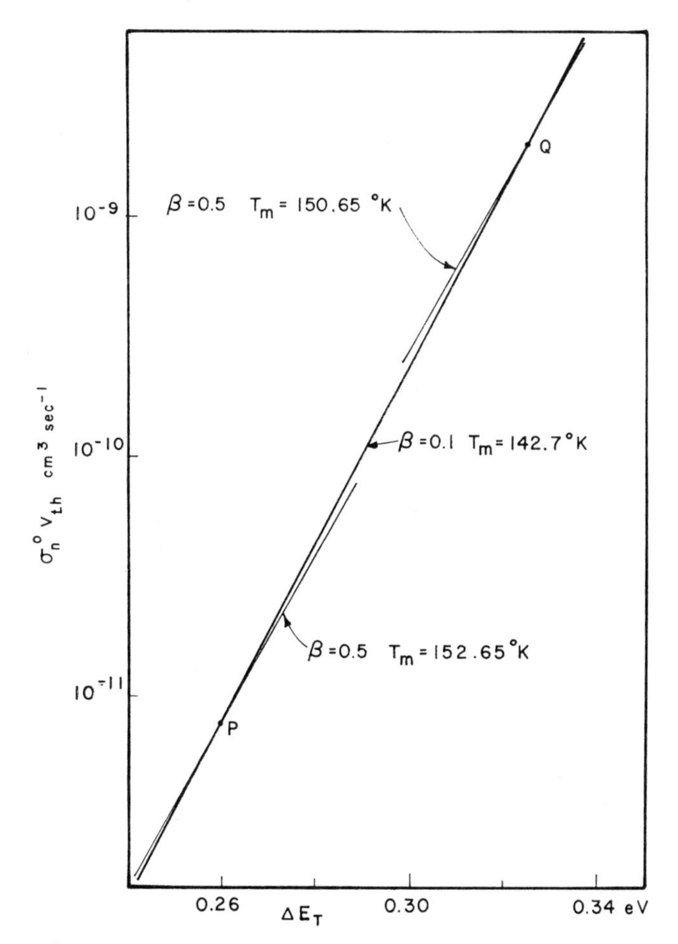

FIG. 9.11 Effect on ΔE_T and $\sigma_n{}^0 v_{th}$ values if there is an error of $\pm 1°$K in determining the temperature T_m for the current peak when the heating rate β is 0.5°K sec^{-1}. *The alternative intersection points for $\pm 1°K$ error are P and Q and represent a considerable spread in the values inferred for ΔE_T and $\sigma_n{}^0 v_{th}$. (After Smith and Milnes, 1972.)*

The intersection of the $\beta = 0.1$ and 0.5 lines now show an uncertainty in ΔE_T of from 0.256 to 0.326 eV, and in $\sigma_n{}^0 v_{th}$ of from 5.6×10^{-12} to 2.2×10^{-9} cm^3 sec^{-1}. The range for the capture coefficient makes this measurement about equal in value to an educated guess. The ΔE_T values, however, deserve more consideration since as a percentage the range is only 20 to 30%. Unfortunately this is not a realistic measure of the way the number is used. In almost any application, ΔE_T will appear in an exponential term of the form exp $(-\Delta E_T/kT)$ when used to determine the trap effects on semiconductor properties. Whether such a range of data is useful depends on the application. In the case of attempting to identify an impurity by its energy level the problem is doubly complex since the experimental ΔE_T can be referenced to either band edge.

In a typical apparatus for thermally stimulated current studies the temperature corresponding to current maximum can be determined to $\pm 1°K$, provided attention is given to the details of the measurement. However, the sample calculation that we have developed shows that the temperature error needs to be held to within 0.1°K if ΔE_T and $\sigma_n{}^0 v_{th}$ of the trap are to be determined independently with good accuracy. The construction of a system with this accuracy was not attempted by Smith and Milnes (1972). However, since they confined their experimental studies to silicon containing an impurity (silver) with a known ΔE_T and $\sigma_n{}^0 v_{th}$ it was possible to confirm the general outline of the theory even with a system of only $\pm 1°K$ accuracy. The silicon, before silver doping, was 50 or 200 Ω-cm n type and the silver was diffused to a uniform concentration for 8 h at either 1100°C (about 5.6×10^{14} Ag cm^{-3}) or 1200°C (about 2×10^{15} Ag cm^{-3}).

The experimental apparatus consisted of a double-wall Pyrex vacuum container with a copper and stainless-steel cold finger mounted inside. The space between the inner Pyrex wall and the cold finger was also evacuated. Power was obtained from two lead–acid storage batteries in order to eliminate power line interference. Heating rates up to 1°K sec^{-1} were obtained in certain ranges. The temperature was monitored by an iron constantan thermocouple attached to the copper block opposite the sample. The sample was connected in series with a carefully shielded 15-V battery and a Keithley 602 MOS electrometer ammeter. Two x-y recorders were used to monitor the current and the thermocouple emf versus time.

The current-versus-temperature recorder produced TSC curves such as that shown in Fig. 9.12. The x axis shows increasing thermocouple voltage to the right, and the y axis is current. This plot was obtained for a sample that had been cooled from room temperature to -6.6 mV (120°K), and then exposed to an intense microscope light which increased the current from less than 10^{-13} to 10^{-4} A for 1 min. After holding the sample in the dark for 5 min it was then heated at 0.37°K sec^{-1} and Fig. 9.12 was obtained. Numerous variations of exposure time, temperature, dark time, and heating rate were tried and some gave different results, which are discussed later. The illumination above produced the most useful data and was used to obtain the results that are presented in detail below.

When the experimental results are converted from thermocouple readings to temperature, it is convenient to plot the current on a logarithm scale

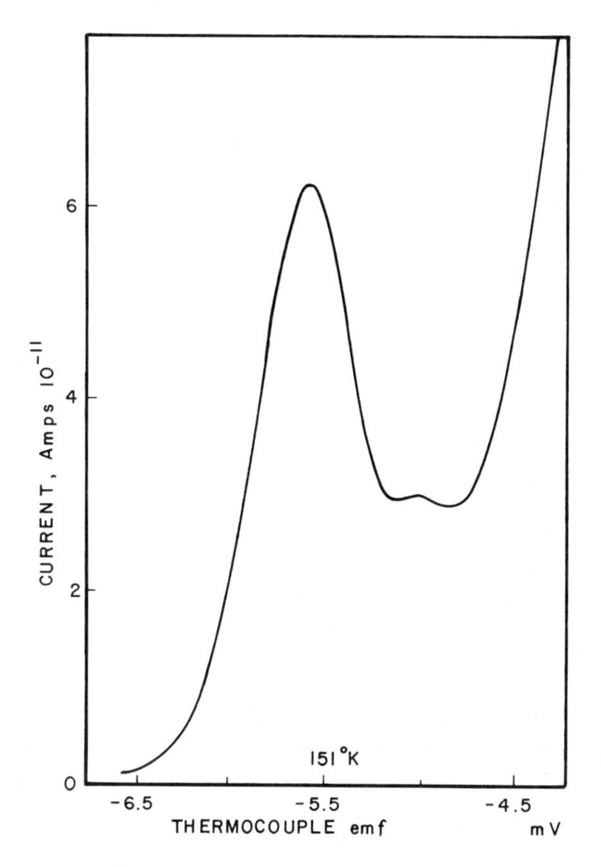

FIG. 9.12 Unprocessed *x-y* recorder plots showing the thermally stimulated current peak for 50 Ω-cm *n*-type silicon diffused with silver at 1200°C (specimen 50N12). *The illumination conditions prior to the run are discussed in the text. The current observed in the absence of preillumination is too small to be seen on this scale. (After Smith and Milnes, 1972.)*

and the temperature on a linear scale. Figure 9.13 shows this done for two runs of different heating rates. It will be noted that the higher heating rate does produce higher peak temperature and amplitude as expected. The initial rise of run 3P1 shows an activation energy of 0.29 eV in agreement with Smith and Milnes (1971) and Thiel and Ghandhi (1970). However, run 4P2 (0.27°K $\sec^{-1}$) shows the effect of exposing the specimen at a temperature of 115°K. The first several points are increased by charge released from a lower temperature trap. If these points were included in the initial rise data, assuming that we did not know that silver at 0.29 eV was

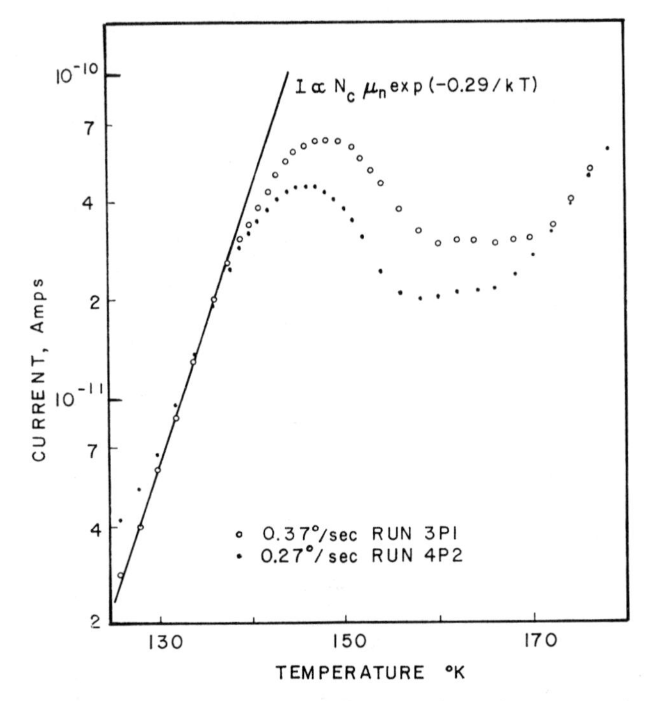

FIG. 9.13 Thermally stimulated current experimental results converted to current-temperature plots showing initial rise and peak shift with heating rate for silver-doped silicon. (After Smith and Milnes, 1972.)

the impurity being sought, some quite invalid interpretation might have resulted.

Figure 9.14 displays the experimental results for $1/T_m$ versus β with each point being the average of several runs. Samples of 50 and 200 Ω-cm n-type silicon silver diffused at 1100 and 1200°C were tested and all results were similar for similar conditions. Thermally stimulated current runs on the undiffused silicon material gave no observable peaks under any conditions because of the very large background currents. Figure 9.14 also includes calculated lines corresponding to (9.24) for trap levels 0.30, 0.29, and 0.28 eV. All the curves are for $\sigma_n{}^0 v_{\rm th} = 10^{-10}$ cm^3 sec^{-1}. If the capture coefficient has another value, the lines shift vertically up or down as may be seen from (9.24). Inspection shows that the slope of the $\Delta E_T = 0.30$ eV line is slightly greater than the other two, and this helps to determine the energy level. The vertical placement defines $\sigma_n{}^0 v_{\rm th}$. This figure demonstrates another approach to the problems discussed in connection with Figs. 9.10 and 9.11.

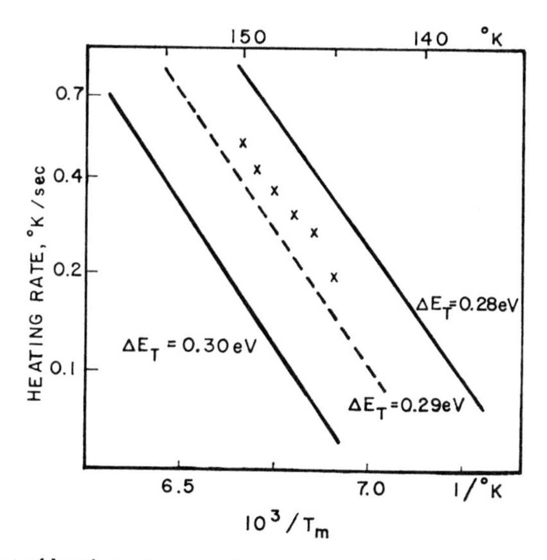

FIG. 9.14 Effect of heating rate on peak temperature shift.
The points (×) show the average of several runs at the same heating rate for the same silver-doped silicon specimen. The straight lines are the calculated results for trap levels of 0.28, 0.29, and 0.30 eV with $\sigma_n{}^0 v_{th} = 10^{-10}$ *cm*^{3} *sec*$^{-1}$*. (After Smith and Milnes, 1972.)*

Without prior knowledge of the sample, one would be reluctant to draw a straight line through the data and compute the trap depth and capture coefficient. Since we are fairly confident of the silver content in the sample from other experiments (Smith and Milnes, 1971) we can state that there is a correspondence between theory and experiment. The scatter in the points is due largely to the temperature measuring problems. From the scale of Fig. 9.13 it is obvious that there is difficulty in determining the exact temperature values for the peaks to better than the nearest degree. By averaging several results the accuracy is improved somewhat but still leaves much to be desired.

In spite of these difficulties one can conclude that a consistent picture has appeared. The peak temperature shifts with heating rate in the direction and to the degree expected. The theory then is supported by the experimental results with these specimens containing a known trapping level. However, if thermally stimulated experimental data are used as the sole sources of information, the results clearly demonstrate that it is easily possible to infer false values for trap depths and capture cross sections.

Smith and Milnes (1972) also discuss results observed for temperature and illumination conditions other than the ones described above. It is rather easy

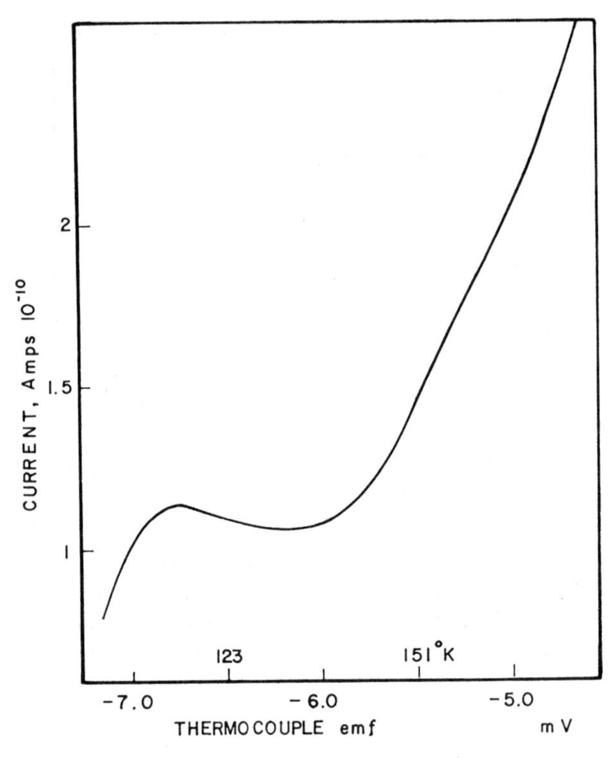

FIG. 9.15 Results of illuminating sample 50N12 at 100°K and heating at 0.645°K sec^{-1}.
Notice the considerable difference from Fig. 9.12, where the illumination was at 120°K and the heating rate was 0.37°K sec^{-1}. (After Smith and Milnes, 1972.)

to miss information completely or arrive at incorrect conclusions if care is not taken to cover a range of conditions. Figure 9.15 was obtained by cooling the sample to approximately 100°K, exposing to room light for 15 sec, and holding in the dark for 5 min before heating. The temperature was then increased at 0.645°K sec^{-1}. One peak appears around 115°K at an amplitude of 10^{-10} A. Faster heating increased the amplitude and shifted the peak. Careful examination of Fig. 9.15 also shows a nonlinearity in the rising portion of the curve following the peak. At faster rates this tends to straighten out.

Figure 9.16 shows the same sample cooled to 100°K and exposed to intense microscope light for 1 min followed by 10 min in the dark. Heating was at 0.51°K sec^{-1}. Now two peaks are clearly evident, one at 120°K and the

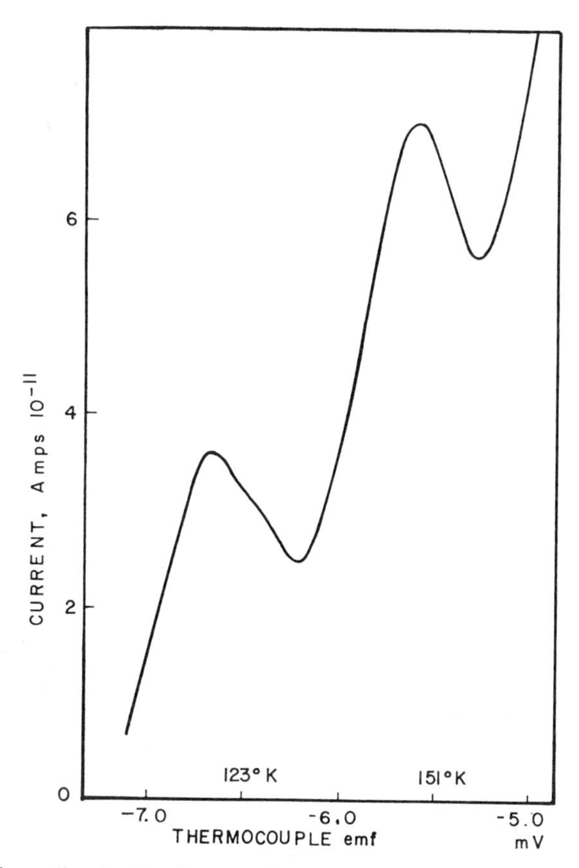

FIG. 9.16 Thermally stimulated current for the same sample as in Fig. 9.15, but with a heating rate of 0.51°K sec^{-1}. (After Smith and Milnes, 1972.)

second near 150°K. The amplitudes are now at 3 and 6 × 10^{-11} A even though the time and intensity of excitation are much larger than previously. Also note the nonlinearity on the falling side of the first peak. Figure 9.17 shows the same sample, again excited by microscope light for 1 min and held in the dark for 10 min before heating at 0.34°K sec^{-1}. The amplitude again drops and the single peak at 120°K of Fig. 9.16 now is clearly a double peak. The peak near 150°K again appears. Any attempt, however, to analyze the peaks according to one of the theories available to assign an energy and capture cross section to a trap would be an extremely questionable action for the reasons of accuracy of interpretation that we have previously discussed.

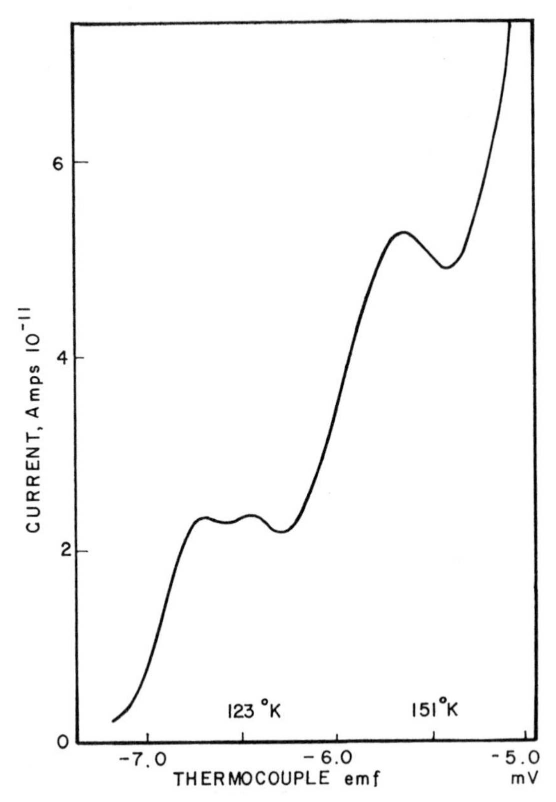

FIG. 9.17 Thermally stimulated current for the same specimen and preillumination as for Fig. 9.16, but with a heating rate of 0.34°K sec^{-1}.
Note the emergence of the twin peak at about 123°K. The traps causing these peaks are unknown. (After Smith and Milnes, 1972.)

A number of other combinations of light and temperature were investigated. The length of time the sample was exposed to light had no apparent effect on the TSC curve. For example, exposing 10 min to intense microscope light and waiting 1 min produced substantially similar results if done at the same temperature and heating rate. In general the only significant effects observed depended on heating rate and temperature of exposure. The exposure temperature determined primarily which peaks were emphasized. All peaks were apparently active regardless of the excitation conditions, although peaks were almost masked in some cases.

In conclusion, TSC experiments can produce considerable data on high-resistivity silicon but interpretation becomes extremely difficult if more than one peak is seen. The samples discussed here have also been studied by examination of the decay of photoinduced conductivity (Smith and Milnes, 1971). These studies showed trap action through the silver center to be dominant, although a long tail on the decay gave some indication of the presence of something unidentified. Nevertheless by careful isolation of the TSC silver peak it is possible to show that a straightforward theory is quantitatively valid for it.

9.9 Other Experimental Thermally Stimulated Current Studies

The II–VI compounds have been studied extensively by thermally stimulated current techniques; see, for example, Bube (1960). The general characteristic of these materials, which makes this method attractive, is their high resistivity. When the traps have emptied, the curves return almost to zero. This produces distinct peaks without interference from dark current. However, problems of interpretation are caused by the presence of more than one trap.

A complex TSC curve for copper-doped high-resistance GaAs is shown in Fig. 9.18. Many peaks are present in this curve and the fine structure from −126 to −96°C is completely reproducible. These curves demonstrate the sensitivity of the method and the complexity of interactions in a supposedly simple situation. It also represents one of the common problems of TSC experiments; that is, the amount of information produced goes far beyond

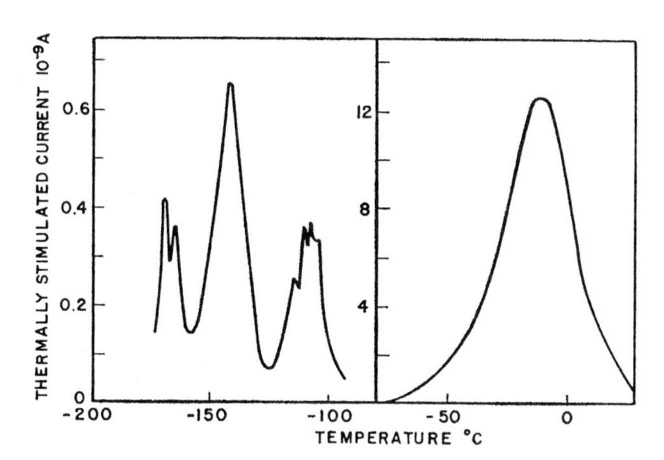

FIG. 9.18 Thermally stimulated current for GaAs crystal with copper doping. (After Blanc et al., 1961.)

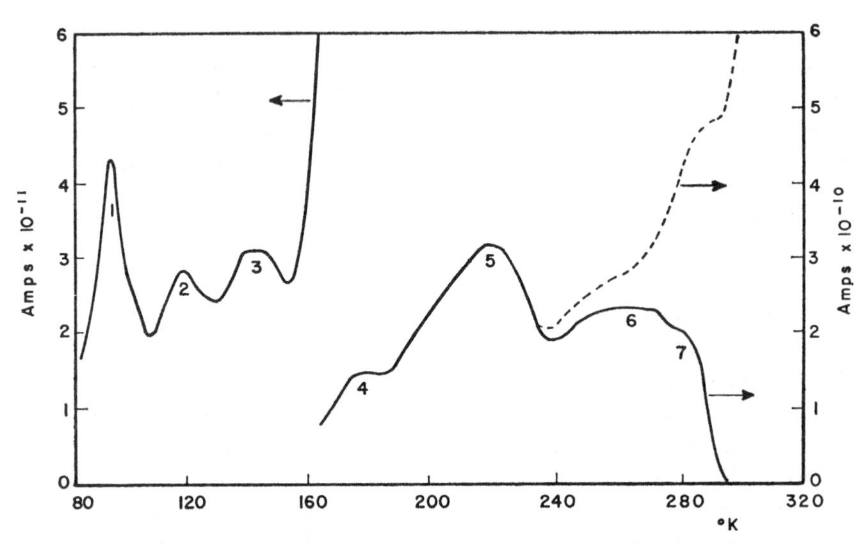

FIG. 9.19 Thermally stimulated current plot for copper-doped GaAs.
The dashed line is without the dark current subtracted. (After Saunders and Jewitt, 1965.)

reasonably simple explanations. Figure 9.19 is another complex TSC plot for copper-doped GaAs. Note that the dark conductivity in this crystal becomes important in resolving peaks 6 and 7. The dashed line is the data as taken and the solid line is with the dark current subtracted.

Thermal cleaning (Hoogenstraaten, 1958) can be used to enhance the peaks, with the results as shown in Fig. 9.20. The procedure is as follows.

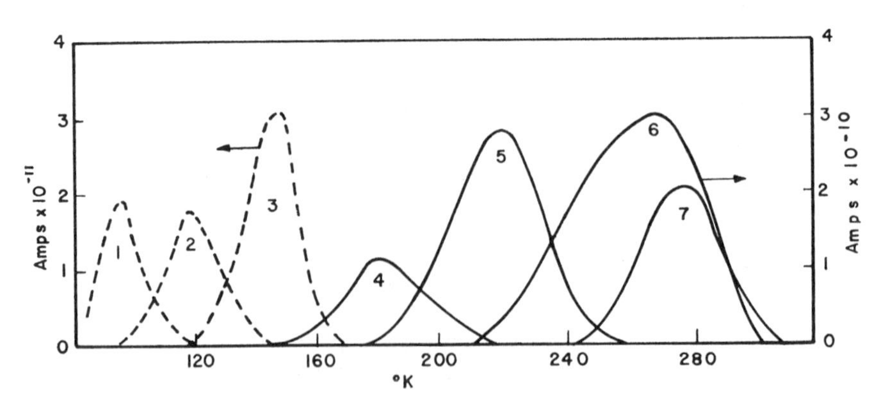

FIG. 9.20 Thermally stimulated current with thermal cleaning applied to the GaAs specimen of Fig. 9.19. (After Saunders and Jewitt, 1965.)

Peak 1 is plotted, but during peak 2 the heating is stopped and the temperature held constant until the current decays close to the dark level. Heating is then resumed and peak 3 plotted. This is continued until all the odd-numbered peaks are plotted. The entire process is then repeated to define the even-numbered peaks. This operation gives the rising part of each peak without interference. The curves are multiplied by a normalizing factor to account for the slight loss in peak height due to the cleaning process, and the rising parts are subtracted from the total to obtain the falling half of each peak. All this makes determining parameters of individual curves more certain.

Figure 9.21 shows a similar technique applied to zinc-doped silicon. It is particularly difficult to observe TSC effects in silicon because of its high

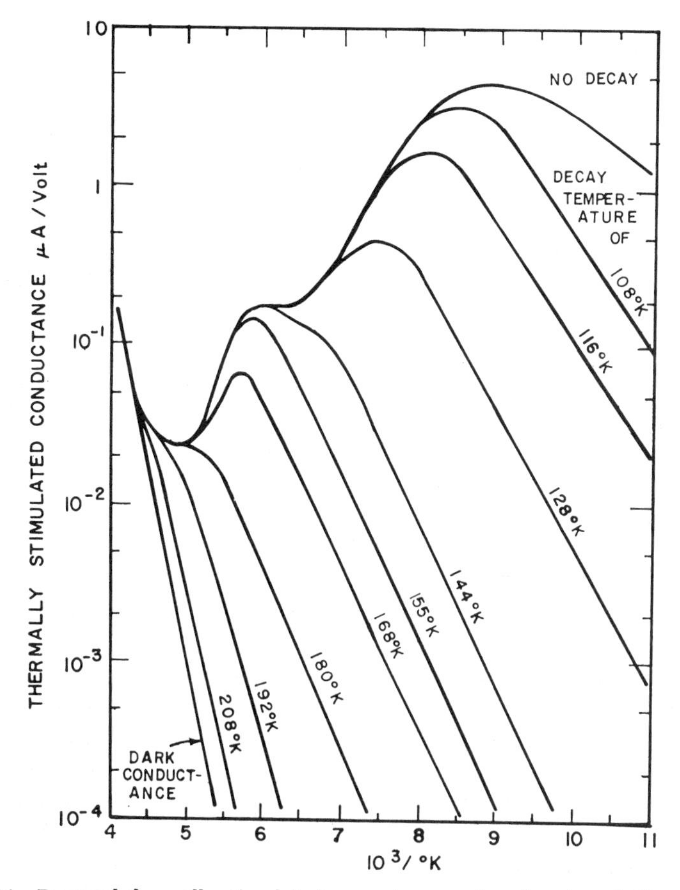

FIG. 9.21 Decayed thermally stimulated current curves for zinc-doped silicon. (After Atalla et al., 1966.)

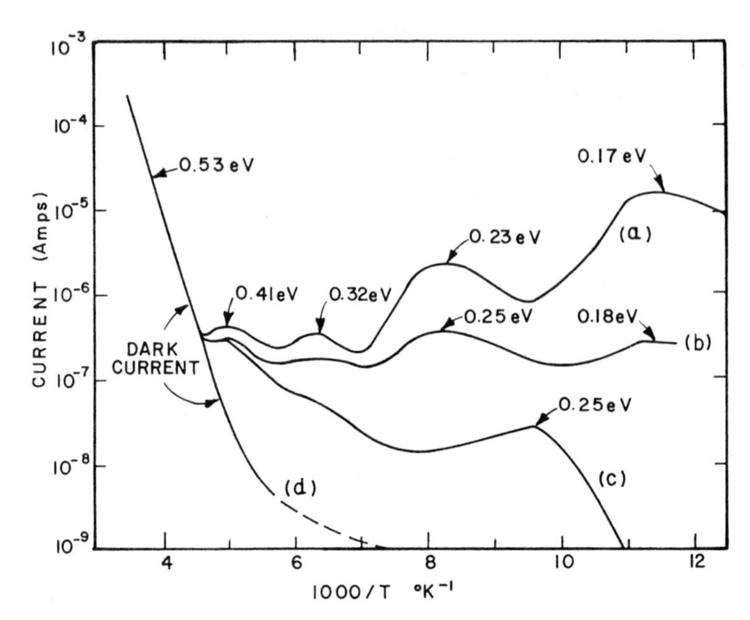

FIG. 9.22 Thermally stimulated current for zinc-doped silicon showing effect of various initial excitation.
(a) after white light excitation; (b) after 1.3 μm illumination; (c) after 1.6 μm illumination; (d) dark current. (After Atalla et al., 1966.)

dark conductivity. The samples studied here were 10^3 to 10^4 Ω-cm at 300°K with electron lifetimes of about 300 μsec. To obtain Fig. 9.21 the sample is cooled and photoexcited, then heated to the temperature shown on the curve, and quickly recooled to a low temperature. Upon reheating without further excitation the curves shown are traced. This process removes carriers from shallow traps and the data can be analyzed by the initial rise method. Care must be taken in this analysis. For example, the slope of the 108 and 116°K lines are identical, whereas the 128°K slope does not agree. The authors attribute this to the presence of other trap depths.

The wavelength of light used to photoexcite the crystal is of considerable importance in the type of curve resulting. Atalla et al. (1966) found that if gold-doped silicon was illuminated with white light and then with extrinsic light, the white light effect was erased. Figure 9.22 is an example of the use of various wavelengths of excitation. Band-gap light in silicon is about 1.1 μm.

Figure 9.23 shows an even more dramatic effect of light. White light excitation produced the solid curve. With band-gap light, $h\nu > 1.25$ eV, three other peaks were found, and the 0.2-eV peak was lost. Presumably the

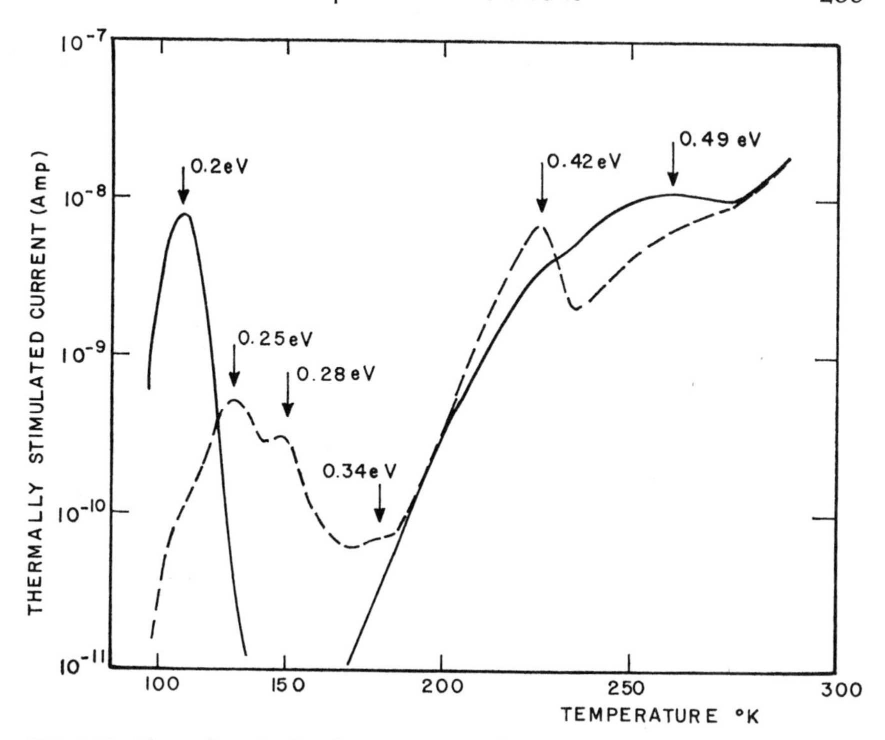

FIG. 9.23 Thermally stimulated current in undoped GaAs, showing the effect of different excitation.

The solid line is the result after white light excitation and the dashed line is after band-gap illumination. (After Schade, 1969.)

0.25 peak is from another source rather than just shifting of the 0.2 peak. One explanation of the difference in behavior for different illumination is based on the position of the quasi-Fermi level. Depending on excitation, the occupancy of the various levels (and the Fermi level) changes, and centers can change their character from recombination centers to traps. Schade (1969) has also suggested that at high trap concentrations trapped carriers can directly pass from the traps into recombination centers without being reactivated into a conduction band and thus being measured as a thermally stimulated current.

Several experimenters have shown that the TSC curves are a function of the initial trap occupancy (Kulshreshtha and Goryunov, 1966). The appearance and disappearance of thermally stimulated current peaks depending on the illumination conditions are also observed for II–VI compound semiconductors such as CdS (Marlor and Woods, 1965; Tscholl,

1966). Evidence is accumulating that photochemical processes involving defects may be taking place in II–VI compounds depending on the temperature and illumination cycles applied (Woods and Nicholas, 1964; Im et al., 1970a).

Figure 9.24*a* shows the effects of various illumination times on the TSC curves of "undoped" high-resistivity GaAs (0.01 ppm of Cu was detected).

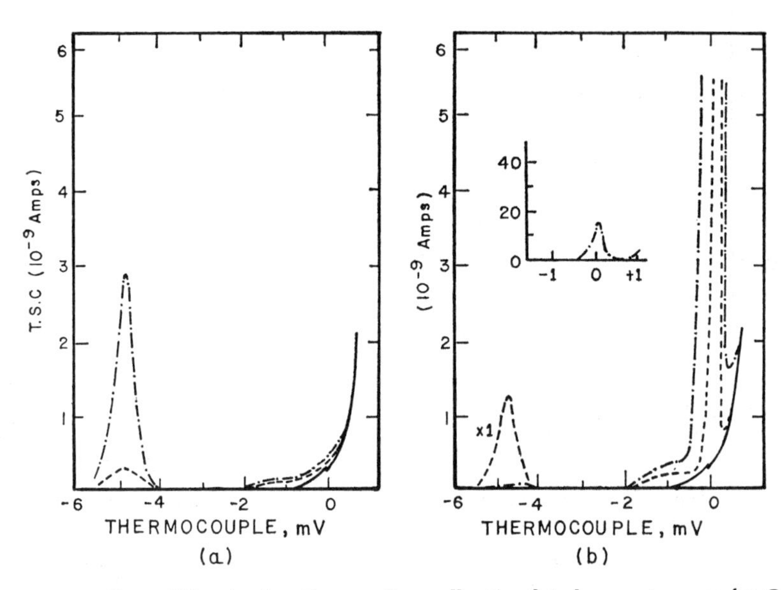

FIG. 9.24 Effect of illumination time on thermally stimulated current curves for GaAs. *(a) ——— dark current, — — — 20 sec illumination, —·—·— 11 min illumination; (b) ——— dark current, — — — 2 h illumination, —·—·— 12 h illumination. The inset figure is for 12 h illumination to a higher scale. The reference junction of the Cu–constantan thermocouple was at 0°C. (After Haisty, 1967.)*

With very weak illumination for times on the order of 5 sec, as many as seven peaks could be observed, and these are attributed to surface states (Haisty, 1967). With stronger illumination the curves shown in Fig. 9.24*b* were recorded. The lower peak at 115°K increased in size up to about 30 min then decreased and shifted slightly to higher temperature. The peak at 278°K continued to grow after 12 h of illumination. All excitation was done with the sample at 77°K. Haisty suggests that the appearance and disappearance of peaks might be due to movement of defects under different states of charge. Blanc et al. (1964) have proposed that 0.2 and 0.5 eV peaks in GaAs correspond, respectively, to paired and unpaired defects. Since the paired defects lose their coulombic attraction during illumination, and since vacancy motion has been observed at low temperatures in semiconductors, they could separate by a few interatomic distances, which is sufficient to change their energy level from 0.2 to 0.5 eV. Upon heating, the trapped charges are released and the defects, now charged, are quickly paired again.

Although in a simple situation the area under a TSC peak may be thought of as related to the total number of traps at a particular energy level if all the traps are initially filled, enough has been said to make it clear that this is not a safe assumption. In most practical studies only very approximate ideas of the trap densities may be inferred from TSC measurements.

Thermally stimulated current peaks are themselves a peaking function of the wavelength of the exciting light since this determines the absorption coefficient and therefore the spatial distribution of the excited traps. This effect has been used by Wakim (1970) to demonstrate the band-gap energy of ZnSe.

9.10 Effect of Electric Field on TSC Measurements

The effects of electric field on the thermally stimulated conductivity process has been studied by Dussel and Bube (1966*c*). In the stationary state, the density of trapped electrons at a given depth is given by

$$n_T(E, \mathscr{E})\, dE = \frac{c(E, \mathscr{E}) N_T(E)\, dE}{c(E, \mathscr{E})n + P(E, \mathscr{E})}\, n \tag{9.25}$$

where $N_T(E)$ is the density of traps at energy depth E; $c(E, \mathscr{E})$ is the capture coefficient for traps of depth E when the applied field is $\mathscr{E}$; $P(E, \mathscr{E})$ is the thermal emission probability for traps of depth E in the presence of field $\mathscr{E}$; and n is the density of free electrons. For $\mathscr{E} = 0$, $P_0 = P(E, 0) = c_0 N_c \exp(-E/kT)$, where N_c is the effective density of states in the conduction band and $c_0 = c(E, 0) = \sigma_n(E, 0)v_{\text{th}}$; $\sigma_n(E, \mathscr{E})$ is the capture cross section of traps with depth E for free electrons in the presence of a field $\mathscr{E}$, and v_{th} is the thermal electron velocity.

For $\mathscr{E} = 0$, the density of free electrons is given by

$$n = \frac{P_0}{c_0} \exp\left(-\frac{E_{\text{F}n} - E}{kT}\right) \tag{9.26}$$

where $E_{\text{F}n}$, like E, is a positive energy measured down from the bottom of the conduction band. For $\mathscr{E} \neq 0$, (9.25) becomes

$$\frac{n_T(E, \mathscr{E})\, dE}{N_T(E)\, dE} = \left[1 + \frac{P(E, \mathscr{E})}{c(E, \mathscr{E})n}\right]^{-1} = \frac{1}{1 + \exp\left[(E_{\text{F}T} - E)/kT\right]} \tag{9.27}$$

where

$$E_{FT} = E_{Fn} + kT \ln \frac{P(E, \mathscr{E})c_0}{P_0 c(E, \mathscr{E})} \tag{9.28}$$

This equation means that if the capture and/or thermal emission probabilities of a trap depend on the applied electric field, the quasi-Fermi level determining the trap occupancy is not the same as that determined from the density of free electrons.

If the distribution of traps to be measured in the TSC is quasicontinuous with a characteristic temperature T^* much larger than the temperatures at which the TSC is measured, the maximum in the TSC is determined by traps with depth $E \approx E_{FT}$, assuming that the trapping parameters are equally affected at different depths. In the following discussion, therefore, we may treat the distribution as if it were effectively a discrete trap with depth E_{FT} and density $N_T = N_T(E_{FT})kT$.

The kinetic equations governing TSC are

$$\frac{dn}{dt} + \frac{dn_T}{dt} = -\frac{n}{\tau} \tag{9.29}$$

and

$$\frac{dn_T}{dt} = cn(N_T - n_T) - Pn_T \tag{9.30}$$

where τ is the lifetime of free electrons. Generally, $n \ll n_T$ and $dn/dt \ll dn_T/dt$, so that

$$\frac{dn_T}{dt} = -\frac{n}{\tau} \tag{9.31}$$

Eliminating dn_T/dt from (9.30) and (9.31) gives

$$n = \frac{\tau P n_T}{1 + \tau c(N_T - n_T)} \tag{9.32}$$

The TSC maximum is given by $dn/dt = 0$ with a linear heating rate, $dT = \beta\, dt$; hence

$$\exp\left(\frac{E}{kT_m}\right)\frac{\beta E}{kT_m{}^2} = cN_c \frac{1 + \tau c N_T}{[1 + \tau c(N_T - n_T)]^2} \tag{9.33}$$

where T_m is the temperature for the maximum TSC. Since strong retrapping is assumed for the traps under consideration, $\tau c(N_T - n_T) \gg 1$ and (9.33) reduces to

$$\exp\left(\frac{E}{kT_m}\right)\frac{\beta E}{kT_m{}^2} = \frac{N_c}{\tau}\frac{N_T}{(N_T - n_T)^2} \tag{9.34}$$

The measured current at the TSC maximum will be determined by (9.32), using the appropriate values, whereas the temperature of the maximum T_m will be determined by (9.34). Both the maximum current and T_m depend on the value of the quasi-Fermi level for traps E_{FT}, which according to (9.28) is determined by the crystal history previous to the TSC measurement.

The predictions of the analysis above have been verified by Dussel and Bube (1966c) for photosensitive CdS–CdSe single-crystal material. The specimen was heated to 400°K and then cooled to 85°K in the dark. Voltage was then applied and the photoconductivity at 85°K was found to increase with time, and the rise was faster the higher the applied voltage, as shown in Fig. 9.25. The effect is due to a decrease in the rate of electron capture

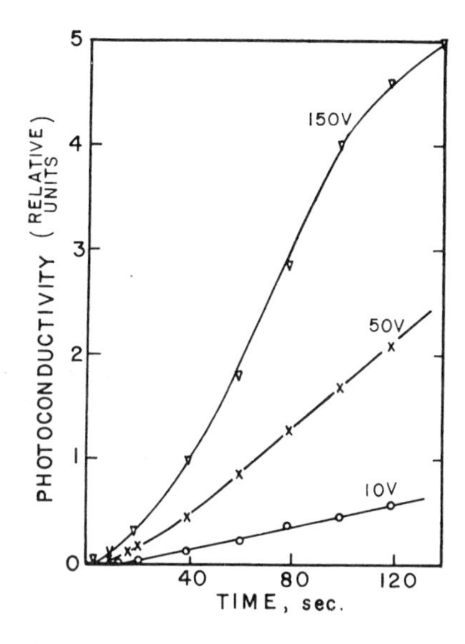

FIG. 9.25 Photoconductivity rise curves at 85°K for CdS–CdSe for an immediate excitation intensity as a function of applied voltage. (After Dussel and Bube, 1966c.)

by traps at increased field strength. When thermally stimulated current measurements were made, after exciting for 30 min with constant excitation intensity with different applied voltages, the results were as shown in Fig. 9.26. The peak of the thermally stimulated current shifts toward higher temperatures and lower conductivities (deeper quasi-Fermi level for traps) with increasing voltages, indicating an increasing difficulty in trap filling.

The general problem of capture cross-section dependence on electric field is discussed in Chapter 5.

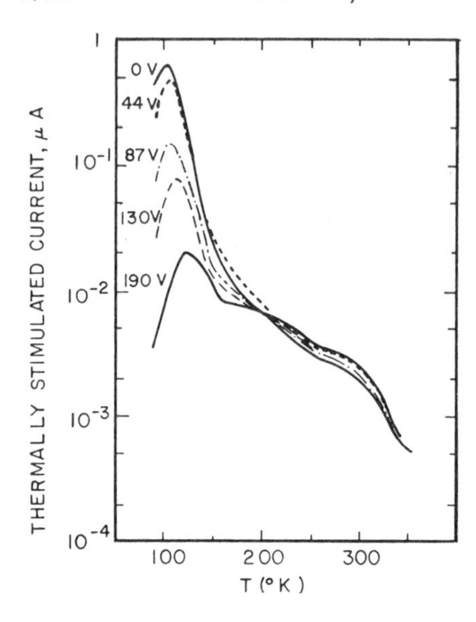

FIG. 9.26 Thermally stimulated conductivity curves (CdS–CdSe) measured with 10 V applied after excitation periods at 85°K during which voltages between 0 and 190 V were applied. (After Dussel and Bube, 1966c.)

9.11 Thermally Stimulated Current in Junction Depletion Regions

As mentioned previously, thermally stimulated current measurements are difficult to make in a semiconductor that is only moderately high in resistance (less than 10^5 Ω-cm) since the dark current tends to be too large. Wysocki (1965), for instance, concludes that TSC measurements are marginal in neutron-irradiated silicon.

The difficulty is overcome if the semiconductor is part of a junction structure. If the junction is reverse biased, the dark current is greatly reduced and emission from traps in the depletion region can be studied. Sah and coworkers (1970b) have made such measurements at constant temperature with a step of applied voltage causing the transient condition, as discussed in Chapter 8. Alternatively, the voltage may be held constant and the temperature increased to provide the thermally stimulated current.

Figure 9.27 shows a TSC curve obtained by Schade and Herrick (1969) for a silicon rectifier (off the shelf). Weisberg and Schade (1968) have studied $GaAs_{0.5}P_{0.5}$ p^+n junctions by this technique. Figure 9.28*a* shows their results for excitation by light at -10 V bias and by a reverse voltage excitation step from 0 to -10 V, both done after cooling. They explain the results by assuming that a neutral deep donor is in nonequilibrium and produces one peak when heated. Excitation by light changes the population of both centers. Figure 9.28*b* shows the effect of reverse bias. As expected

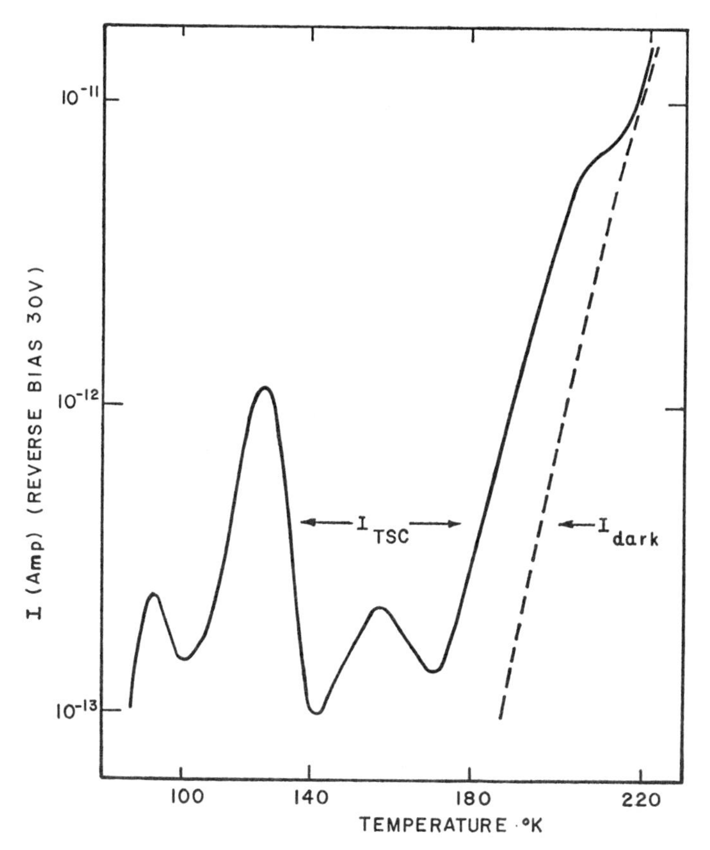

FIG. 9.27 Thermally stimulated current for a reverse-biased silicon diode. (After Schade and Herrick, 1969.)

the curve amplitude increases as the depletion region widens to include more centers. The quantitative relationship was checked by determining the volume of the depletion layer from capacitance measurements. This also demonstrated that the traps were uniformly distributed.

Mirdzhalilova and Paritskii (1967) have studied *p-n* junctions in silicon and germanium. They give the expression

$$i_{\max} \simeq \frac{eL_n N_c}{\tau_n} \exp\left(-\frac{E_T}{kT_m}\right) A\delta \qquad (9.35)$$

where L_n and τ_n are the diffusion length and lifetime of electrons, A is the junction area, and δ is the quantum efficiency of the junction.

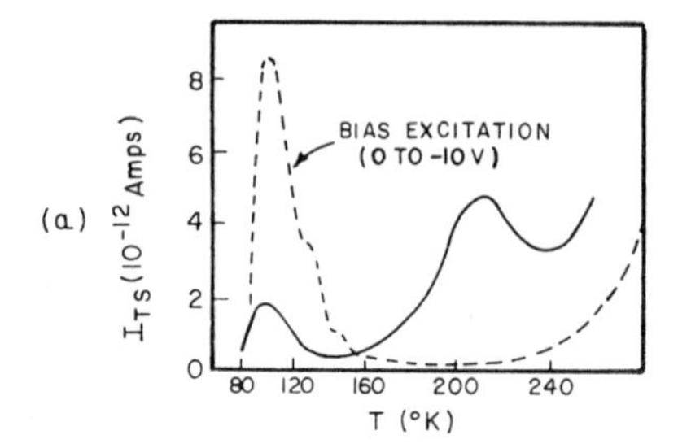

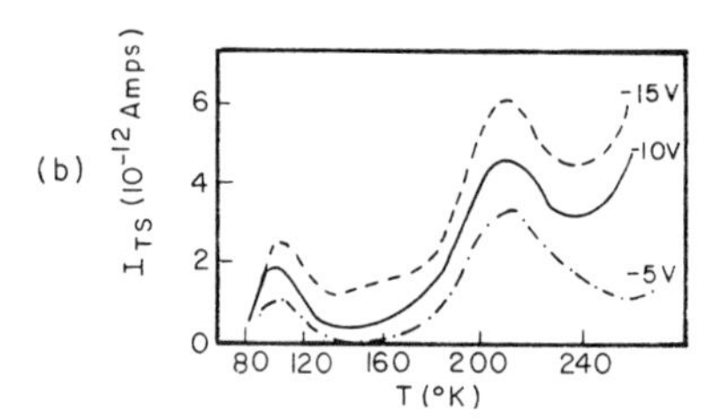

FIG. 9.28 Thermally stimulated current measurements for a p^+n $GaAs_{0.5}P_{0.5}$ diode. *(a) The solid curve was obtained by light excitation with a constant junction bias of -10 V. The dashed curve was obtained after a bias excitation step from 0 to -10 V. (b) Shows the effect of reverse bias on the height of the TSC peaks. (After Weisberg and Schade, 1968.)*

Buehler (1972) has considered the thermal response of *pn* junction current and capacitance where the junction contains deep impurity centers. Energy levels and emission coefficients are found from shifts in either the current maxima or the capacitance slope maxima with heating rate. His theory indicates that the heating rate may be nonlinear and that energy levels may be found independent of the emission coefficient. The analysis accounts for the possibility that only a portion of the space-charge region may contain charged deep centers. This allows the identification of centers in a small segment of the space-charge region. In addition, the analysis allows for the motion of the space-charge width as deep centers discharge. A gold-doped silicon n^+p junction is used to verify the theory.*

Driver and Wright (1963) have studied space charge in CdS diode structures and obtain an expression for the current during heating under reverse bias:

$$i = \tfrac{1}{2}eAdN\nu \exp\left[\frac{-E_T}{kT} - \frac{kT_m{}^2\nu}{\beta E_T}\exp\left(-\frac{E_T}{kT}\right)\right] \tag{9.36}$$

where N is the density of centers in the depletion region, A is the area, and d is the length.

* M. G. Buehler, (1972), Impurity Centers in *p-n* Junctions Determined from Shifts in the Thermally Stimulated Current and Capacitance Response with Heating Rate, *Solid State Electron.*, **15**, 69.

CHAPTER TEN

Lifetime and Capture Cross-Section Measurement Techniques with Respect to Deep Impurities

10.1 General Review of Available Techniques

Most interesting effects of deep level impurities are commonly described in terms of capture cross sections or capture probabilities for holes, electrons, and photons. Carrier capture cross sections are usually obtained from lifetime measurements and photon capture cross sections are deduced from the absorption coefficient at the proper wavelength. These capture cross sections represent suitable statistical averages.

Various physical processes may be involved in the free-carrier capture, each with its particular dependencies. While any one of these processes may be dominant, another may become more important as experimental conditions (e.g., temperature) change. Thus, quite apart from the difficulty of the identification of a recombination process with a particular impurity, the interpretation of data in terms of a physical model is often ambiguous. As a result, reported values of carrier capture cross sections vary widely, sometimes over several orders of magnitude. Even for an extensively studied impurity such as gold in silicon, accepted values may differ by a factor of 10.

In a survey of lifetime measuring techniques Bullis (1968) lists the methods given in Table 10.1.

Photoconductivity decay measurements are widely used for lifetime determinations if the lifetime is in the range from a fraction of a microsecond

TABLE 10.1

Methods of Measuring Carrier Lifetime

Conductivity Decay Methods

1. Photoconductive decay
   - Direct observation of resistivity
   - Q changes
   - Microwave reflection
   - Microwave absorption
   - Spreading resistances
   - Eddy current losses
2. Pulse decay
   - Direct observation of resistivity
   - Microwave absorption
3. Bombardment decay

Conductivity Modulation Methods

4. Photoconductivity
   - Steady state
   - Modulated source
   - Infrared detection, steady state
   - Q changes
   - Microwave absorption
   - Eddy current losses
   - Spreading resistance, modulated source
5. Pulse injection spreading resistance

Magnetic Field Methods

6. Suhl effect (and related effects)
7. Photoelectromagnetic effect
   - Steady state
   - Modulated source
   - Transient decay

Diffusion Length Methods

8. Traveling spot
   - Steady state
   - Modulated source
9. Flying spot
10. Dark spot
11. Sweepout effects
    - Pulse injection
    - Photoinjection
12. Drift field
13. Pulse delay
14. Emitter point efficiency

Junction Methods

15. Open-circuit voltage decay
16. Reverse recovery
17. Reverse current decay
18. Diffusion capacitance
19. Junction photocurrent
    - Steady state
    - Decay
20. Junction photovoltage
21. Stored charges
22. Current distortion effects
23. Current-voltage characteristics

Transistor Methods

24. Base transport
25. Collector response
26. Alpha cutoff frequency
27. Beta cutoff frequency

Other Methods

28. MOS capacitance
29. Charge collection efficiency
30. Noise
31. Surface photovoltage
    - Steady state
    - Decay
32. Bulk photovoltage
    - Steady state
    - Modulated source
33. Electroluminescence
34. Photoluminescence
35. Cathodoluminescence

upward. The difficulties that may be encountered with respect to deep impurities are as follows.

1. Ensuring that the deep impurity under study is really the dominant factor in controlling the lifetime. It is important to vary the deep impurity concentration to observe that the lifetime measured varies as expected. Pokrovskii and Svistunova (1969), for instance, found the variation to be incorrect with indium-doped silicon, as discussed in Chapter 6, and concluded that other centers were dominating the behavior.
2. The problem of providing a sufficiently strong light pulse with an intensity decay time that is considerably less than the expected lifetime decay. High-speed rotating mirror chopped light sources (Garbuny et al., 1957) can provide decay times of the order of 10^{-8} sec. More recently mode-locked laser techniques can provide decay times of a few nanoseconds with reasonable intensity, and there is the possibility of subnanosecond pulses with very special laser techniques.
3. The doping and temperature conditions necessary, to ensure that the lifetime is controlled by the deep impurity, frequently result in a high resistance in the specimen and then unavoidable capacitance in the measuring circuit may result in an *RC* time constant effect that is dominant.

High excitations (10^{22} cm^{-3} sec^{-1}) can be achieved by bombarding the specimen with short electron beam pulses (e.g., 700 keV) from a Van de Graaf generator (Wertheim and Augustyniak, 1956). Short flashes of x-rays have been used by Lindström (1971) for uniform generation of charge carriers.

Another group of techniques is concerned with the observation of conductivity in the presence of a steady-state light source inducing carrier pairs. Measurement of the photoinduced steady-state conductivity $\Delta\sigma = e(\mu_p\Delta p + \mu_n\Delta n)$ is practical in silicon only if $\Delta p = f\tau$ is at least 1% of the dark carrier density. Here f is the rate at which carrier pairs are generated optically and depends on the reflectivity and absorption versus wavelength characteristics of a given specimen, as well as on the light source intensity, which must be accurately known. Such measurements are not very convenient.

Very short decay times require detector bandwidths in the 100-MHz range, which may present serious noise problems. With a sinusoidally modulated light source, however, a narrow-band amplifier can be used to measure the phase shift between photoconductivity and optical excitation (Choo and Heasell, 1962; Staflin, 1965; Zwicker et al., 1971). Optical phase-shift methods are useful in determining lifetime in direct gap radiative semiconductors such as GaAsP (Scifres et al., 1971).

The photoelectromagnetic effect (PEM) is based on the diffusion of carriers which are generated nonuniformly in the semiconductor by a light of known intensity. Pairs generated at the surface diffuse toward the bulk and are separated by the Lorentz force $v \times B$ where B is the magnetic flux in the semiconductor. The PEM theory is well established for equal carrier lifetimes, $\tau_n = \tau_p$, or $\Delta n = \Delta p$: however this is generally not true in the presence of deep impurities. Such trapping effects, when $\Delta n \neq \Delta p$, can be taken into account when the ratio $\Delta n/\Delta p$ is known (Amith, 1959b). Since trapping has a different influence on photoconductive and PEM lifetimes, a careful investigation should reveal τ_n and τ_p. Although surface effects can in principle be eliminated in PEM experiments, exhaustion and accumulation layers at the surface seriously affect the diffusion of the generated carriers.

Combined PEM and PC methods have been extensively used by investigators: see, for instance, Ryvkin, (1964b; Tseng and Li, 1972; Li and Huang, 1972). The Suhl effect is concerned with the behavior of excess carriers deflected by a magnetic field after injection from a point contact.

In the following sections of this chapter, PEM methods, diffusion length methods, junction methods, and noise studies of deep impurities are discussed. The final section presents some capture cross-section values obtained for various deep impurities in silicon. The presentation gives some indication of the scatter that may be expected with choice of measurement method and variation of material control.

The subject index and bibliography of this book should be consulted by readers wishing to study capture cross-section values for other semiconductors such as Ge and GaAs.

10.2 Photoelectromagnetic Effect

The PEM effect may be described as a Hall effect associated with diffusion currents of the optically injected current carriers. It was originally seen in cuprous oxide at low temperature. It has since been observed in many other materials, including Ge, Si, GaAs, InSb, InAs, InP, Bi, Te, PbS, and has been used as a tool for measuring bulk lifetime and surface recombination. However, PEM techniques for lifetime measurements are not in widespread use since their interpretation depends on a number of assumptions that may not always be justified.

Van Roosbroeck in 1956 presented a fundamental treatment of the PEM effect for a slab with strongly absorbed steady radiation on one surface and a parallel, steady uniform magnetic field applied as shown in Fig. 10.1*a*. The voltage developed reverses sign with change of the magnetic field direction as in Fig. 10.1*b*. Two primary quantities that may be measured

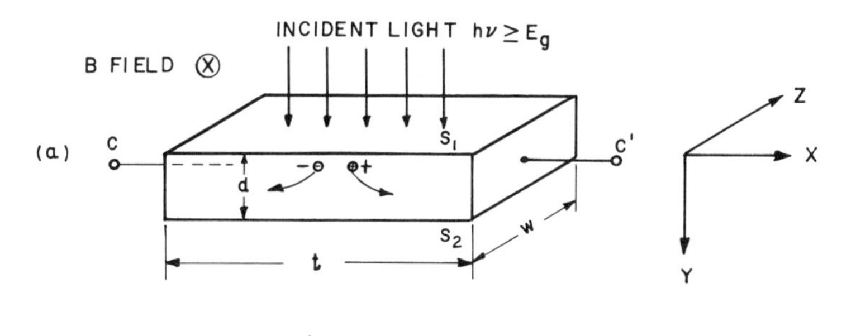

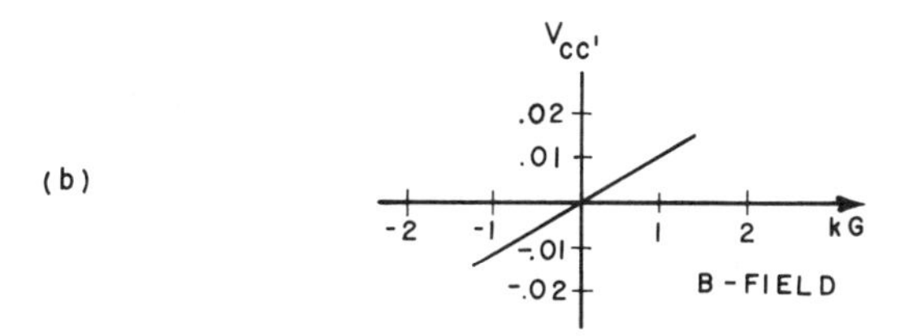

FIG. 10.1 Principle of the photoelectromagnetic effect.
Light absorption at the top surface or in the bulk just below causes diffusion of the excess carriers toward the lower surface. The magnetic field goes into the paper plane and deflects holes toward C′ and electrons toward C.

are the open-circuit voltage V^{oc} and the short-circuit current per unit width, I^{sc}. A photoconductance ΔG can be defined from these with the relationship

$$\Delta G = \frac{I^{sc}}{V^{oc}/t} \tag{10.1}$$

There are some advantages in making PEM measurements at low temperatures (77°K). The dark carrier density is then sufficiently small so that the condition of high injection $\Delta n = \Delta p \gg n_0$ is readily achieved, and useful information can be obtained without knowledge of the illumination intensity. The high carrier mobilities at low temperatures are further advantages of operating under these conditions since the magnitude of the PEM response is correspondingly larger. Also at low temperatures and high mobilities, low and high magnetic field effects may be studied with fields of 10 kG or less. Photoelectromagnetic studies are of course particularly interesting in high-mobility semiconductors such as InSb (Kurnick and Zitter, 1956; Beattie and Cunningham, 1962).

Equipment requirements for the measurement of short-circuit photoelectromagnetic currents have been discussed by Cunningham (1963). A typical arrangement is shown in Fig. 10.2.

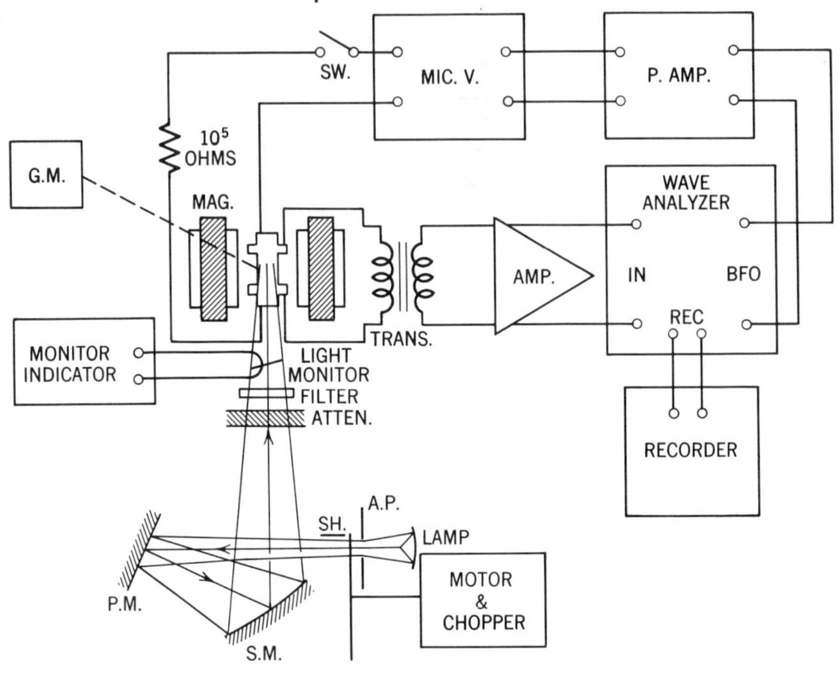

FIG. 10.2 Schematic diagram of a photoelectromagnetic experiment.
The symbols refer to the following: S.M., spherical mirror; P.M., plane mirror; S.H., shutter; A.P., aperture; ATTEN., optical attenuator; MAG., electromagnet; TRANS., transformer; AMP., low-noise preamplifier; G.M., gaussmeter; P.AMP., power amplifier; MIC.V., microvoltmeter; SW., switch; BFO, auxiliary output tuned to input frequency. (After Cunningham, 1963.)

The current flow lines and equipotential lines in a semiconductor slab of finite length are shown in Fig. 10.3. However, the reader is referred to Van Roosbroeck (1956) for further discussion of this. The presentation that follows is confined to a one-dimensional approximation.

10.2.1 Photoelectromagnetic Analysis (One-Dimensional) with Neglect of Trapping Effect

Assume that the slab thickness d is several times the diffusion lengths involved, that the recombination velocity of the dark surface s_2 is much greater than the recombination velocity s_1 of the illuminated surface, and that the excess carriers are created only near the illuminated surface. Trapping is neglected so that $\Delta n = \Delta p$, and both are assumed to be much greater than the dark concentration n_0.

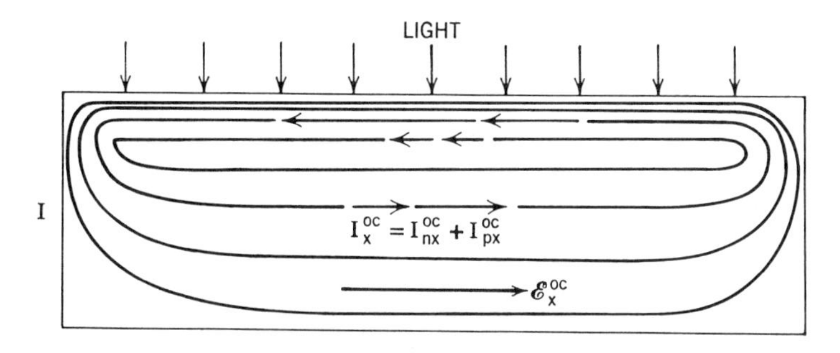

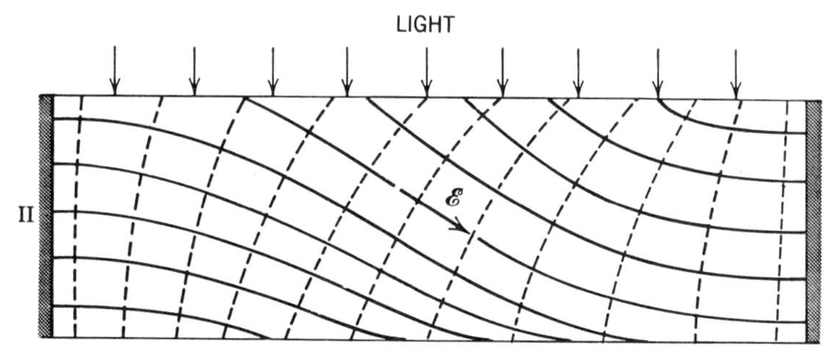

FIG. 10.3 The PEM effect in a semiconductor slab of finite length.
(I) open-circuit current, without electrodes; (II) open-circuit field and equipotentials, with electrodes. (After van Roosbroeck, 1956.)

The equations involved are then

$$J_{py} = -J_{ny} = -eD_p + e\mu_p\Delta p\mathscr{E}_y - \mu_p BJ_{px} \tag{10.2}$$

$$J_{px} = e\mu_p\Delta p\mathscr{E}_x + \mu_p BJ_{py} \tag{10.3}$$

$$J_{nx} = e\mu_n\Delta p\mathscr{E}_x - \mu_n BJ_{ny} \tag{10.4}$$

$$\frac{dJ_{py}}{dy} = -\frac{dJ_{ny}}{dy} = -\frac{e\Delta p}{\tau} \tag{10.5}$$

and

$$\frac{d\mathscr{E}_y}{dy} = \frac{e}{\varepsilon}(\Delta p - \Delta n) = 0 \tag{10.6}$$

Elimination of the currents J_{ny}, J_{nx}, and J_{px} from these equations and substitution of $b = \mu_n/\mu_p$ for the mobility ratio gives

$$\mathscr{E}_y = -\left(\frac{kT}{e}\right)\left(\frac{b-1}{b+1}\right)\frac{1}{\Delta p}\frac{d\Delta p}{dy} - \left(\frac{b-1}{b^{1/2}}\right)(\mu_n\mu_p)^{1/2}B\mathscr{E}_x - \frac{(b-1)(\mu_n\mu_p)^{1/2}B^2}{be\Delta p}J_{py} \tag{10.7}$$

and

$$J_{py} = -\frac{eD_a}{(1+\mu_n\mu_p B^2)}\frac{d\Delta p}{dy} + \left(\frac{\mu_n\mu_p}{1+\mu_n\mu_p B^2}\right)B\mathscr{E}_x\Delta p \tag{10.8}$$

where $D_a = (2b/(b+1))D_p$ is the ambipolar diffusion constant.

For the short-circuit condition $\mathscr{E}_x$ is zero and (10.8) reduces to

$$J_{py} = -\frac{eD}{1+\mu_n\mu_p B^2}\frac{d\Delta p}{dy} \tag{10.9}$$

This is also approximately true for the open-circuit measurement if typical operating conditions are considered.

The assumption of a simple time constant in (10.5) implies that the variation of the excess carriers with distance takes the form

$$\Delta p = \Delta p_0 e^{-y/L^*} \tag{10.10}$$

where $L^* = (D^*\tau)^{1/2}$ is the effective diffusion length and

$$D^* = \frac{(2b/(b+1))D_p}{1+\mu_n\mu_p B^2}$$

The boundary conditions at the surface $y = 0$ are that an injection hole current I_0 is created by the illumination and there is a surface recombination current $s_1\Delta p_0$. The diffusion equation is therefore

$$-D^*\left.\frac{d\Delta p}{dy}\right|_{y=0} = I_0 - s_1\Delta p_0 \tag{10.11}$$

We are now in a position to solve for the PEM short-circuit current per unit width I^{sc}, for $\mathscr{E}_x = 0$, from

$$I^{sc} = \int_0^d (J_{px} + J_{nx})\,dy = \frac{1+b}{b^{1/2}}(\mu_n\mu_p)^{1/2}B\int_0^d J_{py}\,dy \tag{10.12}$$

The result obtained is

$$I^{sc} = \frac{I_0 e(\mu_n\mu_p)^{1/2}B}{(1+\mu_n\mu_p B^2)}\left(\frac{2b^{1/2}D_p\tau}{s_1\tau + L^*}\right) \tag{10.13}$$

At low magnetic fields the short-circuit current is seen to be directly proportional to the magnetic field and to the illumination intensity.

The open-circuit voltage $V^{oc} = \mathscr{E}^{oc}t$ is obtainable from the condition $\int_0^d (J_{px} + J_{nx})\, dy = 0$. Assuming that

$$L^* \gg \left(\frac{\mu_n\mu_p}{1 + \mu_n\mu_p B^2}\right) B\tau\mathscr{E}_x$$

one obtains

$$\mathscr{E}_x^{\,oc} = \frac{2b}{(b+1)} \frac{D_p^{1/2}}{\tau} \frac{B}{(1 + \mu_n\mu_p B^2)^{1/2}} \tag{10.14}$$

Hence the PEM open-circuit voltage is directly proportional to B at low magnetic fields (as shown in Fig. 10.1*b*). It is also independent of the illumination intensity I_0, assuming the condition $\Delta p = \Delta n \gg n_0$ is met.

10.2.2 Photoelectromagnetic and Photoconductivity Studies with Trapping

With trapping present the analysis of the photoelectromagnetic effect becomes more complicated and further assumptions and simplifications are needed. We follow the simple approach of Hilsum and Holeman (1961).

Consider our n-type sample with hole traps illuminated with band-gap light so that the electron concentration increases from n_0 to $\Delta n + n_0$ and the hole concentration from p_0 to $\Delta p + p_0 + \Delta p_T$ where Δp is the increase in the free-hole concentration and Δp_T is the trapped-hole concentration. Then

$$\Delta n = \Delta p + \Delta p_T \tag{10.15}$$

$$\frac{\tau_n}{\tau_p} = 1 + \frac{\Delta p_T}{\Delta p} \tag{10.16}$$

and

$$\frac{\Delta n}{\tau_n} = \frac{\Delta p}{\tau_p} = g \tag{10.17}$$

where g is the rate of generation of charge pairs.

V_{PC}, the photoconductive signal, is the incremental change in voltage across the sample at constant current, and is given by

$$V_{PC} = D\mathscr{E} \frac{d}{tp_0}\left(\frac{b+1}{bc+1}\right)\frac{1}{F} \tag{10.18}$$

where b is the mobility ratio (μ_n/μ_p), c is the concentration ratio (n_0/p_0), D is the photon density, $\mathscr{E}$ is the electric field, d is the sample length, t is the sample thickness, and F is a factor which allows for surface recombination.

The PEM voltage at small magnetic fields is given by

$$V_{PEM} = DHL_a^* \frac{d}{tp_0}\left(\frac{b+1}{bc+1}\right)\frac{1}{F} \tag{10.19}$$

where H is the magnetic field and L_a^* is the ambipolar diffusion length given by

$$\left(\tau_{\text{PEM}}\mu_n \frac{kT}{e} \frac{1 + c}{1 + bc}\right)^{1/2}$$

The surface correction factor is $F = 1 + s_1\tau_{\text{PEM}}/L_a^*$ where s_1 is the surface recombination velocity for minority carriers at the front surface. The sample is considered to be thick compared with the diffusion length. The photoconductive and photoelectromagnetic characteristic times are related to τ_n and τ_p by

$$\tau_{\text{PC}} = \frac{b\tau_n + \tau_p}{b + 1} \tag{10.20}$$

$$\tau_{\text{PEM}} = \frac{\tau_n + c\tau_p}{1 + c} \tag{10.21}$$

For a material such as n-type GaAs where the mobility ratio is large, $\tau_{\text{PC}} \approx \tau_n$ and $\tau_{\text{PEM}} \approx \tau_p$. In most p-type samples τ_{PC} equals $\tau_p/(b + 1)$ and τ_{PEM} equals τ_n.

Measurements of V_{PC} and V_{PEM} therefore provide means for obtaining values of minority and majority lifetimes. The surface correction factor F may be determined in a fairly simple way from spectral response measurements of the photoconductivity.

In this way Hilsum and Holeman obtained values such as 10^{-8} and 10^{-11} sec^{-1} for minority carrier lifetimes in n- and p-type GaAs, respectively. They were also able to conclude that only a small fraction of the minority carriers was free in their experiments. On the other hand, for semiinsulating gallium arsenide at room temperature Holeman and Hilsum (1961) found the lifetimes τ_n and τ_p for a given specimen to be of the same order of magnitude, which suggested that trapping effects were small in this material under these conditions.

From (10.19) the PEM voltage is seen to be proportional to the photon density. On the other hand, for the trap-free model of (10.14) the PEM voltage is independent of the illumination intensity, assuming it is sufficient to provide the $\Delta p = \Delta n \gg n_0$ condition. In Mette and Boatright's (1965) studies of the PEM behavior of 200-μm thick p-type silicon wafers the PEM open-circuit voltage was found to vary sublinearly with illumination intensity. Barker (1966) interpreted this as a $V \propto B(\Delta W \times I)^{1/2}$ dependence where ΔW was the excess photon energy above a threshold (found to be 1.08 eV). To explain this a photodissociation model was proposed.

From the brief review above it is clear that the modeling and interpretation of PEM and PC measurements are not always obvious. However, some progress can be made in terms of interpreting capture cross sections of deep impurities, and PEM techniques have been widely used by investigators in the U.S.S.R. [for instance, see Kalashnikov and Morozov (1961), Ryvkin (1964a, b), and others].

10.3 Diffusion Length and Drift Techniques

Another approach to the determination of lifetime is the measurement of the diffusion length of excess minority carriers. The usual steady-state method of determining diffusion length and hence lifetime from $L = (D\tau)^{1/2}$, is that first used by Goucher (1951). The photocurrent across a reverse-biased *pn* junction is measured when a region close to the junction is illuminated by a slit of light (50 to 400 μm wide). The logarithm of the photocurrent is plotted against the distance of the light from the junction, and the slope of the straight line obtained gives the minority carrier diffusion length. Usually the illumination is chopped at a low frequency to allow an ac amplifier to be used which improves the signal-to-noise ratio of the experiment. If the recombination center controlling the lifetime is a deep impurity, then of course information about the impurity is obtained from the experiment.

Essentially the same technique applies if the induced carrier pairs are created by an electron beam, and this may be as small as 1 μm in diameter (Higuchi and Tamura, 1965). This allows the measurement of diffusion lengths of a few microns. A diffusion length of 10 μm for a carrier of mobility 1000 $cm^2\ V^{-1}\ sec^{-1}$ represents a lifetime of about $4 \times 10^{-8}\ sec^{-1}$.

Diffusion lengths may be measured in a direct gap material, such as gallium arsenide, by observing the dependence of the cathodoluminescence on the voltage of the electron beam (Wittry and Kyser, 1964). Using this technique, and accelerating voltages in the range 5 to 50 kV with a defocused electron beam, Rao-Sahib and Wittry (1969) have observed electron diffusion lengths of 0.6 to 3.2 μm in *p*-type GaAs for carrier concentrations in the range 3.76×10^{19} to $6.9 \times 10^{16}\ cm^{-3}$. Recently Ryan and Eberhardt (1972) have observed very long hole diffusion lengths (200 μm) in high purity epitaxial n-GaAs with a scanning electron microscope.

Another technique for determining diffusion lengths in GaAs is to expose a junction to uniform irradiation that will create electron-hole pairs and observe the short-circuit current assuming the depletion region component to be small. The irradiation may be light, γ-rays, or high-energy (2 MeV) electrons (Aukerman et al., 1967). Correction for the depletion region component may be made in some instances by observing the effect of depletion width increase by varying the voltage across the reverse-biased junction. The junction may be either a *pn* structure or a metal-semiconductor one.

Long minority carrier lifetimes (greater than 5 μsec) may be measured by the Haynes–Shockley experiment (1951) where minority carriers are injected across a junction and swept along a bar (Spitzer et al., 1955; Lemke, 1965).

Drift and diffusion of minority carriers in the presence of trapping have been considered by Jonscher (1957a, b). Direct current diffusion is affected only very little by the presence of trapping in extrinsic material. In the extreme case of heavy trapping in near-intrinsic material the effective diffusion length becomes $L^* = 2^{1/2}L$, where L is the diffusion length for the trap-free case. In the case of sinusoidally variable carrier density, the effective diffusion length is the same as for the trap-free ac case at low frequencies and again at very high frequencies, but may become appreciably less at intermediate frequencies. For a drifting pulse in the presence of traps, Jonscher shows that the attenuation results in an apparent lifetime which is shorter than the true lifetime.

Lifetime determination is also possible by the study of the drift length ($\mu\tau\mathscr{E}$) of minority carrier pulses, in the presence of an applied field $\mathscr{E}$ (Fig. 10.4*a*) where the pulses are generated near a surface by α particles (Davis, 1959), electron beams (McKay, 1948), or light with $h\nu > E_g$ (van Heerden, 1957). The time-integrated current pulse (Fig. 10.4*b*) is

$$A\,\frac{\mu\tau\mathscr{E}}{d}\left[1 - \exp\left(\frac{-d}{\mu\tau\mathscr{E}}\right)\right]$$

where A involves the total number of pairs generated and d is the crystal thickness. By varying the $\mathscr{E}$ field, the quantity $\mu\tau$ can be isolated. Since this method does not observe the carrier density versus time, it is similar in principle to a dc method and requires bombardment-induced conductivity changes of at least 1% of the dark conductivity. Large conductivity changes, on the other hand, cause space-charge buildup, particularly in the presence of traps.

Space-charge injection studies permit capture cross-section measurements in semiinsulating material. Such doping is possible with a deep impurity if it compensates, at low temperatures, the shallow impurities present. Non-equilibrium carriers can be introduced by injection of free carriers which may be trapped by the deep levels and finally recombine. Space-charge injection methods, like most lifetime measuring techniques, show the capture rate of all impurities combined and are therefore sensitive to possible contamination problems.

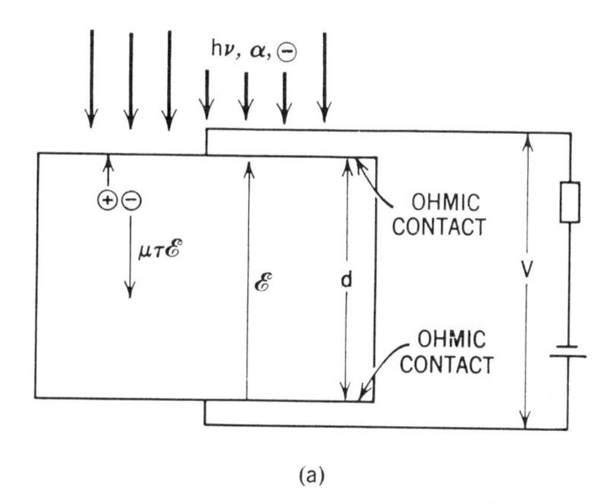

(a)

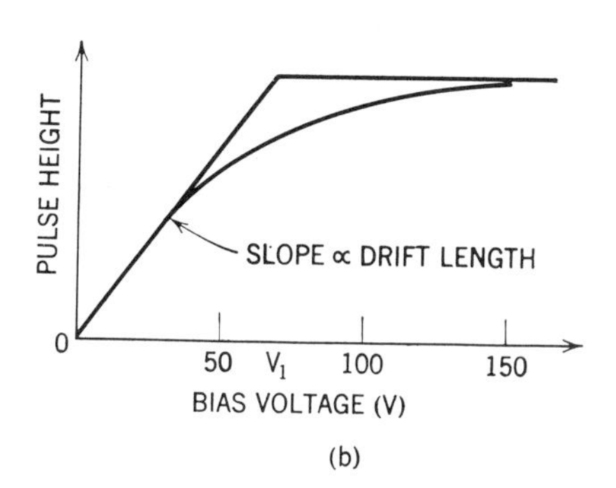

(b)

FIG. 10.4 Drift length experiment for determination of lifetime.
(a) Pulses of electron-hole pairs are generated near one surface. One carrier species (electrons) is swept a drift length $\mu\tau\mathcal{E}$ *through the bulk; the other carriers (holes) recombine at the ohmic contact. (b) The integrated current pulse A reaches its maximum value* A_m *for* $V > V_1$ *when the drift length exceeds the device length d. At small bias, the slope is* $A_m\mu\tau/d^2$. *(After Schibli and Milnes, 1967b.)*

Single-carrier injection studies, as the name implies, require contacts on the specimen that will inject only one carrier and block the other. The injected carrier fills traps and the current through the specimen rises rapidly at a voltage that is termed the trap-filled-limit voltage. Capture cross sections are obtainable from transient single-injection characteristics (Lemke, 1965, 1966a, b).

In $p\pi n$ and $p\nu n$ structures, where junctions of two types are provided, double injection of electrons and holes occurs, and then the capture rate for either carrier species can, under certain conditions, be calculated from two break points in the *I-V* characteristic. These are discussed in Chapter 11.

The large-signal, transient space-charge-limited current in the case of one-carrier injection has a peak at about $0.8t_0$, where t_0 is the space-charge neutral transit time ($t_0 = L^2/\mu V$), which allows determination of τ in low-mobility crystals such as iodine (Many et al., 1962).

Considering again lower resistance semiconductors, various lifetime measurements utilize the time dependence of the current or voltage across a *pn* junction, based on minority charge carrier storage (Bakanowski and Forster, 1960; Melehy and Shockley, 1961; Wilson, 1967).

10.4 Junction Depletion Techniques

A number of other pulse or transient methods of determining lifetime are of interest. Sato et al. (1964) measured the temperature dependence of the relaxation time in conductance of the *p*-type side of silicon *pn* junctions parallel to the junction surface after application of a step of reverse-bias voltage to the junction. The circuit arrangement is shown in Fig. 10.5*a*. When the pulsed reverse voltage is applied to the *p-n* junction, the width of the transition region spreads out toward the lower doped side of the junction, and then decreases as deep traps in the transition region begin to ionize. If the lower doped side of the junction is thin enough to observe the conductivity modulation parallel to the junction due to pulsed field, the relaxation time for conductivity modulation can be measured similarly to the pulsed field effect on a semiconductor surface (Rupprecht, 1961, 1963).

Figure 10.5*b* shows the reciprocal of the relaxation time τ [or more precisely the term $(\tau N_v v_{th})^{-1}$] plotted as a function of reciprocal temperature; N_v is the effective density of states at the valence band edge, and v_{th} is the thermal velocity of holes. Interpretation by Rupprecht's method gives a trap level 0.63 eV above the top of the valence band with a cross section of 5×10^{-14} cm^{-2} for holes. Sato et al. (1964) concluded that these were traps in the bulk near the junction. Measurements of surface effects showed surface traps at $E_v + 0.35$ eV for a dry-air atmosphere and at $E_v + 0.72$ eV for a wet-air or dry-ammonia atmosphere. Other surface studies of silicon by the pulsed field technique of Rupprecht have been made by Sawamoto et al. (1966) and by Sato and Sah (1970).

The energy level of deep impurities may be determined by the study of capacitance transients in metal-semiconductor or *pn* junctions under reverse bias voltage step conditions. Williams (1966) has studied such behavior in *n*-type gallium arsenide with gold Schottky barrier or electrolyte blocking contacts. A reverse bias is applied and the ionization of deep centers in the

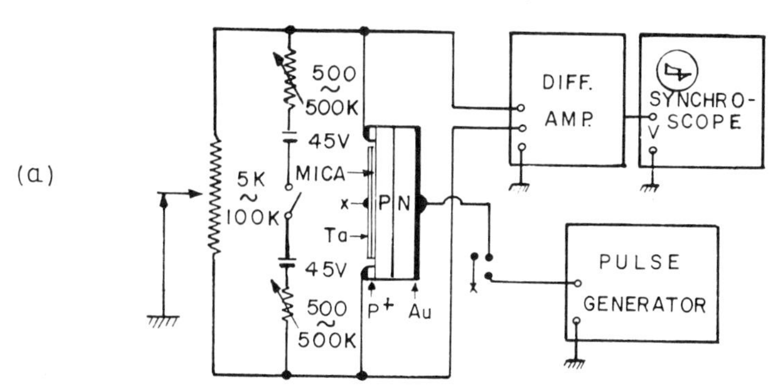

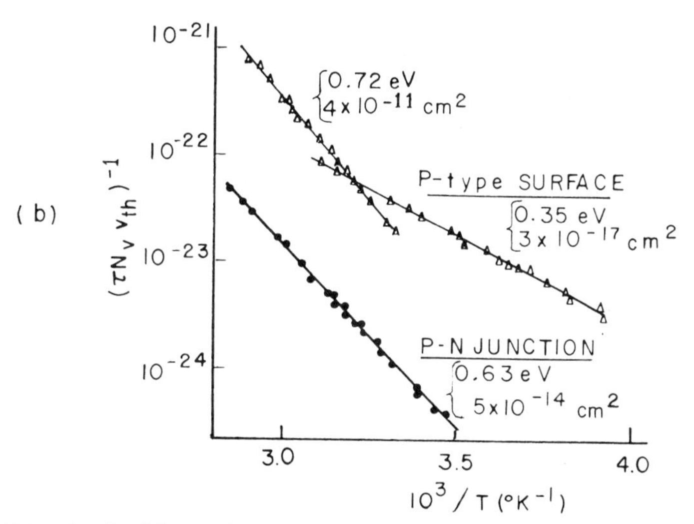

FIG. 10.5 Conductivity modulation relaxation time versus temperature measurements under pulsed conditions in a silicon *pn* **junction showing a bulk trap at 0.63 eV and surface traps at 0.72 and 0.35 eV.**

(a) Diode and circuit arrangement. The tantalum field plate is used for the surface trap studies. (b) Plots of $(\tau N_v v_{th})^{-1}$ *versus reciprocal temperature. (After Sato et al., 1964.)*

dark causes the barrier capacitance to change with time. After this, electrons may be injected into the barrier by strongly absorbed light and captured by the deep centers that have previously ionized. The kinetics of the process can be determined from measurements of the barrier capacitance as function of time. Williams found a donor lying 0.74 eV below the conduction band edge and typically present in concentrations around 10^{16} cm^{-3}. From capture cross-section data it appeared to be negatively charged when occupied.

Recently Sah et al. (1970b) have made many elegant experiments with dark capacitance, photocapacitance, and current transients in gold-doped silicon *pn* junction depletion regions. Such techniques have been discussed briefly in Chapter 8.

10.5 Noise Studies of Deep Impurities

Noise studies may be an effective means of determining cross sections. As a general case, consider a nearly intrinsic semiconductor with both holes and electrons present (Fig. 10.6*a*), in equilibrium with an acceptor and a donor

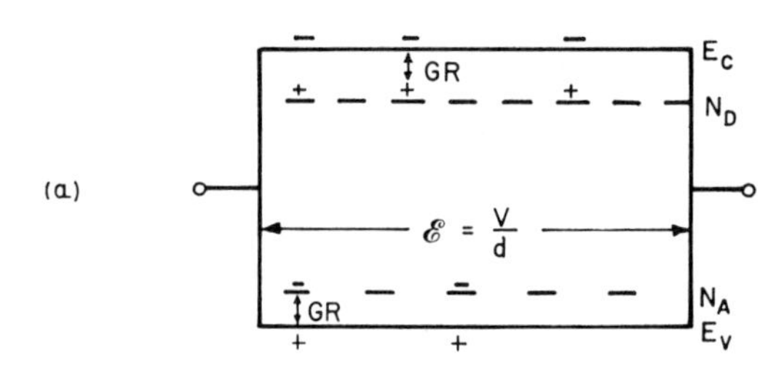

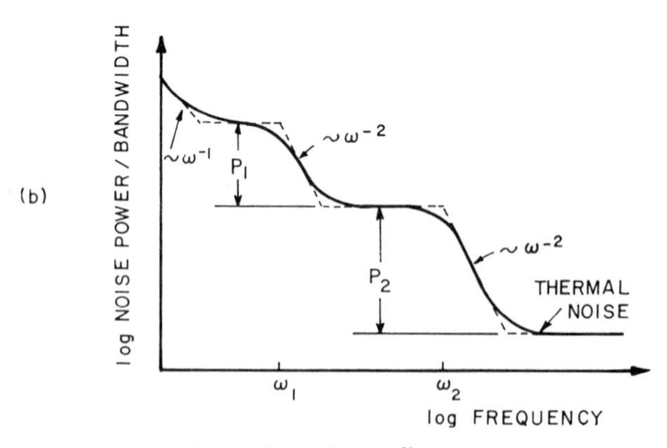

FIG. 10.6 Model of semiconductor for noise studies.
(a) The generation-recombination processes indicated are assumed responsible for the GR noise. (b) Noise spectrum of a semiconductor bar. At low frequencies, 1/f noise dominates; the thermal noise is observed at high frequencies. This GR noise is characterized by two lifetimes $\tau_{1,2} = (\omega_{1,2})^{-1}$. *(After Schibli and Milnes, 1967b.)*

level. Their densities fluctuate independently on account of the generation-recombination (GR) processes between the bands and the nearest level. The frequency spectrum of the GR noise (Fig. 10.6*b*) is then the sum of these fluctuations, and for a specimen of length d (van der Ziel, 1954)

$$r_n = \frac{e}{kT}\frac{V_a^2}{d^2}\left[\frac{\alpha\mu_n\tau_n\sigma_n/(\sigma_p + \sigma_n)}{1 + \omega^2\tau_n^2} + \frac{\beta\mu_p\tau_p\sigma_p/(\sigma_p + \sigma_n)}{1 + \omega^2\tau_p^2}\right] \tag{10.22}$$

where r_n is the ratio of GR noise to thermal noise power, V_a is the applied dc voltage, σ_n and σ_p are the electron and hole conductivities, respectively, and α is the probability that a donor atom is neutral ($\alpha = N_D^0/N_D$) and depends strongly on the donor level $E_c - E_D$ and temperature T. Similarly, $\beta = N_A^0/N_A$ in the second term.

The GR noise spectrum of such a bipolar semiconductor would have two break frequencies, at $\omega_1 = \tau_1^{-1}$ and at $\omega_2 = \tau_2^{-1}$. The amplitudes of these terms (p_1 and p_2 in Fig. 10.6*b*) provide an independent check on the lifetimes. If ω_1 and ω_2 are sufficiently different, then the slope of the noise spectrum approaches -2 (20 dB per decade).

The GR noise of a doped (say *p*-type) semiconductor in the presence of a double acceptor such as zinc, or a donor-acceptor impurity such as gold, shows qualitatively the same spectrum with two breaks (Klaassen, 1962), but the expressions for the amplitudes p_1, p_2 are different from those for the near-intrinsic case considered above.

Generation-recombination noise experiments are limited by factors such as contact ($1/f$) noise and thermal noise (Sah et al., 1966; Yamamoto et al., 1967; Liu, 1967; van Vliet, 1958), by the frequency range of the noise equipment, the resistance of the sample, power dissipation in the device, and by carrier sweepout effects.

The noise spectrum of a specimen of *p*-type gold-doped germanium exhibiting two relaxation times is shown in Fig. 10.7. The correction for $1/f$ noise is shown. The capture cross sections inferred are $\sigma_A = 1.5 \times 10^{-12}$ cm^2 and $\sigma_D = 3.7 \times 10^{-18}$ cm^2 at 100°K. On the other hand, Besfamil'naya and Ostroborodova (1965), also from noise measurements, observed a value of the order of 10^{-14} cm^2 for σ_D at 24°K.

Knepper (1969); and Knepper and Jordan (1972a, b), in studies of $p^+\pi n^+$ double-injection diodes where the π region is indium-doped silicon, have found noise measurements a suitable technique for determining the electron lifetime. From these studies the capture cross section for electrons by neutral indium-trapping centers, σ_n^0, is inferred to be 1.8×10^{-16} cm^2 at 77°K (see Sections 6.3 and 10.6.2 for further discussion of σ_n^0 for indium in silicon).

Copeland (1971) has used noise as a tool to determine the presence of deep impurity levels in *n*-type GaAs used in n^+nn^+ Gunn and LSA oscillators.

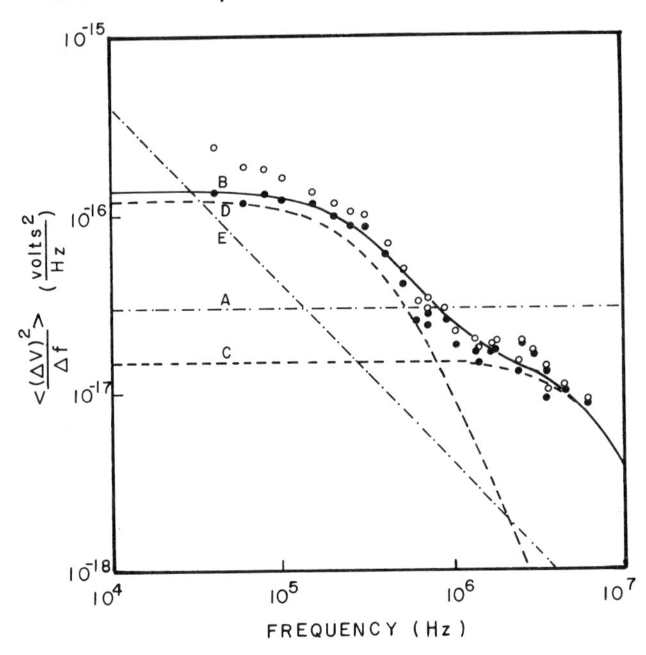

FIG. 10.7 Generation-recombination noise spectrum at 100°K for gold-doped germanium with the Fermi level lying midway between the 0.041-eV donor level and the 0.145-eV acceptor level, so that the noise spectrum simultaneously exhibits relaxation times characteristic of both levels. (After Neuringer and Bernard, 1961a.)

The average charge of these levels changes by impact ionization or hole capture when Gunn domains form in the device. Hence they can cause current runaway or breakdown at low voltage just above the Gunn threshold (3000 V cm^{-1}). The noise measuring circuit used is shown in Fig. 10.8. The device is in a temperature-controlled holder. The dc bias current I from a storage battery is controlled by a variable series resistor which is several times larger than the device resistance. A low-noise low-impedance grounded-base transistor preamplifier is used before the input of a Hewlett–Packard 3590A wave analyzer. A simple external waveform generator circuit is used to obtain an exponential frequency versus time sweep. The wave analyzer has logarithmic outputs so that log-log plots of noise current versus frequency are drawn by an x-y plotter. The system is calibrated before each plot by injecting a known ac current across the device with the dc bias current flowing. With 10 Ω samples, noise signals $\langle i^2\rangle/\Delta f$ as low as 3×10^{-21} A^2 Hz^{-1} can be plotted.

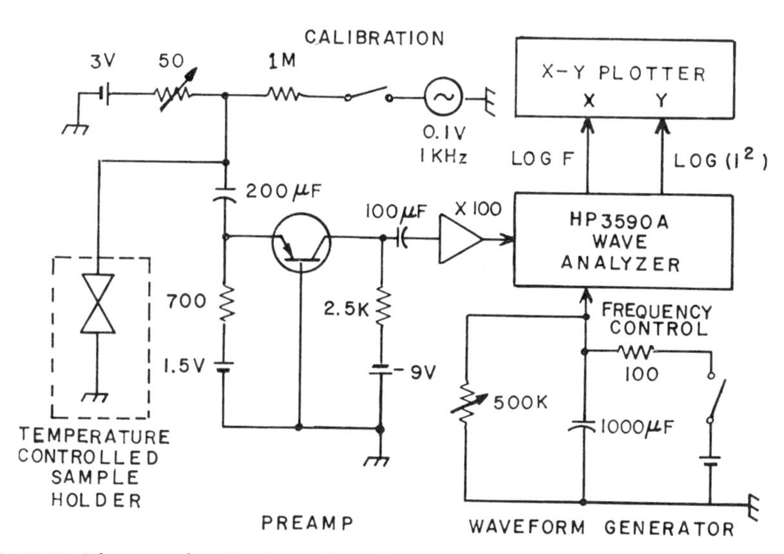

FIG. 10.8 Diagram of a circuit used to plot noise current squared versus frequency. (After Copeland, 1971.)

The measurements show a well-defined time constant, as in Fig. 10.9, for GaAs containing a deep impurity. By studying the noise as a function of temperature Copeland (1971) inferred that the deep level was 0.175 eV below the conduction band edge with an acceptor-like degeneracy factor of 4. It was present with a density of 2.5×10^{14} cm^{-3} and was attributed to a lattice defect produced during the growth process of the epitaxial layer. The capture cross section for electrons was 4×10^{-18} cm^{-2} at 300°K.

Electron fluctuation noise in zinc-doped high-resistivity GaAs has been studied by Anderson and van Vliet (1971). In thermal equilibrium (300°K) and with a small applied field the noise is sometimes extremely low, with a relative variance $(\Delta n^2)/(n)$ of the order of 0.01. The spectra have much structure in the range 10 Hz to 1 MHz with relaxation modes near 40 Hz, 10 kHz, and one or more modes in the middle range. The Hall effect and resistivity data suggest that oxygen donors at 0.8 eV below the conduction band act as the main centers. The spectra give a cross section for electron capture of $\sigma_n \approx 2.5 \times 10^{-15}$ cm^2. Other defect states with cross sections of the order of 10^{-17} and 10^{-19} cm^2 were also observed.

Measurements at 77°K under optical excitation gave different noise spectra. With light of energy near that of the band gap, much structure was observed. Positive and negative modes (the latter characterized by peaks)

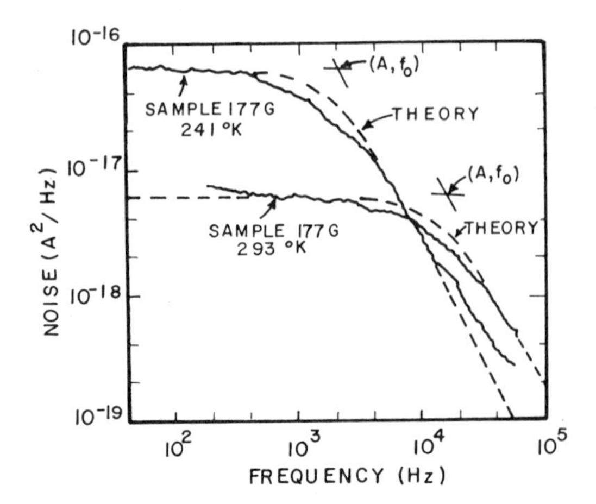

FIG. 10.9 Noise spectra of *n*-type GaAs at 241 and 293°K.
The dashed curves are of the form $A/[1 + (f/f_0)^2]$ *and are fitted to the data. The corners of the experimental curves are not as sharp as those of the theoretical curves because the time constants of all the traps are not exactly the same. (After Copeland, 1971.)*

of rather low magnitude occur. Both for long wavelengths and for strongly absorbed light the noise is high and the spectra are structureless. These properties were attributed to surface traps. The interpretation of noise data for a material such as GaAs is therefore not usually as simple as the Copeland experiment would suggest.

The complexity of interpreting noise measurements is further revealed by the studies of Mantena and Loebner (1971) of high-gain zinc-compensated silicon photoconductors. Six ground-state electronic energy levels have been identified in this material. These levels have been assigned to a donor (arsenic); the three zinc levels Zn^{2-}, Zn^{-}, Zn^{0}; an *A* center; an *E* center; and a silicon vacancy, as shown in Fig. 10.10. The $Zn^{-3/2}$ level is the photoconductivity sensitizing center with an electron capture cross section for Zn^{-} ions of about 10^{-22} cm^2 and a hole capture cross section for Zn^{2-} ions of between 10^{-14} and 10^{-13} cm^2. However, there are at least 20 possible capture and emission transitions of electrons and holes to and from the levels shown in Fig. 10.10. Mantena and Loebner show that any model based on a one-defect recombination level, communicating with both the conduction and valence bands, is inadequate to explain their observed noise characteristics. They suggest that a two-defect-level model is the minimum in complexity needed to produce a better match between theory and experiments. This would be an undertaking of considerable magnitude.

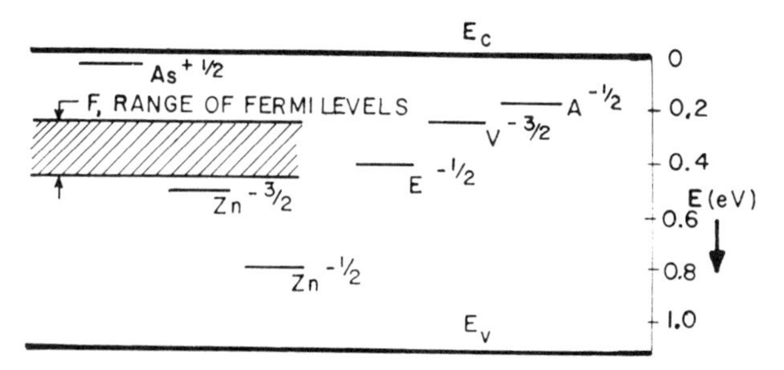

FIG. 10.10 Energy level diagram concentrations and carrier cross sections of studied defects in zinc-counterdoped silicon.
*In the labeling of levels by fractions the Sah-Shockley notation [Phys. Rev., **109**, 1163 (1958)] is followed.*

Identity	Donor	Zn^{2-} Zn^- Zn^0	*E* Center	Vacancy	*A* Center
Symbol	As_{Si}	Zn_{Si}	$[V_{Si}\text{–}As_{Si}]^{pair}$	V_{Si}	$[V_{Si}\text{–}O_I]^{pair}$
Con-centrations	4×10^{15}– 2×10^{16}	2–6 $\times 10^{15}$	$\sim 8 \times 10^{14}$	$\sim 2 \times 10^{15}$	$\sim 6 \times 10^{15}$
Cross sections					
σ_n		10^{-22}– 10^{-19} 10^{-15}	10^{-16}– 10^{-15}	10^{-21}– 10^{-19}	10^{-16}– 10^{-15}
σ_p		10^{-13} 10^{-14}– 10^{-13}	10^{-16}– 10^{-15}	10^{-14}– 10^{-13}	10^{-16}– 10^{-15}

(After Mantena and Loebner, 1971.) (Courtesy of E. M. Pell. Ed., Proceedings of the Third International Conference on Photoconductivity, Pergamon, 1971.)

10.6 Capture Cross Sections in Silicon

Experimental work on carrier capture cross sections of deep impurities in silicon has concentrated mainly on gold, indium, and zinc.

10.6.1 Gold in Silicon

Gold has an acceptor level at $E_c - 0.55$ eV and a donor level at $E_v + 0.35$ eV, and as discussed in earlier chapters it is important in silicon for lifetime control. In spite of considerable interest in gold in silicon, the capture cross-section values spread over as much as a decade (Table 10.2).

TABLE 10.2

Capture Cross Sections of Gold in Silicon[a] *(After Bullis, 1966)*

Temperature (°K)	Acceptor Level		Donor Level		Ref.
	$\sigma_n{}^0$ (cm^2)	$\sigma_p{}^-$ (cm^2)	$\sigma_n{}^+$ (cm^2)	$\sigma_p{}^0$ (cm^2)	
300	5×10^{-16}	1×10^{-15}	3.5×10^{-15}	$\geq 10^{-16}$	Bemski, 1958b
300	1.7×10^{-16}	1.1×10^{-14}	6.3×10^{-15}	2.4×10^{-15}	Fairfield and Gokhale, 1965
77	3×10^{-15}	1×10^{-13}	6×10^{-14}	3×10^{-15}	Davis, 1959
77	5×10^{-16}	2.3×10^{-13}	1×10^{-13}		Bullis, 1966

[a] The convention adopted is that σ_n represents the capture of electrons and σ_p that of holes. The superscript 0, −, or + indicates the charge state of capturing center before capture. Thus $\sigma_n{}^0$ indicates the capture of an electron by a neutral center. The thermal velocity assumed was 10^7 cm sec^{-1} for the values in the table.

The neutral capture cross sections $\sigma_p{}^0$ and $\sigma_n{}^0$ are the same at 77°K, and within an order of magnitude at 300°K. They are roughly a factor of 10 smaller than the coulomb-attractive cross sections $\sigma_p{}^-$ and $\sigma_n{}^+$, as expected; $\sigma_p{}^-$ and $\sigma_n{}^+$ have a much stronger temperature dependence $T^{-2.5}$ to T^{-4} than the neutral cross sections T^0 to T^{-1}.

Hole and electron emission data for the gold acceptor level in silicon have been studied by Parrillo and Johnson (1972). Certain anomalies are resolved if it is assumed that as the temperature varies the energy level of the trap is not locked onto one band edge or the other but remains in the same relative position with respect to both edges. At 300°K the acceptor level is at $E_c - 0.45$ eV with this assumption, and $e_n = 1600$ sec^{-1} and $e_p = 17$ sec^{-1}.

Gold in germanium has one donor level ($E_v + 0.04$ eV) and three acceptor levels ($E_v + 0.16$, $E_c - 0.20$, and $E_c - 0.04$ eV), but only the two lowest levels can be compared with those in silicon. The hole capture cross section $\sigma_p{}^0$ of the 0.04-eV donor level (10^{-18} cm^2) is considerably smaller than $\sigma_p{}^0$ of gold in silicon (2.4×10^{-15} cm^2). But since these levels are 0.04 and 0.35 eV deep, respectively, not too close an agreement would be expected. The smaller value for $\sigma_n{}^0$ (5×10^{-16} cm^2) of gold in silicon, on the other hand, is reasonably close to the corresponding cross section in germanium (2×10^{-16} cm^2) (Ashley and Milnes, 1964). The levels for these processes are both 0.55 eV from the conduction band. The energy differences for the attractive hole capture by the acceptor levels are 0.16 and

0.62 eV in germanium and silicon, respectively, and the respective capture cross sections are 0.5×10^{-14} and 1 to 2.3×10^{-13} cm^2. These simple observations would suggest larger capture cross sections for deeper levels. The spread, particularly of the germanium data, makes a more meaningful comparison difficult.

10.6.2 Indium in Silicon

Indium has a single acceptor level, at $E_v + 0.16$ eV (Morin et al., 1954), and serves therefore as a simple model of a deep impurity. The value of the hole capture cross section (Table 10.3) of the negatively ionized indium atom σ_p^- is about 7×10^{-14} cm^2 at 77°K and decreases with temperature as T^{-2}, according to Wertheim (1958).

The electron capture cross section σ_n^0 of neutral indium has been measured under many conditions; the results spread over five orders of magnitude at 77°K. The main experimental conditions are therefore included in Table 10.3. The author's personal inclination is toward accepting a value of 10^{-16} to 10^{-17} cm^2 for σ_n^0. However, this cannot be reconciled with the results of the careful studies in the U.S.S.R. which suggest values of the order of 10^{-21} cm^2 ($c_n = 10^{-14}$ cm^3 sec^{-1}). The matter therefore still remains open to further examination.

Although photoconductive decay measurements are standard techniques, even the ideal situation of a uniformly irradiated specimen with no surface effects is quite involved as suggested by the brief discussion of the work of Blakemore and Sarver (1968) given in Chapter 7. Photoionization cross sections for indium in silicon are shown in Fig. 10.11. Process 1 has a maximum cross section for light of $h\nu = 2(E_{\mathrm{In}} - E_v)$, in agreement with theory (Lucovsky, 1965), and 40% of this maximum for $h\nu = 6(E_{\mathrm{In}} - E_v)$. Hence monochromatic light of $h\nu \approx 1$ eV would cause mainly transition 1, if the indium level is 50% occupied. The cross section for transition 2 is very small and hence difficult to measure (Chikovani and Pokrovskii, 1967) because of

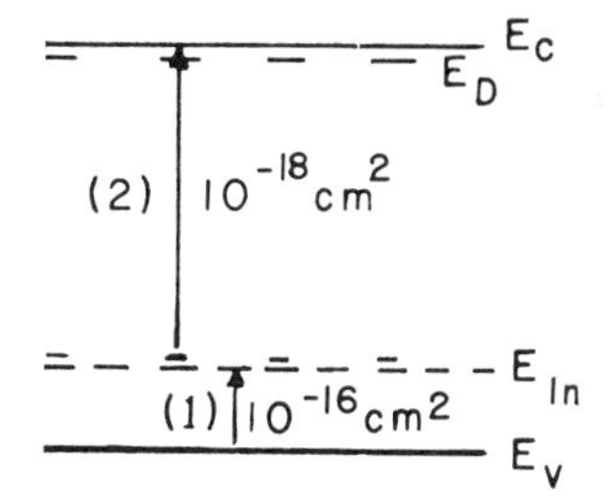

FIG. 10.11 Photon capture cross sections of indium in silicon.

The absorption process involves transfer of a hole to the valence band or of an electron to the conduction band. (After Schibli and Milnes, 1967b.)

TABLE 10.3

Capture Cross Sections of Indium in Silicon *(After Schibli and Milnes, 1967)*

Quantity	Method[a]	Value (σ in cm^2, c in $cm^3\ sec^{-1}$)		Temperature Dependence	Indium Density, N_{In} ($10^{16}\ cm^{-3}$)	Shallow Donor Density, N_D ($10^{16}\ cm^{-3}$)	Ref.
$\sigma_n{}^0$	P	3.7×10^{17}	300°K	T^{-1}	14	$< 10^{-4}$	Wertheim, 1958
$\sigma_n{}^0$	PCD	$< 10^{-19}$	> 125°K		0.5–1	2–3	Pokrovskii and Svistunova, 1961
$\sigma_n{}^0$	PCD	4×10^{-22}	80°K	T^{-3} (60–77°K)	0.14–6	0.72–9	Pokrovskii and Svistunova, 1962
$c_n{}^0$	PC	2×10^{-12}	30°K	exp (0.035 eV/kT)	0.45	2.7	Pokrovskii and Svistunova, 1964a
$c_n{}^0$	PC	8×10^{-15}	16–60°K	Independent	10	350	Pokrovskii and Svistunova, 1965a
$c_n{}^0$	PhX	8×10^{-8}	1.1–4.2°K	Independent	0.3	0.6	Loewenstein and Honig, 1966
$\sigma_n{}^0$	DI	9×10^{-17}	77°K	$T^{-1.3}$ (64–100°K)	2	0.004	Wagener and Milnes, 1964
$\sigma_n{}^0$	N	1.8×10^{-16}	77°K		0.2	0.04	Knepper, 1969
$\sigma_n{}^0$	Ind	10^{-18}	77°K		2–6	0.004	Barrera, 1967
$c_n{}^0$	PC	1.5×10^{-14}	25–30°K		0.5–6	$> N_{In}$	Chikovani and Pokrovskii, 1967
$\sigma_p{}^-$	P	1.7×10^{-14}	300°K	T^{-2}	14	$< 10^{-4}$	Wertheim, 1958
$\sigma_p{}^-$	PCD	$> 10^{-18}$					Pokrovskii and Svistunova, 1969
$c_p{}^-$	PC, N	$2–4 \times 10^{-6}$	77°K		0.5–10	0.03–0.15	Godik and Pokrovskii, 1965
$c_p{}^-$	N	1.5×10^{-6}	78°K				Proklov et al., 1965
$\sigma_p{}^-$	DI	10^{-13}	77°K		2	0.004	Wagener and Milnes, 1964
$\sigma_p{}^-$	N	7×10^{-14}	77°K		0.2	0.03	H. Preier, private communications
$\sigma_p{}^-$	N	7×10^{-14}	77°K		2–6	0.004	Barrera, 1967
$c_p{}^-$	PC	10^{-6}	77°K	exp (0.055 eV/kT)	5.1	0.085	Blakemore, 1956

[a] PCD, photoconductive decay method; N, noise method; PC, steady state photoconductivity; Ind, indirect; DI, double-injection method; P, pulse method; PhX, photoexcitation.

the vicinity of other transitions (shallow levels); this reported photon capture cross section at 28°K shows an increase from 1.5×10^{-19} cm^2 at 1.03 eV to 9×10^{-19} cm^2 at 1.12 eV.

10.6.3 Zinc in Silicon

Zinc has two deep acceptor levels, at $E_v + 0.31$ and $E_c - 0.55$ eV, and its effects are therefore correspondingly complex. The lower level ($E_v + 0.31$ eV) is observed in *p*-type material, when $N_D < N_{Zn}$. As compensation increases, the lower level is filled completely, and the upper level becomes electrically active when $N_{Zn} < N_D < 2N_{Zn}$. Table 10.4 shows that the measured carrier capture cross sections are almost independent of temperature from 80°K up to room temperature. The maximum doping level was 5×10^{16} cm^{-3} and impurities like aluminum, boron, and manganese, which are known to form pairs with zinc (Fuller and Morin, 1957; Ludwig and Woodbury, 1961), were absent in those experiments. The large cross section for hole capture by singly negative-charged zinc atoms agrees with other attractive cross sections (σ_p^- for indium, σ_n^+ for gold) and is explained by the phonon cascade model. A very similar value was obtained from double-injection experiments on rather short devices (Kornilov, 1966a). However, no electron capture cross section σ_n^0 was deduced from the lower breakpoint in the *I-V* curve, and no negative-resistance region was observed, while early double-injection experiments (Holonyak, 1962) on zinc-doped silicon did produce such a region. The double-injection theory in the presence of deep levels has been applied for single-level impurities, including field-dependent parameters and gradients in either the shallow or deep level density as discussed in Chapter 11. Analyses have been made recently of space-charge flow in the presence of multilevel impurities (Weber and Ford, 1970; Weber et al., 1971). However, the treatment tends to become involved and assumptions have to be made that make double-injection studies rather indirect as a means of determining capture cross sections for multilevel impurities.

Recently the thermal capture rates of electrons and holes at zinc centers in silicon have been obtained by Herman and Sah in reverse biased n^+p and p^+n junctions from junction capacitance transients due to capture of photogenerated and injected carriers.* The electric field dependence of the thermal capture rates of holes at singly ionized zinc centers is $E^{-1.1}$, whereas that at doubly ionized zinc centers is $\exp(-E/E_0)$. The temperature dependence of the thermal capture rate of holes at doubly ionized zinc centers is small. The thermal capture rate of electrons at neutral zinc centers

* J. M. Herman, and C. T. Sah (1973), Thermal Capture of Electrons and Holes at Zinc Centers in Silicon, *Solid State Electron.*, **16** (in process).

TABLE 10.4

Capture Cross Sections of Zinc in Silicon *(After Schibli and Milnes, 1967)*

Quantity	Method	Value (cm^2)	N_{Zn} (10^{16} cm^{-3})	N_D (10^{15} cm^{-3})	Ref.
σ_n^0	PC	10^{-15} (80–200°K)	1–5	0.3–1	Glinchuk et al., 1964
σ_p^-	PC	10^{-13} (80–200°K)	1–5	0.3–1	Glinchuk et al., 1964
σ_n^-	PC	10^{-16} (90–200°K)	1–2	20–30	Glinchuk and Litovchenko, 1964a
σ_n^-		$\leq 10^{-18}$ (80°K)	$1–2 \times N_D$	0.7–7	Kornilov, 1965a
σ_n^-	PC and PCD	10^{-18}–10^{-20} (100–300°K)	0.5	7	Kornilov, 1965c
σ_p^-	DI	3×10^{-14} (77°K)	1	8	Kornilov, 1966a
σ_p^-	PC	10^{-14} (80–200°K)	0.01–0.5	$< N_{Zn}$	Kornilov, 1966b
σ_n^0	PC	5×10^{-16} (80–200°K)	0.01–0.5	$< N_{Zn}$	Kornilov, 1966b

is 2.0×10^{-9} cm^3 sec^{-1} ($\sigma_n{}^0 \sim 2 \times 10^{-16}$ cm^2) with very little field dependences, whereas that at singly ionized zinc centers is 5.0×10^{-11} cm^3 sec^{-1} ($\sigma_n{}^- \sim 5 \times 10^{-18}$ cm^2) with very little field and temperature dependence below 170°K. The auger-impact emission rate of electrons trapped at doubly ionized zinc centers by electrons is 3.5×10^{-10} cm^3 sec^{-1} with very little field and temperature dependence below 170°K.

The hole capture cross section $\sigma_p{}^{2-}$ of the upper, double-attractive level at $E_c - 0.55$ eV, is expected to be quite large; and according to Herman and Sah (1973) is about 7×10^{-14} cm^2 at 10^4 V cm^{-1} and is quite field dependent. The phonon cascade model hardly applies to the capture of electrons by neutral zinc atoms (Glinchuk et al., 1964) because the polarizability, and hence the number of excited levels, may be insufficient (Bonch-Bruevich and Glasko, 1962).

The observed electron capture cross section $\sigma_n{}^-$ of the repulsive, singly charged zinc atoms is probably near 10^{-18} to 10^{-20} cm^2. The recent photoconductivity quenching studies of Sklensky and Bube (1972) give 10^{-20} cm^2 at 50°K and below. An earlier observed value (Glinchuk and Litovchenko, 1964a) of 10^{-16} cm^2 is much larger than $\sigma_n{}^-$ for zinc in germanium.

In the charge exchange experiment (Kornilov, 1965a) electrons are excited optically from the lower zinc level to the conduction band and subsequently captured by neutral and singly ionized zinc atoms. As a result, the electron population on the upper zinc level will be greatly above the equilibrium value. Suitable infrared radiation, or thermal stimulation, will release this excess, and the resulting current surge is a measure of $\sigma_n{}^-$. The temperature independence of $\sigma_n{}^-$ has been explained by a tunneling process of the free electron through the coulomb barrier.

The photon capture cross sections (Fig. 10.12) are not very different (according to Kornilov, 1964a) for the $E_c - 0.55$ eV level, but the transfer of an electron to the lower level, $E_v + 0.31$ eV, has only one-tenth the probability of the transfer of an electron from the lower zinc level to the

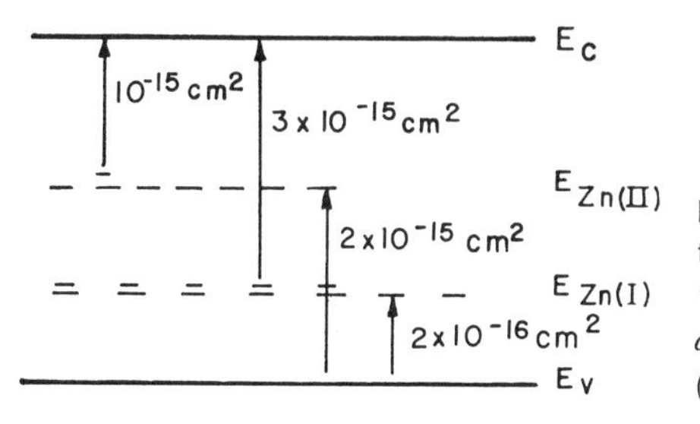

FIG. 10.12 Photon capture cross sections of zinc in silicon.
The absorption process involves transfer of a carrier from an impurity level to a band. (After Schibli and Milnes, 1967b.)

conduction band. The two-level nature and the low solubility make measurements of the spectral dependence of these photon capture cross sections difficult. Zavadskii and Kornilov (1971) have studied the photon-capture cross section (σ_n^{2-}) of the negative doubly charged Zn center, involving the excitation of an electron from the $E_c - 0.55$ eV level to the conduction band. This cross section was found to be about 10^{-17} cm^2 for $h\nu = 1.05$ eV. This cross section is smaller than predicted by the Lucovsky model (1965) for the photoionization cross section of a deep center, and the difference is attributed to coulomb interaction effects.

In a recent study by Herman and Sah† the spectral dependences of the photoionization cross sections for holes at neutral and singly ionized zinc centers have been measured using photocapacitance transient techniques in n^+p junctions. Lucovsky's delta function potential model fits the results well. The photoionization cross section for holes at singly ionized zinc centers is much greater than the photoionization cross section for electrons at doubly ionized zinc centers. A large electric field dependence and a small temperature dependence are observed near the absorption edge of the photoionization cross section for holes at singly ionized zinc centers. The temperature dependence indicates that a possible two-step photothermal process has taken place when the center is photoionized.

The Lucovsky expression for the photoionization cross section is

$$\sigma(h\nu) = \frac{1}{n}\left[\frac{\mathscr{E}_{\text{eff}}}{\mathscr{E}_0}\right]^3 \left[\frac{8e^2h}{3m^*c}\right](E_I)^{1/2}\frac{(h\nu - E_I)}{(h\nu)^3}, \qquad (10.23)$$

where n is the optical index of refraction, $\mathscr{E}_{\text{eff}}/\mathscr{E}_0$ is the effective field ratio, E_I is the observed binding energy, and m^* is the effective mass of the trapped carrier.

The presence of two deep levels allows the quenching of intrinsic photoconductivity ($h\nu > 1.1$ eV) by impurity light ($h\nu < 1.1$ eV) in compensated silicon containing zinc. The impurity light essentially transfers electrons from the lower to the upper zinc level. As a result, the electron lifetime becomes shorter, thus decreasing the photoconductivity. The modulating action of the impurity light has a time constant of the order of tens of milliseconds. The photoconductivity of n-type zinc-compensated silicon has recently been discussed by Mantena and Loebner (1971) who show that the problem is one of considerable complexity.

Since ionized impurity scattering is significant at 80°K and doping levels

†J. M. Herman, and C. T. Sah (1973), Photoionization Cross Sections of Holes at Zinc Centers in Silicon, *J. App. Phys.*, **44**, 1259.

in excess of 10^{16} cm^{-3} (Kornilov, 1966a), the additional scattering (depending on Z^2) by doubly charged impurities ($Z = 2$) upon optical excitation should also be considered in principle (Chéroff et al., 1961; Bube et al., 1961; Fowler, 1961).

The spectral sensitivity of a reverse-biased zinc-doped *pn* junction has a sharp dip at 1.65 eV; this is not understood in detail (Kornilov, 1964b). Other unusual photoconductivity spectra of some silicon specimens doped with gold or with zinc show an unexpected increase in sensitivity above 105°K (Glinchuk et al., 1966a), possibly caused by a small density of other deep centers. Such interpretation, however, would require reexamination of earlier work; for example, the unusually large σ_n^- value (10^{-16} cm^2) could be due to such deep extraneous centers.

Preliminary impact-ionization studies (Kornilov, 1964c) on p^+pp^+ structures indicate a critical field of about 2×10^4 V cm^{-1} for the lower zinc level, with $T = 77°K$, $N_{Zn} = 3 \times 10^{15}$ cm^{-3}, $N_D = 7 \times 10^{14}$ cm^{-3}, and a device length of 30 μm.

10.6.4 Other Impurities: Fe, W, Cu, B, Al, Ga, P, As, Sb, Bi, and S

The hole capture cross section, σ_p^0, of the iron donor level ($E_v + 0.40$ eV) from pulsed space-charge-limited current flow in p^+pp^+ structures at 300°K has the value $7 \times 10^{-16} \times 1/g_{Fe}$ cm^2 where g_{Fe} is the degeneracy of the iron level. Lemke's measurement (Table 10.5) is independent of the iron doping level, a distinct advantage for low-solubility impurities with a tendency to precipitate. This result agrees well with Collins' findings in his original studies of iron in silicon (see Table 10.5).

Tungsten in silicon has three acceptor levels (Zibuts et al., 1964), at $E_c - 0.22$, $E_c - 0.30$, and $E_c - 0.37$ eV, and two donor levels, at E_v + 0.31 and $E_v + 0.34$ eV, have also been reported (Zibuts et al., 1964). The photon and electron capture cross sections of the three acceptor levels have been obtained (Table 10.5) from the photoconductivity kinetics of tungsten-doped silicon (10^{15} cm^{-3}). The $E_c - 0.22$ eV level gave rise to two time constants, which was interpreted as this level having two capture cross sections. Such behavior would set tungsten apart from all other known deep impurities. Furthermore, if the three levels are associated with the same impurity atoms, then the electron capture by the highest level ($E_c - 0.22$ eV) is doubly repulsive capture and one might expect a smaller value than 10^{-16} cm^2, as observed. Such large values for doubly repulsive capture have been observed for σ_n^{2-} of gold in germanium (Ridley and Pratt, 1965) at high fields (more than 1000 V cm^{-1}) at 20°K, but the low-field value is about three orders of magnitude smaller.

TABLE 10.5

Capture Cross Sections of Iron, Tungsten, Copper, and Shallow Impurities in Silicon[a] *(After Schibli and Milnes, 1967b)*

Impurity	Quantity	Value (σ in cm^2, c in cm^3 sec^{-1})	Temperature (°K)	Method[b]	Ref.
Fe	σ_p^{0}	3×10^{-16}	250–350	PC	Collins and Carlson, 1957
Fe	σ_n^{+}	15×10^{-16}	250	PC	Collins and Carlson, 1957
Fe	σ_p^{0}	$7 \times 10^{-16} \times 1/g_{Fe}$	250–300(?)	Trans. SCLC	Lemke, 1966b
W	σ_n^{0}	2.5×10^{-18}			
W	σ_n^{-}	6×10^{-17}			
W	σ_n^{2-}	$0.6–1 \times 10^{-16}$			
W	q^{-}	4.5×10^{-18}		PCD	Zibuts et al., 1967
W	q^{2-}	5×10^{-17}			
W	q^{3-}	$1–1.5 \times 10^{-16}$			
Cu	σ_p^{0}	3.5×10^{-20}			
Cu	q^{+}	5×10^{-18}			
B	c_p^{-}	2.2×10^{-6}	4.2	S injection	Gregory and Jordan, 1964
B	c_n^{0}	2×10^{-12}[b]	30; exp $(+0.038/kT)$	PCD	Pokrovskii and Svistunova, 1964b
B	c^{-}	1.5×10^{-4}	20	PC, N	Godik and Pokrovskii, 1965
B	c_p^{-}	7×10^{-5}	20	N	Proklov et al., 1965
B	σ_p^{-}	8×10^{-12}	4.2; 1000 V cm^{-1}	DI	Brown and Jordan, 1966
B	σ_n^{0}	2×10^{-14}	4.2; 667 V cm^{-1}	DI	Brown and Jordan, 1966
B	c_n^{0}	5×10^{-8}	1.1–4.2	PhX	Loewenstein and Honig, 1966
Al	c_n^{0}	4×10^{-8}	1.1–4.2	PhX	Loewenstein and Honig, 1966
Ga	c_n^{0}	1×10^{-12}[c]	30; $\sim$exp $(0.035/kT)$	PCD	Pokrovskii and Svistunova, 1964a
Ga	c_n^{0}	7×10^{-8}	1.1–4.2	PhX	Loewenstein and Honig, 1966
P	c_n^{+}	6.9×10^{-6}	1.1–4.2	PhX	Loewenstein and Honig, 1966
P	σ_n^{+}	8×10^{-12}	4.2	DI	Brown and Jordan, 1966
P	σ_p^{0}	2×10^{-14}	4.2	DI	Brown and Jordan, 1966
As	c_n^{+}	$5–8.6 \times 10^{-6}$	1.1–4.2	PhX	Loewenstein and Honig, 1966
Sb	c_n^{+}	$1.6–2.6 \times 10^{-6}$	1.1–4.2	PhX	Loewenstein and Honig, 1966
Bi	c_p^{0}	6×10^{-11}	30; $\sim$exp $(0.036/kT)$	PCD	Pokrovskii and Svistunova, 1964b
Bi	c_n^{+}	$1–3 \times 10^{-5}$	1.1–4.2	PhX	Loewenstein and Honig, 1966

[a] c is the capture coefficient ($c = \sigma v_{th}$), and q is the photon capture cross section. The superscript indicates the charge state of the center before the capture process. g_{Fe} is the degeneracy of the iron level.
[b] PC, steady state photoconductivity; PCD, photoconductive decay method; Trans. SCLC, transient space charge limited current method; N, noise method; DI, double injection; PhX, photoexcitation.
[c] Recombination between impurity levels probably occurred in these experiments.

The lower tungsten acceptor levels, $E_c - 0.30$ and $E_c - 0.37$ eV, have smaller capture cross sections, 6×10^{-17} and 2.5×10^{-18} cm^2, respectively. The photon capture cross sections of the three tungsten levels are 10^{-16}, 5×10^{-17}, and 5×10^{-18} cm^2, respectively, and are thus smaller for the deeper levels. These values were measured at 4.5, 3.5, and 2.5 μm, respectively; but they are probably functions of wavelength.

The donor level of copper ($E_v + 0.24$ eV) in silicon has a very small hole capture cross section, $\sigma_p{}^0 = 3.5 \times 10^{-20}$ cm^2, and the photon capture cross section is 5×10^{-18} cm^2 (Zibuts et al., 1967). The photoionization cross section of electrons from neutral sulfur centers in silicon is 2×10^{-16} cm^2 and from singly ionized sulfur centers is about 10^{-16} cm^2, for the photon energy range 0.7 to 1.0 eV (Rosier and Sah, 1971). Thermal emission rates from Au, Ag, and Co traps in silicon have been measured by Yau and Sah (1972).

Capture cross sections of shallow impurities at low temperatures are also included in Table 10.5 for comparison. The neutral capture coefficients are about 4×10^{-8} cm^3 sec^{-1} from 1.1 to 4.2°K, and the coulomb-attractive coefficients are about a hundred times larger.

CHAPTER ELEVEN

Space-Charge-Limited Current Studies in Semiconductors Containing Deep Impurities

Single- and double-injection currents in insulators and semiinsulators are of considerable interest, since the *V-I* characteristics of associated devices are usually rich in structure due to various electronic properties of the device material. Single-injection devices are those in which either electrons or holes are injected from a suitable contact into the bulk of the device and are collected by a contact on the other side of the bulk. Currents are limited by space charge (much as in a vacuum diode) and by trapping of injected carriers by defect states in the bulk. This subject has been treated extensively in theory by Rose (1955b, 1964), Lampert (1964), and others.

The current-voltage characteristics obtained are confined to a triangular region in a log-log plot. This triangle is bounded by Ohms law, by the $I \propto V^2$ line appropriate for the space-charge effect in a trap-free insulator, and by a vertical rise of current at a voltage which is known as the traps-filled-limit voltage, V_{TFL}. For a semiconductor containing traps Lampert has shown that the transition from the ohmic line to the V_{TFL} line is along an $I \propto V^2$ line which is determined in position by the trap parameters and the thermal equilibrium free-carrier density. This is illustrated in Fig. 11.1. Experimental studies of single injection have been made by many workers, including Gregory and Jordan (1964), Henderson and Ashley (1966) and Henderson et al. (1972). However, it is not easy to obtain purely ohmic contacts, and studies of space-charge flow in semiconductors have tended to

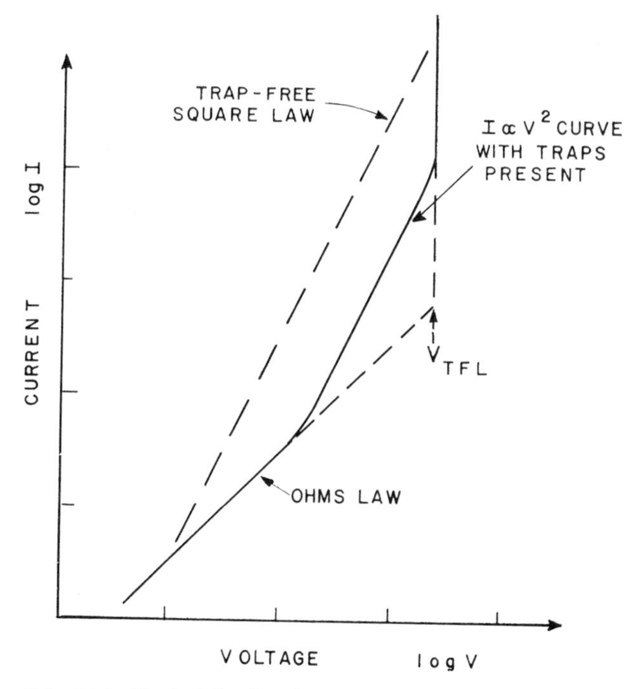

FIG. 11.1 Single-injection characteristic triangle.

be done more with double-injection devices than with single-injection structures.

Double-injection devices are those in which electrons are injected into the bulk from one contact and holes are injected from another contact on the other side of the bulk. The simplest of these devices is the much studied *p-i-n* structure. Space charge and trapping also limit the currents in these devices, but additional complexities are introduced since both electrons and holes are present, and the currents can be further limited by recombination and space charge of both polarities. The topic of double injection in high-resistance semiconductors has been reviewed recently by Baron and Mayer (1970). There are many different conditions to be considered which depend on temperature, device length, capture cross-section ratios, and so on. The present chapter touches on only a few aspects of these problems. The conditions treated in detail illustrate many deep impurity effects in space-charge-limited (SCL) current flow, but are not exhaustive. For instance, diffusion effects will be neglected, although these may be quite significant. Regional approximation methods may be the most convenient way of analyzing certain situations (Lampert and Mark, 1970; Antognetti et al., 1971). Quasineutrality assumptions may or may not be allowable (Baranenkov and Osipov, 1971), and the multivalent nature of the trapping centers may have to be considered (Zwicker et al., 1970).

The reader with a trap-controlled space-charge effect to explain, therefore, may well have to go beyond the confines of this chapter: for instance to the books by Lampert and Mark (1970) and by Baron and Mayer (1970).

In general in this chapter we will be concerned with steady-state dc characteristics. Trap-controlled moving domains and recombination wave oscillations in high-resistance semiconductors are considered in Chapter 12. Transient space-charge-limited current effects have been considered by Many and Rakavy (1962), Fazakas et al. (1969), R. H. Dean (1969), and Tove and Andersson (1973).

11.1 The Ashley-Milnes Space-Charge Regime

Lampert (1956, 1962) has analyzed double injection in a perfectly compensated semiconductor. Ashley (1963, 1964), using the model of Fig. 11.2*a*, has expanded this treatment to include partial compensation and arbitrary occupancy of the trapping centers. The shallow donor centers play no role other than providing compensating electrons for the trapping centers. These centers can either be neutral, in which case they are electron traps, or singly negatively charged, in which case they are hole traps. The capture cross section for holes by the hole traps is greater than the capture cross section for electrons by the electron traps, and hence the electron lifetime is greater than the hole lifetime. In the Lampert model of perfect compensation the electron lifetime is essentially infinite, so that at lowest fields the current is a space-charge-limited electron current and is proportional to the square of the voltage. At some critical voltage, holes can begin to transit the bulk, resulting in filling of some of the hole traps. This increases the hole lifetime, decreases the electron lifetime, allows holes to transit at lower voltages, and a negative-resistance region ensues. When the electron and hole lifetimes become equal, the current becomes two-carrier, recombination limited, and is proportional to the square of the voltage. Ultimately the injection rates become greater than the recombination rate, and the current becomes two-carrier, space-charge limited, and proportional to the cube of the voltage. Figure 11.2*b* shows the complete qualitative *V-I* characteristic for this model.

Also shown in Fig. 11.2*b* is the *V-I* characteristic resulting from the modification of the Lampert model by Ashley, which more closely represents practical situations. This modification considers a high-resistivity extrinsic

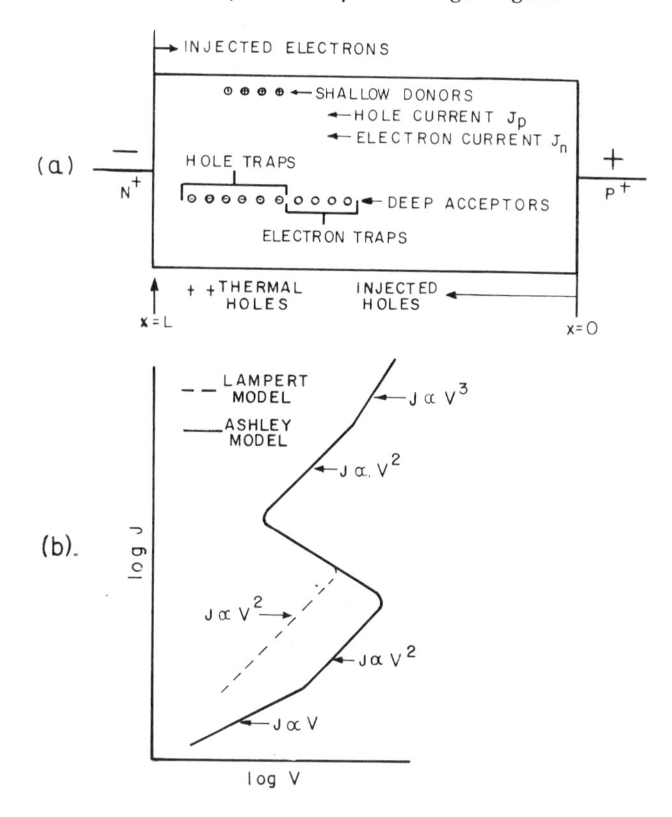

FIG. 11.2 Double injection in $p^+ \pi n^+$ structures.
(a) Model of the π semiinsulating region (Ashley-Milnes). (b) Expected I-V characteristics. (After Wagener and Milnes, 1965.)

semiconductor, so that at lowest values of electric field the behavior is simply ohmic. Initially injected electrons, like the holes, are trapped, so that no electron current flows and the current is composed entirely of thermal holes. When the field becomes great enough so that electrons can transit the bulk, a recombination-limited electron current dominates the thermal hole current, and the current becomes proportional to the square of the voltage, as in the Lampert model. An important difference between the two models in this region, however, is the fact that the space charge in the Ashley case is predominantly that of trapped injected electrons rather than the free electrons themselves. This has become known as the Ashley–Milnes regime. At higher fields when the injected holes can transit the bulk a negative resistance ensues, and for higher currents the Ashley model gives results very similar to those of Lampert. However, Ashley's calculations show that the breakdown voltage (beginning of negative-resistance region) can be considerably greater than expected from the Lampert model due to the influence of the space charge trapped on the impurity centers.

Among the information about the properties of the device material that can be obtained from the characteristics of Fig. 11.2*b* are values for the capture cross sections that the impurity centers offer to holes and electrons. At the voltage where the Ohm's law region intersects the square-law region of the Ashley characteristic (V_Ω) the electron transit time equals its lifetime. Thus measurement of this voltage yields the value of capture cross section that neutral impurity centers offer to electrons ($\sigma_n{}^0$). Similarly, measurement of the breakdown voltage (V_B) yields the value of capture cross section that singly negatively charged impurity centers offer to holes ($\sigma_p{}^-$). Double injection therefore is a tool for measuring lifetimes and capture cross sections in semiconductor materials. Lifetimes in material containing deep impurities are typically quite small (about 10^{-8} sec) and difficult to measure with the standard photoconductive decay technique. The double-injection method employs steady-state measurements and hence is attractive. Moreover, since two different lifetimes are obtained with the same measurement, the double-injection method is doubly productive. Also, since only electrical connections to the device are necessary (incident radiation, magnetic fields, etc., not used) the apparatus needed for making measurements as a function of temperature is simple. On the other hand, the validity of the method depends on the degree to which the model truly represents the physics of the specimen. Imperfect contacts, diffusion effects, thermal effects, mobility and capture cross-section dependencies on field strength, and multiple trapping actions may present problems in the application of the model. However, the Ashley–Milnes regime has been shown to represent experimental observations at 77°K for gold-doped germanium (Ashley and Milnes, 1964; Ham and Ashley, 1970), for indium-doped silicon (Wagener and Milnes, 1965), for zinc-doped silicon (Kornilov, 1966a), for gold-doped silicon (Zwicker et al., 1970a), and for manganese-doped GaAs (Weiser and Levitt, 1964).

However, it is not found to be very effective in predicting the threshold voltage for the onset of the negative-resistance region. For this one must turn to the studies of Keating (1964), Dumke (1964), Saunders (1968), Wright and Ibrahim (1968), and Ashley et al. (1973).*

11.1.1 Analysis of the Regime, Neglecting High-Field Effects

A single recombination center is assumed, which, in equilibrium, is partially occupied by electrons. The recombination-center level is arbitrarily located with respect to the valence and conduction band edges except that it is

* K. L. Ashley et al., GaAs:Mn Double-Injection Devices, *Solid State Electron.*, **16,** in process.

sufficiently deep for the density of thermally excited carriers (in particular, holes) to be much less than the density of recombination centers. Also, there is assumed an infinite reservoir of electrons and holes at the cathode and anode, respectively. In line with this assumption, the electric field is taken as always being equal to zero at both electrodes. Finally, if the voltage-current characteristic of the model is to contain a negative-resistance region (or a breakdown), it must be assumed that the capture cross sections of the recombination centers are different for holes and electrons. In particular, it is assumed that $\sigma_p \gg \sigma_n$ where σ_p is the capture cross section of the recombination center for holes and σ_n is the capture cross section of the recombination center for electrons.

Both types of carriers are able to recombine at the lowest injection levels because the recombination-center level contains both filled and unfilled states. Therefore in the absence of relaxation processes, upon application of a voltage, injected electrons would build up on the recombination centers in the vicinity of the cathode and the recombination centers would become deficient of electrons (corresponding to hole trapping) on the anode side of the device. This shift of the equilibrium recombination-center occupation level would result in a space-charge barrier to current flow. This barrier would prevent the single-carrier square-law behavior of the Lampert model, which assumes all recombination centers full in equilibrium.

If, however, a certain small number of thermally induced holes is assumed to be present, a degree of charge relaxation is provided. Consequently, a large shift of the recombination-center equilibrium is prevented; and the barrier to electron flow is reduced and a square-law behavior is permitted. The characteristic for voltages below the breakdown voltage then consists of a square-law segment and an Ohm's law segment (Fig. 11.2*b*). The Ohm's law segment of the characteristic prevails when the injection level of electrons is substantially below the thermal hole carrier concentration.

In the square-law segment of the characteristic, the current is carried by injected electrons, n cm^{-3}, and thermally induced holes, p_0 cm^{-3}, and $n > p_0$. As the injection level of n is increased (with increasing voltage), the value of p remains near p_0 because of the "recombination barrier" to hole injection. This assumes that $\sigma_p^- \gg \sigma_n^0$, where σ_p^- is the capture cross section for holes and σ_n^0 is the capture cross section for electrons. Thus the current is (neglecting diffusion terms)

$$J = (\mu_n n + \mu_p p_0) e\mathscr{E} \tag{11.1}$$

Since the recombination charge predominates over the free-carrier charge, Poisson's equation is

$$\frac{\varepsilon}{e}\frac{d\mathscr{E}}{dx} = p_r - p_{r0} \tag{11.2}$$

where p_r is the density of holes on the recombination center and p_{r0} is the equilibrium value. To represent the recombination of carriers, use is made of the particle conservation equation. Now

$$\frac{1}{e}\frac{dJ_p}{dx} = \mu_p p_0 \frac{d\mathscr{E}}{dx} \tag{11.3}$$

But $-(1/e)(dJ_p/dx)$ is equal to the rate of recombination n/τ_n or

$$-\frac{1}{e}\frac{dJ_p}{dx} = \frac{n}{\tau_n} = np_r\sigma_n v_{th} \tag{11.4}$$

where $\tau_n = (p_r\sigma_n{}^0 v_{th})^{-1}$ and v_{th} is the thermal velocity for electrons. Thus, equating the right sides of (11.3) and (11.4)

$$\frac{\varepsilon}{e}\frac{d\mathscr{E}}{dx} = -\frac{\varepsilon n}{e\mu_p p_0 \tau_n} \tag{11.5}$$

and from (11.2), the right side of (11.5) must be δp_r or

$$\delta p_r = -\frac{\varepsilon n}{e\mu_p p_0 \tau_n}$$

But $\varepsilon/e\mu_p p_0$ is the dielectric relaxation time of the material, so

$$\delta p_r = -\left(\frac{\tau_{rel}}{\tau_n}\right) n \tag{11.6}$$

Hence the charge density is proportional to the free-electron density, but is larger than n by the ratio of the dielectric relaxation time to the electron lifetime.

Obtaining the *V-I* characteristic, then, consists of eliminating n between (11.1) and (11.5) (now neglecting the current contribution of holes) and solving for $\mathscr{E}$. The integration of $\mathscr{E}$ to determine voltage is then performed with $\mathscr{E}$ set equal to zero at the cathode ($x = 0$). The resulting relationship between current and voltage is

$$J_{SCL} = \frac{9}{8}\frac{\tau_n}{\tau_{rel}}\varepsilon\mu_n\frac{V^2}{L^3} \tag{11.7}$$

This expression is the same as for the trap-free insulator except for the modification factor τ_n/τ_{rel}. It is evident from the dependence on τ_{rel} that, although the current is carried principally by electrons, its magnitude is directly proportional to p_0. Since the magnitude of the current in the Ohm's law region is, of course, also directly proportional to p_0, a variation of p_0 has the effect of shifting the entire characteristic (below the breakdown voltage) accordingly.

The voltage as a function of distance through the semiinsulating region is given by a three-halves power law:

$$V(x) = \left(\frac{8}{9}\frac{J}{\sigma_n{}^0\tau_n\mu_n}\right)^{1/2} x^{3/2} \tag{11.8}$$

This has been verified by Ham and Ashley (1970) for gold-doped germanium diodes.

As previously indicated, for $n < p_0$, Ohm's law predominates and for applied voltages below the square-law regime,

$$J_\Omega = e\mu_p p_0 \frac{V}{L} \tag{11.9a}$$

The two regions, Ohm's law and square law, intersect where $J_\Omega = J_{\mathrm{SCL}}$ or (omitting a $\frac{9}{8}$ factor) the intersection voltage is

$$V_\Omega \simeq \frac{L^2}{\mu_n\tau_n} \tag{11.9b}$$

Thus the intersection occurs approximately when the lifetime and transit time, $L^2/\mu_n V_\Omega$, of the electron are equal. If V_Ω is determined by experiment, (11.9b) gives τ_n and therefore the capture cross section $\sigma_n{}^0$ may be determined.

Double injection in long *p-i-n* diodes with one trapping level has been studied by Deuling (1970) who obtains the Ashley–Milnes regime as a limiting case of a more general treatment. The treatment has been further extended by Zwicker et al. (1970b) and Weber and Ford (1970) to cover the condition of a semiinsulating region containing multivalent trapping centers.

11.1.2 Analysis Including High-Field Effects

The analysis that has just been presented has been extended by Wagener (1964) to include high-field effects.

In the prebreakdown space-charge region the current is predominately due to injected electrons, and therefore (diffusion being assumed negligible) is given by

$$J = e\mu_n n\mathscr{E} \tag{11.10}$$

where n is the free-electron density. This density is obtained from the conditions of continuity in the steady state. Hence,

$$\frac{n}{\tau_n} = \frac{1}{e}\frac{dJ_n}{dx} = -\frac{1}{e}\frac{dJ_p}{dx} \tag{11.11}$$

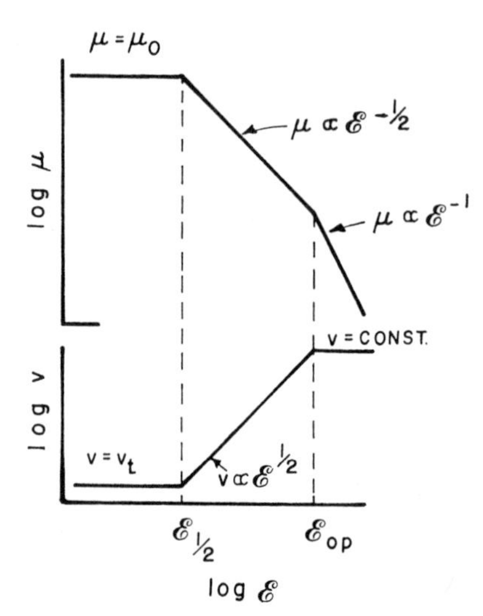

FIG. 11.3 Idealized field dependence of mobility and carrier velocity. (After Wagener and Milnes, 1965.)

where x is the distance variable, and τ_n is the electron lifetime given by

$$\tau_n = (\sigma_n{}^0 v_{\text{th}} N_{T_e})^{-1} \tag{11.12}$$

where v_{th} is the thermal velocity and N_{T_e} is the density of electron traps.

It is well known that high electric field strengths cause carrier mobilities to be lowered, and these effects become appreciable in silicon at 77°K for field strengths in excess of a few hundred volts per centimeter. Because double-injection structures may be operated in this range the effects of nonconstant mobilities on the device characteristics must be considered.

An idealized picture of variation of carrier mobility and carrier velocity with electric field $\mathscr{E}$ is shown in Fig. 11.3. In this first-order approximation the carrier mobility μ is constant at its low-field value of μ_0 until some critical field $\mathscr{E}_{1/2}$ is reached. In this range the carrier velocity v is also constant at its thermal value v_{th}. The actual distribution of carrier velocity is neglected in these calculations, and all of the carriers are assumed to have the same velocity—namely that of the appropriate statistical average. The value of $\mathscr{E}_{1/2}$ is that field strength at which the carrier drift velocity approximately equals the thermal velocity, and for greater fields the drift velocity and carrier velocity are approximately the same. Under these conditions, the mobility decreases as $\mathscr{E}^{-1/2}$ as the velocity increases as $\mathscr{E}^{1/2}$. At some higher critical field $\mathscr{E}_{\text{op}}$ optical phonons are excited and the velocity saturates.

The calculated value of $\mathscr{E}_{1/2}$ for holes in silicon is about 600 V cm^{-1} and about 7000 V cm^{-1} for $\mathscr{E}_{\text{op}}$, in the 77°K temperature range. Using single-injection devices, $\mathscr{E}_{1/2}$ for holes is observed to be about 650 V cm^{-1}. No direct observation of $\mathscr{E}_{\text{op}}$ has been attempted, but a field-induced breakdown has been noted to occur at about 8700 V cm^{-1}. Thus the range over which $\mu \propto \mathscr{E}^{-1/2}$ is quite appreciable, and an important portion of the *V-I* characteristics can occur in this range for devices of appropriate length. Under low-field conditions, the breakdown voltage is proportional to the square of the device length, and since the field at which mobilities become field dependent is a constant, there exists a certain critical device length such that devices shorter than this break down before mobilities become field dependent, and for longer devices mobilities become field dependent before breakdown occurs.

In addition to the mobilities and carrier velocities being field dependent at high fields, the capture cross sections also are field dependent. In the region where $\mu \propto \mathscr{E}^{-1/2}$ Lax (1960) calculates the capture cross section σ to vary approximately as $\mathscr{E}^{-2}$. More generally, the field dependencies of the capture cross sections are given by

$$\sigma_n^{\ 0} \propto \mathscr{E}^{-\alpha_n} \tag{11.13}$$

$$\sigma_p^{\ -} \propto \mathscr{E}^{-\alpha_p} \tag{11.14}$$

where α_n and α_p are constants.

J_p is due entirely to thermal holes, and in the high-field region becomes

$$J_p = ep_0\mu_{p0}(\mathscr{E}_{1/2p}\mathscr{E})^{1/2} \tag{11.15}$$

where μ_{p0} is the low-field mobility. Combining (11.10), (11.11), and (11.15), n becomes

$$n = -\frac{p_0\mu_{p0}(\mathscr{E}_{1/2p})^{1/2}}{2\sigma_{n0}v_{\text{th}}N_{T_e}(\mathscr{E}_{1/2n})^{(\alpha_n-1/2)}}\,\mathscr{E}^{(\alpha_n-1)}\frac{d\mathscr{E}}{dx} \tag{11.16}$$

Combining (11.10) and (11.16)

$$J = K_1\mathscr{E}^{(\alpha_n-1/2)}\frac{d\mathscr{E}}{dx} \tag{11.17}$$

where K_1 is a constant and is given by

$$K_1 = \frac{ep_0\mu_{p0}\mu_{n0}(\mathscr{E}_{1/2p})^{1/2}}{2\sigma_{n0}v_{\text{th}}N_{T_e}(\mathscr{E}_{1/2n})^{(\alpha_n-1)}}$$

Integrating (11.17),

$$\mathscr{E}^{(\alpha_n+1/2)} = \frac{\alpha_n+\frac{1}{2}}{K_1}Jx + K_2$$

and K_2 is seen to be zero if the boundary condition $\mathscr{E} = 0$ for $x = 0$ is imposed. Therefore,

$$\mathscr{E} = \left[\frac{\alpha_n+\frac{1}{2}}{K_1}Jx\right]^{[1/(\alpha_n+1/2)]} \tag{11.18}$$

and integrating (11.18) to obtain the device voltage,

$$V = \left[\frac{\alpha_n + \frac{1}{2}}{K_1} J\right]^{[1/(\alpha_n + 1/2)]} \left(\frac{\alpha_n + \frac{1}{2}}{\alpha_n + \frac{3}{2}}\right) L^{(\alpha_n + 3/2)/(\alpha_n + 1/2)}$$

and

$$J = KL^{-(\alpha_n + 3/2)} V^{(\alpha_n + 1/2)} \tag{11.19}$$

where L is the device length.

It is not difficult to extend the calculation to include the most general form of field dependence of mobility, that is,

$$\mu_p \propto \mathscr{E}^{-1/\beta_p}$$
$$\mu_n \propto \mathscr{E}^{-1/\beta_n} \tag{11.20}$$

where β_p and β_n are constants between 1 and ∞. The carrier velocities are then generalized, and the *V-I* characteristics become

$$J = K_3 L^{-(\alpha_n + 2 - 1/\beta_p)} V^{(\alpha_n + 1 - 1/\beta_p)} \tag{11.21}$$

where K_3 is a constant and is given by

$$K_3 = \left[\frac{\alpha_n + 2 - 1/\beta_p}{\alpha_n + 1 - 1/\beta_p}\right]^{(\alpha_n + 1 - 1/\beta_p)} \times \left\{\frac{e p_0 \mu_{p0} \mu_{n0} (\mathscr{E}_{1/2p})^{1/\beta_p}}{\left[\frac{\beta_p(\alpha_n + 1 - 1/\beta_p)}{(\beta_p - 1)}\right] \sigma_{n0} v_{\text{th}} N_{T_e} (\mathscr{E}_{1/2n})^{(\alpha_n - 2/\beta_n)}}\right\} \tag{11.22}$$

From (11.7) and (11.19) a V^2 region in the prebreakdown characteristic should transform into a $V^{(\alpha_n + 1/2)}$ region if the electron mobility becomes field dependent at some voltage V_F before breakdown. Measurement of the *V-I* characteristic in this region, and determination of the slope, therefore provides an experimental determination of the field dependence of the capture cross section. A log-log plot of J as a function of L at constant voltage and temperature in the high region also gives an experimental determination of α_n.

The transition voltage V_Ω is at all times given by the relation

$$V_\Omega = L^2 \frac{\sigma_n{}^0 v_n N_{T_e}}{\mu_n} = \frac{L^2}{\mu_n \tau_n} \tag{11.23}$$

and is correct in the high-field region when the appropriate high-field quantities are used for σ_n, v_n, and μ_n. When breakdown occurs before mobilities become field dependent, the breakdown voltage is given by

$$V_B = L^2 \left(\frac{e}{4\pi\varepsilon} \frac{\sigma_p{}^0 v_{\text{th}}}{\mu_p}\right)^{1/2} N_{T_h} \tag{11.24}$$

where ε is the dielectric constant and N_{T_h} is the density of hole traps. When mobilities become field dependent before breakdown, (11.24) becomes,

$$V_B = L^{[(\alpha_p+3)/(\alpha_p+1)]}\left[\frac{e}{4\pi\varepsilon}\frac{\sigma_{p0}v_{\text{th}}}{\mu_{p0}}\right]^{[1/(\alpha_p+1)]}\left[N_{T_h}{}^2(\mathscr{E}_{1/2p})^{(\alpha_p-1)}\right]^{[1/(\alpha_p+1)]} \tag{11.25}$$

Equation 11.25 indicates that the field dependence of the capture cross section for holes can be experimentally determined by finding V_B as a function of L for devices in which mobilities become field dependent before breakdown occurs.

The electron capture cross section is determined by observation of V_Ω since (11.23) gives

$$\sigma_n{}^0 = \frac{\mu_n}{v_{\text{th}}N_{T_e}}\left(\frac{V_\Omega}{L^2}\right) \tag{11.26}$$

Similarly (11.24), when used for short devices where mobilities do not become field dependent, gives, with observation of V_B, the hole capture cross section

$$\sigma_p{}^- = \frac{4\pi\varepsilon\mu_p}{ev_{\text{th}}N_{T_h}{}^2}\left(\frac{V_B}{L^2}\right)^2 \tag{11.27}$$

For the longer devices this becomes

$$\sigma_p{}^- \approx \frac{4\pi\varepsilon}{eN_{T_h}{}^2}\frac{V_B}{L^3} \tag{11.28}$$

where $\sigma_p{}^-$ in (11.28) is the high-field value of the capture cross section. Thus using (11.28) in conjunction with a series of devices of different lengths (or the same device at different temperatures), $\sigma_p{}^-$ can be found as a function of field strength under high-field conditions, and this result should check with α_p determined by the method of (11.25).

Diffusion effects in such devices have been examined recently by C.-H. Tsao, who finds that the influence on the I-V characteristic shape can be similar to that of the field dependent effects discussed above.*

11.1.3 Experimental Studies of Indium-Doped Silicon *p*-π-*n* Structures

Devices of *p*-π-*n* structure were fabricated from silicon containing 2×10^{16} atoms cm^{-3} of indium and compensated with about 4×10^{13} atoms cm^{-3} of shallow donors. The devices were made by standard diffusion of boron and phosphorus into opposite surfaces of wafers of various thicknesses. The

* C.-H. Tsao, "Computer Simulation Studies of Semiconductor Device Behavior—Application to PIN Diodes", Ph.D. Thesis, Carnegie-Mellon University, April 1973.

resulting surfaces were degenerate p and n, respectively, and provided excellent injection.

The V-I characteristics for all of the devices were taken from the Ohm's law region to breakdown and corrected for junction voltage. Where heating was not significant dc measurements were used, and 100 μsec pulses, with a duty cycle of 10^{-4}, were used at the higher levels. The characteristics shown in Fig. 11.4 are for a device typical of the shorter structures. For this device breakdown occurs before the electric field becomes high enough to cause mobilities to become field dependent, and it exhibits an Ohm's law region followed by a perfect square-law region. Nearing breakdown the characteristic departs from the square-law behavior, and the current increases rapidly until breakdown occurs. Breakdown is indicated by the small arrows pointing to the upper left. The temperature dependence of both the Ohm's law and square-law regions is as predicted by the Ashley model, and in particular the current density is directly proportional to the thermal hole density p_0 in both regions. Inspection of Fig. 11.4 shows that V_Ω varies only slightly with

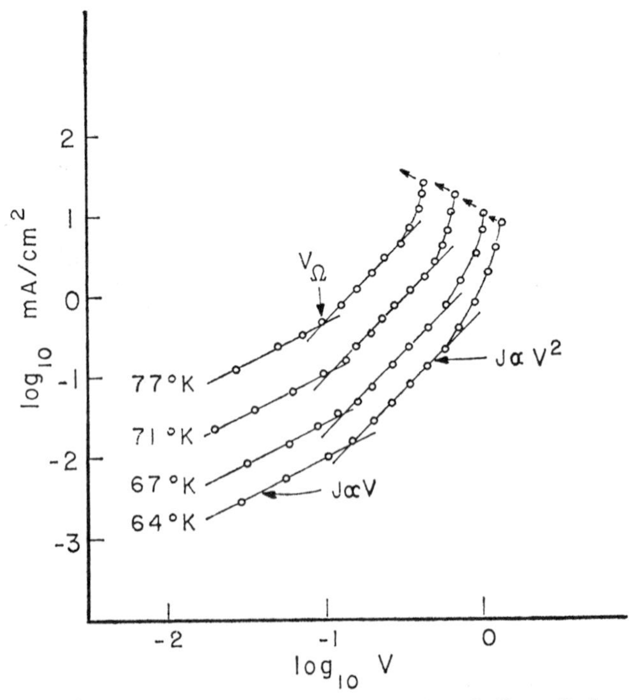

FIG. 11.4 ***V-I* characteristics of 1.6-mil-thick indium-doped silicon diode as a function of temperature.**

The experimental points were taken under dc conditions. (After Wagener and Milnes, 1965.)

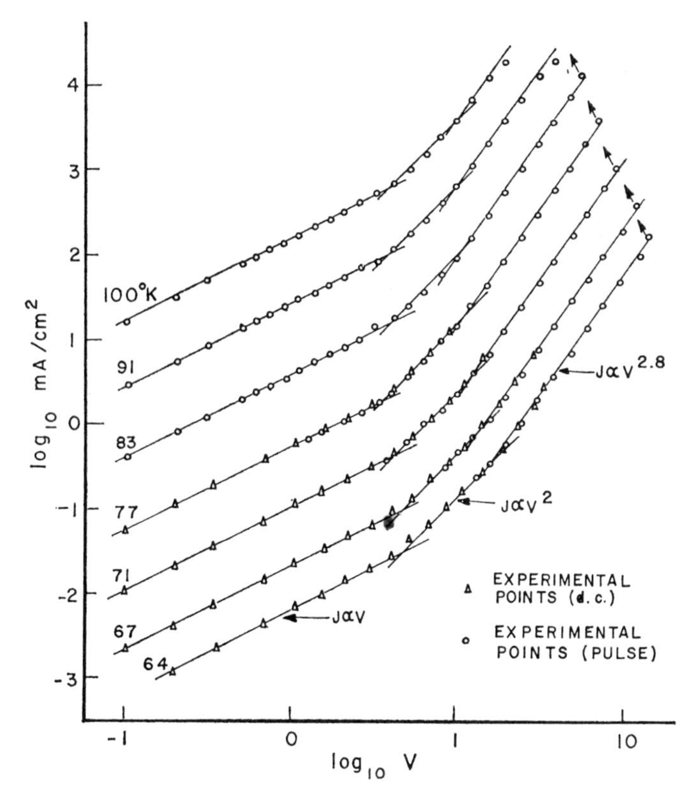

FIG. 11.5 *V-I* **characteristics of 15.9-mil-thick indium-doped silicon diode as a function of temperature.**

Circles indicate experimental points taken with pulse techniques; triangles are experimental points taken with direct current. (After Wagener and Milnes, 1965.)

temperature, whereas the breakdown voltage varies quite appreciably with temperature.

The characteristics shown in Fig. 11.5 are typical of the longer devices. Three distinct regions are apparent in these data. The Ohm's law and square-law regions are followed by what is very close to a $V^{2.8}$ region, which extends over almost an order of magnitude in voltage before breakdown. This is the region in which the mobilities are field dependent. Applying (11.19) to these results, α_n is seen to be 2.3 or $\sigma_n \propto \mathscr{E}^{-2.3}$. This of course assumes, perhaps too readily, that diffusion effects are not contributing to the curve shape seen. The Ohm's law, square-law intersection occurs in the low-field region, so the same general temperature dependence of V_Ω as in Fig. 11.4 is observed. The breakdown occurs in the field-dependent mobility region, so that V_B is both temperature and field dependent.

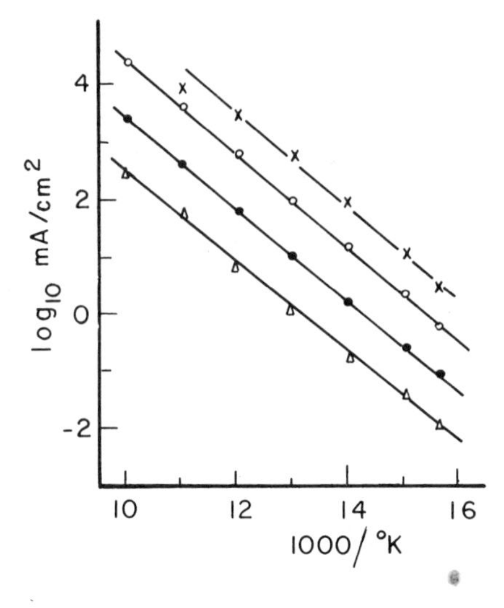

FIG. 11.6 Current as a function of temperature at constant voltage for the different regions of the characteristics of indium-doped silicon diodes.
The straight lines correspond to an energy level 0.158 ± 0.002 eV above the valence band: (×) *11.6 mil device,* $V = 20$ V ($J \propto V^{3.0}$); (○) *15.9 mil device,* $V = 20$ V ($J \propto V^{2.8}$); (●) *15.9 mil device,* $V = 8$ V ($J \propto V^2$); (△) *15.9 mil device,* $V = 2$ V ($J \propto V$). *(After Wagener and Milnes, 1965.)*

The theory predicts that the current in all regions of the prebreakdown characteristic should be proportional to the thermal hole density. That this is observed experimentally is shown in Fig. 11.6 by a plot of current density as a function of temperature at constant voltage. The bottom line is for the Ohm's law region and the next to bottom line for the square-law region of the device of Fig. 11.5. The next to the top line is for the high-field region in this device, and the top line is for the high-field region of a slightly shorter device. The lines are of a slope corresponding to the hole activation energy of the indium level.

In the high-field region of the characteristics the current should vary as $L^{-(\alpha_n+3/2)}$ at constant voltage and temperature, as given by (11.19), and a log-log plot of current versus device length under these conditions should have a slope equal to $-(\alpha_n + \frac{3}{2})$. Figure 11.7 is such a plot for two different values of voltage and temperature. The slopes of the two curves shown are -4.0 and -4.3, making $\alpha_n = 2.5$ and 2.8, respectively. This compares with the value for α_n of 2.3 obtained by simply measuring the slope of the V-I characteristic.

For the set of devices used in this investigation the transition to the space-charge region occurs before the mobilities become field dependent. For this case V_Ω, as derived by Ashley and given in (11.23), should be directly proportional to L^2. The value for this voltage is experimentally obtained by the intersection of the straight lines of slope 1 and slope 2 in the characteristic. A plot of this voltage as a function of device length is given in Fig. 11.8, where it is seen that the L^2 dependency does indeed exist.

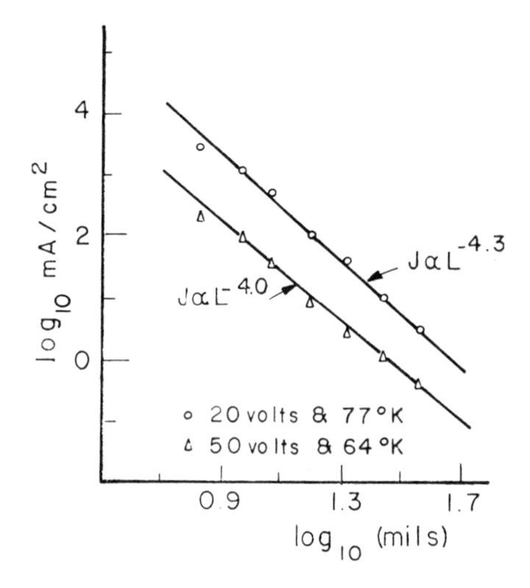

FIG. 11.7 Current as a function of $p^+\pi n^+$ device length (indium-doped silicon) at constant voltage and temperature in the high-field regime. (After Wagener and Milnes, 1965.)

Knowing μ_n, v_n, and N_{T_e}, and obtaining the quantity V_Ω/L^2 from any point on the straight line of Fig. 11.8, (11.26) can be used to obtain the electron capture cross section. The values for $\sigma_n{}^0$ have been calculated in this manner in the temperature range 64 to 100°K. The results indicate a temperature dependence of $T^{-1.3}$, and $\sigma_n{}^0 = 2.4 \times 10^{-14} T^{-1.3}$ most accurately describes the low-field electron capture cross section in this

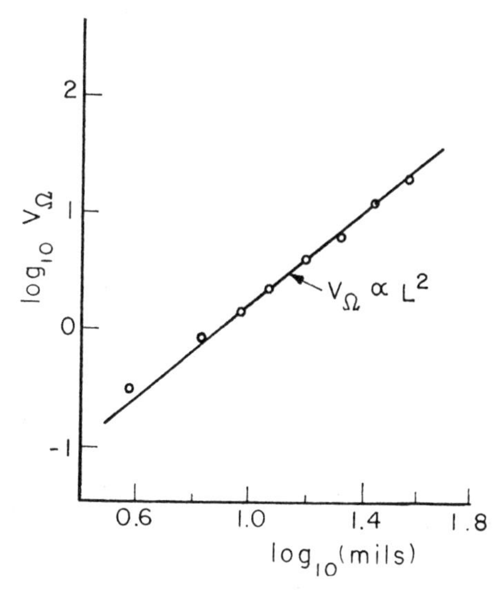

FIG. 11.8 V_Ω as a function of device length at 77°K for indium-doped $p^+\pi n^+$ structures. (After Wagener and Milnes, 1965.)

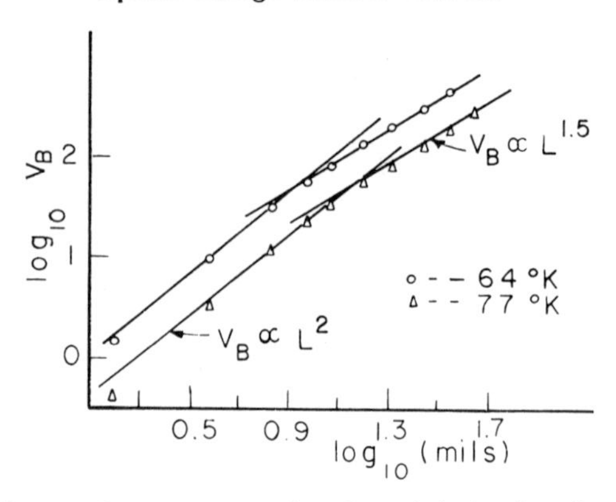

FIG. 11.9 Breakdown voltage, V_B, as a function of device length ($p^+\pi n^+$ indium-doped silicon). (After Wagener and Milnes, 1965.)

temperature range. This agrees quite well with the results of Wertheim (1958), but is much larger than the capture cross sections observed by Pokrovskii and Svistunova (1969) (as discussed in Chapter 6).

An alternative method of arriving at the electron capture cross section is to determine the voltage variation across the device by probing. This gives, from (11.8), $V(x) \propto x^{3/2}$, and $\sigma_n{}^0$ can be determined from the slope of this line. Applying this technique to gold-doped germanium, Ham and Ashley (1970) obtain capture cross sections that are in good agreement with other determinations. Although the probing is somewhat cumbersome to set up, the method has the advantage of avoiding junction voltage corrections.

Let us turn now to consideration of the threshold voltage for onset of the negative-resistance region. Equation 11.24 describes the breakdown voltage for low fields, and indicates that V_B should be proportional to L^2. The relationship of V_B to L under high-field conditions is given in (11.25) and is strongly dependent on the high-field dependency of the capture cross section for holes. Figure 11.9 shows V_B plotted as a function of L for two different temperatures. For the shorter devices the $V_B \sim L^2$ dependency is observed, since breakdown occurs before the high-field regime sets in. However, for the longer devices where high-field effects are present, it is seen that V_B varies very closely as $L^{1.5}$ rather than as L^2. Applying (11.25) to this portion of Fig. 11.9, $(\alpha_p + 3)/(\alpha_p + 1) = 1.5$, which gives $\alpha_p = 3.0$. It is apparent from Fig. 11.9 that as the temperature is lowered, shorter devices enter the high-field regime before breakdown. This is expected, of course, since the breakdown voltage rapidly increases as the temperature is decreased.

Equation 11.27 is available to determine the low-field magnitude and temperature dependence of σ_p^- if V_B/L^2 is taken from Fig. 11.9. The value of σ_p^- obtained for 77°K is about 10^{-13} cm^2, which is in good agreement with other determinations. For further discussion see Wagener and Milnes (1965), Wright and Ibrahim (1968), and Saunders (1968).

11.2 Effects of a Density Gradient of the Deep Impurity

The purpose of this section is to study the prebreakdown condition for $p^+\pi n^+$ structures where the contacts (p^+, n^+) permit double injection and the semiinsulating π region contains a diffused gradient of the deep recombination-center impurity. Attention is concentrated on the shift of the V_Ω point and the Ashley V^2 regime with the graded impurity density. The structures are such that injection of electrons (from the n^+ contact) occurs at the low end of the impurity gradient or, alternatively, at the high end.

The semiinsulator model considered (Fig. 11.10) is similar to Ashley's but with the recombination-center density N_R an exponential function of the

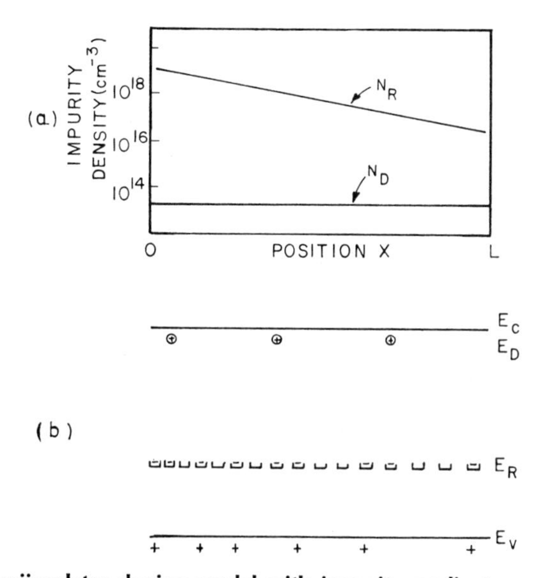

FIG. 11.10 Semiinsulator doping model with impurity gradient.
(a) Assumed doping profile, with exponential gradient of the deep impurity recombination-center density N_R, and a constant shallow donor density N_D. (b) Energy band diagram for equilibrium. (After Schibli and Milnes, 1967c.)

position x. The doping levels are such that $N_R > N_D \gg p_0$, where p_0 is the thermal hole density. The electric field E is assumed zero at both injecting contacts, which are located at $x = 0$ and $x = L$. In Fig. 11.10*b*, all the shallow donor atoms (E_D, N_D) have lost one electron to the deep acceptor level (E_R, N_R). The presence of a few thermal holes is shown in the valence band E_v, the electrons corresponding to these holes having also gone to the deep acceptor level. However, the thermal hole density is low because low-temperature conditions are assumed. (In indium-doped silicon at 77°K, the resistivity can be in excess of 10^4 Ω-cm for these reasons.)

In the theoretical studies (Schibli and Milnes, 1967c) the net recombination-center density ($N_R - N_D$) at the low doped end of the device is held constant and the gradient specified determines the density at the high doped end. (The analysis could also be readily applied to represent the condition where the shallow donor density N_D varies exponentially and the recombination-center density N_R is constant, and $N_R > N_D$. However, to maintain a clear exposition this condition is not considered here.)

Some features of the I-V curve of the structure under consideration may be deduced by simple physical reasoning. At very low currents the injected carriers recombine close to the contacts and the bulk is in quasiequilibrium. The current is carried by thermal holes and the I-V relation is ohmic. In this regime the magnitude of the current at a given voltage does not depend on its direction with respect to the doping gradient.

As the voltage is raised further, the structure shows an $I \propto V^2$ characteristic where the current is carried principally by injected electrons and the V^2 dependence arises from the space-charge conditions that develop. The space charge is determined by the recombination kinetics, which control the device behavior. If the n^+ contact is at the highly doped end of the device, the injected electrons meet a high recombination-center density in the vicinity of the cathode, before penetrating deeply into the device. If the n^+ contact is at the low doped end of the device, the electrons injected find relatively few recombination centers until they approach the anode. Near the anode, however, holes are readily available and the space-charge buildup is less than before. Hence V for a given current is less than before.

The argument may be reexpressed in a slightly different way as follows. Electron injection at the highly doped end causes a large negative space-charge density near the cathode, whereas the compensating positive space charge (to maintain overall neutrality) is distributed throughout the device. The distance between the centers of these charge distributions is a measure of the voltage and is relatively large. Electrons injected at the low doped end build up only a moderate space-charge density in front of the cathode because the recombination-center density available is small, and as a consequence the negative space-charge region extends deep into the device.

The centers of these space-charge distributions are less far apart and hence the voltage is less.

These physical arguments are a little sweeping in a situation as complicated as the one under consideration. However, they do provide some insight into the phenomena under consideration.

The state of charge of the recombination centers under injection conditions has some effect on the lifetime of free carriers which therefore is position and injection dependent. Keating (1964) has discussed the role of lifetime modulation in the double-injection region. For the Ashley V^2 region the effect is not of major significance for the class of semiinsulating materials under consideration (indium-doped silicon and gold-doped germanium). However, the matter needs examination whenever a space-charge problem of this kind is studied.

The analytical treatment for space-charge flow in the presence of a density

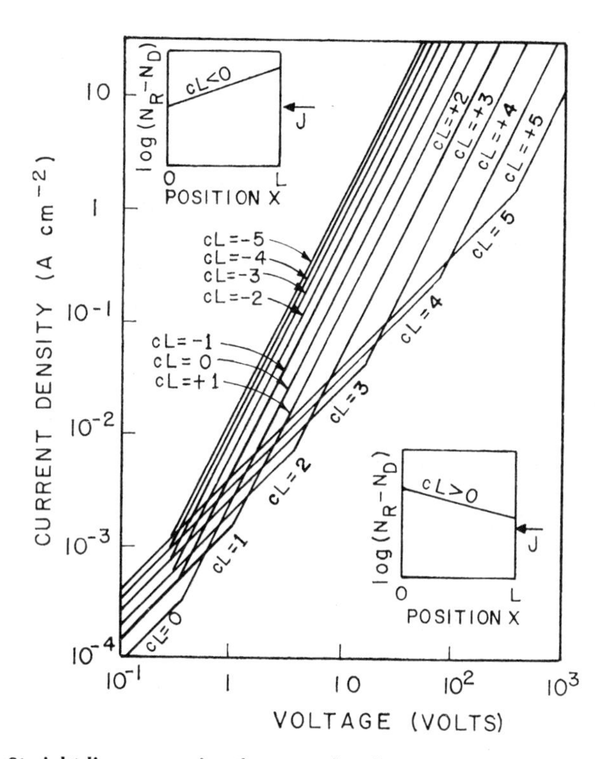

FIG. 11.11 Straight-line approximations to the theoretical *I-V* characteristics for a $p^+\pi n^+$ structure containing a density gradient of the deep impurity.

The parameter cL represents the doping gradient. Note the shift of the V_Ω intersection point with doping gradient. (After Schibli and Milnes, 1967c).

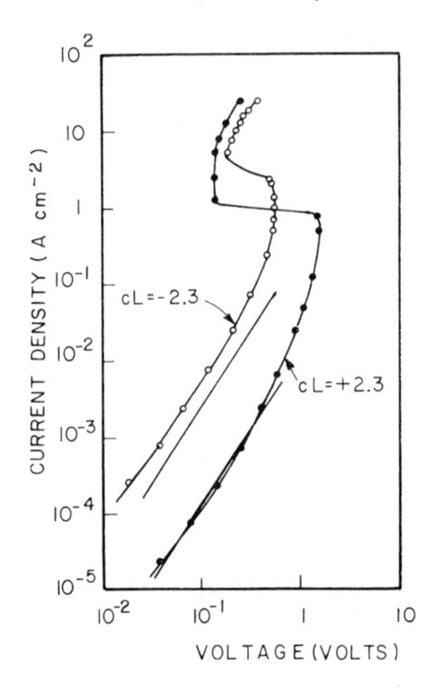

FIG. 11.12 ***I-V*** **characteristics of experimental $p^+\pi n^+$ indium-doped silicon devices at 77°K. The recombination-center density varies from 10^{17} to 10^{18} cm^{-3} ($cL = \pm 2.3$). For $cL = 2.3$ electrons are injected at the indium-rich end, and for $cL = -2.3$, at the indium-poor end. (After Schibli and Milnes, 1967c.)**

gradient requires computed solutions. The equations and a detailed discussion (Schibli and Milnes, 1967c) therefore will not be developed here. However, a very brief discussion follows.

The deep impurity density is assumed to vary with distance as e^{-cx}, hence cL is an important parameter characterizing each device. If the equations are approximated to give piecewise linear solutions, one can follow the shift of the V_Ω intersection point of the Ashley–Milnes regime. The theoretical results are shown in Fig. 11.11. If $cL < 0$, and the current flow is in the direction shown, the electrons are injected at the low doped end and penetrate closer to the anode end than for $cL > 0$, before encountering holes and recombination centers in quantity. Hence the space-charge buildup is less and the current flow for a given voltage across the device is greater for $cL < 0$ than for $cL > 0$ in the high-current portion of the regime.

The curves computed from the equations of the full analytical treatment are reasonably similar in form to those of Fig. 11.11, although of course lacking the abrupt corners of the linearized piecewise treatment.

To verify the treatment, $p^+\pi n^+$ diodes of indium-doped silicon were made with the deep impurity concentration varying from 10^{18} to 10^{17} cm^{-3} over a distance of 35μm. The parameter cL was ± 2.3, depending on which sides the p^+ and n^+ contacts were placed. The experimental results, after subtraction of junction voltages, are shown in Fig. 11.12 together with the square-law lines calculated from the approximation equations. The agreement between Figs. 11.11 and 11.12 is seen to be reasonably satisfactory, except that, at high currents, double-injection breakdown is seen, which is not provided for in the approximate treatment but does show in the computed solutions of the more complete analysis. Other studies of the effects of non-uniform distributions of traps include those of Hwang and Kao (1972) and Choo (1971).

11.3 Filament Formation in the Double-Injection Regime

At high current levels in high-resistance semiconductors a negative-resistance regime develops for reasons connected with lifetime changes and charge redistribution. At voltages and currents large enough for the holes to be making transit, the recombination centers become less significant than the injected carriers in determining the space charge and the current flow. The lifetime of the holes therefore becomes effectively larger and the space charge is partially relaxed. These effects result in the voltage decreasing as the current rises, and a region of negative resistance ensues. This negative-resistance regime is followed by a $J \alpha V^2$ double-injection region (Fig. 11.2) that has been modeled by Lampert (1964) and others. At even higher voltage levels this may be expected to develop into a V^3 region because of field-dependent effects.

Although the Lampert models are conceptually elegant and comforting to have, there is some doubt that such bulk treatments represent the behavior of practical semiconductor structures such as *p-i-n* diodes effectively. In most structures that have been carefully studied, the negative-resistance region appears to be associated with the development of a current filament that occupies only a limited portion of the cross section of the device. As the current is increased, the filament grows in diameter until it occupies the whole cross section of the device. Sometimes a second filament develops before the first filament fully occupies the device.

Effects of this kind were observed by a number of workers prior to 1962. One detailed study was that of Melngailis and Milnes (1962) who examined the filament development in the impact-ionization breakdown of compensated germanium at 4.2°K.

The *I-V* characteristics observed in double injection and in impact ionization are of the current-controlled form (as in Fig. 11.2*b*, where specifying the current completely determines the voltage). This can be distinguished from the voltage-controlled class of negative-resistance curves (obtainable from Fig. 11.2*b* by interchanging the *J* and *V* axes). These two classes of curves have been considered by Ridley (1963) in a discussion that invokes the principle of minimum entropy production. In the case of the

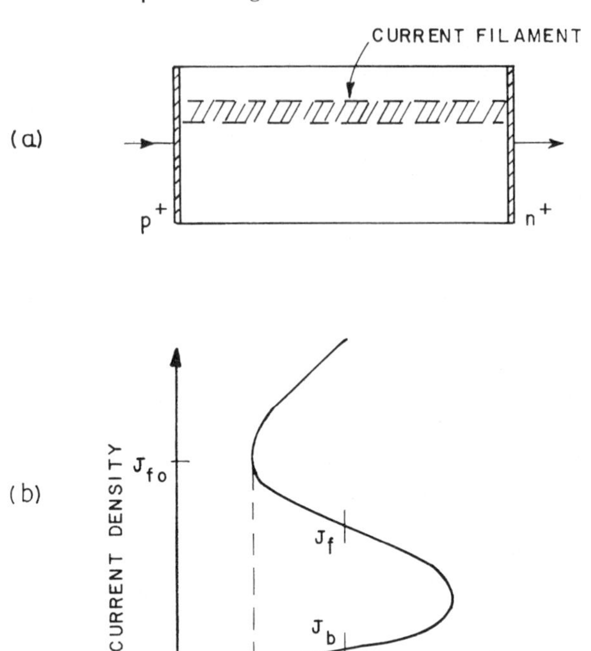

FIG. 11.13 Formation of a current filament in a structure with a current-controlled negative resistance.

voltage-controlled negative resistance, moving domains of high electric field would be expected to develop. These have now become familiar as Gunn domains in GaAs. Domains are also associated with certain oscillations of current that are caused by other physical phenomena besides valley transfer.

For the current-controlled negative-resistance characteristics, Ridley postulates, on minimum entropy production or joule-heating considerations, that the current density across the device does not remain constant. Instead a filament of high current density develops as shown in Fig. 11.13*a*. If this filament is of current density J_f, if the current density in the bulk is J_b, and if the field strength is V/L, then the power dissipation per unit volume is

$$W = \frac{V}{L} J_b(1 - a) + J_f a \tag{11.29}$$

where a is the filament area as a fraction of the cross-sectional area of the devices. With a defined in this way the average device current density is of course

$$J = J_b(1 - a) + J_f a \tag{11.30}$$

W may be thought of as the rate of entropy production and it will be a minimum when V is as low as possible, namely the value V_0 in Fig. 11.13*b*. The current densities will then become J_{b0} and J_{f0}. If one hypothesizes that these densities remain relatively unchanged as the device current is increased, the device voltage will remain at V_0 and the filament area will increase with J according to

$$a = \frac{J - J_{b0}}{J_{f0} - J_{b0}} \tag{11.31}$$

Thus the filament will grow at constant voltage until the device current density is J_{f0} and the filament then fills the device. Further increase of current then requires increase of the device voltage.

Although we may feel that the approach above is a considerable oversimplification, and neglects the detailed physics of the situation, regimes of current rise at almost constant voltage are indeed found in double-injection diodes. Figure 11.14 shows this for a $p^+\pi n^+$ indium-doped silicon structure

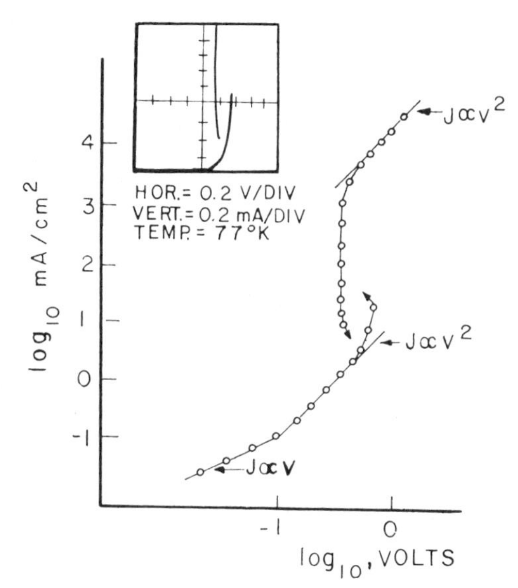

FIG. 11.14 Complete *V-I* characteristic typical of filamentary breakdown for a $p^+\pi n^+$ silicon device at 71°K.

The circles are experiments points, where the current density was obtained by dividing the current by the total device area. Inset is an oscillograph picture of a double-injection device showing the breakdown and hysteresis in which the maximum prebreakdown current exceeds the minimum postbreakdown current. (After Wagener and Milnes, 1964.)

at 71°K. The inset is an oscillograph showing the drastic hysteresis (minimum postbreakdown current less than the maximum prebreakdown current) and the characteristic near-constant value of the postbreakdown voltage. In this picture the drive to the device increases until breakdown occurs at about 1.1 mA. The trace then jumps to a current of about 1.5 mA on the post-breakdown portion, and subsequent increase in drive increases the current at constant voltage. Subsequently decreasing the drive causes the current to decrease to nearly 0.4 mA at near-constant voltage before jumping back to the prebreakdown state. The main part of Fig. 11.14 shows the characteristics of this device, after correcting for junction voltage. The prebreakdown region conforms to the Ashley theory. Immediately after breakdown a near-constant voltage is observed, followed by a square law at the highest currents observed.

The development and growth of current filaments in indium-doped semiinsulating silicon structures have been observed by recombination radiation photography (Barnett and Milnes, 1966). The $p^+\pi n^+$ diodes examined were of stepped construction as shown in Fig. 11.15*a*, so that there would be no possibility that the aluminum contact wire would act as a nucleation source for the filaments. The recombination radiation was observed through the p^+ (2 μm) skin. The current-voltage characteristic observed for a particular device is shown in Fig. 11.15*b*. The sensitivity of the photographic system did not allow filaments to be observed at currents much below 10 mA. Figure 11.16 shows photographs of the filament for currents of 10, 20, and 40 mA. Included in Fig. 11.16 is a photograph of the device taken through a microscope and also a photograph taken through the image-converter optical system (with the device removed from the Dewar). The vertical band of light in the photograph represents reflections from the edge of the step in the device thickness (Fig. 11.15*a*). The white circle added to the photograph indicates the approximate nucleation point of the filament on the device.

The formation of a filament occurs when a critical injection condition is reached, voltage V_B, at which the hole traps are filled along a path between the contacts. The voltage then drops to a lower value V_M, but rises again as the current is increased and the filament further develops. According to the photographic evidence the filament is continuing to grow even after the constant voltage period of current rise is complete.

With increasing current the device voltage again begins to approach the value V_B, and this may encourage the onset of a second filament displaced from the original. Figure 11.17 shows a second filament present in the device with 80 mA of current flowing. At 160 mA a bridge is forming between the two filaments. Interaction between filaments is, of course, to be expected by carrier diffusion, trap-filling mechanisms, crystal inhomogeneities, heating,

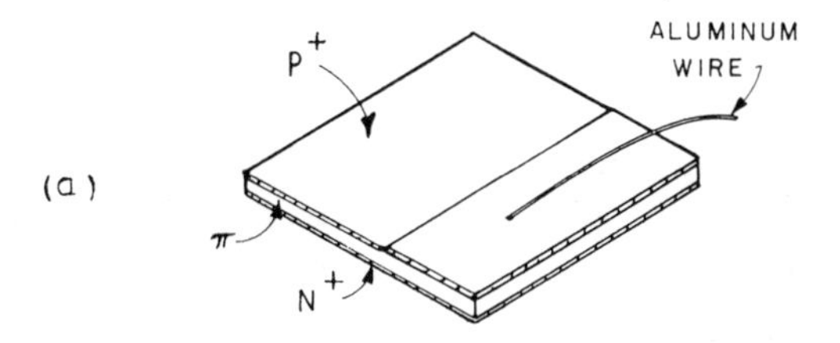

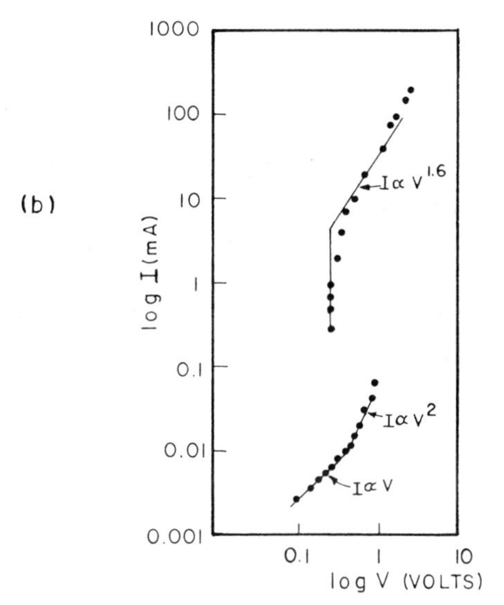

FIG. 11.15 Silicon $p^+\pi n^+$ diode for filament observation and its *I-V* characteristic (77°K). (After Barnett and Milnes, 1966.)

and perhaps photon-radiation mechanisms. Onset of a second filament may be accompanied by a small discontinuity in the *I-V* characteristic and/or by a change in the slope of the characteristic. There are signs of such effects in Fig. 11.15*b* at currents above about 60 mA.

The inhomogeneity that results in the nucleation of the filament is not known. A slight nonuniformity of device thickness would be one possible factor, although care was taken to minimize such effects. Another factor causing filament nucleation might be a small doping variation. The condition that is normally assumed for onset of breakdown is that the hole transit time becomes less than or equal to its lifetime. This is equivalent to

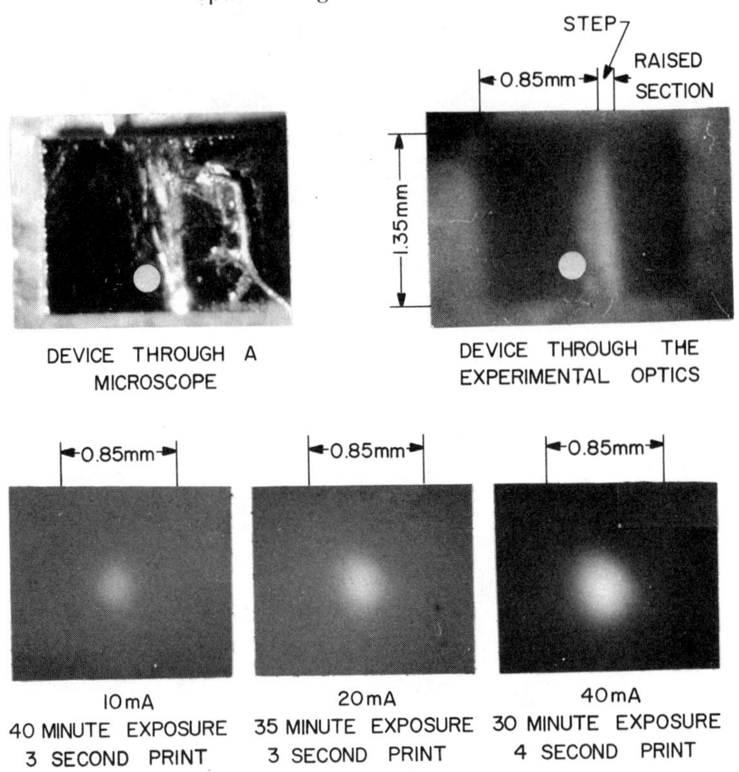

FIG. 11.16 Photographic evidence for filament formation in device (#25-1) at 10, 20, and 40 mA.
The white circles superimposed on the device photographs indicate the filament location. (After Barnett and Milnes, 1966.)

assuming that the holes at the far boundary from their injecting contact are sufficient in number to neutralize the charged recombination centers (hole traps) effectively. The breakdown voltage is strongly dependent on the concentration of negatively charged deep acceptors. As the thermal hole density is small, this approximately equals N_D, the compensating shallow donor density. Thus it may be hypothesized that variations in N_D, particularly in the vicinity of the n-type contact, could be the inhomogeneity factor responsible for the development of the filament.

Characterization of the filament and of the conditions controlling its growth is a formidable problem. The attempts that have been made up to now have involved assumptions that limit their validity. The difficulties are discussed by Barnett (1966), Knepper (1969), Varlamov and Poltoratskii

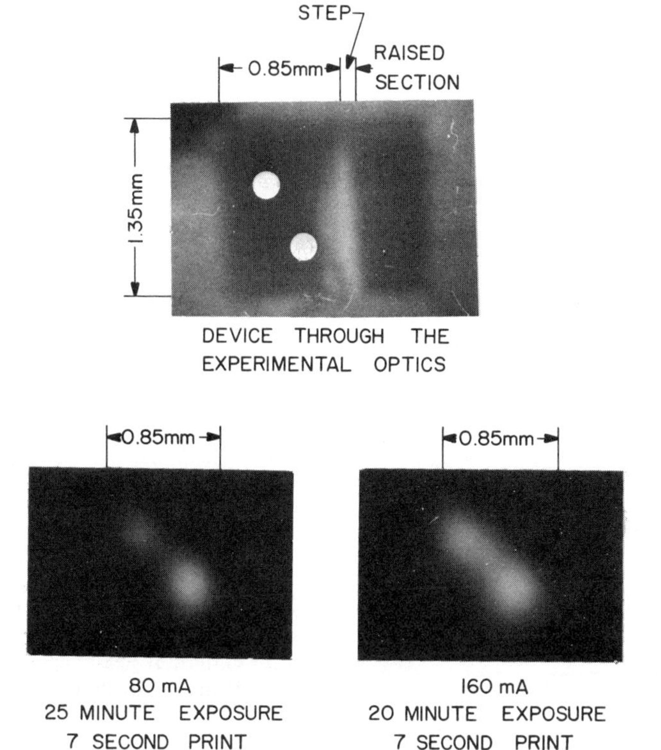

FIG. 11.17 Photographs showing the development of a second filament (80 and 160 mA). (After Barnett and Milnes, 1966.)

(1969), Muller and Guckel (1968), and Carlson (1969). The effect of a magnetic field on a current filament, in gold-doped silicon, has been studied by Homma et al. (1972).

Filamentary action has been observed in oxygen-doped semiinsulating GaAs at room temperature (Barnett and Jensen, 1968; Ferro and Ghandhi, 1971). The characteristic of such a diode is shown in Fig. 11.18. The current rise at almost constant voltage is seen to be several decades. Such diodes have been considered for use as part of a solid-state display mosaic that can be *x-y* addressed. The bistable action of the negative-resistance device with a load line provides a storage function (Barnett, 1969).

One consequence of the discovery that filamentary current flow is normal in double-injection structures is that considerable caution must be exercised in interpreting the results of experiments with such devices.

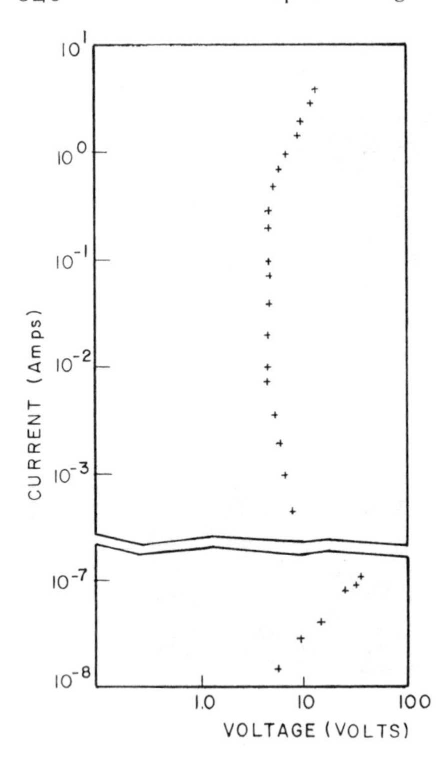

FIG. 11.18 Voltage-current characteristics for GaAs device, at 300°K, showing rise of current at nearly constant voltage during filament growth. (After Barnett and Jensen, 1968.)

11.4 Optical Feedback as a Factor in the Double-Injection Threshold in Gallium Arsenide

In a study of GaAs p^{+}-π-n^{+} structures where the intrinsic region is produced by manganese doping, Weiser and Levitt (1964) find that the prebreakdown region is in general agreement with the Ashley–Milnes model. However, the model does not effectively predict the threshold voltage for double-injection breakdown and the onset of the negative-resistance regime. The model suggests that the threshold voltage should be given by

$$V_{\text{Th}} = \frac{L^2}{2(\pi)^{1/2}} \left(\frac{e\sigma_p^{-} v_{\text{th}_p} P_{R0}}{\varepsilon \mu_p N_R} \right)^{1/2} n_{R0} \tag{11.32}$$

where L is the width of the π region, σ_p^{-} is the capture cross section of negatively charged manganese centers for holes, v_{th_p} is the thermal velocity of holes, μ_p is the hole mobility in the π region, and P_{R0} is the number of neutral manganese centers. Also N_R is the total number of manganese centers, n_{R0} is the number of negatively charged manganese centers, e is the electronic charge, and ε is the static dielectric constant of GaAs. With

$L = 4 \times 10^{-3}$ cm, $P_{RO}/N_R \sim 1$, $n_{RO} \sim 3 \times 10^{17}$ cm^{-3}, $\mu_p \sim 500$ cm^2 V^{-1} sec^{-1}, $v_{\mathrm{th}_p} \sim 10^7$ cm sec^{-1}, and $\varepsilon \sim 10^{-12}$ F cm^{-1} we find $V_{\mathrm{Th}} \sim 8 \times 10^{10}$ $(\sigma_p^-)^{1/2}$. Using the reasonable value of 10^{-14} cm^2 for σ_p^- the value obtained for V_{Th} is about 8000 V. The experimentally observed value, however, is only of the order of 5 V. Similarly, the current in the post-negative-resistance regime occurs at a much lower value than predicted by Lampert (10 A cm^{-2} compared with 10^4 A cm^{-2}).

The threshold voltage for the onset of the negative-resistance regime has been reexamined recently by Ashley et al.* for a $p^+\pi n^+$ structure in which the thermal hole density p_0 in the π region is high, namely comparable with the donor density N_D in the region. Under these conditions the $J \alpha V^2$ region of the prebreakdown characteristic is raised (in a way similar to that suggested by the dotted line in Fig. 11.2) and then the threshold knee occurs at a much lower voltage than given by (11.32).

This modification of the Ashley–Milnes theory should be relevant to many double-injection structures. For semiconductors with recombination of a highly radiative nature, however, another factor may have to be considered. Dumke (1964) has suggested that an optical feedback effect could modify the double-injection characteristics. In particular, the possibility exists that the intense light emission at the $p^+\pi$ boundary before the onset of the negative resistance may be reabsorbed in the π region. Such a reabsorption together with the superlinearity of the light intensity with current before breakdown could contribute to the negative resistance.

A recent study of the Dumke model is that of Selway and Nicolle (1969), who examined its applicability to chromium-doped GaAs *p-i-n* diodes. The chromium was determined to be a deep acceptor at about 0.64 eV above the valence band if inserted by diffusion.

Dumke's model for optical feedback is developed as follows. If some of the recombination occurring at the p/i junction is radiative, then for a given current the light flux produced is $F = \gamma_J J$, where γ_J is the quantum efficiency that depends on J. If a fraction ϕ of this light is reabsorbed in the semiinsulating region, it will produce a photocurrent given by

$$\Delta J = \phi \gamma_J \tau_n \mu_n \frac{V}{L^2} \tag{11.33}$$

The photocurrent and recombination-limited current may be added and the relationship between current and voltage is

$$J = aV^2 + b\gamma_J JV \tag{11.34}$$

where $b = \phi \tau_n \mu_n / L^2$, a constant.

* K. L. Ashley et al. (1973). GaAs:Mn Double-Injection Devices, *Solid State Electron.*, **16** (in process).

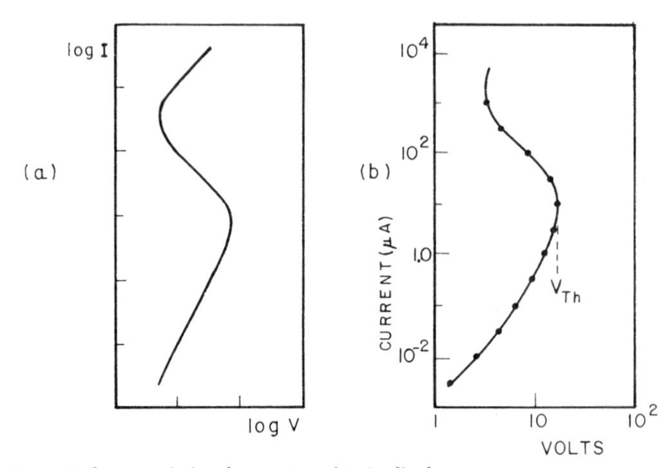

FIG. 11.19 *I-V* **characteristics for GaAs** $p^{+}in^{+}$ **diodes.**
(a) Characteristic predicted from theory of optical feedback, with $\gamma_I \propto I$. *(b) Characteristic observed for chromium-doped GaAs diode (with i region about 25 μm wide). (After Selway and Nicolle, 1969.)*

If we consider a particular case in which the quantum efficiency is proportional to the current, $\gamma_J = cJ$, then (11.33) becomes

$$J = aV^2 + bcJ^2V \qquad (11.35)$$

At low currents the first term predominates and yields a square law, but at high currents the second term increases and eventually yields $VJ =$ constant, a negative-resistance region. At even higher currents the series resistance of the diode will terminate the negative-resistance region with a positive ohmic region. A typical characteristic is given in Fig. 11.19*a*, and there is a clear similarity between this and the experimental characteristic of Fig. 11.19*b*.

In a more general case, if $\gamma_J \propto J^n$, where n is a positive number, then the negative-resistance region is described by

$$V \propto \left(\frac{1}{J}\right)^n \qquad (11.36)$$

If the series resistance R is appreciable, as may be the case in the chromium-diffused diodes, in which a partly compensated n region exists near the n/i junction, then we may describe the high current part of the characteristic by

$$V = \frac{1}{b\gamma_I} + IR \qquad (11.37)$$

where I is the current rather than current density.

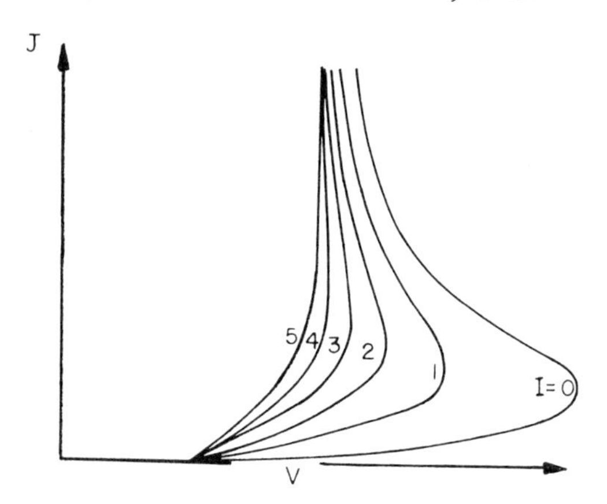

FIG. 11.20 Theoretical behavior of *J-V* characteristics of *p-i-n* diode under progressively more intense external illumination, *I*. (After Dumke, 1964.) (Courtesy of Dunod, Paris and Academic Press, New York.)

Dumke considered a quantum efficiency expression of the form

$$\gamma = \frac{J^n}{J^n + \text{const}} \tag{11.38}$$

with n in the range 1 to 4. Under external illumination the optical feedback should have less effect and the negative-resistance region should become less significant. The Dumke predictions in this respect are shown in Fig. 11.20 which was derived for $n = 2$ in (11.38). The curve $I = 0$ is for no external illumination and the other curves represent equally spaced steps of illumination. Such effects have been seen in GaAs p^+ in^+ structures.

Selway and Nicolle (1969) measured the quantum efficiency versus current curves of their diodes at room temperature and obtained values in the range 0.21 to 0.88. The negative-resistance slopes of the *V-I* curves for the diodes were then determined and good agreement was found with the measured values for n as shown in Table 11.1. As the temperature was decreased the light output observed from the diodes increased, and the negative resistance became more pronounced, as shown in Fig. 11.21.

Selway and Nicolle observed uniform light emission in their structures, and therefore believe the current flow in the devices did not form filaments. On the other hand, Barnett and Jensen (1968) found filament formation for oxygen-doped GaAs diodes. More recently Ferro and Ghandhi (1970) have reported current filaments in chromium-doped GaAs structures. Whether or not filaments form in GaAs is probably related to factors such as

TABLE 11.1

Comparison of the Slope of the Negative-Resistance Region of the Current–Voltage Characteristic of GaAs Diodes with the Slope Predicted from the Variation of Quantum Efficiency with Current

Diode Number	Variation of Quantum Efficiency with Current ($n = d \ln \gamma I / d \ln I$)	$1/n$	Slope of Negative Resistance ($d \ln V / d \ln I$)
1	0.88	1.14	1.18
2	0.73	1.37	1.29
3	0.21	4.8	4.7
4	0.84	1.19	2.0

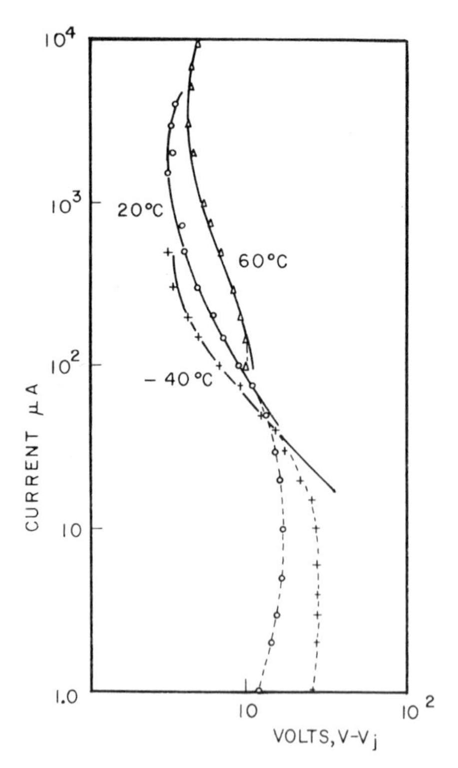

FIG. 11.21 Experimental and predicted characteristics for a GaAs (Cr) p^+in^+ structure with radiative feedback. (After Selway and Nicolle, 1969.)

the uniformity of the material, its efficiency as a light emitter and reabsorber, and other details of the fabrication.

Other characteristics of GaAs double-injection diodes have been examined by Saunders (1968), including the threshold voltage variation with temperature, and the peak-to-valley ratio. Both of these parameters appear to be affected by the concentration of shallow minority carrier traps. This might be expected from the analysis of Keating (1964).

The extent to which the Dumke optical feedback effect may be present in germanium and silicon *p-i-n* diodes at low temperatures is not clear. The radiative quantum efficiencies expected for these materials is lower than for GaAs, and other direct gap semiconductors. Other factors must be considered such as reabsorption efficiency, deep impurity concentration, carrier lifetimes, and transit times. Dumke (1964) estimates that an internal quantum efficiency of the order of 10^{-3} might be sufficient to affect the breakdown voltage and the negative-resistance regime.

11.5 Possible Uses of Double-Injection Negative-Resistance Characteristics

A current-controlled negative-resistance device if connected to a constant voltage V_{dc} in series with a load resistance, as shown in Fig. 11.22, can be made to switch between a low current condition P and a high current condition Q. This is normally accomplished by the application of a pulse of voltage that either aids or opposes the circuit battery voltage V_{dc}, depending on whether turn-on or turn-off is required. The transitions take place with slopes determined by the load resistance. The switching takes place in a time that is usually between 10^{-6} and 10^{-9} sec. Numerous investigators therefore, on encountering such switching actions, imply that they may be useful. In fact, two-terminal negative-resistance switching is not very convenient to use in practice, and almost no applications are found in working circuits.

As discussed previously, illumination has some effect on the shape of the negative-resistance curve and on its peak and valley voltages. It is therefore possible to fabricate a double-injection diode which is triggered from high to low conductivity by incident photons of one energy and from low to high conductivity by photons of another energy. This has been described for silicon *p-π-n* structures (77°K) by Marsh et al. (1965).

In some instances the switching of a double-injection space-charge-limited device can result in the generation of microwave oscillations. Hagenlocher and Chen (1969) have reported that nickel-doped silicon *n-π-n* structures at 300°K, pulsed for 500 μsec at a repetition rate of 100 Hz, have given outputs of 2 W at 1.68 GHz with an efficiency of 0.2%.

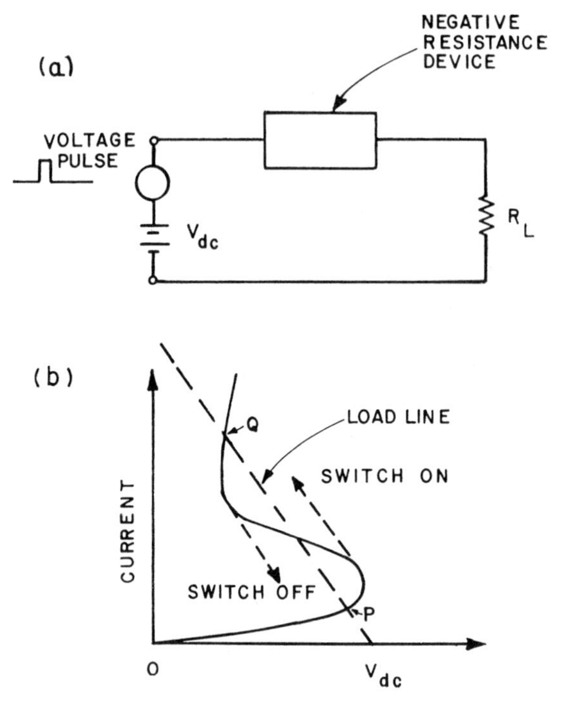

FIG. 11.22 Switching of a current-controlled negative-resistance diode. *(a) typical circuit; (b) characteristic with load line added.*

The possible use of GaAs light-emitting space-charge-limited diodes for optical display with inherent memory has already been mentioned in Section 11.3, and other references are given in the subject index.

CHAPTER TWELVE

Trap-Controlled Domain Propagation and Other Oscillations in High-Resistance Deep Impurity Semiconductors

Oscillations may occur in high-resistance semiconductors for many reasons, and the frequency range may be from tens of gigahertz to less than one hertz, depending on the type of oscillation and on the semiconductor properties.

The highest frequency oscillations are those caused by intervalley transfer of electrons. Avalanche-type diode oscillations will not be considered here since this chapter is primarily concerned with bulk phenomena and in particular with phenomena associated with deep impurities. Space-charge negative resistance due to the lifetime changes involved in double injection is another cause of oscillation. Since the lifetimes involved tend to be in the 10^{-5} to 10^{-8} sec range, the oscillations are usually in the low megahertz region. Other modes of oscillation to be considered are acoustoelectric domains, recombination wave oscillations (Antognetti et al., 1971), and low-frequency moving-domain oscillations caused by field enhancement of carrier trapping.

Intervalley Transfer of Electrons. This effect, proposed by Watkins and Ridley (1961) and later verified experimentally by many investigators, is responsible for the Gunn effect in GaAs. The negative resistance is caused by the transfer of hot electrons from the high-mobility central valley to a secondary minimum in the conduction band of low mobility. Copeland

(1966) has shown that the domains formed travel at about the velocity of the electrons outside of the domain. This velocity is of the order of 10^7 cm sec^{-1}. Therefore if the device length is 10^{-3} cm, the oscillation frequency is about 10 GHz. There is also a mode of oscillation known as limited space-charge accumulation (LSA), that does not involve full domain propagation and results in an even higher frequency of oscillation (Copeland, 1968).

Gunn-type oscillations in GaAs and other semiconductors are the subject of an extensive literature (Bulman et al., 1972) and need not further concern us here. A useful introduction to Gunn oscillations and to avalanche diode oscillators may be found in the book by Sze (1969).

Acoustoelectric Effect. In materials which possess piezoelectric properties, coupling between acoustic waves and electrons in the conduction band can take place (Haydl et al., 1967). The result is a simultaneous buildup of acoustic waves and high-field domains. The domains travel at velocities of approximately those of shear waves in the semiconductor, which is of the order of 10^5 cm sec^{-1}. Such oscillations have been seen in CdS (Hutson and White, 1962; Haydl and Quate, 1965), in GaSb (Sliva and Bray, 1965), in GaAs (Hervouet et al., 1965), in Te (Ishiguro et al., 1965), and in InSb. A perspective on acoustoelectric instabilities has been provided by Bray (1969).

Space-Charge-Limited Negative-Resistance Oscillations. Space-charge-limited flow diodes, particularly those that are really trap limited, have been considered in the steady state in Chapter 11. The small-signal operation of space-charge-limited diodes has been the subject of study by Dascalu (1968). Some of these devices, according to Dascalu, offer high-frequency negative resistance (the low-frequency differential resistance being always positive). In these devices transit time effects play an important role.

The basis of Dascalu's analysis is the separation of the alternating current into components associated with carrier density and carrier velocity modulation. Since each component is connected with the variation of a physical quantity, such as space-charge density, carrier velocity, and electric field, the mechanism of the diode response can be examined at all frequencies, including those comparable to or greater than the reciprocal carrier transit time.

Since there are a wide variety of space-charge-limited structures, depending on the contact behavior, the trapping conditions, and the field strengths involved, it is left to the reader to pursue special interests from the starting point provided by the subject index to the bibliography (under the heading "Oscillations"). However, in Section 12.3, the oscillations seen by Holonyak and coworkers are examined briefly. The class of oscillations that have been termed recombination waves is also discussed qualitatively.

Field-Enhanced Electron Trapping Causing Oscillations. Ridley (1966) has

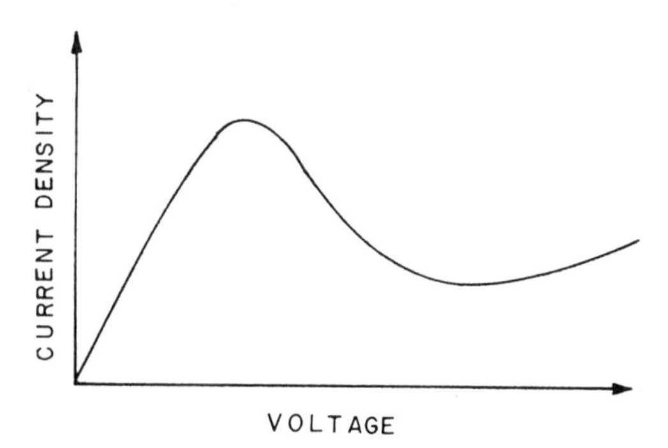

FIG. 12.1 Voltage-controlled differential negative-resistance characteristic.

shown that if a semiconductor has a deep impurity with an electron capture coefficient that increases with electric field, the material may possess a bulk differential negative resistance (Fig. 12.1). A material that possesses a voltage-controlled differential negative resistance may be expected to develop high- and low-field regions (Ridley, 1963) and the domain so formed propagates between the contacts. The average J versus V characteristic observed for the material flattens and drops with the onset of domain action instead of following a V^2 relationship as in space-charge-limited flow. The domain travels at a velocity that is a function of the emission and capture rate of the impurity involved, and is usually in the range 1 to 100 cm sec^{-1}, corresponding to oscillation frequencies of 10 to 10^3 Hz for a device of typical length, 10^{-1} cm.

Such oscillations have been seen in compensated silicon containing zinc (Kornilov and Anfimov, 1967) and, at low temperatures (15 to 77°K) in germanium containing gold (Ridley and Pratt, 1965; Kurova et al., 1967, 1969) and in germanium containing copper (Zhdanova et al., 1966b). They have also been seen in GaAs (Northrop et al., 1964), in CdS (Boer, 1971), in ZnS, and in ZnSe (Bube and Lind, 1958).

While considerable success has been had in predicting domain velocities and characteristics connected with the Gunn effect (Butcher and Fawcett, 1966; Butcher et al., 1966), similar efforts connected with the trap-controlled model have not been as fruitful. Ridley and Wisbey (1967) have proposed a nonlinear analysis for the trap-controlled case which predicts for germanium about the correct value for the current in the presence of a domain, but does not adequately give the domain velocities observed. Sacks and Milnes (1970), in the analysis that is about to be described, have made some further progress.

12.1 Analysis of Domains Caused by Field-Enhanced Trapping

Consider a bulk semiconductor that has been compensated to a high resistivity and that contains a multiply charged deep level. For explicitness and to match later experimental observations the semiconductor is assumed to be GaAs. Figure 12.2 shows the band diagram with the impurity and the

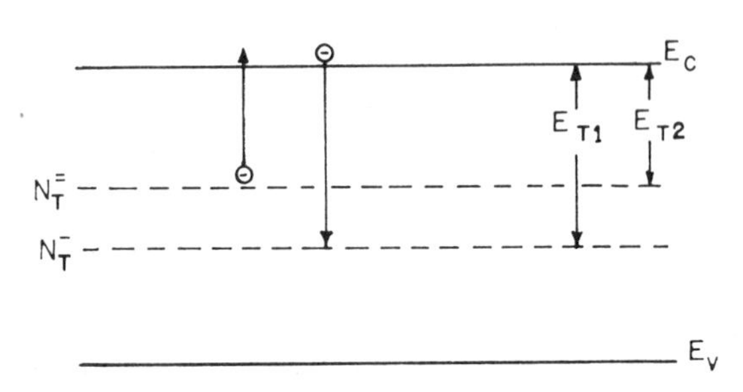

FIG. 12.2 Band diagram of GaAs showing electron transitions for a deep impurity with two charge states: N_T^{-} and N_T^{2-}.

transitions being considered. The basic concept is that the electron capture cross section of the N_T^{-} center, which may be copper in oxygen-doped GaAs, involves a coulombic repulsion barrier and that electrons in the conduction band more easily surmount this barrier and become trapped if they are raised in energy by an electrical field. Therefore if the stability of the material is disturbed by a local rise of field strength, the increase in the rate of capture by the N_T^{-} center reduces the free-electron concentration. This local reduction in electron concentration further increases the field, and the free-electron concentration is further lowered.

At the same time, as Bagaev et al. (1968) comment, a region with a positive space charge is formed a little closer to the anode because the influx of electrons into this region from the domain, where the electron density is low, is less, at least for some time, than the flow of electrons away from this region in the direction of the anode. In the steady state the domain moves toward the anode, because an influx of electrons from the cathode side of the domain is matched by an outflow on the anode side. In the equations that follow, for Fig. 12.2, it will be assumed that transitions to and from the valence band can be neglected; and the occupation of all levels, other than the multiply charged one, is constant. We can then write down the basic set of equations:

$$J = -env + eD\frac{\partial n}{\partial x} \tag{12.1}$$

$$\frac{\partial \rho}{\partial t} + \frac{\partial J}{\partial x} = 0 \tag{12.2}$$

$$\frac{\partial n}{\partial t} = \left.\frac{\partial n}{\partial t}\right]_g + \frac{1}{e}\frac{\partial J}{\partial x} \tag{12.3}$$

$$\left.\frac{\partial n}{\partial t}\right]_g = e_1 N_T^{2-} - c_2 n N_T^{-} + I \tag{12.4}$$

$$\frac{\rho}{e} = -N_{T0}^{-} + n_0 + N_T^{-} - n \tag{12.5}$$

$$\frac{\partial \mathscr{E}}{\partial x} = \frac{\rho}{\varepsilon} \tag{12.6}$$

where J is the current density (A cm^{-2}), n is the free-electron density (cm^{-3}), v is the electron drift velocity (cm sec^{-1}), D is the electron diffusion coefficient (cm^2 sec^{-1}), ρ is the charge density (C cm^{-3}), $\mathscr{E}$ is the electric field (V cm^{-1}), $\partial n/\partial t]_g$ is the net electron generation rate (cm^{-3} sec^{-1}), e_1 is the emission rate from the filled level (sec^{-1}), c_2 is the electron capture rate for the empty level (cm^{-3} sec^{-1}), N_T^{2-} is the density of the occupied level (cm^{-3}), N_T^{-} is the density of the unoccupied level (cm^{-3}), and I is the electron generation due to light (cm^{-3} sec^{-1}). Zero subscripts refer to thermal equilibrium quantities. The electric field and the current are assumed to be in the positive x direction.

Implicit in (12.1) and (12.4) are several assumptions: under conditions of high field, GaAs has two classes of electrons with different mobilities and so the actual current equation should read

$$J = J_1 + J_2 = e\left(-n_1 v_1 + \frac{\partial(D_1 n_1)}{\partial x} - n_2 v_2 + \frac{\partial(D_2 n_2)}{\partial x}\right) \tag{12.7}$$

where the subscripts 1 and 2 refer to parameters of the central and secondary valleys, respectively. Kroemer (1966) has pointed out that for GaAs at room temperature and above the expressions $v_1 = \mu_1 \mathscr{E}$ and $v_2 = \mu_2 \mathscr{E}$ (where μ_1 and μ_2 are independent of field) are valid approximations. Since he has determined that the equilibrium time between the two classes of electrons is of the order of 10^{-13} sec, it is valid to say that n_1 and n_2 are instantaneous functions of the field. Hence in the absence of a diffusion term the conduction current may be written

$$\begin{aligned} J_c &= env \\ &= en\left(\frac{\mu_1 n_1}{n_1 + n_2} + \frac{\mu_2 n_2}{n_1 + n_2}\right)\mathscr{E} \end{aligned} \tag{12.8}$$

where μ_1 and μ_2 are the lower and upper valley mobilities, respectively (cm^2 V^{-1} sec^{-1}), and $n = n_1 + n_2$.

Now assume for mathematical simplicity that $D_1 = D_2 = D =$ a constant. (This is a rather sweeping assumption that could be dispensed with in any later development of the theory.) Then addition of a diffusion term to (12.8) gives (12.1) in form. For the two-valley model, (12.4) becomes the sum of two equations:

$$\left.\frac{\partial n_1}{\partial t}\right]_g = e_{11} N_T^{2-} - c_{21} n_1 N_T^{-} + I \tag{12.9}$$

$$\left.\frac{\partial n_2}{\partial n}\right]_g = e_{12} N_T^{2-} - c_{22} n_2 N_T^{-} \tag{12.10}$$

where e_{11} is the emission rate from the occupied level to the lower valley (sec^{-1}), e_{12} is the emission rate from the occupied level to the upper valley (sec^{-1}), and c_{21} and c_{22} are the capture rates for the lower and upper valley electrons, respectively (cm^3 sec^{-1}). The required form of (12.4) is then derived by adding (12.9) and (12.10) together, which gives $e_1 = e_{12} + e_{22}$ and $c_2 = (c_{21} n_1 + c_{22} n_2)/(n_1 + n_2)$.

Equations 12.1 through 12.6 modified in this way describe the model with the exception of the functional forms of e_1, c_2, and v. These equations could be used to compute the formation and propagation characteristics of a domain in the bulk material by adding the conditions at the boundaries to the equations above. However, for Gunn effect computations, contact conditions and random doping variations have profound effects on the formation of the high-field domain. Hence for specimens of finite length there is some uncertainty in making proper allowance for the actual contact conditions. The present treatment is therefore confined to the simpler case of a constant velocity domain far from the contacts.

Under these circumstances it is convenient to apply the variable transformation

$$y = x - v_{\text{dom}} t$$

where x and t are the independent variables for (12.1) through (12.6) and v_{dom} is the domain velocity in the positive x direction (cm sec^{-1}). Then we have the following transformation equations:

$$\frac{\partial}{\partial x} = \frac{d}{dy} \times \frac{\partial y}{\partial x} = \frac{d}{dy} \tag{12.11}$$

$$\frac{\partial}{\partial t} = \frac{d}{dy} \times \frac{\partial y}{\partial t} = -v_{\text{dom}} \frac{d}{dy} \tag{12.12}$$

Using (12.11) and (12.12) in (12.1) through (12.6) reduces the initial set of partial differential equations into a set of ordinary nonlinear differential equations. Before solving these equations, we must discuss the explicit functional forms of electron drift velocity and capture cross sections as a function of field.

12.1.1 Velocity-Field Characteristics

The form of the drift velocity versus field curve used is taken from Kroemer's (1966) analysis, where v is given as

$$v = -\mu_0 \mathscr{E}_1 \left(\frac{\mathscr{E}/\mathscr{E}_1 + B(\mathscr{E}/\mathscr{E}_1)^{K+1}}{1 + (\mathscr{E}/\mathscr{E}_1)^K} \right) \tag{12.13}$$

$\mathscr{E}_1$, B, K, and μ_0 are determined by trial and error to match a curve of v versus $\mathscr{E}$. The curve we have chosen to match, Fig. 12.3, is the theoretical

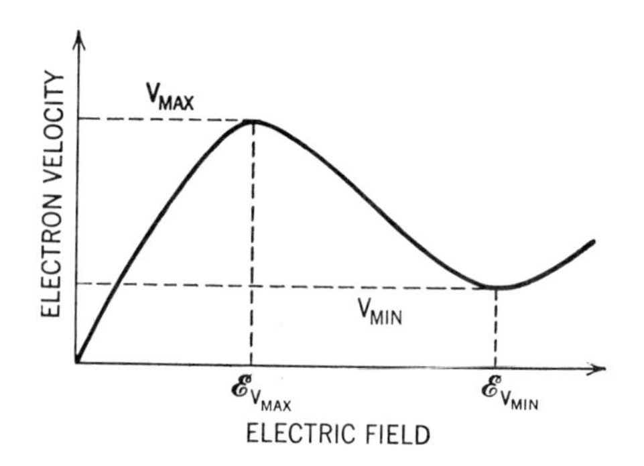

FIG. 12.3 Velocity-field characteristic for electrons in GaAs. (After Sacks and Milnes, 1970.)

one derived by McCumber and Chynoweth (1966). Using their values of $\mathscr{E}_{v_{max}}$ and $\mathscr{E}_{v_{min}}$, and an initial slope of 5000 cm^2 V^{-1} sec^{-1}, the values of the parameters obtained are

$$\mu_0 = 5000 \text{ cm}^2 \text{ V}^{-1} \text{ sec}^{-1} \qquad B = 9.137 \times 10^{-3}$$

$$\mathscr{E}_1 = 2454 \text{ V cm}^{-1} \qquad K = 1.676$$

This gives

$$v_{max} = -6.34 \times 10^6 \text{ cm sec}^{-1} \qquad \mathscr{E}_{v_{max}} = 3200 \text{ V cm}^{-1}$$

$$v_{min} = -3.57 \times 10^6 \text{ cm sec}^{-1} \qquad \mathscr{E}_{v_{min}} = 3.100 \times 10^4 \text{ V cm}^{-1}$$

12.1.2 Capture Coefficient Field Dependence

The field dependence of the electron capture cross section comes from the concept that a negatively charged impurity is surrounded by a potential barrier of height ϕ (Fig. 12.4). For an electron in the conduction band to be captured by this center it must either go over the barrier or tunnel through it. Let $P(E, \mathscr{E})$ be the distribution function for the electron energies as a function of electric field $\mathscr{E}$, and let $N(E)$ be the density of state function.

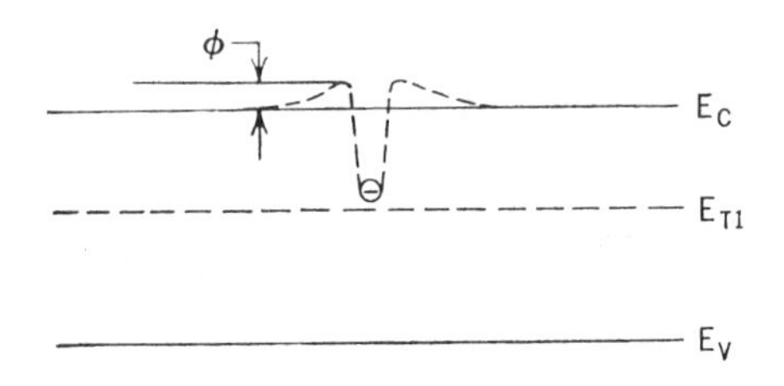

FIG. 12.4 Electrostatic potential for an electron in the neighborhood of a negatively charged impurity (After Sacks and Milnes, 1970.)

The cross section of an electron of energy greater than ϕ may be assumed to be the cross section of a neutral center σ_0. Then the field dependence of the cross section, following Barraud (1967), may be given by

$$\sigma(\phi, \mathscr{E}) = \sigma_0 \left(\frac{\int_\phi^\infty N(E)P(E, \mathscr{E})\, dE}{\int_0^\infty N(E)P(E, \mathscr{E})\, dE} \right) \tag{12.14}$$

This expression is difficult to use because the electron energy distribution function for GaAs is not simple in form (Conwell and Vassel, 1966). An approach which does permit a calculable answer is to assume that $P(E, \mathscr{E})$ is a shifted Boltzmann distribution with an effective temperature T_e, where T_e is computed as a function of the field from the equations of McCumber and Chynoweth (1966). Equation 12.14 then takes the form

$$\begin{aligned} \sigma(\phi, \mathscr{E}) &= \sigma_0 \frac{\int_\phi^\infty E^{1/2} \exp(-E/kT_e)\, dE}{\int_0^\infty E^{1/2} \exp(-E/kT_e)\, dE} \\ &= \sigma_0 \left(1 - \frac{2}{\pi^{1/2}} \int_0^{\phi/kT_e} x^{1/2} \exp(-x)\, dx \right) \end{aligned} \tag{12.15}$$

Then the capture coefficient becomes

$$c_2(\mathscr{E}) = \sigma(\phi, \mathscr{E}) v_{\text{th}}(T_e) \tag{12.16}$$

where $v_{\text{th}}(T_e)$ is the thermal velocity at temperature T_e.

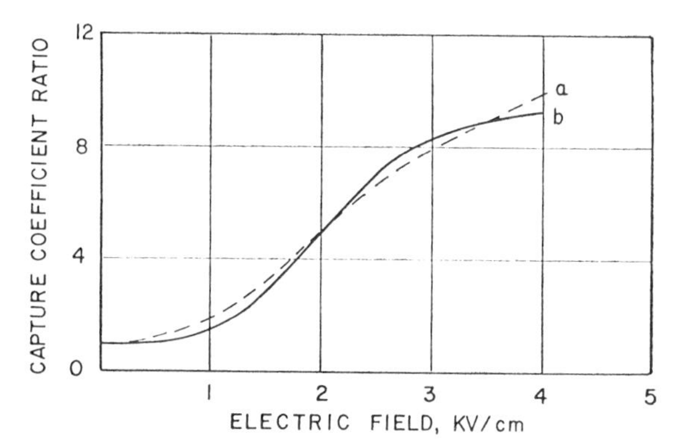

FIG. 12.5 Capture coefficient ratio of a negatively charged impurity as a function of electric field.

Ordinate is ratio of coefficient at field $\mathscr{E}$ to zero-field coefficient: (a) $\phi = 0.080$ eV; (b) simple mathematical function used for computation. (After Sacks and Milnes, 1970.)

Barraud (1967) has found that ϕ equal to 0.080 eV fits well with his experiments. The computation of this curve is quite time-consuming, and it is convenient to represent it by a simple mathematical function:

$$c_2(\mathscr{E}) = c_{20}\left(1 + 9 \times \frac{(\mathscr{E}/2100)^4}{1 + (\mathscr{E}/2100)^4}\right) \tag{12.17}$$

where c_{20} is the capture coefficient at zero field and $10c_{20}$ is the capture coefficient of an electron once it has crossed the barrier. Figure 12.5 shows curves plotted where a is (12.16) and b is (12.17). The curves have been normalized to 1.0 at $\mathscr{E}$ equal to zero. The value of the zero-field cross section computed from (12.15) is 1×10^{-17} cm^2, where we have assumed a value of 1×10^{-16} cm^2 for σ_0. Equation 12.16 approaches infinity as $\mathscr{E}$ approaches infinity, clearly a nonphysical situation. In fact, for electrons with large amounts of energy, we would expect c_2 to decrease, since to be captured, the electron would have to lose its energy through multiple phonon interactions (Ridley and Pratt, 1965). The simple form (12.17), is more reasonable in this respect since the capture cross section approaches a constant as $\mathscr{E}$ approaches infinity.

12.1.3 Steady-State J versus $\mathscr{E}$ Characteristics

Solving (12.1) through (12.6) for zero space charge (ρ) and $\partial\mathscr{E}/\partial X = 0$ gives the current-field characteristic for the homogeneous electric field condition (Fig. 12.6). The ratio of the maximum to the minimum current,

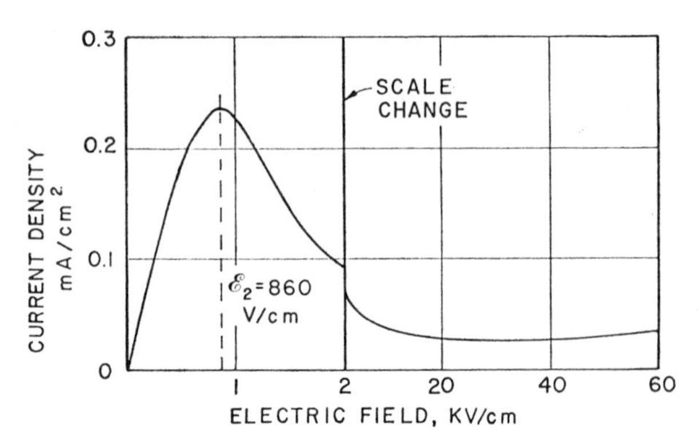

FIG. 12.6 Computed current-field characteristics. (After Sacks and Milnes, 1970.)

J_{max}/J_{min}, is 8.2, and the field for domain nucleation is 860 V cm^{-1}. For pure intervalley transfer effect, this ratio is 1.8 and the field for domain nucleation is 3200 V cm^{-1}. It may be concluded therefore that, for the trap conditions assumed, the cross-section dependence on the field dominates the effects of intervalley transfer in the GaAs. Since the capture coefficient in our model approaches a constant at high electric fields, the minimum current, although reduced by the factor $n(\mathscr{E} = 0)/n(\mathscr{E} = \infty)$, still occurs at 32 kV cm^{-1}.

This solution does not give the current versus field curve that would be measured, because the fields, in general, are not uniform (Barraud, 1967; Northrop et al., 1964). To arrive at the correct characteristics, the spatial variations in (12.1) to (12.6) must be left in and solved together with the appropriate boundary conditions outside the domain, as described elsewhere (Sacks and Milnes, 1970).

12.1.4 Discussion of the Computed Results

Figure 12.7 shows a typical domain shape for a trap density N_T of 10^{17} cm^{-3} and n_0 of 5×10^8 electrons cm^{-3}. The velocity of propagation is 0.575 cm sec^{-1} and the current density is 59.4 μA cm^{-2}. The excess domain voltage is $V_{exc} = \int_{-\infty}^{+\infty} (\mathscr{E} - \mathscr{E}_0)\, dy = 344$ V. The figure shows that the domain is nearly symmetric in contrast to Gunn domains in low-resistivity GaAs. In the latter, the domain consists of a narrow accumulation layer of electrons and a broad depletion layer in which $n < n_0$. In the trap-controlled case, the charge in the domain is almost wholly associated with trap occupancy rather than with electrons, and to a good approximation $n = J_0/ev(\mathscr{E})$; that is, diffusion effects are small.

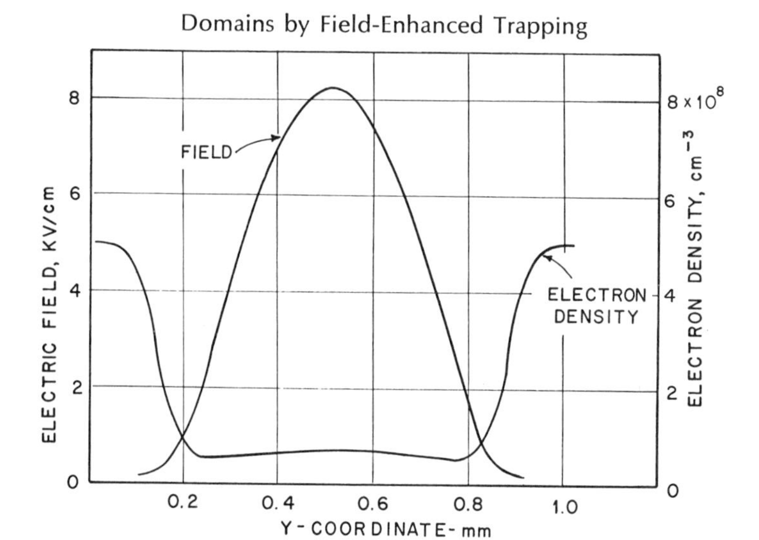

FIG. 12.7 Computed high-field domain, showing electric field, and electron density. The current is 59.4 μA cm^{-2}, and the excess domain voltage is 344 V. Propagation velocity is 0.575 cm sec^{-1}. (After Sacks and Milnes, 1970.)

In addition to the boundary conditions (i.e., $\mathscr{E} \to \mathscr{E}_0$ as $y \to \pm\infty$) the solutions must also satisfy the relation $V_a = V_{\text{exc}} + \mathscr{E}_0 L$ where V_a is the applied device voltage, and V_{exc} has been defined above as $V_{\text{exc}} = \int_{-\infty}^{+\infty} (\mathscr{E} - \mathscr{E}_0)\, dy$.

Figure 12.8 shows the excess domain voltage as a function of $\mathscr{E}_0$, the field outside the domain, for two cases of illumination: (*a*) dark ($n = 5 \times 10^8$); and (*b*) $n = 2.5 \times 10^{11}$. Curve (*c*) is effectively a load line for the device, and is a plot of

$$V_{\text{exc}} = V_a - \mathscr{E}_0 L$$

for $V_a = 1000$ V and $L = 0.5$ cm. The intersection of this curve with the appropriate device curve gives the operating point of the device. For example, in the case above for no illumination, the intersection of the load line with curve (*a*) gives the field outside the domain as 108 V cm^{-1} and the excess domain voltage as 946 V. From Fig. 12.6, the device current is found to be 43 μA cm^{-2}.

As $\mathscr{E}_0$ approaches $\mathscr{E}_2 = 860$ V cm^{-1}, the device curves approach straight lines. The slope of these lines defines a critical length, which for the case of no illumination is 0.24 cm. A device longer than this can have a load line which intercepts the device curves at two points. However, such a device

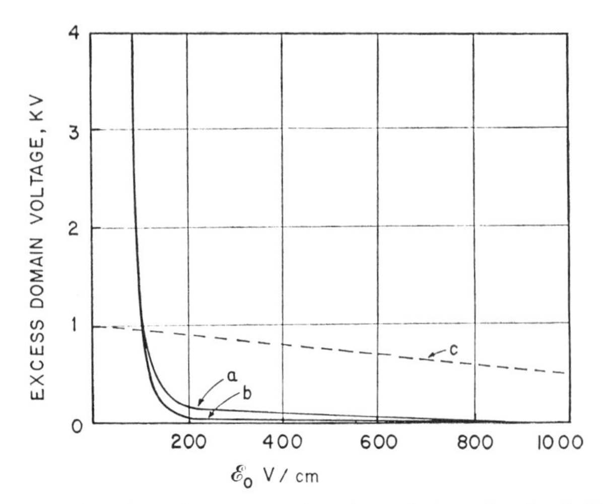

FIG. 12.8 Excess domain voltage as a function of the electric field outside the domain: *(a)* **dark curve;** *(b)* **illumination to produce 2.5×10^{11} electrons cm^{-3}. The dashed curve (c) is computed for a 0.5-cm device with 1000 V applied. (After Sacks and Milnes, 1970.)**

will have a voltage across it such that the average field, V_a/L, is less than $\mathscr{E}_2$, so that a domain already formed in the device can continue to propagate, but a new one will not nucleate when the original domain disappears at the anode. The field $\mathscr{E}_2$ corresponds to the maximum of the J versus $\mathscr{E}$ curve as shown in Fig. 12.6.

The J versus V curve has been plotted in Fig. 12.9 for a 0.5-cm-long device. For simplicity, we have assumed a homogeneous field until the critical field for domain nucleation occurs. It is recognized from the calculations of McCumber and Chynoweth (1966) that the actual field configuration may be nonuniform and dependent on the contacts. However, with our simplification Fig. 12.9 shows that the current follows the static characteristic until 430 V. After this point the current is found by using a load line as demonstrated in Fig. 12.8. The dashed curve of Fig. 12.9 is the current (from Fig. 12.6) that would flow if the field were uniform. In the ideal case, with domain length small compared with device length, the current would rise rapidly to this value along some load line when the domain disappeared at the anode. The current would then fall to its previous value along the same load line when the domain nucleated again at the cathode. The upper and lower curves in Fig. 12.9 then form the envelope for the current oscillations in the device. Figure 12.10 shows a plot of

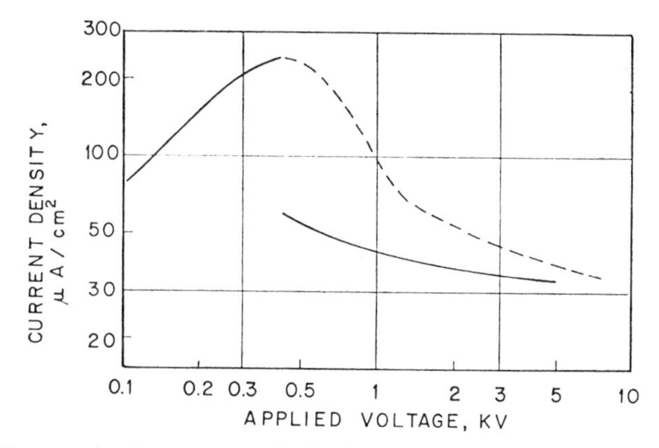

FIG. 12.9 Current density versus applied voltage computed for a 0.5-cm-long device. *The dashed portion of the curve is the current that would exist if the field were homogeneous. The threshold voltage for domain nucleation is 430 V. (After Sacks and Milnes, 1970.)*

envelope amplitude as a function of bias, assuming a load resistor small compared to the dc device impedance.

Again looking at Fig. 12.8, it is seen that the excess domain voltage approaches infinity as the outside field, $\mathscr{E}_0$, approaches $\mathscr{E}_1 = 82.5 \text{ V cm}^{-1}$, corresponding to a minimum current of 32.9 $\mu\text{A cm}^{-2}$. However, while the voltage across the domain is unbounded, the maximum field is not. For no illumination, the situation within the device is shown in Fig. 12.11. The

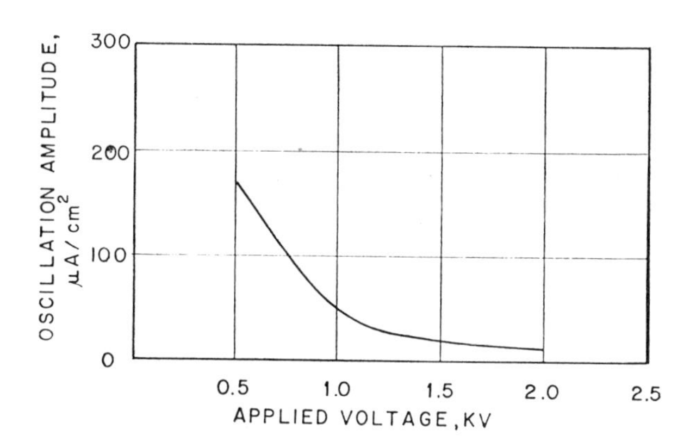

FIG. 12.10 Current oscillation amplitude computed as a function of applied bias.

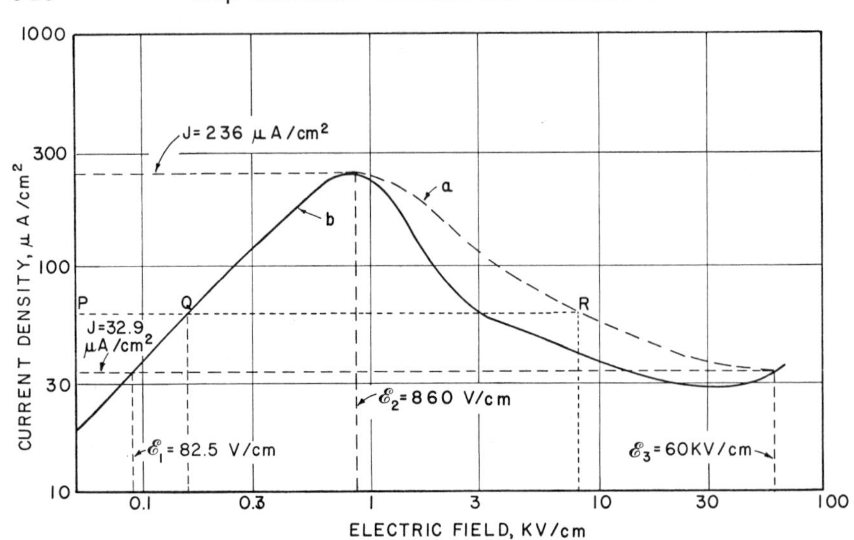

FIG. 12.11 Maximum domain field characteristic as a function of device current. *(a) Maximum field curve. (b) Current-field curve. (After Sacks and Milnes, 1970.)*

dashed curve (*a*) is the maximum field within the domain plotted as a function of device current. For example, the line *PQR* represents a device current of 60 μA cm^{-2}. The intersection point *Q* with the current-field characteristic (*b*) shows that for 60 μA cm^{-2} to flow, the outside electric field must be 153 V cm^{-1}. Simultaneously, point *R* specifies that the maximum domain field is 8 kV cm^{-1}. As the excess domain voltage increases and $\mathscr{E}_0$ approaches $\mathscr{E}_1$, the current decreases toward 32.9 μA cm^{-2} and the maximum domain field approaches 60 kV cm^{-1}. Since V_{exc} is unbounded, the result is a flat-topped domain, with a 60 kV cm^{-1} peak field, which increases in width as the applied voltage increases. For the case of $\mathscr{E}_0 > \mathscr{E}_2$ or $\mathscr{E}_0 < \mathscr{E}_1$, dipole domain solutions do not exist.

Comparing this result with a pure Gunn domain, we find that Copeland (1966) computed a curve similar to Fig. 12.11 and found a maximum field of 64 kV cm^{-1}.

Figures 12.12 and 12.13 show the computed properties of the domains. The domain velocity at constant applied bias increases considerably with illumination and the width decreases slightly. Figure 12.12 shows that for low illumination levels, the domain velocity decreases with increasing field while at sufficiently high illumination, this situation is reversed. In both

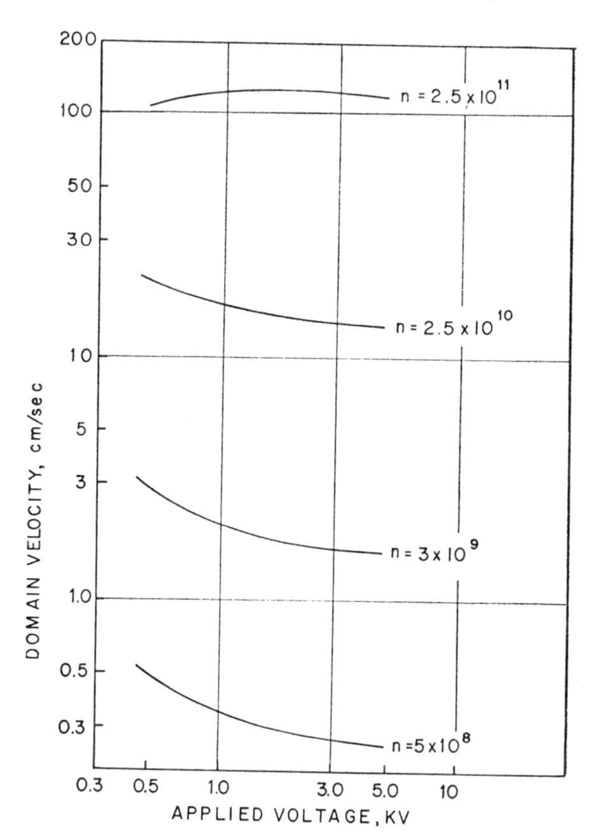

FIG. 12.12 Domain velocity calculated as a function of the applied voltage for various electron densities produced by illumination.
The device length assumed was 0.5 cm. (After Sacks and Milnes, 1970.)

cases, the velocity saturates at high applied bias as $\mathscr{E}_0$ approaches 82.5 V cm^{-1}. The domain width, defined as the distance between the points on the electric field curve where the field has risen to 10% above $\mathscr{E}_0$, decreases slightly with illumination and increases with increasing bias.

The values of $\mathscr{E}_1$ and $\mathscr{E}_3$ arrived at by the computations are a consequence of the J versus $\mathscr{E}$ curve shape, and so depend on the electron velocity field characteristic as well as the capture coefficient. For the intervalley transfer effect with constant diffusion coefficient, values could be found geometrically by the equal areas rule (Butcher et al., 1966). This is not possible in our problem since no analytic solution exists.

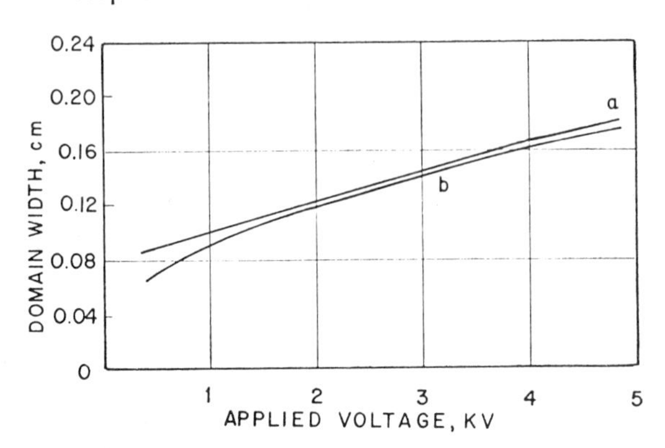

FIG. 12.13 Domain width calculated as a function of the applied voltage for two illumination levels.
(a) $n = 5 \times 10^8$; (b) $n = 2.5 \times 10^{11}$ cm^{-3}. The device length assumed was 0.5 cm. (After Sacks and Milnes, 1970.)

12.2 Relation to Experimental Observations of Domain Oscillations

Table 12.1 summarizes most of the available experimental results of studies of trap-controlled low-frequency oscillations in GaAs.

The computed results of Sacks and Milnes (1970) are in qualitative agreement with experimental observations at 300°K and many of the domain velocities and widths observed are within a factor of 2 of the computed quantities. The domain velocity of Tokumaru (1969), reference h, is an exception since it was taken on material six orders of magnitude lower in resistivity than our computer model. Since computation time is long it has not been possible to examine the effects of extensive parameter changes. A few preliminary calculations were done with $n_0 = 1 \times 10^8$ (a factor of 5 lower than the initial computations). For these computations the domain velocity was found to be reduced by a factor of 25.

The qualitative agreement, however, is good in many respects. For instance, decrease of oscillation amplitude with increase of applied voltage, as predicted by Figs. 12.9 and 12.10, has in fact been observed by Barraud (1963).

It is also of interest to compare the computed curves of Fig. 12.12 with the data taken by Ridley and Pratt (1965) for gold-doped germanium at 20°K. Figure 12.14 is a reproduction of their graph which in general is quite similar to Fig. 12.12. Copper-doped germanium exhibiting static and moving high-field domains has been studied by Kagan et al. (1972). Magnetic field effects on domain velocity in gold-doped n-type germanium have been examined by Vrana et al. (1971).

TABLE 12.1

Experimental Observations for Low-Frequency Oscillations in GaAs

Ref.[a]	Resistance (Ω-cm)	Temperature (°K)	Threshold Field (V cm^{-1})	Domain Velocity[b] (cm sec^{-1})	Domain Width (cm)	Dependence on Increasing Illumination[c]		Dependence on Increasing Applied Bias[c]	
						Velocity	Width	Velocity	Width
a	1×10^{10}	80	1200	2×10^{-3}	1×10^{-1}	Incr.	Decr.	—	Incr.
b, c	1×10^{10}	300	800	3×10^{-1}	1×10^{-1}	—	—	Incr.	—
d	2×10^{3}	229	—	2×10^{2}*	4×10^{-2}	Incr.	—	Incr.	—
e	1×10^{8}	300	200–400	1×10^{-1}	2×10^{-2}	Incr.	—	—	—
f	1×10^{5}	300	1000	1×10^{-1}	2×10^{-2}	Incr.	—	—	—
g	1×10^{8}	300	—	1×10^{-1}	3×10^{-2}	Incr.	—	—	—
h	1×10^{2}	300	1500	3×10^{4}*	—	Incr.	—	Incr.	—

[a] (a)Bagaev et al., 1968; (b) Barraud, 1963; (c) Barraud, 1967; (d) Dorman, 1968; (e) Northrop et al., 1964; (f) Shirafuji, 1968; (g) Smith, 1965; (h) Tokumaru, 1969.

[b] The asterisk (*) indicates that the measurement was taken with the device illuminated.

[c] "Incr." indicates that the parameter is an increasing function. "Decr." indicates that the parameter is a decreasing function.

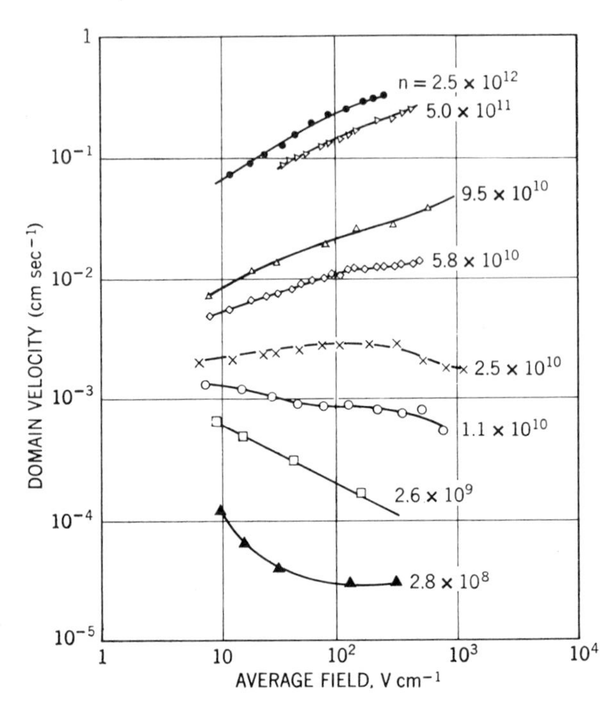

FIG. 12.14 Domain velocity versus field, measured for gold-doped germanium at 20°K.

The average electric field was obtained from the applied voltage divided by the device length. The electron concentration was varied by illuminating the specimen. (After Ridley and Pratt, 1965.)

The experimental studies of Sacks and Milnes (1971) were with copper- and oxygen-doped semiinsulating GaAs. Figure 12.15*a* shows the current waveform during the oscillation. When the specimen was illuminated with a thin slit of light the current increased when the domain crossed this section of the bar, as shown in Fig. 12.15*b*. Consider what must happen when the domain crosses this region. By creating additional carriers in the high-field region, the current would be constrained to increase if the field remains the same. But since the current must be continuous, the field in the domain must decrease. However, this implies that the voltage across the domain decreases, and so the outside field must increase. This gives an increase in sample current. It is conceivable that as the domain field decreases, the domain width increases to compensate, and so the sample current could remain the same. However, this turns out not to be the case. Figure 12.15*c* shows the result of moving the slit image. The picture is actually a double

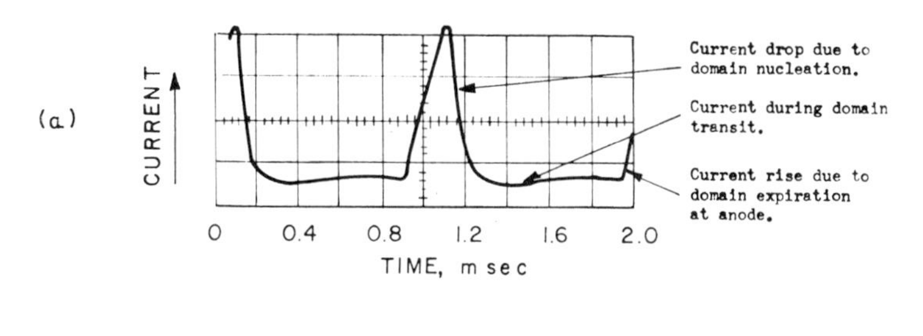

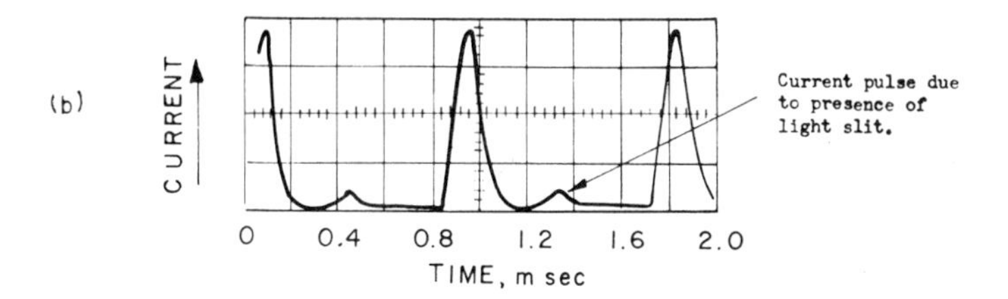

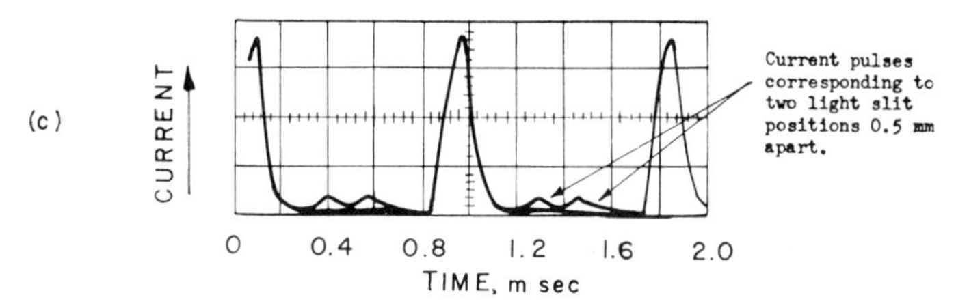

FIG. 12.15 Effect of light slit on current waveform during domain propagation in GaAs.

(a) no slit; (b) with slit; (c) double exposure with slit moved 0.5 mm. (After Sacks and Milnes, 1971.)

exposure. The first exposure was taken with the slit image near the cathode. This produced a current pulse at $t = 0.40$ msec. The second exposure was taken after the slit was moved 0.50 mm toward the anode. The current pulse has now moved to a position $t = 0.47$ msec. Hence by moving the slit further from the cathode, the current pulse occurs later in time. This indicates that our domain is moving from cathode to anode. The propagation velocity can be found by dividing the time displacement of the two current pulses in Fig. 12.15*c* by the distance the slit was moved. This is found to be 310 cm sec^{-1} for the particular operating conditions in this illustration. We can compare this to the average domain velocity by dividing the

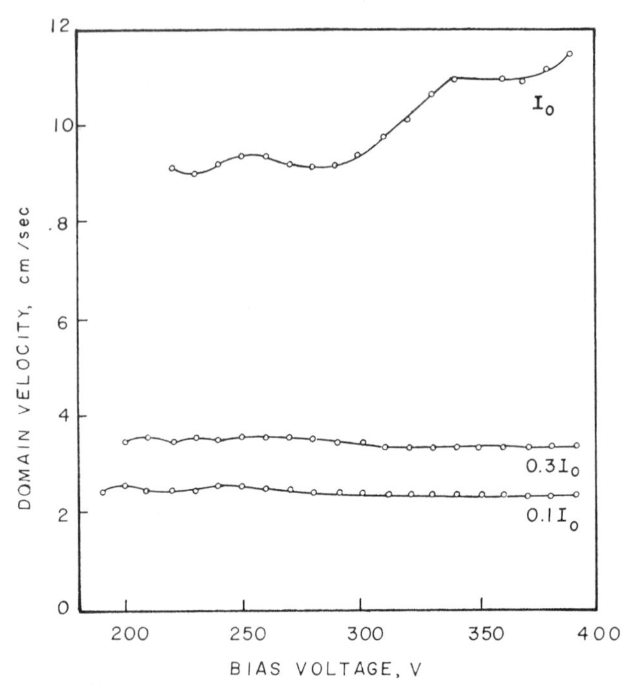

FIG. 12.16 Domain velocity as a function of bias for three illumination levels. (After Sacks and Milnes, 1971.)

distance between contacts (0.288 cm) by the total transit time (0.70 msec). This gives $v_{\text{dom}} = 330 \text{ cm sec}^{-1}$.

The domain velocity as a function of device voltage is shown in Fig. 12.16 for three intensities of illumination. If the wavelength of the light was longer than 1.10 μm (1.1 eV) or shorter than 0.85 μm (1.45 eV), no effect on the oscillations was found. The frequency of oscillation had two maximums: one at 0.89 μm and the other at 0.99 μm. The first maximum corresponds to the band-gap energy of GaAs, while the second is equivalent to an energy of 1.25 eV. If we assume that this maximum is due to electron generation from an acceptor, then we obtain an energy of 0.18 eV from the valence band, which is reasonably compatible with the energy level of copper as a singly charged acceptor in GaAs.

For any prediction of the effects of temperature on domain velocity, it is necessary to have an analytical expression for velocity in terms of the basic quantities of the material. Gunn (1969) has derived an expression for the propagation velocity of a domain for two classes of electrons in a model that can be conceived of as including a trap-controlled negative-resistance situation. The expression he obtained was

$$v_{\text{dom}} = \frac{J}{eN}\left[1 + \frac{e}{\varepsilon}\frac{N\mu' s_2}{(s_1 + s_2)^2}\right]^{-1} \tag{12.18}$$

where the meanings of the terms are given in his paper. Gunn's treatment is sufficiently different than our development of the specific trap-controlled situation so that it is not readily possible to use his expression. However, taking (12.18) as a starting point, it is found that the expression

$$v_{\text{dom}} = \frac{J}{eN_{TO}^{-}} \tag{12.19}$$

gives domain velocities that agree closely with the computed results (Sacks and Milnes, 1971). Here J is the current density during domain propagation and N_{TO}^{-} is the equilibrium density of the singly charged acceptor.

Equation 12.19 may be explained physically as follows. The trailing edge of the domain is a region where electrons are accumulating on the traps by the process $N_{TO}^{-} + e \rightarrow N_{TO}^{2-}$ and the leading edge is one where the inverse of this process is taking place, as suggested by Fig. 12.17. For a current density J, the number of electrons per unit area entering the accumulation region in time δt is $(J/e)\delta t$. If we assume that each N_{TO}^{-} center in the accumulation layer is converted to a N_{TO}^{2-} center by the capture of an electron, then the change in width (δx) of the accumulation layer in time δt must be such that

$$\delta x \cdot N_{TO}^{-} = \delta t \frac{J}{e} \tag{12.20}$$

Rearranging terms gives

$$\frac{\delta x}{\delta t} = \frac{J}{eN_{TO}^{-}} \tag{12.21}$$

But the ratio $\delta x/\delta t$ is actually the domain velocity, and so we have derived (12.19).

This is encouraging, but the assumption that every N_{TO}^{-} center is converted is not representative of the computed values for the N_{TO}^{-} density as a function of position in the domain. Computed values of the domain parameters indicate that the occupancy of the N_{TO}^{-} centers changes very little from the accumulation edge to the depletion edge. For example, a typical solution gave 10^{17} cm^{-3} for N_{TO}^{2-} and 7.94×10^{14} cm^{-3} for N_{TO}^{-} with a maximum deviation of N_{T}^{-} from its equilibrium value of $\pm 5.64 \times 10^{12}$ cm^{-3}. Therefore, while the reasoning above gives some physical insight as to the mechanism of domain propagation, the computed

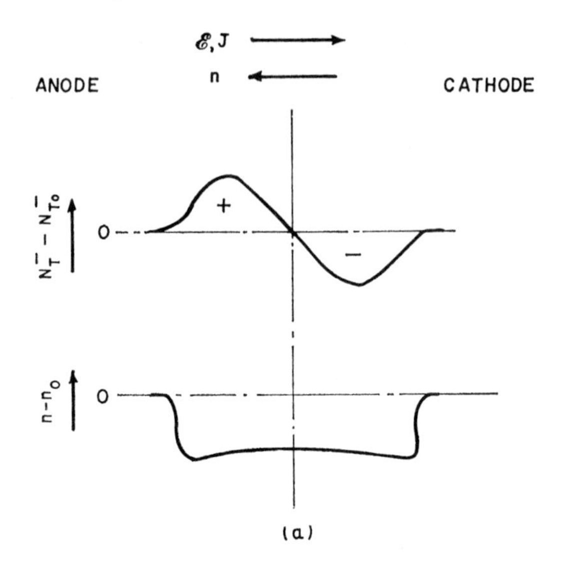

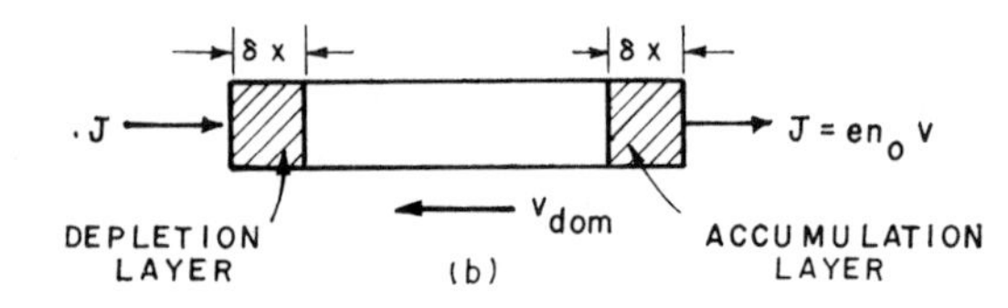

FIG. 12.17 Illustration of mechanism for domain propagation.
(a) Sketch of computed curves for changes in N_T^- and n across domain. (b) Representation of domain with depletion and accumulation layers. (After Sacks and Milnes, 1971.)

results show that the assumption that the depletion layer involves the conversion of virtually all the N_T^- centers to N_T^{2-} is not valid. Further study therefore may be needed to examine more thoroughly the validity of (12.19). However, for present purposes we will accept it as an empirical relationship and examine the temperature dependence it suggests for v_{dom}. In equilibrium, in the absence of illumination, the balance of emission and recombination terms gives

$$\left.\frac{dn}{dt}\right|_g = e_1 N_T^{2-} - c_2 n_0 N_{TO}^- = 0 \tag{12.22}$$

Hence

$$N_{TO}^- = \frac{e_1 N_T^{2-}}{c_2 n_0} \tag{12.23}$$

where e_1 is the emission rate from the N_T^{2-} center ($\sec^{-1}$), c_2 is the capture coefficient of the N_{T0}^- center at zero field ($\text{cm}^3 \sec^{-1}$), n_0 is the equilibrium electron density (cm^{-3}), and N_T^{2-} is the doubly charged acceptor concentration (cm^{-3}) and is almost the same as N_T, the total impurity concentration, if the Fermi level is high. Rewriting (12.19) gives

$$v_{\text{dom}} = v_n \frac{n_0^2 c_2}{e_1 N_T^{2-}} \tag{12.24}$$

where J has been written as $en_0 v_n$. From the earlier analysis we have that

$$\frac{e_1}{c_2} \propto \exp\left(-\frac{E_{T2}}{kT}\right) \tag{12.25}$$

where E_{T2} is the energy between the N_T^{2-} center and the conduction band.
The equilibrium electron density varies with temperature according to

$$n_0 \propto \exp\left(-\frac{E_n}{kT}\right) \tag{12.26}$$

where E_n, the electron activation energy, is determined primarily by the energy level of the deep donor present (oxygen in GaAs). Assuming that the electron velocity, v_n, is not a significant function of temperature, we obtain from (12.24)

$$v_{\text{dom}} \propto \exp \frac{(E_{T2} - 2E_n)}{kT} \tag{12.27}$$

Equation 12.27 represents the activation energy to be expected for the domain velocity.

A survey of impurity levels in GaAs shows the existence of a Cu^{2-} level 0.99 eV below the conduction band edge. The measured conductivity activation energy (E_n) of the material was 0.67 eV. Using these two values in (12.27) gives an activation energy of -0.35 eV for the domain velocity.

Figure 12.18 shows a plot of the empirical formula (dashed line) together with experimental results. Equation 12.27 applies for the case of no illumination, and it can be seen that as the light intensity decreases, the theoretical curve is approached. In this particular GaAs therefore it seems clear that copper is the trap responsible for the domain formation.

Oscillations in chromium-doped semiinsulating GaAs have been reported by Viehmann (1969). The trapping center responsible was not identified.

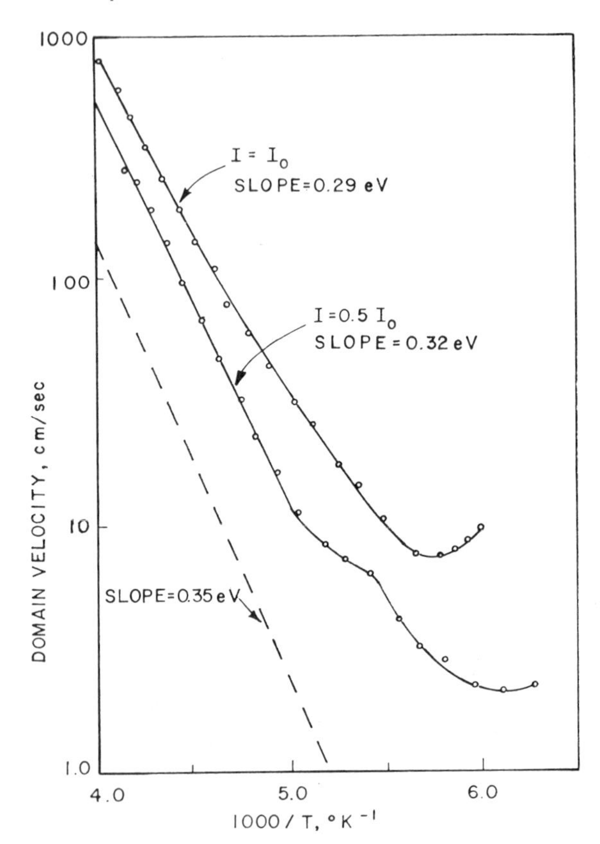

FIG. 12.18 Temperature dependence of domain velocity for two illumination levels. *Dashed curve is plot of (12.27). (After Sacks and Milnes, 1971.)*

12.3 Oscillations in Silicon *p-i-n* Structures due to Deep Levels

Oscillations have been observed in silicon *p-i-n* diodes containing deep impurities, such as gold, cobalt, and zinc, by Holonyak and Bevacqua (1963) and coworkers, Moore et al. (1967), and Streetman et al. (1967, 1969).

Consider the oscillations, which are approximately sinusoidal, that are seen in gold-compensated silicon *p-i-n* diodes. In these devices, below the threshold for oscillation, the 300°K forward-biased dc *J-V* characteristics exhibit current densities that vary as the cubed power of the applied voltage. Such a dependence follows from space-charge-limited injection theory for certain conditions of injection in trap-controlled high-resistance semiconductors. Excess holes and electrons are injected at the *p-i* and *n-i* junction, respectively, and recombine in the *i* region of the diodes.

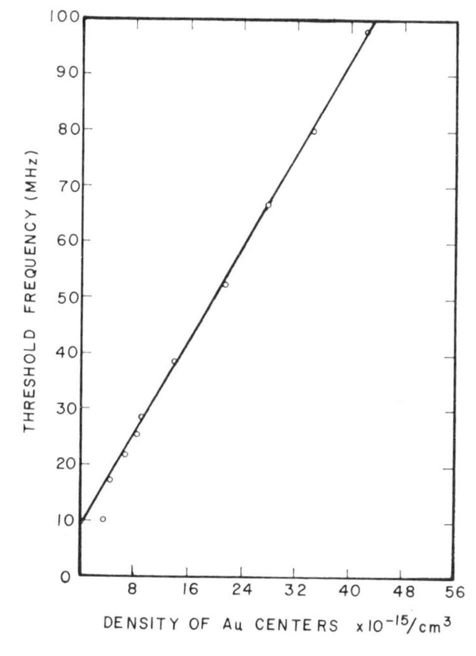

FIG. 12.19 Frequency at the threshold of current oscillations as a function of the density of gold centers in silicon. (After Moore et al., 1967.)

The frequency of oscillation at threshold is found to depend only on the gold density and the temperature. At 300°K the results obtained are as shown in Fig. 12.19, and the frequency is given by

$$f_t = 2.1 \times 10^{-9} N_{\mathrm{Au}} + 8.1 \times 10^6 \tag{12.28}$$

where N_{Au} is the density of gold centers per cubic centimeter in the i region. The oscillation is not a transit time effect because there is no dependence of frequency on length, although the threshold voltage varies as the square of the length of the i region.

Fairfield and Gokhale (1966) have found the capture cross section $\sigma_n{}^0$ for electrons by neutral gold centers to be 1.7×10^{-16} cm^2. The term that depends on N_{Au} in (12.28) is then very nearly the reciprocal of the lifetime of electrons against capture by neutral gold centers, $1/\tau_n = N_{\mathrm{Au}} v_{\mathrm{th}} \sigma_n{}^0$, where v_{th} is the thermal velocity of electrons.

The oscillations in n-type zinc deep level doped p-i-n structures can be several tens of percent of the applied dc bias. The model that appears most able to explain the oscillations is one developed by Zwicker (1972). Essentially the concept is that the incremental negative resistance in the p-i-n diode (the low- to high-level switching) is not caused by an abrupt shift in the conductivity across the entire i region at breakover, but rather results from the gradual growth of the fraction of the total length of the *i* region which has been "switched" to the high conduction state. Thus there is in effect a negative-resistance element present within the *p-i-n* device at all bias levels below breakover. For instance, in an *n*-type zinc-doped *p-i-n* structure it may be that for a given applied bias voltage perhaps 5% of the *i* region located nearest the *p* contact is "switched" to the high conduction state by the interaction of the injected carriers and the traps, while the rest of the *i* region is in the low conduction state. The steady-state recombination kinetics in the two portions of the *i* region are fundamentally different, that is, in the low-level ("unswitched") portion the emission rates necessary for the Ashley–Milnes regime are important, while in the high-level ("switched") portion only carrier capture processes are involved; we therefore expect, qualitatively, a similar difference in the transient properties of each portion.

The complete device may now be thought of as two *p-i-n* diodes in series, one with a very small breakover voltage and one with a nearly normal, large one. For currents where the first section is in breakover and the second section is biased well below breakover, the diode is indeed a negative resistance in series with a positive one. Relaxation oscillations are therefore possible if the transient incremental resistance of the positive ("unswitched") load resistor happens to be sufficiently smaller than its computed steady-state value such that the total device resistance becomes negative. The negative-resistance portion of the device is simply viewed as the active element in a relaxation oscillator.

Zwicker's qualitative explanation of why we might expect the transient incremental resistance of the "unswitched" portion of the *i* region to be well below its computed steady-state value is as follows. For a current oscillating below some certain critical frequency (perhaps on the order of a carrier capture frequency, or roughly 10^6 Hz), the "switching" portion of the *i* region should be able to follow its computed steady-state J-V curve faithfully. However, in the "unswitched" portion, where the response is retarded by the slow emission processes, the critical frequency for which the steady-state solution is roughly valid should be much lower than the first. For a current oscillating above this second frequency, the trapped charge (on the deep levels) will be unable to rearrange itself rapidly enough to follow the varying current. Here instead of following the steady-state curve faithfully, the voltage across the low-level, "unswitched" portion of the *i* region will have a tendency to remain constant. In effect, then, the transient incremental resistance of the "load" resistor will indeed be much smaller than its steady-state value, as is required for oscillations.

The maximum amplitude of such oscillations may be determined by assuming that the "switching" portion of the *i* region oscillates fully from "on" to "off" while the voltage across the positive-resistance load remains constant. This shows that the oscillation amplitude increases with applied voltage, as is observed in experimental studies. The shape of the experimental oscillation waveform changes with varying injection, from sinusoidal at low levels to pulselike near breakover. This is also in accord with the proposed model where, near breakover, the entire device is assumed to be in the process of switching fully from "on" to "off" (subject to external load resistances) and the oscillation waveform is expected to be nonsinusoidal. The temperature dependence of the oscillation frequency is roughly proportional to the dominant thermal emission rate, for instance, varying as $\exp[-0.3/(kT)]$ for p-type zinc-doped devices and as $\exp[-0.5$ or $0.6/(kT)]$ for n-type devices.

The computation of the actual oscillation frequency is a difficult problem. The model indicates only that the frequency should be related to the transient response of the "unswitched" portion of the i region. Therefore we might expect the carrier thermal emission times, rather than a carrier lifetime, to be the important frequency-determining quantities. The trapping centers in the low-level portion of the i region are viewed as the capacitor in a relaxation oscillator. These obviously provide for charge storage and for delayed (quasiexponential) transient response owing to the long, thermally activated, charge release times. (The thesis of Zwicker and related papers should be studied for a more detailed examination of the model.)

Typically the frequencies observed for such oscillations are in the range 10 kHz to 100 MHz, although up to 400 MHz has been observed for heavily doped specimens. However, it is difficult to obtain electron lifetimes below 1 or 2 nsec for silicon by deep impurity doping.

It appears that this class of oscillations is therefore not likely to go high into the gigahertz range. Neither are there any indications that this mode offers any serious competition to avalanche diode or Gunn oscillators in terms of efficiency or power-handling ability. Hagenlocher and Chen (1969) have observed for nickel-doped n-π-n silicon diodes oscillations at 1.68 GHz with an output power of 2 W, but at an efficiency of only 0.2%.

12.4 Recombination Wave Oscillations

Certain recombination instabilities that are possible, theoretically, in semiconductors have been examined by Bonch-Bruevich and Kalashnikov (1965), Konstantinov and Perel' (1965), Karageorgii-Alkalaev and Leiderman (1968), and Bisio and Chiabrera (1970).

Basically the models assume that the semiconductor contains two kinds of traps. At least one of these kinds, called traps of the first type, must have a capture cross section for electrons that depends on the field strength due to electron heating effects. The capture cross section of the second kind of trap need not depend on the field strength. The oscillation action then, as summarized by Bonch-Bruevich (1969), is physically as follows. Assume that as a result of a fluctuation, the electron density in the band is lower and the density at the traps of the first type is higher than their equilibrium values. In the absence of traps of the second type this would either lead to the usual process of relaxation or to a fluctuation instability (due to the dependence of the capture coefficient on the field intensity, which increases in the case of a fluctuation such as the one considered). In the presence of traps of the second type, however, the equilibrium between these traps and the bands also breaks down. As a result, the equilibrium value of the carrier density in the band is also reestablished by liberation of carriers from the second impurity. The decrease in the field intensity connected with this effect leads to liberation of some of the electrons captured by the first impurity. This, in turn, induces a recombination flux from the conduction band to the traps of the second type, and so on. If certain conditions between the parameters of the system are satisfied, this process can be oscillatory with a frequency that depends on a recombination time. Illumination may be helpful in establishing the oscillation conditions.

Adequate comparison of theory with experiment is difficult because of the complexity of the system and difficulties of eliminating contact-injection effects. It is not certain that true recombination wave oscillations have yet been observed.

12.5 Other Oscillatory Modes

Oscillations that are related to plasma effects and are influenced by magnetic fields rather than by trapping actions have been observed in semiconductors.

One class of semiconductor plasma instability observed by Ivanov and Ryvkin (1958) has been termed oscillistor action (Larrabee and Steele, 1960). Such action has been observed in *n*- and *p*-type germanium (300 to 77°K), in *p*-type silicon (77°K), in *n*- and *p*-type indium antimonide (77°K), and in other semiconductors. An oscillistor essentially consists of a semiconductor bar with one ohmic and one injecting (*p*-*n* junction) contact connected in series with a load resistance across a dc or pulsed power supply. If a magnetic field (typically a few thousand gauss) is applied so that it is approximately parallel to the electric field along the bar produced by the contacts, oscillations develop and the percentage modulation of the dc component of current observed is up to 70%. The oscillation frequencies tend to be in the range from a few kilohertz to a few tens of megahertz. The reason for the oscillation is that in a longitudinal magnetic field, a helical perturbation of the current develops that is unstable (Glicksman, 1961, 1962).

Plasma action is also involved in the emission of microwave radiation from InSb (mainly at 77°K) when placed in electric and magnetic fields in a cavity structure. The frequency range is from some megahertz up to about 100 GHz, and the power levels are from nanowatts to a few microwatts. A review of the two-stream plasma and acoustoelectric mechanisms possibly involved has been given by Glicksman (1969). Helicon and Alfvén waves in semiconductors have been studied by Baynham and Boardman (1971).

Oscillations in impact-avalanche transit time diodes (IMPATT diodes) and in Gunn diode structures have been reviewed by Sze (1969).

CHAPTER THIRTEEN

Electric Field Impact Ionization of Deep Impurities

Impact ionization of shallow impurities in germanium at low temperatures (about 4°K) has been studied by McWhorter and Rediker (1959), Koenig and Gunther-Mohr (1957), Sclar and Burstein (1957), Lambert (1962), Kachlishvili (1968), and others.

The electric field strength for impact ionization of shallow impurities in germanium tends to be in the range 20 to 100 V cm^{-1} depending on the particular dopant and its density. At the critical field the conductivity of the germanium, at 4°K, increases rapidly by typically five to eight orders of magnitude with very little further change of field strength. If the germanium is partially compensated with a second impurity (for instance, antimony-doped germanium compensated with indium, or vice versa), the voltage-current plot after breakdown shows a negative-resistance region with a sustaining voltage of several times less than the voltage for turn-on. McWhorter has suggested that the source of this negative resistance is an inelastic scattering action formed by the sharing of an electron, in *n*-type material, by two donor atoms, one of which has been previously ionized by a compensating acceptor. The breakdown path is filamentary in nature in compensated material and can be controlled by a reverse-biased junction gate. This provides a three-terminal structure with interesting switching circuit properties (Melngailis and Milnes, 1961, 1962). Devices based on impact ionization of germanium have also been proposed as shunting elements to limit safely the voltage rise in superconducting magnet coils that may be tending to become resistive because of incorrect operation. Impact ionization of shallow states has also been considered as a technique for increasing photoconductor sensitivity at low temperatures (Crandall, 1971).

This brief review of shallow impurity impact ionization is sufficient for our present purposes, and we now turn to a consideration of impact ionization of deep impurities in the temperature range 65 to 300°K. In particular, impact ionization of indium, nickel, and gold in silicon is discussed.

Impurities that lie deep in the band gap, say from 0.1 to 0.5 eV deep, can be impact-ionized by hot carriers at field strengths of 10^3 to 10^4 V cm^{-1}. Such effects have been observed in the temperature range 65 to 300°K by a number of experimenters (Reddi, 1963; Kornilov, 1964c; Ghandhi et al., 1965, 1966; Schairer and Stath, 1972; and others) in silicon and other semiconductors. However, the modeling of deep impurity impact ionization until recently has been less satisfactory than that of shallow impurity low-temperature impact ionization. This chapter presents theoretical and experimental studies of impact ionization in silicon (McCombs, 1971). The studies relate to the $E_v + 0.16$ eV acceptor level of indium, the $E_c - 0.35$ eV acceptor level of nickel, the $E_v + 0.35$ eV donor level of gold and the $E_c - 0.54$ eV acceptor level of gold. The model that is developed and verified by the experiments with silicon should be applicable to deep impurity ionization in many other semiconductors.

13.1 Modeling of Impact Ionization for Indium, Nickel, and Gold in Silicon

Consider a semiconductor containing N_T atoms cm^{-3} of a deep acceptor trap impurity level, $E_c - E_T$, and partially compensated by a shallow donor of density N_D atoms cm^{-3} where $N_D < N_T$. The semiconductor will be of high resistance ν type ($n \gg p$, but small n) provided $E_c - E_T \gg kT$. The density of the deep impurity atoms that have the acceptor level empty of electrons will be about $(N_T - N_D)$ cm^{-3}. If ohmic contacts are provided and a low electric field is applied, the current flow is small because the free-electron density is low. Increasing the electric field causes heating of the free electrons and a few acquire enough energy, equal to or exceeding $E_c - E_T$, to impact-ionize some of the deep impurity atoms and so raise the free-carrier density. The effect is that the current increases with voltage by many orders of magnitude (Fig. 13.1*a*) at a certain field strength (the breakdown field for impact ionization). In experimental studies it is desirable that the ohmic contacts are provided with circular guard rings as shown in Fig. 13.1*b*, to prevent surface effects from disturbing the observed current-voltage relationship of the main $n^+ \nu n^+$ section.

Comparing impact ionization with avalanche is helpful. Impact ionization, as considered here, is illustrated in Fig. 13.2*a* for electrons. A free

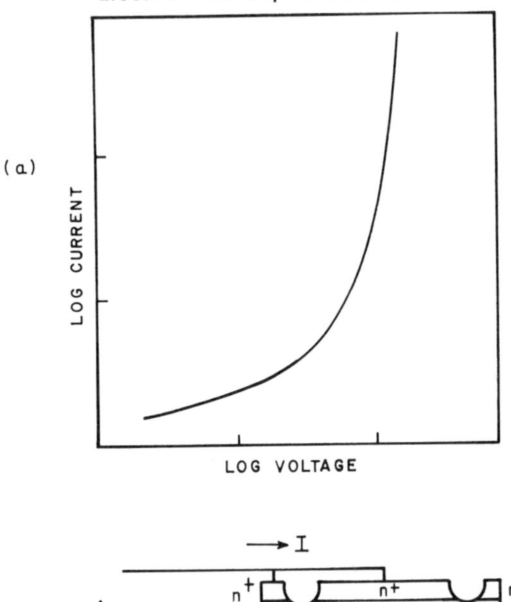

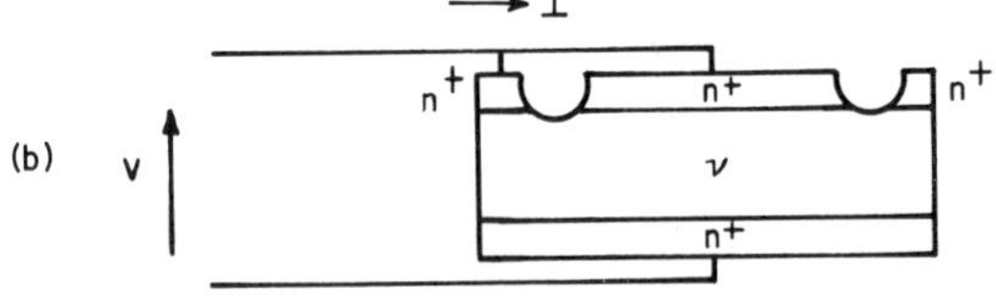

FIG. 13.1 Typical current versus applied voltage characteristic with impact ionization of a deep impurity.

The current-voltage curve is shown in (a). The structure shown in (b) is an $n^+\nu n^+$ *device with a circular guard ring. (After McCombs, 1972.)*

high-energy electron impacts a trapped electron, freeing the trapped electron. In Fig. 13.2*b*, avalanche is shown in which a free electron impacts a valence band electron into the conduction band producing a free electron-hole pair. Avalanche ionization is band-to-band impact ionization, whereas the process shown in Fig. 13.2*a* is trap-to-band impact ionization. Impact ionization as used here will refer to the process in Fig. 13.2*a* or to the analogous process shown in Fig. 13.2*c* in which a hot hole impact-ionizes a hole from a deep level. The kind of Auger process in which a majority carrier impact-ionizes a minority carrier from a deep impurity is not considered here. This implies that in our model the energy level to be impacted must be closer to the majority carrier band than to the minority carrier band.

The minimum energy for impact-ionization processes is lower than for avalanche processes. The minimum free-electron energy necessary for impact ionization will be assumed to be the trap depth (E_T). As would be expected, the impact ionization reported here occurs at fields lower than the 2 to 10 $\times$ 10^5 V cm^{-1} associated with avalanche.

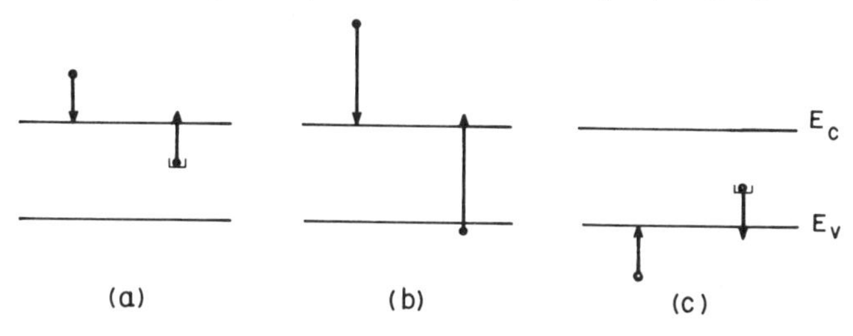

FIG. 13.2 Impact ionization processes.
(a) The hot electron impact ionization of a trapped electron. (b) The hot electron avalanche ionization of an electron from the valence band. (c) A hot hole impact ionizing a trapped hole. (After McCombs, 1971.)

The free-carrier density at field strengths where impact ionization is significant is determined by the balance between capture rate, the thermal generation rate, and the impact-ionization rate. Consider electrons as the majority carrier and let the concentration be n cm^{-3}. The capture rate is usually described in terms of a capture cross section, such that the free-electron capture rate is $n\sigma_{\text{cap}}v_{\text{th}}N_{\text{cap}}$, where σ_{cap} is the capture cross section, v_{th} is the thermal velocity $(2kT_0/m^*)^{1/2}$, T_0 is the lattice temperature, and N_{cap} is the concentration of capture centers. The impact ionization can be described similarly in terms of an impact cross section. The impact-ionization rate is

$$\frac{dn}{dt} = n\left(\frac{n''}{n}\right)\sigma_{\text{imp}}v_{tt}N_{\text{gen}} \tag{13.1}$$

where n''/n is the fraction of free electrons with energy greater than the trap depth, σ_{imp} is the impact cross section, v_{tt} is the velocity appropriate to the trap depth $(2E_T/m^*)^{1/2}$, and N_{gen} is the concentration of centers with trapped electrons available to be impact-ionized. The fraction of free electrons with sufficient energy to impact-ionize the trapped electrons is a rapidly increasing function of the applied electric field. A similar method of expressing the impact rate has been used previously (Reddi, 1963). However, an effective-temperature Maxwell–Boltzmann energy distribution was used in Reddi's calculations. The use of such an energy distribution involves inconsistencies between low- and high-field mobilities. Furthermore, a Maxwell–Boltzmann effective-temperature distribution does not conform with the energy distribution observed experimentally by Pinson and Bray (1964) in measurements that relate to p-type germanium. The energy distributions used in the discussion that follows are those obtained from Monte Carlo calculations for silicon by McCombs and Milnes (1968). The hole-lattice scattering constants

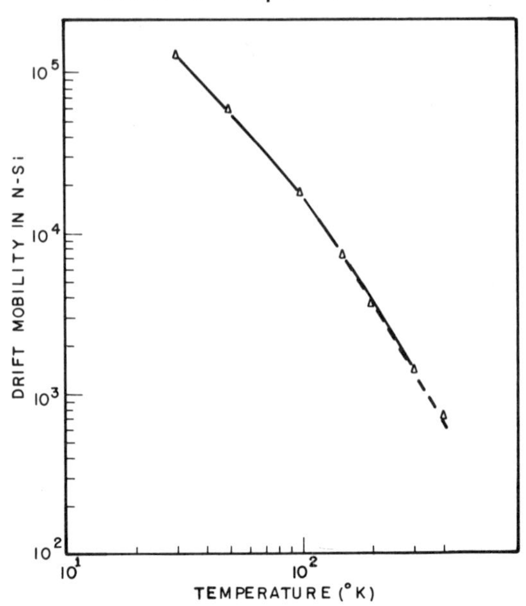

FIG. 13.3 The fit of the calculated to the measured low-field drift mobility for electrons in silicon versus temperature.
(———) measurement of Long (1960); (-----) measurement of Ludwig and Watters (1956); (△) calculated points, with the scattering constants producing best fit (see Table 13.1). (After McCombs, 1971.)

used were those determined from the low-field hole mobility versus temperature dependence. The electron-lattice scattering constants were obtained in a similar manner by the comparison of theory and experiment as shown in Fig. 13.3. The constants used in obtaining this fit are given in Table 13.1. Details of this kind of approach are discussed elsewhere (McCombs, 1971).

13.1.1 Prebreakdown Region

Several effects other than impact ionization can cause the majority carrier concentration to increase with the electric field. Majority carrier space-charge injection is one such effect. For deep traps, this injection is small up to the "traps-filled-limit voltage" (Lampert, 1956):

$$V_{\mathrm{TFL}} = \frac{eN_{\mathrm{cap}}L^2}{2\varepsilon} \tag{13.2a}$$

TABLE 13.1

Constants Used for Electron Scattering in Silicon[a]

Acoustical:	$w_{a\perp} = 2.3 \times 10^8\ \text{sec}^{-1}$	with $T_R = 1$,	and $w_{a\parallel} = 1.5 w_{a\perp}$
Optical:[b]			
$i =$	1	2	3
$E_{\text{opt}_i} =$	0.025	0.048	0.060 eV
$w_i =$	0.3	0.14	$2.0 \times w_{a\perp}$
Intervalley:	g	f	f

[a] These constants are determined from low-field mobility and are used to calculate high-field quantities.
[b] The phonon characteristic energies are those obtained by Folland (1970). The scattering constants (w_i) have been modified from those used by Folland in order to improve the calculated low-field electron mobility. The electron scattering constants used here produce better agreement with the pressure-saturated relative conductivity measurements of Aubrey et al. (1963) than do Folland's values.

where e is the electronic charge, L is the device length, and ε is the dielectric constant. Expressed in terms of an apparent electric field, the "traps-filled-limit field" is

$$\mathscr{E}_{\text{TFL}} \equiv \frac{V_{\text{TFL}}}{L} = \frac{eN_{\text{cap}}L}{2\varepsilon} \tag{13.2b}$$

For the devices in the McCombs–Milnes study the concentrations of capture centers (N_{cap}) were large enough that impact-ionization breakdown occurred at fields less than $\mathscr{E}_{\text{TFL}}$ so that majority carrier space-charge injection effects were small.

The majority carrier concentration also depends on the thermal generation rate and this increases with the electric field (as a result of the Poole–Frenkel effect). The applied electric field changes the potential energy maximum for the thermal generation of the trapped carrier by

$$\Delta E = -e\left(\frac{e\mathscr{E}}{\pi\varepsilon}\right)^{1/2} \tag{13.3}$$

where $\mathscr{E}$ is the applied electric field. This is for the empty center attractive to the majority carrier. This would be the condition for an In^- center and a trapped hole that is being emitted during the process $\text{In}^0 \rightarrow \text{In}^- + h$. For π- and ν-type gold-doped silicon the processes are $\text{Au}^+ \rightarrow \text{Au}^0 + h$ and $\text{Au}^- \rightarrow \text{Au}^0 + e$, respectively. The empty trap is therefore neutral with respect to the majority carrier. The change in the potential energy maximum with field is then (Reddi, 1963)

$$\Delta E = -1.65e\mathscr{E}^{0.8}A_L^{0.2} \tag{13.4a}$$

where

$$A_L \simeq 2 \times 10^{-31}(0.5/E_T) \text{ V-cm}^4 \tag{13.4b}$$

from Lax's (1960) approximation for the potential around a neutral center. In the one-dimensional model, considering thermal generation only, this would give

$$\frac{n(\mathscr{E})}{n(\mathscr{E} = 0)} = \exp\left(-\frac{\Delta E}{kT_0}\right) \tag{13.5}$$

Hartke (1968) has considered this in three dimensions and obtains, for the attractive case,

$$\frac{n(\mathscr{E})}{n(\mathscr{E} = 0)} = 0.5 + 0.5\int_0^{\pi/2} \exp\left(-\frac{\Delta E \cos^{0.5}\theta}{kT}\right)\sin\theta\, d\theta \tag{13.6}$$

For the empty trap neutral with respect to the emitted carrier the term $\cos^{0.5}\theta$ should be replaced by $\cos^{0.8}\theta$.

The capture cross section may also be a function of electric field. For a Maxwell–Boltzmann distribution the capture cross section is

$$\sigma_{\text{cap}}(\mathscr{E}) = \frac{\int_0^\infty E\sigma(E)f(E)\, dE}{\int_0^\infty Ef(E)\, dE} \tag{13.7}$$

where $\exp(-E/kT_e)$ is used for $f(E)$, and E is the free-carrier energy. In the present treatment $\sigma_{\text{cap}}(\mathscr{E})$ is calculated using for $f(E)$ the distribution obtained from the Monte Carlo calculations of McCombs and Milnes (1968). Lax's (1960) calculations for an attractive capture involving optical phonons give

$$E\sigma(E) \propto \frac{(1 + 0.85(E_{\text{opt}} - E)/E_{\text{opt}})^{0.5}}{(1 - E/E_{\text{opt}})^{2.5}}[1 - (1 + y + 0.5y^2)\exp(-y)] \tag{13.8}$$

where E_{opt} is the optical phonon characteristic energy, and y is $(E_{\text{opt}} - E)/kT_0$. For neutral capture involving optical phonons, Lax obtains

$$E\sigma(E) = \begin{cases} \text{constant} & \text{for } E < E_{\text{opt}} - kT_0 \\ 0 & \text{for } E > E_{\text{opt}} - kT_0 \end{cases} \tag{13.9}$$

From the equations presented above, the effects of the electric field on the capture and thermal generation rates have been computed for silicon with indium, nickel, or gold doping. For ν-type gold-doped silicon the majority

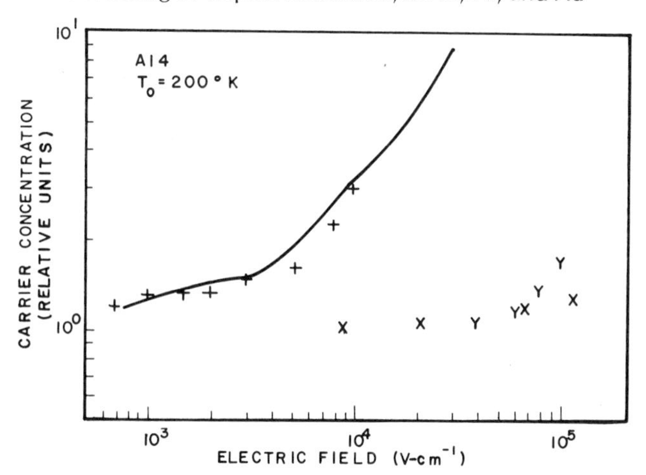

FIG. 13.4 The effect of the field-dependent capture and the thermal generation rates on the electron concentration in the prebreakdown region of ν-type gold-doped silicon devices.

The capture rate contribution is calculated using the strongest phonon (0.060 eV) only. (+) measured field dependence of the electron concentration; (———) calculated contribution of the capture rate; (X) calculated contribution of the thermal generation rate; (Y) thermal generation rate measured by Tasch and Sah (1970). (After McCombs, 1971.)

carrier concentration relative to the low-field value is shown in Fig. 13.4. This relative electron concentration was obtained from

$$\frac{n(\mathscr{E})}{n(\mathscr{E}=0)} = \frac{I(\mathscr{E})}{I(\mathscr{E}=0)} \frac{v_d(\mathscr{E}=0)}{v_d(\mathscr{E})} \tag{13.10}$$

by observing the prebreakdown current as a function of field and allowing for the field dependence of the drift velocity.

For this material the effect of field dependence on the capture rate (the solid line in Fig. 13.4) is seen to be much greater than the effect of the field on the thermal generation rate (points X in Fig. 13.4). The measurements of Tasch and Sah (1970), shown as points Y, support the conclusion that the effect of the field on the thermal generation for ν-type gold-doped silicon is small relative to the effect on the capture rate at these fields. The field dependence of the thermal generation rate is an exponential-like function of $\Delta E/kT_0$ where

$$\frac{\Delta E}{kT_0} = 0.074 \left(\frac{\mathscr{E}}{10^4}\right)^{0.8} \left(\frac{300°\text{K}}{T_0}\right) \qquad \text{(neutral)} \tag{13.11}$$

and

$$\frac{\Delta E}{kT_0} = -0.84 \left(\frac{\mathscr{E}}{10^4}\right)^{0.5} \left(\frac{300°\text{K}}{T_0}\right) \qquad \text{(attractive)} \tag{13.12}$$

Hence the exponent for neutral generation (gold in silicon) is much smaller than for the attractive case (such as for indium in silicon). Calculations for indium in silicon (McCombs, 1971) show that the effect of the electric field on the thermal generation rate is the dominant effect on the increase of carrier concentration with field in the prebreakdown region before impact ionization becomes significant.

13.1.2 Impact Ionization

The rate of majority carrier concentration change, considering electrons as the majority carrier, is given approximately by

$$\frac{dn}{dt} = -nc_{\text{cap}}N_{\text{cap}} + n_1 c_{\text{gen}}N_{\text{gen}} + nc_{\text{imp}}N_{\text{gen}} \tag{13.13}$$

where n is the electron density in the conduction band and n_1 is numerically equal to the density of the electrons that would be in the conduction band if the Fermi function was equal to 0.5 at the trap level. In our case at steady state dn/dt is equal to zero. Then from (13.13),

$$n = \frac{n_1 c_{\text{gen}}N_{\text{gen}}}{c_{\text{cap}}N_{\text{cap}} - c_{\text{imp}}N_{\text{gen}}} \tag{13.14}$$

from which

$$n = \frac{n_1 (c_{\text{gen}}/c_{\text{cap}})R_{\text{comp}}}{1 - (c_{\text{imp}}/c_{\text{cap}})R_{\text{comp}}} \tag{13.15}$$

where $R_{\text{comp}} = N_{\text{gen}}/N_{\text{cap}}$.

At low fields c_{imp} is negligibly small and (13.15) reduces to

$$n(\mathscr{E} = 0) = n_1 \frac{c_{\text{gen}}}{c_{\text{cap}}} R_{\text{comp}} \tag{13.16}$$

At a higher field ($\mathscr{E}$) the carrier concentration is given by

$$\frac{n(\mathscr{E})}{n(\mathscr{E} = 0)} = \frac{c_{\text{gen}}(\mathscr{E})}{c_{\text{gen}}(\mathscr{E} = 0)} \frac{c_{\text{cap}}(\mathscr{E} = 0)}{c_{\text{cap}}(\mathscr{E})} \left(\frac{1}{1 - r(\mathscr{E})R_{\text{comp}}}\right) \tag{13.17}$$

where

$$r(\mathscr{E}) = \frac{c_{\text{imp}}(\mathscr{E})}{c_{\text{cap}}(\mathscr{E})} = \frac{n''(\mathscr{E})}{n} \frac{\sigma_{\text{imp}} v_{tt}}{\sigma_{\text{cap}}(\mathscr{E}) v_{\text{th}}} \tag{13.18}$$

From (13.17) impact-ionization generation of a large number of carriers occurs when $r(\mathscr{E})R_{\text{comp}}$ approaches unity. The electric field for this condition is termed the field for impact-ionization breakdown.

From (13.18)

$$r(\mathscr{E})R_{\text{comp}} = \frac{n''}{n} \left(\frac{\sigma_{\text{imp}}}{\sigma_{\text{cap}}(\mathscr{E})}\right) \frac{v_{tt}}{v_{\text{th}}} R_{\text{comp}} \tag{13.19}$$

In order for impact ionization to affect the steady-state majority carrier concentration the left side of (13.19) must approach unity. The fields measured for impact-ionization breakdowns are in the range 10^3 to 10^4 V cm^{-1} for the specimens studied. The fraction n''/n of electrons (or p''/p for holes) with enough energy to cause impact ionization is much less than unity (for example, on the order of 10^{-4} for gold-doped silicon). The compensation ratio R_{comp} ($=N_{gen}/N_{cap}$) may be near unity for typical specimens, and the velocity ratio v_{tt}/v_{th} is only moderately greater than unity. From (13.19) it follows that the impact cross section tends to be large compared with the capture cross section.

13.2 Experimental Studies

The specimens of indium-, nickel-, and gold-doped silicon studied were of the doping densities and compensation ratios shown in Tables 13.2 through 13.4. The contacts were ohmic and guard rings were provided to eliminate surface effects.

TABLE 13.2
Indium-Doped Silicon Materials

Material No.	Active Indium Centers (cm^{-3})	Shallow Donors (cm^{-3})	Hole Generation Centers (cm^{-3})	Hole Capture Centers (cm^{-3})	R_{comp} ($=N_{gen}/N_{cap}$)
MI	6×10^{16}	4×10^{13}	6×10^{16}	4×10^{13}	1.5×10^3
MIII	2.4×10^{15}	4×10^{14}	2×10^{15}	4×10^{14}	5.0

TABLE 13.3
Nickel-Doped Silicon Materials

Material No.	Original Room-Temperature Resistivity (Ω-cm)	Diffusion Temperature for 1 h (°C)	Room-Temperature Resistivity after Diffusion (Ω-cm)	R_{comp}
N3	273.5	1000	1.1×10^4	0.0117
N4	273.5	950	2.6×10^3	0.072

TABLE 13.4a

Gold-Doped (π-Type) Silicon Materials

Material No.	Original Room-Temperature Resistivity (Ω-cm)	Diffusion Temperature and Time	Room-Temperature Resistivity after Diffusion (Ω-cm)	Active Gold Centers (cm^{-3})	Shallow Acceptor Centers	R_{comp} ($= N_{gen}/N_{cap}$)
A17	15.4	1015°C 34 h	292.0	1.73×10^{15}	8.5×10^{14}	0.966
A19	51.6	1015°C, 34 h	1035.0	1.29×10^{15}	2.6×10^{14}	0.252
A20	1.34	1200°C, 34 h	182.0	2.01×10^{16}	1.3×10^{16}	1.83

TABLE 13.4b

Gold-Doped (ν-Type) Silicon Materials

Material No.	Original Room-Temperature Resistivity (Ω-cm)	Diffusion Temperature and Time	Room-Temperature Resistivity after Diffusion (Ω-cm)	Active Gold Centers (cm^{-3})	Shallow Donor Centers	R_{comp}
A13	0.515	1193°C, 13 h	1.05×10^{5}	1.48×10^{16}	1.2×10^{16}	4.29
A14	3.51	1002°C, 58 h	8.04×10^{4}	1.82×10^{15}	1.4×10^{15}	3.33

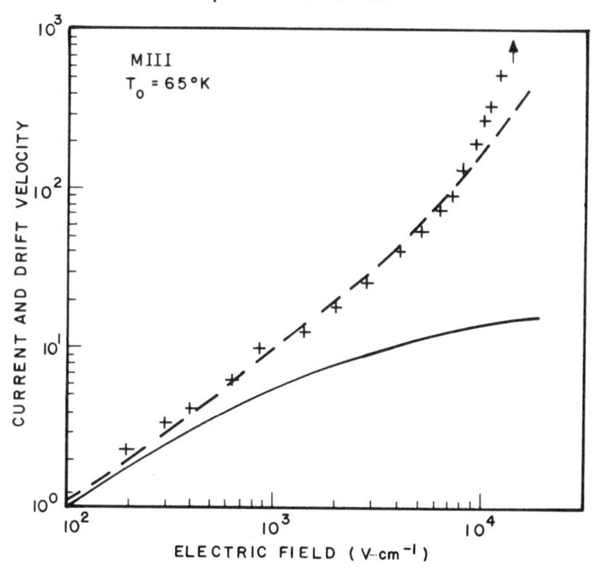

FIG. 13.5 Typical current dependence on electric field for indium-doped silicon. *The three-dimensional Poole-Frenkel field-dependent generation rate multiplied by the field-dependent drift velocity and the drift velocity are shown also: (+) measured current; (-----) the Poole-Frenkel effect; (———) the drift velocity. (After McCombs, 1971.)*

Voltage pulse techniques were found to be unsatisfactory since time-delay effects were observed. A possible explanation is the contact-delay model of Lampert and Rose (1959). Pulse measurements therefore created difficulties of interpretation and were abandoned in favor of low-temperature dc measurements.

The current versus electric field curve for indium-doped silicon (indium density 2.4×10^{15} cm^{-3}, material MIII of Table 13.2) is shown in Fig. 13.5. The calculated drift velocity is also shown. Because of the effect of the electric field on the thermal generation rate, the current is seen to be larger than would exist from a direct dependence of a constant carrier density on the drift velocity. Just before breakdown the impact ionization causes the current to rise rapidly. This breakdown region current is described by the $1/(1 - r(\mathscr{E})R_{\text{comp}})$ term in (13.17) (modified for holes as the majority carrier). The prebreakdown region current can be described approximately by the three-dimensional Poole–Frenkel effect.

The prebreakdown region is the region before the impact ionization affects the current. The breakdown region of the current versus electric field characteristic is the section where the impact ionization affects the carrier concentration. The relative impact-ionization rate $[r(\mathscr{E})]$ is obtained from the current using

$$J = ev_d(\mathscr{E})p_0 \frac{c_{gen}(\mathscr{E})}{c_{gen}(\mathscr{E} = 0)} \left(\frac{1}{1 - r(\mathscr{E})R_{comp}}\right) \tag{13.20}$$

The breakdown is complete at the field where $r(\mathscr{E})R_{comp}$ equals unity. As $r(\mathscr{E})R_{comp}$ increases to approach unity, the measured current increases by orders of magnitude. The theory involving the $1/(1 - r(\mathscr{E})R_{comp})$ term assumes that R_{comp} is independent of field. The compensation ratio (R_{comp}) is equal to the concentration of hole generation centers (N_{gen}) divided by the concentration of hole capture centers (N_{cap}). This is contant with field as long as the hole concentration is much less than both of the trap concentrations above.

The field at which the impact ionization dominates can be varied by examining materials with various compensation ratios. The impact-ionization rate as a function of field can then be obtained over a range of fields. The calculated impact-ionization rate (Fig. 13.6) for a chosen value of the ratio $\sigma_{imp}/\sigma_{cap}$ is seen to conform well with the measured ionization rates

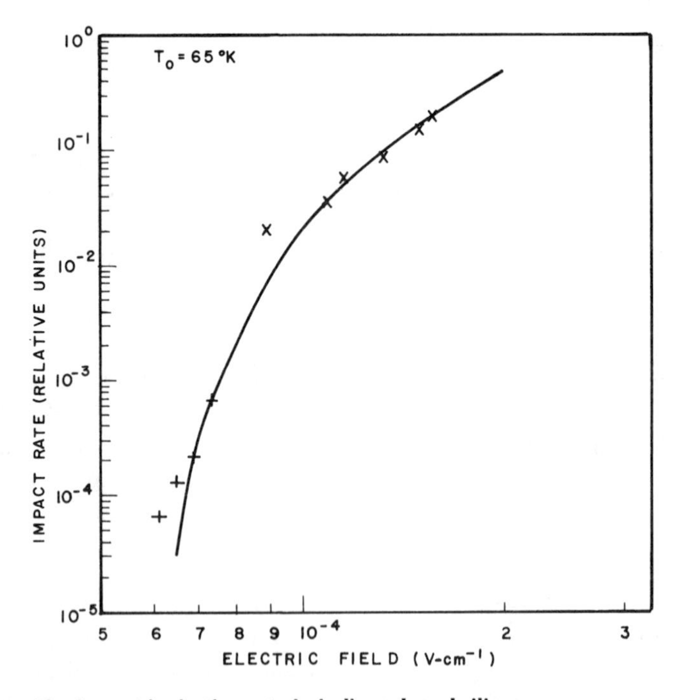

FIG. 13.6 The impact-ionization rate in indium-doped silicon: *(+) rate measured in material MI; (×) rate measured in material MIII; (———) calculated impact-ionization rate. (After McCombs, 1971.)*

for two specimens of indium-doped silicon with quite different dopings and compensation ratios. The ratio $\sigma_{imp}/\sigma_{cap}$ used in the calculation was 98 and the value of σ_{cap} was assumed independent of the field strength. With σ_{cap} taken as 7×10^{-14} cm^2, the impact cross section is 6.9×10^{-12} cm^2.

For nickel-doped silicon the process involved is the impact ionization of an electron from the upper acceptor level, $E_c - 0.35$ eV, by a high-energy conduction band electron. The nickel-doped materials showed breakdown at lower fields (about 10^3 V cm^{-1}) than the indium- and gold-doped silicon. Field-dependent effects on the thermal generation and capture rates were therefore not important in the prebreakdown region, and the current variation followed the drift velocity dependence on electric field until impact ionization became significant. The dependence of the impact-ionization rate on the electric field is shown in Fig. 13.7 where the calculated line is for $\sigma_{imp}/\sigma_{cap}$ taken as 5.3×10^9.

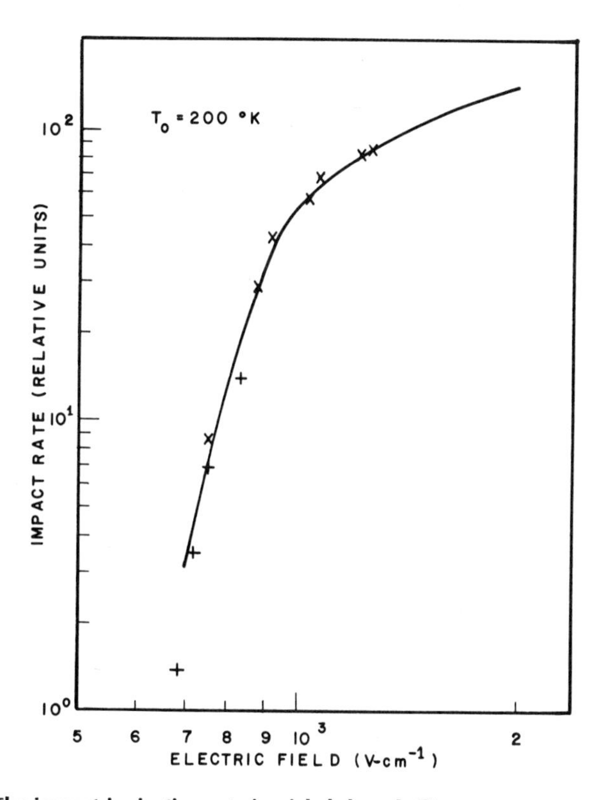

FIG. 13.7 The impact-ionization rate in nickel-doped silicon:
(+) rate measured in material N4; (×) rate measured in material N3; (———) calculated rate. (After McCombs, 1971.)

Lax (1960) refers to the cross sections of repulsive centers as being "exceedingly small—in the range 10^{-24} cm^2 to 10^{-21} cm^2." Hence for the $\sigma_{imp}/\sigma_{cap}$ ratio determined, σ_{imp} would be in the range of 5×10^{-15} to 5×10^{-12} cm^2. The breakdown field for the nickel level ($E_c - 0.35$ eV) is 8.34×10^2 V cm^{-1} for material N4 and 1.26×10^3 V cm^{-1} for material N3 at 200°K. The material with the lower ratio of impact centers to capture centers (i.e., lower R_{comp}) has the higher field for breakdown, as expected and as similarly observed for indium-doped silicon.

Pulse measurements have been made by Ghandhi et al. (1966) on nickel-doped silicon microwave limiter devices at room temperature. They report breakdown for a field of 1 to 2×10^4 V cm^{-1} at room temperature. However, they are mainly concerned with holes in $n^+\pi n^+$ devices, instead of electrons in $n^+\nu n^+$ devices with which we are concerned. They report a sharp change in the resistivities of some of their samples at fields below their breakdown. This is probably due to the impact center (electrons on the upper acceptor level) that is studied in this chapter. It appears that this

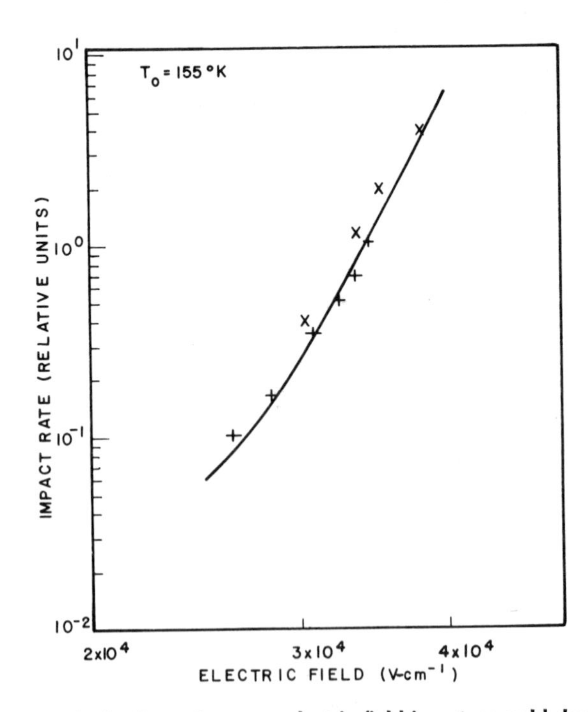

FIG. 13.8 Impact-ionization rate versus electric field in π-type gold-doped silicon: *(×) rate measured in material A19; (+) rate measured in material A17; (———) calculated rate. (After McCombs, 1971.)*

center's concentration in their samples is small enough compared to the free-carrier concentrations that the impact ionization and hole injection empties it.

For π-type gold-doped silicon the impact process involved is a hot hole impact-ionizing a trapped hole from an Au^+ atom at an energy level $E_v + 0.35$ eV. This may be expressed alternatively as the excitation of the electron from the valence band onto an Au^+ site which becomes neutral. For ν-type gold-doped silicon the process is a hot electron impact-ionizing an electron from the $E_c - 0.54$ eV acceptor level of gold.

The measured impact-ionization rates for the π-type materials A17 and A19 (Table 13.4a) are shown in Fig. 13.8. The impact breakdown fields for A17 and A19 are 3.45 and 3.81 $\times$ 10^4 V cm^{-1}, respectively, at 155°K. The theoretical line is that calculated with $\sigma_{\text{imp}_\pi}/\sigma_{\text{cap}}$ taken as 1.2×10^5 at 3×10^4 V cm^{-1}, where the variation of σ_{cap} with field is allowed for and where σ_{imp_π} is assumed independent of field strength.

The measured impact-ionization rates for the ν-type material A14 are shown in Fig. 13.9. The impact breakdown field is 1.4×10^4 V cm^{-1} at 175°K. The theoretical line is that calculated with $\sigma_{\text{imp}_\nu}/\sigma_{\text{cap}}$ taken as 1.54×10^5 at 1.4×10^4 V cm^{-1}, where σ_{cap} varies with field as shown in Fig. 13.4, and again σ_{imp_ν} is assumed independent of field strength.

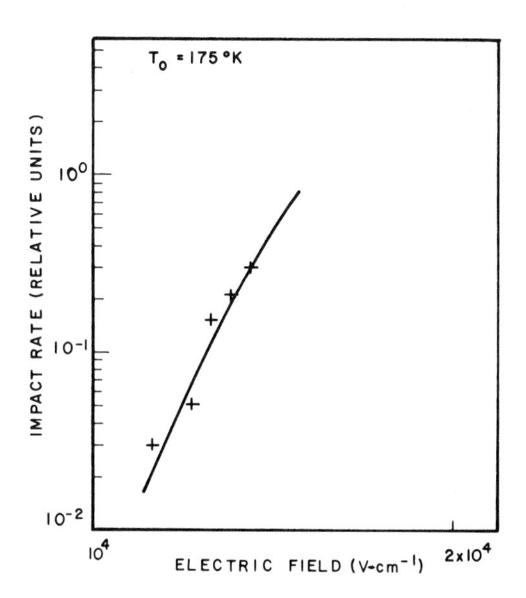

FIG. 13.9 Impact-ionization rate versus electric field in ν-type gold-doped silicon: *(+) rate measured in material A14; (———) calculated rate. (After McCombs, 1971.)*

Measured hole neutral capture cross sections for the gold donor range from 2.4×10^{-15} to at least 10^{-16} cm^2 at 300°K, and 3×10^{-15} cm^2 is observed at 77°K. Values for electron neutral capture by the gold acceptor are 5×10^{-16} to 1.7×10^{-16} cm^2 at 300°K, and 3×10^{-15} to 5×10^{-16} cm^2 at 77°K. If we assume 3×10^{-15} cm^2 for the low-field hole capture cross section and allow for the reduction expected at the field strength 3×10^4 V cm^{-1}, then from the $\sigma_{imp_\pi}/\sigma_{cap}$ ratio of 1.2×10^5 determined at this field strength, it follows that σ_{imp_π} is 1.5×10^{-10} cm^2. Similarly for the ν material by using 5×10^{-16} cm^2 for the low-field electron capture cross section, we obtain σ_{imp_ν} as 2×10^{-11} cm^2.

13.3 The Overlap Effect at Large Densities of Impact Centers

The impact cross sections obtained from experiments are large enough that in samples with large densities of impact centers the impact cross sections for the centers overlap. The impact cross section suggests the distance over which the center is effective in being "seen" by the free carrier for impact ionization. When a free carrier of energy greater than the trap energy "sees" an impact center, the free carrier gives its energy to the carrier trapped on the center. However, if a hot free carrier "sees" more than one impact center, it may be assumed that it tries to give its energy to more than one trapped carrier. Because the energy distribution decreases rapidly with energy, most of the free carriers which cause impact ionization have energies not much greater than the trap energy. A free carrier which has barely enough energy to ionize one trapped carrier cannot afford to expend energy trying to ionize more than one trapped carrier at the same time. Therefore we may hypothesize that hot carriers in regions of space where the impact areas overlap are inefficient in achieving impact excitation of trapped carriers. If the measured impact-ionization rates for gold-doped silicon, with large densities of impact centers, are interpreted without a correction based on this hypothesis, then there is an apparent lowering of the impact cross section with increase of the impact center density. From the doping density at which the onset of this "lowering" effect occurs it is possible to infer the impact cross-sectional area.

The problem of impact cross-section overlap is a quite complicated physical one if it is to be considered in detail. Therefore only a simple geometrical model is presented by McCombs (1971), but the corrections it indicates are of the right order of magnitude.

The concentration of trapped majority carriers is N_{gen}. Assume that these impact centers are uniformly spaced, each at a distance L_{geom} from its neighbors. The actual spacing of deep impurities and shallow impurities is probably very different from this, but we will ignore this in our simple approach. The volume per center is L_{geom}^3 or $1/N_{gen}$ and so L_{geom} is $(1/N_{gen})^{1/3}$. The area per center is L_{geom}^2 in a two-dimensional view. Let the impact cross-sectional area (σ_{imp}) be represented by L_{imp}^2, that is, a square with side of length L_{imp}. When L_{geom} is greater than L_{imp}, none of the impact cross-sectional area will overlap. When L_{geom} is less than L_{imp} and greater than $\frac{1}{2}L_{imp}$, some of the impact cross-sectional area will overlap. Let the density of impact centers (N_{gen}) for which L_{geom} is just equal to L_{imp} be defined as N_{crit}. This means that

$$\sigma_{imp} = L_{imp}^2 = L_{geom}^2 = \left(\frac{1}{N_{crit}}\right)^{2/3} \tag{13.21}$$

The gold-doped specimens A20 and A13 of Tables 13.4a and 13.4b are found from the measurements to have impact breakdown fields that are higher than predicted from the equations of Section 13.1. From this it may be inferred that N_{crit} is approximately 2×10^{15} centers cm^{-3} for π-type gold-doped silicon and 1.6×10^{15} cm^{-3} for ν-type material. From (13.21) then σ_{imp_π} is about 6×10^{-11} cm^2, and σ_{imp_ν} about 7×10^{-11} cm^2. These values agree within a factor of 3 or 4 with the impact cross sections derived from the ratios $\sigma_{imp}/\sigma_{cap}$ obtained from the analysis of the measured ionization rates for the lightly doped specimens. This agreement is better than might have been expected when one considers the gross simplifications of the overlap model used. The doping densities of the indium- and nickel-doped silicon materials were low enough so that corrections for overlap were not required.

13.4 Conclusions

The model that has been discussed for impact ionization of deep impurities is a more complete treatment than previously available. In particular, the hot carrier energy distributions used are obtained from Monte Carlo treatments of the scattering processes rather than from effective-temperature Maxwell–Boltzmann approximations.

Experimental results for indium-, nickel-, and gold-doped silicon are found to conform with the model. Impact cross sections in the range 1.5×10^{-10} to 5×10^{-12} cm^2 are obtained. These cross sections are defined with reference to the fraction of hot carriers participating in impact ionization at high electric fields.

Impact ionization of traps may be a factor in switching and memory actions observed in certain semiconductor structures, such as ZnSe–Ge heterojunctions (Hovel and Urgell, 1971).

CHAPTER FOURTEEN

Impurity-Band Conduction and Hopping Associated with Deep Impurities

Shallow dopants are normally throught of in terms of a hydrogen-like model. At low and medium impurity concentrations, in the case of a donor, conduction occurs only if the lattice temperature is high enough for the excess electron to be excited to the conduction band. At low temperatures the electron is still bound to its atom. The wave functions of these electrons do not overlap if the doping concentration is low and conduction does not occur (providing the crystal is free of compensating acceptor impurities). However, as the donor doping is increased, significant wave-function overlap develops because of the closer spacing of the impurities. The discrete levels of the donor electrons then become bands which broaden with increased doping. The concept of the impurity band is, in its simplest form, an approximately gaussian density of states centered at the original impurity energy level. With further increase of doping concentration, the band may become wide enough to overlap the adjacent conduction band. This merging of the bands then eliminates the activation energy previously needed for conduction and one loosely speaks of the semiconductor as having metallic-like conduction temperature behavior.

Actually this is an oversimplified picture, for several reasons. To begin with, it neglects the fact that the levels forming the impurity band, at concentrations such as 0.01%, must be drawn from the conduction band levels, and therefore, in principle, the simple concept of the superposition of an impurity band on the low-concentration intrinsic band is not correct (Long, 1968). Secondly, the donor impurities are randomly distributed and this has some effect on the width of the impurity band. However, the principal objection to the model presented is the neglect of impurity compensation. Such compensation leads to a hopping action that is the dominant mode of conduction at low temperatures for all except the very highest doping concentrations.

Consider at low temperatures an n-doped semiconductor with compensating acceptors to which some of the donors have lost their electrons and become $N_D{}^+$. There is now a conduction mechanism available in which electron hopping occurs from a neutral donor $N_D{}^0$ to an ionized donor $N_D{}^+$. This has been studied by Fritzsche (1958), Mott (1956), Conwell (1956), Pollak and Geballe (1961), and many others. However, before considering hopping, we very briefly review the simplest possible impurity band concepts.

14.1 Impurity Band Conduction

Modeling of an impurity band is simplest if it is assumed that the impurities are on a regular lattice. The atomic polyhedra around each impurity are represented by spheres of radius r_s where $(4\pi/3)r_s{}^3 = 1/N_I$ for an impurity density N_I cm^{-3}. The bound-electron wave function and energy within each sphere are assumed to satisfy the Wannier effective mass equation. The bottom and top of the bands are characterized by periodic and antiperiodic wave functions, respectively. The energy values satisfying the resulting boundary conditions, and therefore presumably representing the band edges, have been determined in this way by Baltensperger (1953) and are shown plotted in Fig. 14.1 as a function of r_s (normalized in units of $\varepsilon a_{\mathrm{H}}^*$, where a_{H}^* is the first Bohr orbit for an electron of mass m^*). As expected, when the distance between impurities decreases the hydrogenic levels of successively lower excited levels form bands and overlap. We are principally interested in the impurity band derived from the ground state (1s). Conwell (1956) shows that for r_s equal to 4 or 5 $\varepsilon a_{\mathrm{H}}^*$ the width of this band is about the value of kT at 10°K, and so at the impurity concentration corresponding to this, the band may be thought of as formed significantly. From Fig. 14.1 for increased concentration the 1s impurity band is seen to have overlapped with the excited-state bands when r_s is somewhat smaller than $2\varepsilon a_{\mathrm{H}}^*$, and can be considered definitely merged with the conduction band at $r_s \simeq \varepsilon a_{\mathrm{H}}^*$. The impurity concentrations corresponding to these conditions are shown in Table 14.1 for germanium and silicon.

Impurity band action manifests itself in Hall measurements as an apparent failure of the semiconductor to deionize completely in carrier density as the

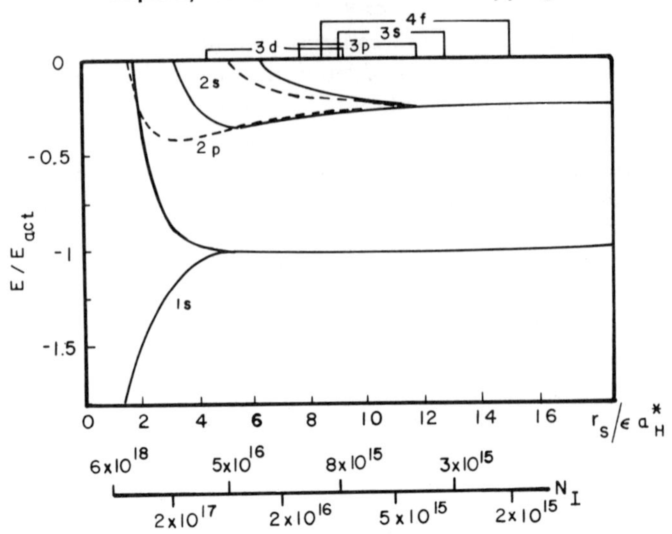

FIG. 14.1 Baltensperger's results (1953) for energy levels of impurities versus impurity spacing.

The lower abscissa scale, representing impurity concentration for a relative dielectric constant of 16 and effective mass of $\frac{1}{4}m_0$*, has been added for illustrative purposes. At the top are indicated the radii of the atomic cell for which the edges of some further bands cross the value* $E = 0$*. (After Conwell, 1956.)*

TABLE 14.1

Estimated Impurity Concentrations for "Formation of an Impurity Band," and for Its Overlap with the Conduction Band *(After Conwell, 1956)*

Case	m^*/m_0 of Hydrogenic Approximation	N_I for Overlap (cm^{-3})	N_I for "Band Formation" (cm^{-3})
n–Ge	0.13	$>9 \times 10^{17}$	$>1 \times 10^{16}$
p–Ge	0.2	$<3 \times 10^{18}$	$<5 \times 10^{16}$
n–Si	0.27	$>2 \times 10^{19}$	$>1 \times 10^{17}$
p–Si	0.5	$<1 \times 10^{20}$	$<1 \times 10^{18}$

temperature is lowered. The carrier densities $(1/R_H e)$ for specimens of n- and p-type germanium are shown in Fig. 14.2. The lightest doped p-type germanium, for instance, is seen to deionize from 4×10^{16} cm^{-3} to only about 6×10^{15} cm^{-3} as the temperature is lowered. This is associated with

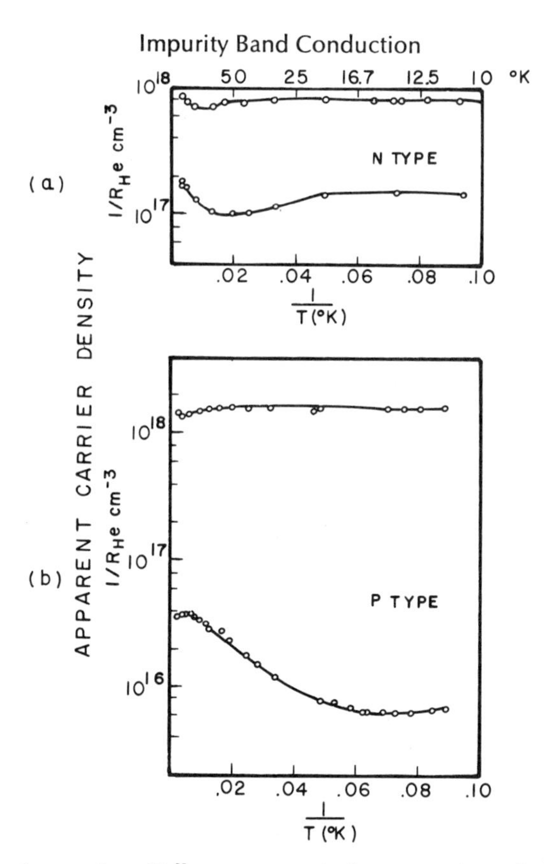

FIG. 14.2 Low temperature Hall measurements for germanium showing failure to deionize.

(a) Reciprocal of product of Hall constant and charge on the electron versus reciprocal of absolute temperature for some n-type germanium samples. (b) Reciprocal of product of Hall constant and charge on the electron versus reciprocal of absolute temperature for some p-type germanium samples. The magnetic field intensity was about 3000 O. The data were taken by P. P. Debye. (After Conwell, 1956.)

effective mobilities (from $R_H\sigma$) of 1600 cm^2 V^{-1} sec^{-1} at 300°K, 4000 cm^2 V^{-1} sec^{-1} at 50°K, and 400 V^{-1} sec^{-1} at 12°K. For the more heavily doped *p*-type germanium specimen (1.5×10^{18} cm^{-3}) the carrier density varies very little with temperature in the range 300 to 10°K and the mobility (500 cm^2 V^{-1} sec^{-1}) is also relatively independent of temperature. Hence these curves are displaying the features expected of impurity band conduction. However, in relatively lightly doped specimens the low-temperature conductivity is more than expected from the impurity band concept and is found to be very dependent on the compensating doping present. Shallow level or deep level compensating impurities considerably increase the conduction observed, which becomes of a hopping nature.

Conwell (1956) in an elementary approach to dc hopping conductivity considers a donor concentration N_D (total), of which $N_D{}^+$ are donors without a bound electron as a result of partial compensation. An electron bound to one donor ion can tunnel over to an adjacent empty one. To obtain the jump frequency, the system of the two neighboring donor ions and an electron is treated like a hydrogen molecule ion. For this system the rate at which the electron oscillates between the ions is given by $1/h$ times the energy difference, ΔE, between the symmetric and antisymmetric combination of wave functions localized on each ion. The diffusion constant for the process is then $4r_s{}^2\Delta E/h$. Using the Einstein relation, Conwell obtains a conductivity

$$\sigma = N_D{}^+ \frac{e^2}{kT} \frac{4r_s{}^2\Delta E}{h} \tag{14.1}$$

In this form the expression is valid for either n- or p-type material. For the large r_s values of interest here,

$$\Delta E \simeq 8\left(\frac{r_s}{\varepsilon a_{\mathrm{H}}^*}\right) E_{\mathrm{act}} \exp\left(-\frac{2r_s}{\varepsilon a_{\mathrm{H}}^*}\right) \tag{14.2}$$

where E_{act} denotes the activation energy. If we neglect the volume occupied by the minority impurities, and assign unity to a constant factor that is not very different from 1, we obtain the following for the impurity band resistivity:

$$\rho \simeq \frac{hkT}{16\varepsilon E_{\mathrm{act}}^2}\left(\frac{N_D}{N_D{}^+}\right) \exp\left(\frac{2r_s}{\varepsilon a_{\mathrm{H}}^*}\right) \tag{14.3}$$

For germanium, at 4°K, this gives a resistivity in ohm-centimeters:

$$\rho \simeq 5 \times 10^{-5} \frac{N_D}{N_D{}^+} \exp\left(\frac{2r_s}{\varepsilon a_{\mathrm{H}}^*}\right) \tag{14.4}$$

Equation 14.3 shows directly the importance of compensation in the factor $N_D/N_D{}^+$.

Since this early work of Conwell, more refined theoretical studies and experimental studies of hopping conductivity have been made. Table 14.2 lists some of the materials in which hopping conductivity has been studied experimentally.

TABLE 14.2

Impurity Concentrations (Experimental) for Hopping Studies and for Overlap of the Impurity Band with the Conduction Band

Semi-conductor	Dopants	Impurity Concentration for Hopping Studies (cm^{-3})	Impurity Concentration for Overlap with Conduction Band (cm^{-3})	Ref.[b]
n–Ge	Sb	2×10^{15}–10^{17}		Fritzsche, 1958
n–Ge	Sb, radiation defects	7×10^{15}–2×10^{17}		Davis and Compton, 1965
(S.I.)Ge	n shallow: Cu	3.5×10^{15}[a]		Dobrego and Ryvkin, 1964
(S.I.)Ge	Au	2×10^{16}[a]		Beglov et al., 1969
p–Ge	Ga:As	4×10^{14}–8×10^{16}[a]		Dobrego, 1970
n–Si	In or B; P or As	$\sim 10^{16}$–10^{17}		Pollak and Geballe, 1961
GaSb	Zn, Te	$10^{20}, 10^{18}$		Aliev et al., 1967
n–GaAs	Si	10^{16}		Emel'yanenko et al., 1961, 1965
n–GaAs	Si	$>5 \times 10^{15}$	3.5×10^{16}	Basinski and Olivier, 1966
n–GaAs		2–8×10^{17}		Baranskii and Glushkov, 1970
n-p–GaAs	Cu	$>6 \times 10^{18}$		Morgan et al., 1965
n–GaP	S	$\leq 6 \times 10^{16}$	2.1×10^{18}	Hara and Akasaki, 1968
p–GaP	Zn	$\sim 10^{18}$		Cohen and Bedard, 1968
InSb				Putley, 1960
n–InAs		1–4×10^{16}		Zotova et al., 1964
n–InP		$\sim 8 \times 10^{16}$		Kesamanly et al., 1964
n GaP–ZnSe				Yu and Glicksman, 1972
SiC				Patrick and Choyke, 1959
n–CdS	Cd (excess) (0.021 eV)			Crandall, 1968
n–CdSe	(0.014 eV)			Burmeister and Stevenson, 1967

[a] Hopping photoconductivity.

[b] The appropriate literature should be inspected for other information, since only a selection of the references available are cited in this table.

14.2 Hopping Conductivity and Its Frequency Dependence

Prior to about 1960, studies of hopping conductivity were concerned with behavior under dc conditions. In 1961, Pollak and Geballe showed that the ac hopping conductivity tends to be orders of magnitude larger than the dc hopping conductivity. Their studies were of the complex conductivity in n-type silicon, with various doping and compensation levels, for frequencies between 10^2 and 10^5 Hz and temperatures between 1 and 20°K.

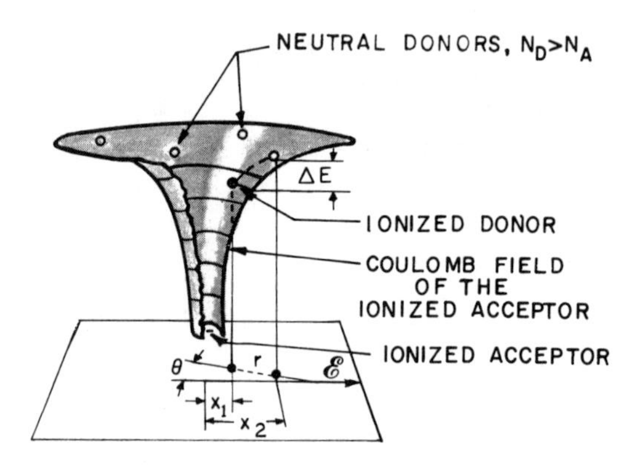

FIG. 14.3 A two-dimensional representation of the Mott-Conwell model for impurity conduction.

The upper half of the figure shows a random distribution of majority donor atoms in the coulombic potential of the negatively charged acceptor. The charge distribution can respond to an applied field ℰ by an electron hop (with energy increase ΔE) from an adjacent uncharged donor to the charged donor as indicated by the dashed line. The lower half of the figure shows the impurity atoms projected on the plane containing the field ℰ and illustrates the notation used in the text. Ionized donors and acceptors are represented by plus and minus signs; neutral donors by open circles. (After Pollak and Geballe, 1961.)

Figure 14.3 shows the coulomb field around an ionized acceptor in a semiconductor where donors are the major dopant. The state of lowest energy is achieved when the donor nearest to the acceptor is ionized. On application of a steady electric field a current can flow only if thermal energy is sufficient to overcome the coulombic potential around the minority impurity. On the other hand, even if there is insufficient thermal energy, the potential is altered by the applied electric field and a new equilibrium has to be established. This will take a time of the order of the hopping time and the net polarization can be detected as a current in an ac experiment.

Figure 14.4 shows the variation of the conductivity (real and imaginary parts) with frequency for silicon containing 1.4×10^{16} cm^{-3} phosphorus

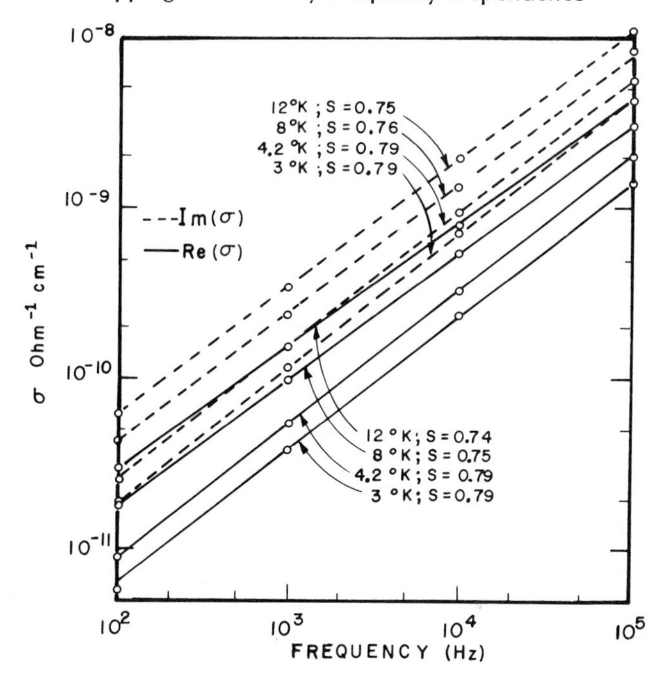

FIG. 14.4 Frequency dependence of the real and imaginary parts of the conductivity for *n*-type silicon ($N_D = 1.4 \times 10^{16}$ and $N_A = 0.8 \times 10^{15}$ cm^{-3}). (After Pollak and Geballe, 1961.)

and 0.8×10^{15} cm^{-3} boron doping. The variation with frequency is a power law of the form

$$\sigma_{ac} = \sigma - \sigma_{dc} = A\omega^S \tag{14.5}$$

where S is about 0.8 and A is complex and depends on temperature, donor concentration, and degree of compensation. From the Kramers–Kronig relation

$$\frac{\text{Im}(A)}{\text{Re}(A)} = \tan\frac{\pi S}{2} \tag{14.6}$$

The real part of the conductivity as a function of temperature and frequency for a specimen of n-type silicon is presented in Fig. 14.5. The activation energy slope (0.006 eV) for the dc impurity conduction is clearly seen. The increased slope at the high-temperature end represents ionization of the donor electrons to the conduction band. The ac impurity conductivity is always appreciably larger than the dc conductivity. The dependence of the ac conductivity on the majority dopant concentration depends on the relative contributions from impurity band conduction and from hopping.

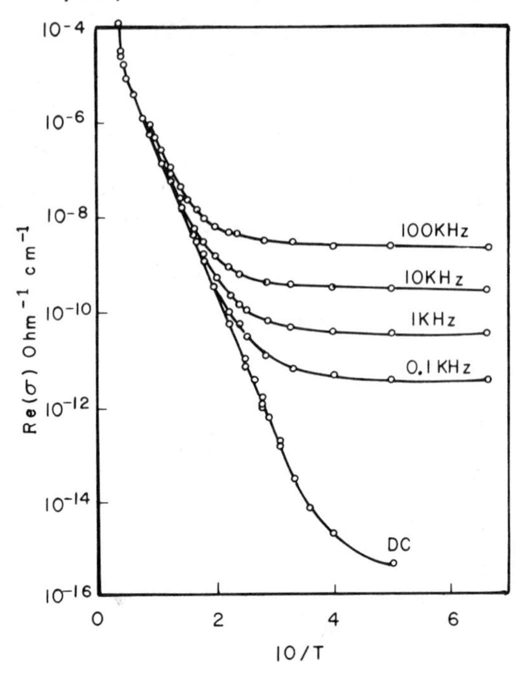

FIG. 14.5 Inverse temperature dependence of the real part of the conductivity, and the dc conductivity for *n*-type silicon ($N_D = 2.7 \times 10^{17}$ cm^{-3} phosphorus atoms and 0.8×10^{15} cm^{-3} boron atoms). (After Pollak and Geballe, 1961.)

At temperatures below 4 or 5°K, the ac conductivity is fairly independent of the majority carrier concentration. On the other hand, increase of the minority concentration increases the ac conductivity, as shown in Fig. 14.6, throughout the measured temperature range. Figure 14.7 shows that the conductivity is almost directly proportional to the minority impurity concentration and also shows a slightly lower conductivity for arsenic as the majority dopant.

Consider an electron (or hole) jumping between two states, with a certain statistical rate. At frequencies much lower than the jumping rate equilibrium can be established faster than the field changes. Hence the polarization will keep pace with the applied field without appreciable phase shift, and the time derivative of the polarization will have a very small in-phase part; that is, the real part of the conductivity will be very small. The magnitude of the polarization will gradually decrease with increasing frequency as it has less and less chance to keep up with the field variation. This decrease will never be faster than inversely proportional to frequency (i.e., proportional to the rate of change of field). This inverse dependence will be approached

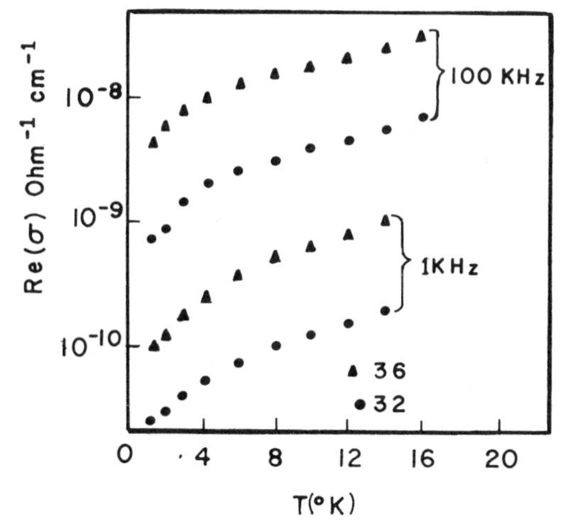

FIG. 14.6 The effect of varying minority concentration on the temperature dependence of the real part of the conductivity for silicon with different boron and similar antimony concentrations.

(Sample No. 36, $N_A = 5.2 \times 10^{15}$, $N_D = 1.5 \times 10^{16}$ cm $^{-3}$; Sample No. 32, $N_A = 1.1 \times 10^{15}$, $N_D = 1.4 \times 10^{16}$ cm^{-3}.) (After Pollak and Geballe, 1961.)

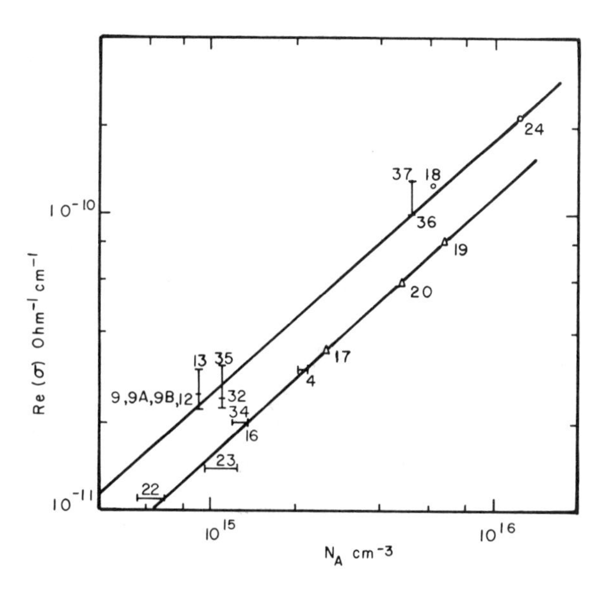

FIG. 14.7 Dependence of the real part of the silicon conductivity at 10^3 Hz on minority concentration at 1.2°K.

The numbers indicate different samples. The upper curve is for specimens with phosphorus or antimony as the majority dopant and the lower curve is for arsenic as the majority dopant. The slope of both curves is about 0.85. (After Pollak and Geballe, 1961.)

at frequencies very high compared to hopping time, when the zero-field equilibrium hardly has time to be disturbed. Because the conductance is proportional to ω times the polarization, the magnitude of the conductance has to be a nondecreasing function of ω. As the importance of the real part relative to the imaginary part becomes more at higher frequencies, the real part of the conductance must also be an increasing function of frequency.

The hopping model used by Pollak and Geballe (1961) is of the form

$$J = \frac{d}{dt}\left(V^{-1} \sum_i e x_i f_i\right) \tag{14.7}$$

where J is the current, V is the volume over which the summation extends, x_i is the projection along the field direction of the separation of the ith majority impurity and the nearest minority impurity, and f_i is the occupancy probability (for a hole in n type) of the ith majority impurity, and is given by the solution of

$$f_i = \sum_j W_{ij} f_i - \sum_i W_{ji} f_i \tag{14.8}$$

together with the normalizing condition

$$V^{-1} \sum_j f_j = N_A \tag{14.9}$$

The transition rates W_{ij} have been calculated by Miller and Abrahams (1960). After development and simplification of (14.8) the expression obtained for the real part of the conductivity is

$$\mathrm{Re}(\sigma) = \frac{N_A N_D}{12} \frac{e^2}{kT} \int r^2 \tau^{-1} \frac{(\omega\tau)^2}{1 + (\omega\tau)^2} 4\pi r^2 \, dr \tag{14.10}$$

where r is the spacing of the pairs between which the hopping takes place and τ is a function of the spacing and the jump energy ΔE:

$$\tau \simeq 5 \times 10^{-13} \left(\frac{r}{a}\right)^{-3/2} \exp\left(\frac{2r}{a}\right) \tanh\left(\frac{\Delta E}{2kT}\right) \tag{14.11}$$

Numerical solution of these equations gives results that are in acceptable agreement with the behavior observed.

In further work Pollak (1962, 1964) has considered the intermediate region between hopping and impurity band conduction, and has extended the theory to very low temperatures and to highly compensated material. Pollak and Watt (1969) have applied the model of the ac hopping conduction to a study of the paring of lithium and boron in silicon.

Compensation dependence of low-temperature conduction in n-type germanium (7×10^{15} to 2×10^{17} antimony atoms cm^{-3}) has been studied by Davis and Compton (1965) with the compensation controlled by fast-neutron irradiation.

The conductivity can in general be expressed by

$$\sigma = \sum_{i=1}^{3} \sigma_i^{(0)} \exp\left(-\frac{\varepsilon_i}{kT}\right) \tag{14.12}$$

where ε_i is the activation energy of the ith conduction process and $\sigma_i^{(0)}$ is the extrapolated value of σ_i for $1/T \rightarrow 0$; ε_i is the familiar donor ionization energy and is observed in all samples with donor concentrations less than 10^{18} cm^{-3}. This energy decreases with increasing donor concentration as expected with the formation of the impurity band; ε_i is also found to increase with compensation.

For an intermediate concentration range ($2 \times 10^{17} > N_D > 2 \times 10^{16}$ cm^{-3}) a second activation energy is observed. This increases with decreasing donor concentration and is stress sensitive. It appears to be associated with the thermal excitation of electrons from the donor ground state to a band that arises from the interaction of negatively charged donors.

For an even lower concentration range a third activation energy [about 1 to 1.5×10^{-3} eV for germanium (antimony doped plus irradiation-induced compensating defects)] is seen. This is the energy associated with the transition of an electron from an occupied to an unoccupied donor site. Although the transition is considered to take place by tunneling, an activation energy exists because of the need to overcome the coulomb barriers associated with the compensating impurities. For concentrations such that the resonance energy of the electron exchange process is less than that associated with the coulomb field (i.e., $N_D < 5 \times 10^{14}$ antimony atoms cm^{-3}), the theories of Miller and Abrahams (1960) and others correctly predict the dependence of the third activation energy on donor separation and compensation. In this concentration range the energy increases with donor concentration and has a minimum value for a given donor concentration when the compensation is one-half.

Impurity conduction has some effects on magnetoresistance. At low concentrations there is a positive magnetoresistance associated with a decrease in size of overlapping wave functions or a field-induced change in phase difference between wave functions on neighboring donors (Mikoshiba, 1962). At high concentrations negative magnetoresistance is observed. This has been explained by Toyazawa (1962) in terms of scattering by localized spins of electrons of partially isolated impurity atoms. This has been observed in many semiconductors including Ge, GaAs, InSb, and InAs (Zotova et al., 1964).

14.3 Association with Deep Impurities

From the discussion of hopping conductivity that has been presented, it is clear that deep impurities may form the compensation that is needed to provide the ionized majority dopants. However, no systematic studies of this are known to the author, except for the irradiation defect level studies of Davis and Compton (1965).

From the impurity band discussion, and Table 14.1, doping concentrations in excess of 10^{16} cm^{-3} in germanium and 10^{17} cm^{-3} in silicon are necessary if the wave-function overlap is to be enough to exhibit impurity band conduction. These densities are derived from a hydrogenic model and conceptually might have to be larger for the tighter wave functions of deep impurities. A search for impurity band action in copper- and gold-doped silicon has been made by Aitchison (1960). The maximum concentrations that he achieved were about 10^{16} cm^{-3}, and the results were negative.

However, the literature does contain a few accounts of effects that are attributed to deep impurity bands. Morgan et al. (1965), for instance, in luminescence emission studies infer the existence of impurity conduction in the copper donor band of graded GaAs junctions containing large concentrations of copper (more than 6×10^{18} cm^{3}). Mixed conduction in chromium-doped GaAs has been reported by Inoue and Ohyama (1970). Also in photodielectric studies of compensated *p*-type germanium and silicon Oksman and Smirnov (1969) find effects that they attribute to hopping along donor levels, as opposed to the normal hopping process that involves the majority dopant.

In general therefore effects that are clearly due to banding of deep impurities are rare. However, effects such as interimpurity recombination (Dobrego and Ryvkin, 1964) and hopping photoconductivity (Beglov et al., 1969) are reasonably well established.

Symbols

Symbol	Definition	Dimensions
A	Cross sectional area	cm^2
A	Also used as constant in various equations	—
a	Current filament fractional area in space-charge-limited flow	—
a^*	Effective Bohr radius	
a_H^*	Radius of first Bohr orbit for electron of mass m*	
a_n	Radius of orbit in hydrogen model of an impurity center	
a_0	Radius of first Bohr orbit for hydrogen	
B	Magnetic field strength	
B	Bar dimension	cm
B	Constant in (12.13)	—
b	Ratio of electron to hole mobility	—
b	Exponent in $\sigma_T \propto T^{-b}$	—
C	Capacitance of a diode	F
$\bar{C}$	Capacitance in complex form	F
$\bar{C}_f$	Frequency-dependent part of complex capacitance	F
C_{AC}	Alternating-current value of capacitance	F
C_D	Flat band normalizing capacitance when all donors ionized	F
C_{DC}	Direct-current value of capacitance at zero signal frequency	F
C_{HF}	High-frequency value of capacitance	F

Symbol	Definition	Dimensions
C_{LF}	Low-frequency value of capacitance	F
C_1	Impurity concentration in the liquid	cm^{-3}
C_P	Parallel equivalent circuit capacitance	F
C_S	Impurity concentration in the solid	cm^{-3}
$C(\omega, \psi_S)$	Capacitance at frequency ω for a variable barrier with barrier height ψ_S	F
c	Carrier concentration ratio n_0/p_0, (10.18)	—
c	Constant relating light production to current, (11.35)	
c	Factor in the impurity density gradient $\exp(-cx)$, Fig. 11.11	cm^{-1}
c_{cap}	Capture coefficient for a trap	$cm^3 \ sec^{-1}$
$c_n{}^0$	Capture coefficient for electrons by a neutral center	$cm^3 \ sec^{-1}$
$c_A{}^D$	Interimpurity recombination coefficient, (6.41)	$cm^3 \ sec^{-1}$
c_p	Capture coefficient for holes	$cm^3 \ sec^{-1}$
D	Diffusion coefficient for an impurity	$cm^2 \ sec^{-1}$
D	Diffusion constant for minority carriers	$cm^2 \ sec^{-1}$
$D_{n,p}$	Diffusion constant for electrons, holes	$cm^2 \ sec^{-1}$
D_a	Ambipolar diffusion constant, $(2b/(b + 1)) D_p$	$cm^2 \ sec^{-1}$
D	Defined by (4.22)	
D	Photon density, (10.18)	cm^{-2}
D^*	Detectivity of a photodetector	$cm \ (Hz)^{1/2} \ W^{-1}$
d	Specimen dimensions	cm
E_A	Energy level of an acceptor, measured either from an arbitrary reference level or with respect to the energy of the valence band edge, E_v	eV
E_{act}	Activation energy for electron hopping between two donors	eV
E_c	Energy of the conduction band edge with respect to an arbitrary level	eV
E_D	Energy level of a donor, measured either from an arbitrary level or with respect to the energy of the conduction band edge, E_c	eV
E_f	Fermi energy level with respect to an arbitrary level	eV

Symbol	Definition	Dimensions
E_{fN}, E_{fP}	Fermi energy level in N-type or P-type material	eV
$E_{G,g}$	Energy gap of the semiconductor	eV
E_{In}	Energy level of indium acceptors in silicon	eV
E_i	Fermi energy level for the intrinsic semiconductor condition, where $p = n = n_i$	eV
E_n	Binding energy from hydrogen model, (1.2)	eV
E_T	Energy level of a trap with respect to a band edge	eV
E_0	Electron energy above the conduction band edge	eV
E_{opt}	Optical phonon energy	eV
E_R	Energy level of a recombination center	eV
E_T^*	Equality Fermi level in (6.51)	eV
E_v	Energy at the valence band edge with respect to an arbitrary reference level	eV
$\mathscr{E}$	Electric field	V cm^{-1}
$\mathscr{E}_{1/2}$	Electric field at which mobility begins to vary as the square root power of the field	V cm^{-1}
$\mathscr{E}^{\text{oc}}$	Electric field observed for open-circuit conditions	V cm^{-1}
e	Electronic charge	C
$e_{n,p}$, $e_{n,p}^t$	Thermal emission rate for electrons or holes from a trap	sec^{-1}
$e_{n,p}^0$	Emission rate of electrons or holes from a trap, caused by optical excitation	sec^{-1}
F	Force, (1.1)	
F	Freezeout factor in capacitance analysis, (8.65)	
F	Factor associated with surface recombination, Section 10.2.2	
F	Photon flux in double-injection threshold analysis, Section 11.4	
f	Signal frequency	Hz
f	Rate at which carrier pairs are generated optically	cm^{-3} sec^{-1}
f_B	Break frequency in linearized capacitance frequency diagrams, (8.53)	Hz

Symbol	Definition	Dimensions
f_F	Probability of ionization of a deep level, (8.64)	—
f_i	Probability of ionization of a deep level, (8.63)	—
f_i	Occupation probability of the ith majority impurity, (14.8)	—
f_0	Break frequency in linearized noise frequency characteristics, Fig. 10.9	Hz
f_T	Fermi-Dirac electron distribution function, (6.3)	—
$f_{T_{MB}}$	Maxwell–Boltzmann approximation for the electron distribution function, (6.4)	—
f_t	Frequency of oscillation at threshold voltage for gold-doped silicon	Hz
G_0	Rate of optical generation of holes, (7.5)	$cm^{-3} sec^{-1}$
G_p	Parallel equivalent conductance, (8.38)	Ω^{-1}
g	Thermal generation rate	$cm^{-3} sec^{-1}$
$g_{A,D}$	Degeneracy factor associated with an acceptor or donor impurity level	—
g_{Fe}	Degeneracy factor of the iron level	—
g_L	Generation rate produced by band-gap light, (7.48)	$cm^{-3} sec^{-1}$
g_{ph}	Generation rate produced by photons, (7.17)	$cm^{-3} sec^{-1}$
H	Hamiltonian, (1.6)	
H	Magnetic field, (10.19)	
h	Hole	
h	Planck's constant	
$\hbar$	Planck's constant/2π	
$\hbar\omega$	Optical phonon energy, (5.16)	
I, I_0	Light intensity	photons $cm^{-2} sec^{-1}$
I	Current	A
$I^{oc,sc}$	Open-circuit or short-circuit current, per unit width	$A\ cm^{-1}$
I_B	Direct-current base current in a transistor	A
I_E	Direct-current emitter current in a transistor	A
I_f	Forward current in a diode	A
I_0	Leakage current in a diode	A

Symbol	Definition	Dimensions
I_0	Hole current created by illumination (10.11)	A
I_r	Reverse current in a diode	A
I_{TSC}	Thermally stimulated current	A
Im	Imaginary part of an equation	—
J	Current density	A cm^{-2}
J_b	Bulk current density, (11.29)	A cm^{-2}
J_d	Diffusion-limited current in Shockley model, (8.6)	A cm^{-2}
J_f	Filament current density, (11.29)	A cm^{-2}
$J_{p(x,y)}$	Current density caused by hole flow in x and y directions, (10.2)	A cm^{-2}
J_r	Recombination current density in a diode, (8.5)	A cm^{-2}
J_{SCL}	Space-charge-limited current density, (11.7)	A cm^{-2}
J_{Ω}	Current density in ohmic low-current regime before space-charge-limited flow, (11.9a)	A cm^{-2}
$J_{1(2)}$	Component of reverse current density caused by generation in the $p(n)$ side of a junction	A cm^{-2}
K	Constants defined by their respective equations	—
k	Boltzmann's constant	eV $°K^{-1}$
$k°$	Distribution coefficient, C_s/C_1, at the melting point, Section 2.3	
L	Device length	cm
L^*	Effective diffusion length, (10.10)	cm
L_a^*	Ambipolar diffusion length	cm
L_{geom}	Geometric spacing of impact centers assumed uniformly distributed	cm
L_{imp}	Spacing defined by the cross section for impact ionization	cm
$L_{n,p}$	Diffusion lengths for electrons and holes	cm
L_0	Diffusion length used when $L_n = L_p$	cm
ln	Logarithm to the base e	—
l_c	Mean free path for an acoustical phonon collision, (5.16)	cm

Symbol	Definition	Dimensions
m^*	Effective electron mass in the conduction band of a semiconductor	g
m_0	Mass of an electron in free space	g
m_V^*	Effective hole mass in the valence band	g
N	Density of impurity centers	cm^{-3}
N_A	Density of acceptor impurity centers	cm^{-3}
N_{A_i}	Density of acceptor impurity centers containing i excess electrons	cm^{-3}
N_{A_r}	Density of acceptor impurity centers containing exactly r excess electrons	cm^{-3}
N'_{A_r}	Density of acceptor impurity centers containing at least r excess electrons	cm^{-3}
$N_{\mathbf{Ag}}$	Density of silver centers	cm^{-3}
$N_{\mathbf{Au}}$	Density of gold centers	cm^{-3}
N_c	Effective density of states at the conduction band edge	cm^{-3}
N_D	Density of donor impurity centers	cm^{-3}
$N_D^{0,+}$	Density of neutral or positively charged donor impurity centers	cm^{-3}
N_{gen}	Density of impurity centers with trapped electrons (holes) available to be impact-ionized	cm^{-3}
N_I	Density of impurities in impurity band modeling	cm^{-3}
$N_{\mathbf{In}}$	Density of indium impurity centers	cm^{-3}
N_0	Density of uncharged impurity centers	cm^{-3}
$N_R^{0,-}$	Density of recombination centers, neutral or negatively charged	cm^{-3}
N_r	Surface recombination centers per unit area	cm^{-3}
N_s	Density of impurity centers which contain s excess electrons, (4.16)	cm^{-3}
N_s	Density of shallow donors, (8.62)	cm^{-3}
$N_T(E)$	Density of traps at energy E	cm^{-3}
$N_T^{0,-,+}$	Density of traps with neutral, negative, or positive charge	cm^{-3}
N_{T0}^{-}	Density of traps with negative charge under equilibrium conditions, (7.19)	cm^{-3}
N_T^{2-}	Density of traps with double negative charge	cm^{-3}

Symbol	Definition	Dimensions
N_v	Effective density of states at the valence band edge	cm^{-3}
n	Electron density in the conduction band	cm^{-3}
n'	Excess electron density above equilibrium density, (7.30)	cm^{-3}
n''	Density of conduction band electrons with energy greater than the impact center trap depth	cm^{-3}
n^*	Equality electron density, (6.50a)	cm^{-3}
n^m	Electron density in the conduction band at the temperature T_m	cm^{-3}
n_b, n_{eq}	Equilibrium electron density in bulk material	cm^{-3}
n_i	Electron density for intrinsic semiconductor	cm^{-3}
n_i	Density of electrons at the ith trap level, (9.4)	cm^{-3}
$\hat{n}_i$	Initial occupation density of the ith trap level at temperature T_0, (9.9)	cm^{-3}
n_0	Electron density in the conduction band in thermal equilibrium	cm^{-3}
n_{p0}	Equilibrium electron density in p material	cm^{-3}
n_s	Electron density in the conduction band at a semiconductor surface, (6.44)	cm^{-3}
n_{s1}	Electron density at the surface if the Fermi level is at the recombination center level	cm^{-3}
$n_T(E, \mathscr{E})$	Density of trapped electrons at an energy E for a field strength $\mathscr{E}$	cm^{-3}
n_1	Electron density in the conduction band if the Fermi level is at the impurity level E_T, (6.7)	cm^{-3}
$n_{1,2}$	Refractive index air, silicon, (7.3)	—
$P(E)$	Probability of the energy level E being occupied by two electrons with paired spins, (4.11)	
$P(U)$	Sticking probability, fraction of electrons that are captured with binding energy U that remain captured	
p	Hole density in the valence band	cm^{-3}

Symbol	Definition	Dimensions
p^*	Equality hole concentration, (6.50b)	cm^{-3}
$p_{b,0}$	Equilibrium hole density in the bulk	cm^{-3}
p_{n0}	Equilibrium hole density in n material	cm^{-3}
p_r	Hole density on the recombination center, (11.2)	cm^{-3}
p_{r0}	Hole density on the recombination center in equilibrium, (11.2)	cm^{-3}
p_s	Hole density in the conduction band at a semiconductor surface, (6.44)	cm^{-3}
p_{s1}	Hole density at the surface if the Fermi level is at the recombination center level	cm^{-3}
p_1	Hole density in the valence band if the Fermi level is at the impurity level E_T, (6.8)	cm^{-3}
$p_{1,2}$	Amplitudes of noise plateaus, Fig. 10.6	
Q	Quantum efficiency, carrier pairs per photon	
q	Charge increment, Figs. 8.14 and 8.15	C
R	Reflectivity fraction at surface of a semiconductor, (7.3)	—
R	Ratio of shallow to deep donor concentrations, (8.62)	—
R_{comp}	Ratio of generation to capture centers in impact ionization, (13.15)	—
R_H	Hall coefficient ($1/en$ or $1/ep$)	C^{-1} cm^3
R_N	Ratio of trapped electron to trapped hole concentration, Fig. 6.9	—
Re	Real part of an equation, (8.48) and Fig. 14.5	—
r	Distance from an ion, (1.1)	cm
r	Impurity pair separation distance, (3.2)	cm
r	Spacing of pairs between which hopping occurs, (14.10)	cm
r	Recombination, (6.1)	cm^{-3} sec^{-1}
r_e	Recombination rate in equilibrium	cm^{-3} sec^{-1}
r_0	Geometrical radius of a center at which the potential energy is about $2kT$ below the maximum, Section 5.3.1	cm
r_s	Recombination rate per unit area at a surface, (6.44)	cm^{-2}

Symbol	Definition	Dimensions
r_s	Radius of sphere representing the atomic polyhedra around an impurity atom in the impurity band conduction model, Section 14.1	cm
S	Elastance, (8.60)	F^{-1}
S	Power law dependence of ac hopping conductivity on frequency, (14.5)	—
s	Surface recombination velocity, (6.45)	cm sec^{-1}
T	Absolute temperature	°K
T_B	Temperature of break point on linearized capacitance-temperature plots	°K
T_D	Temperature at which an impurity is diffused, Fig. 8.5	°C, °K
T_J	Temperature at which a junction transient measurement is made, Fig. 8.22	°K
T_m	Temperature at which the conductivity maximum is observed in a thermally stimulated emission experiment, (9.10)	°K
T_0	Lattice temperature, (13.9)	°K
$T_{1/2}$	Temperature at half-maximum on the rising portion of a thermally stimulated current curve, (9.15)	°K
t	Thickness of specimen	cm
t	Time	sec
t_{off}	Turn-off time for photocurrent decay, Fig. 8.20	sec
t_r	Total diode recovery time, (8.11)	sec
$t_{\text{I,II}}$	Plateau and decay times in diode recovery transient, Fig. 8.4	sec
U	Binding energy of capture state in the Lax model, (5.9)	eV
$U_{C,0}$	Bloch function at the minimum of the conduction band, (1.9)	
u	Electrostatic potential, (1.4)	eV
V	Volume, (14.7)	cm^3
$V(r)$	Periodic potential, (1.6)	eV
V, V_a	Applied voltage	V
V_B	Breakdown voltage at which the current rises rapidly	V
V_D	Diffusion voltage of a diode	

Symbol	Definition	Dimensions
V_{exc}	Excess voltage across the high-field domain, Section 12.1.4	V
V_{Ga}	Concentration of gallium vacancies	cm^{-3}
V_i	Concentration of vacancies, (3.1)	cm^{-3}
V_J	Voltage across junction in transient measurements, Fig. 8.22	V
V_R	Reverse voltage across the junction, Fig. 8.23	
V_{TFL}	Traps-filled-limit voltage in space-charge-limited flow, Fig. 11.1	V
V_{Th}	Threshold voltage for onset of negative-resistance regime, (11.32)	V
V_{Ω}	Transition voltage in space-charge-limited flow, (11.23)	V
v	Symbol for $\exp(-eV/kT)$ in (8.45)	—
$v_{1,2}$	Electron drift velocity for two valleys, (12.7)	$cm\ sec^{-1}$
$v_d(\mathscr{E})$	Effective carrier drift velocity at field $\mathscr{E}$, (13.10)	$cm\ sec^{-1}$
v_{dom}	Domain velocity, (12.19)	$cm\ sec^{-1}$
v_n	Electron drift velocity, (12.24)	$cm\ sec^{-1}$
v_0	Thermal velocity of electrons of energy E_0, (5.2)	$cm\ sec^{-1}$
v_{th}	Thermal velocity of carriers	$cm\ sec^{-1}$
$v_{th_{n,p}}$	Thermal velocity of electrons or holes	$cm\ sec^{-1}$
v_{tt}	Velocity of electrons having an energy equal to the trap depth, (13.19)	$cm\ sec^{-1}$
W	Illuminating power available, (7.54)	W
W	Power dissipation per unit volume, (11.29)	$W\ cm^{-3}$
W	Width of the space-charge layer, (8.1)	cm
W_{ij}	Transition rates in hopping analysis, (14.8)	
w	Squared ratio of optical to acoustical matrix elements, (5.11)	
$w_{a_{\perp,\parallel}}$	Acoustical scattering constants, perpendicular and parallel, Table 13.1	sec^{-1}
w_i	Optical scattering constants, Table 13.1	sec^{-1}
x	Distance	cm
x_m	Maximum molar solid solubility	cm^{-3}

Symbol	Definition	Dimensions
y	Distance, Fig. 8.12	cm
y	Dimensionless parameter in (5.14)	—
y	Distance transformation, (12.11)	
Z	Charge, in electron charge units, on a trap, (5.12)	—
α	Optical absorption coefficient, (7.2)	cm^{-1}
$\alpha_{n,p}$	Power law dependence of capture cross section on electric field for electrons, holes, (11.13) and (11.14)	—
β	Rate of temperature rise in thermally stimulated current experiments	$°K\ sec^{-1}$
$\beta_{p,n}$	Involved in power law dependence of mobility on electric field, (11.20)	—
γ	State degeneracy ratio	—
γ_J	Quantum efficiency for emission of light at a current density J, (11.33)	—
ΔF	Root mean square photon flux, (7.4)	photons $cm^{-2}\ sec^{-1}$
ΔG	Change of conductance caused by light, (7.52)	Ω^{-1}
Δn_i	Excess electron density during turn-on phototransient, (7.20)	cm^{-3}
Δn_{ss}	Excess electron density under steady-state illumination	cm^{-3}
ΔS	Elastance difference, (8.60)	F^{-1}
$\Delta\sigma$	Change of conductivity produced by light, (7.51)	Ω^{-1}
$\Delta\phi_{PF}$	Barrier lowering induced by electric field in Poole–Frenkel emission, (5.18)	eV
$\Delta\phi_S$	Barrier lowering induced by electric field in Schottky emission, (5.19)	eV
δ	Loss angle for the capacitance of a junction containing a deep impurity, Fig. 8.16	rad
ε	Dielectric constant (permittivity) of a semiconductor	$F\ cm^{-1}$
ε_0	Dielectric constant (permittivity) of free space	$F\ cm^{-1}$
η	Factor in exp (eV/kT)	—
η	Relative occupancy, (9.7)	—
λ	Normalizing factor, (8.68)	—

Symbol	Definition	Dimensions
μ_I	Mobility limited by scattering from ionized impurities, (3.3)	$cm^2 V^{-1} sec^{-1}$
$\mu_{n,p}$	Mobility of electrons, holes	$cm^2 V^{-1} sec^{-1}$
μ_0	Mobility at low electric fields	sec^{-1}
ν	Thermal generation vibration frequency	—
$\nu_{1,2}$	Factors in (6.33)	—
Π	Multiply successive terms	—
π	Lightly doped *p*-type material	
ρ	Charge density, (5.37)	$C cm^{-3}$
$\rho(\Psi)$	Charge density as a function of Ψ, (8.61)	$C cm^{-3}$
σ	Conductivity	$\Omega^{-1} cm^{-1}$
σ'	Kinetic capture cross section, Section 5.1.4	cm^2
σ_{cap}	Capture cross section of a trap	cm^2
σ_i	Photoionization cross section for photon absorption, (7.2)	cm^2
σ_i	Capture cross section of the *i*th level for electrons, (9.4)	cm^2
σ_{imp}	Impact-ionization cross section, (13.1)	cm^2
$\sigma_n^{0,-,+}$	Electron capture cross section for neutral, repulsive, and attractive processes	cm^2
$\sigma_p^{0,-,+}$	Hole capture cross section for neutral, attractive, and repulsive processes	cm^2
σ_T	Capture cross section of a trap	cm^2
τ	Exponential time constant or lifetime	sec^{-1}
τ_B	Lifetime in the bulk, (6.42)	sec^{-1}
τ_d	Decay time constant produced by turn-off of illumination, (7.21a)	sec^{-1}
τ_d	Transit time in a drift field	sec^{-1}
τ_g	Regeneration time, (6.49)	sec^{-1}
τ_i	Initial (fast) time constant for photoconductivity decay transient, (6.28)	sec^{-1}
τ_n	Electron lifetime as an excess minority carrier	sec^{-1}
τ_{n0}	Electron lifetime, minimum value, when all recombination centers are neutral and therefore available for recombination	sec^{-1}
τ_{obs}	Observed lifetime if both surface and bulk recombination are significant, (6.42)	sec^{-1}

Symbol	Definition	Dimensions
τ_{on}	Time constant of turn-on transient produced by light, (7.20)	$\sec^{-1}$
τ_p	Hole lifetime, (7.29)	$\sec^{-1}$
τ_{p0}	Hole lifetime, minimum value when all recombination centers are active	$\sec^{-1}$
τ_{rel}	Dielectric relaxation time $(\varepsilon/e\mu_p p)$	$\sec^{-1}$
τ_s	Surface recombination lifetime, (6.43)	$\sec^{-1}$
τ_T	Hole-trapping time, (6.49)	$\sec^{-1}$
τ_t	Principal (slow) time constant for photoconductive decay transient, (6.29)	$\sec^{-1}$
ϕ	Energy difference between the bottom of the conduction band and the emission center, (5.24)	eV
ϕ	Potential barrier between a semiconductor and a metal, Fig. 8.18	V
ϕ_b	Energy barrier between Si and SiO_2, Fig. 8.25	eV
ϕ_S	Electrostatic potential between surface states and center of band gap, Fig. 6.6	V
ψ_S	Electrostatic potential at a metal-semiconductor interface, Fig. 8.18	V
$\psi(r_1, r_2)$	Wavefunction, (1.9)	
ω	Angular signal frequency $(2\pi f)$, (8.37)	rad $\sec^{-1}$

Bibliography

1 Aanas'ev, V. C., L. G. Paritskii, N. F. Prikot, and S. M. Ryvkin (1964), Effect of Trapping Levels on the Lux-Ampere Characteristics of Silicon, *Sov. Phys. Solid State,* **5,** 2326.

2 Abrahams, M. S., and C. J. Buiocchi (1970), Mechanism of Thermal Annihilation of Stacking Faults in GaAs, *J. Appl. Phys.,* **41,** 2358.

3 Adachi, E. (1967), Slow Photoconductivity Relaxation in Oxygen-Doped *n* Germanium, *J. Appl. Phys.,* **38,** 1972.

4 Adamic, J. W., Jr., and J. E. McNamara (1964), Paper 153, Washington, Meeting of the Electrochemical Society, October 1964.

5 Adams, H. D., W. J. Beyer and R. L. Petritz (1961), Photoconductivity Research on Cu–Ge, *J. Phys. Chem. Solids,* **22,** 167.

6 Adamsky, R. F. (1969), Effect of Deposition Parameters on the Crystallinity of Evaporated Germanium Films, *J. Appl. Phys.,* **40,** 4301.

7 Adirovich, É. I. (1961), A New Method of Determining the Effective Cross Sections of Local Centers, *Sov. Phys. Solid State,* **2,** 2006.

8 Adirovich, É. I., Sh. A. Mirsagatov, and V. V. Morozkin (1967), Negative Differential Resistance and Inductive Effects in Silicon Carbide, *Sov. Phys. Solid State,* **8,** 2692.

9 Afanas'ev, V. F. (1971), Carrier Lifetime in Gallium Arsenide Counters, *Sov. Phys.—Semicond.,* **4,** 1171.

10 Afromowitz, M. A., and M. Di Domenico, Jr. (1971), Measurement of Free-Carrier Lifetimes in GaP by Photoinduced Modulation of Infrared Absorption, *J. Appl. Phys.,* **42,** 3205.

11 Agraz-Guerna, J., and S. S. Li (1970), Steady State Recombination and Trapping Effects in Gold and Phosphorus-Doped Silicon, *Bull. Am. Phys. Soc.,* **15,** 314.

12 Agraz-Guerna, J., and S. S. Li (1970), Low Temperature Photomagnetoelectric Properties of Gold-Doped *n*-Type Silicon, *Phys. Rev.,* **2B,** 1847.

13 Agraz-Guerna, J., and S. S. Li (1970), Recombination and Trapping Processes of the Injected Carriers in Gold-Doped Silicon at Low Temperatures, *Phys. Rev.,* **2B,** 4966.

14 Aitchison, R. E. (1960), *An Experimental Investigation of the Energy Levels of Copper and Gold in Silicon and a Search for Accompanying Impurity Band Conduction,* Stanford University Electronics Laboratories Report 211–2, Document AD 244 171, National Technical Information Service, Springfield, Virginia.

15 Akimchenko, I. P., V. S. Vavilov, and A. F. Plotnikov (1963), Spectra and Kinetics of Photoconductivity Related to Simple Structural Defects in Germanium Single Crystals, *Sov. Phys. Solid State,* **5,** 1031.

16 Akimchenko, I. P., V. S. Vavilov, V. A. Vdovenkov, and A. F. Plotnikov (1968), Photoeffect Due to Generation of Excitons in Semiconductors, *Sov. Phys. Solid State,* **10,** 770.

17 Aladinski, V. K. (1967), Impact Ionization in Germanium, *Sov. Phys.—Semicond.,* **1,** 801.

18 Alekseeva, V. G., and N. G. Zhdanova (1972), Nonlinearity of the Current-Voltage Characteristics of Manganese-Doped Germanium, *Sov. Phys.—Semicond.,* **6,** 492.

19 Alekseeva, V. G., I. V. Karpova, and S. G. Kalashnikov (1961), Recombination at Gold Atoms in *p*-Germanium, *Sov. Phys. Solid State,* **3,** 699.

20 Alekseeva, V. G., N. G. Zhdanova, M. S. Kagan, S. G. Kalashnikov, and E. G. Landsberg (1970), Temperature Dependence of the Electron-Capture Coefficient of the Upper Level of Copper in Germanium, *Sov. Phys.—Semicond.,* **3,** 1179.

21 Alekseeva, V. G., N. G. Zhdanova, M. S. Kagan, S. G. Kalashnikov, and E. G. Landsberg (1972), Capture of Hot Electrons by Cu^{2-} Centers in Germanium, *Sov. Phys.—Semicond.,* **6,** 270.

22 Alferov, Zh. I., and N. S. Zimogorova (1969), Photoelectric Properties of *p-n* Junctions in Silicon-Doped Gallium Arsenide, *Sov. Phys.—Semicond.,* **3,** 385.

23 Alferov, Zh. I., D. Z. Garbuzov, and E. P. Morozov (1967), Effect of Thermal Annealing on the Photoluminescence of Gallium Arsenide, *Sov. Phys. Solid State,* **8,** 2589.

24 Alferov, Zh. I., V. I. Ivanov-Omskii, L. G. Paritskii, and V. Ya. Frenkel (1969), Investigations of Semiconductors at the Leningrad Physicotechnical Institute, *Sov. Phys.—Semicond.,* **2,** 1169.

25 Alferov, Zh. I., D. Z. Garbuzov, E. P. Morozov, and D. N. Tret'yakov (1969), Radiative Recombination in Gallium Arsenide *p-n* Structures with *p*-Type Regions Doped with Germanium, *Sov. Phys.—Semicond.,* **3,** 600.

26 Alferov, Zh. I., D. Z. Garbuzov, E. P. Morozov, I. I. Protasov, and D. N. Tret'yakov (1969), The Energy Spectrum of Gallium Arsenide Doped with Silicon, *Sov. Phys. Solid State,* **10,** 2260.

27 Alferov, Zh. I., V. K. Ergakov, V. I. Korol'kov, V. G. Nikitin, D. N. Tret'yakov, and A. A. Yakovenko (1971), Investigation of *S*-Type Diodes Made of Chromium-Doped Semi-Insulating GaAs, *Sov. Phys.—Semicond.,* **4,** 1748.

28 Alferov, Zh. I., D. Z. Garbuzov, O. A. Ninua, and V. G. Trofim (1972), Influence of Uniaxial Deformation on the Photoluminescence Spectra of $Al_xGa_{1-x}As$ Solid Solutions of Compositions Corresponding to the Transition from Direct to Indirect Band Structure, *Sov. Phys.—Semicond.,* **5,** 1228.

29 Aliev, M. I., G. I. Safaraliev, and S. G. Abdinova (1967), About the Role of Minority Carriers in the Impurity Conduction of GaSb, *Phys. Status Solidi*, **22,** 741.

30 Alieva, B. S., and V. I. Tagirov (1971), Determination of the Depth of Impurity Levels in Semiconductors, *Sov. Phys.—Semicond.*, **4,** 1872.

31 Allen, H. A., and G. A. Henderson (1968), Efficient Red-Emitting *p-n* Junctions Formed in GaP by Solution Growth of Thick Layers of *p*-Type Material on Vapor-Grown *n*-Type Substrates, *J. Appl. Phys.*, **39,** 2977.

32 Allen, J. W. (1964), Energy Levels of Transition Metal Impurities in semiconductors, *Physics of Semiconductors, Proceedings of the 7th International Conference, Paris, 1964,* Academic Press, New York and London, Dunod, Editeur, Paris, p. 781.

33 Allred, W., G. Cumming, J. Kung, and W. G. Spitzer (1968), *Site Distribution of Silicon in Silicon-Doped Gallium Arsenide, International Symposium on Gallium Arsenide, Dallas, Texas, 1968,* Institute of Physics, London, 1968.

34 Ambroziak, A. (1968), *Semiconductor Photoelectric Devices,* Iliffe, London.

35 Amelio, G. F. (1972), A New Method of Measuring Interface State Densities in MIS Devices, *Surface Science,* **29,** 125.

36 Amitay, M., and M. Pollak (1966), An Experimental Investigation of the Hall Effect in the Hopping Region, *Proceedings of the International Conference on the Physics of Semiconductors, Kyoto, Japan, 1966, J. Phys. Soc. Japan, Suppl.* **21,** 549.

37 Amith, A. (1959a), New Parallel Photoelectromagnetic Effect, *Phys. Rev.*, **116,** 330.

38 Amith, A. (1959b), Photoconductive and Photoelectromagnetic Lifetime Determination in Presence of Trapping. I. Small Signals, *Phys. Rev.*, **116,** 793.

39 Amsterdam, M. F. (1970), The Anomalous Behavior of Schottky Barrier Diodes Made on Lightly Doped GaAs, *Metall. Trans. AIME,* **1,** 643.

40 Ancker-Johnson, B., and W. P. Robbins (1971), Dynamic and Steady-State Injection of Electron-Hole Plasma in *p*-Type InSb, *J. Appl. Phys.*, **42,** 762.

41 Ancker-Johnson, B., W. P. Robbins, and D. B. Chang (1970), Transient High-Density Injection in a Semiconductor with Traps, *Appl. Phys. Lett.*, **16,** 377.

42 Anderson, C. L., and C. R. Crowell (1972), Threshold Energies for Electron-Hole Pair Production by Impact Ionization in Semiconductors, *Phys. Rev.*, **B5,** 2267.

43 Anderson, L. H., and K. M. van Vliet (1971), Electron Fluctuations in Zn-Doped GaAs in Thermal Equilibrium and under Optical Excitation, *Proceedings of the 3rd International Conference on Photoconductivity, Stanford University, 1969,* E. M. Pell, Ed., Pergamon Press, New York.

44 Andrews, A. M., and N. Holonyak, Jr. (1972), Properties of *n* Type Ge-Doped Epitaxial GaAs Layers Grown From Au-Rich Melts, *Solid-State Electron,* **15,** 601.

45 Andrews, A. M., H. W. Korb, N. Holonyak, C. B. Duke, and G. G. Kleiman (1972), Photosensitive Impurity-Assisted Tunneling in Au-Ge-Doped $Ga_{1-x}Al_xAs$ *p-n* Diodes, *Phys. Rev.*, **B5,** 4191.

46 Angelova, L. A., R. Bindemann, V. S. Vavilov, V. M. Grachev, L. I. Marina, and A. E. Yunovich (1969), Photoluminescence of Gallium Phosphide Doped with Zinc and Oxygen, *Sov. Phys.—Semicond.*, **3,** 269.

47 Anisimova, I. D., N. A. Ivashneva, and Y. D. Mozzhorin (1970), Characteristics of the Long-Wave-Length Electroluminescence Spectrum of *p-n* Junctions in Indium Arsenide, *Sov. Phys.—Semicond.,* **3,** 1412.

48 Antell, G. R., and A. P. White (1968), Double Diffused Gallium Arsenide Transistor, *International Symposium on Gallium Arsenide, Dallas, Texas, 1968,* Insti-stitute of Physics, London, 1968.

49 Antognetti, P., A. Chiabrera, and S. Ridella 1971a), Negative Resistance in Silicon Doped with Gold, *Solid-State Electron,* **14,** 1119.

50 Antognetti, P., A. Chiabrera, and S. Ridella (1971b), Recombination Wave Diode, *Solid-State Electron,* **14,** 1123.

51 Antognetti, P., A. Chiabrera, and S. Ridella (1972), Recombination Instabilities in Semiconductors Doped with Deep Two Level Traps, *J. Appl. Phys.,* **43,** 4676.

52 Antonov, A. S. (1966), On the Mechanism of Lithium Ion Drift in the Electrical Field of a *p-n* Junction in Silicon, *Phys. Status Solidi,* **16,** 761.

53 Antonov-Romanovskii, V. V. (1963a), Effective Trapping and Recombination Cross Sections of Free Charges in Solids, *Sov. Phys. Solid State,* **5,** 975.

54 Antonov-Romanovskii, V. V. (1963b), Electric Field Effect on the Effective Recombination and Trapping Cross Sections, *Sov. Phys. Solid State,* **5,** 979.

55 Antonov-Romanovskii, V. V. (1966), *Kinetics of the Photoluminescence of Crystal Phosphors,* Document AD 674 713, National Technical Information Service, Springfield, Virginia.

56 Arimura, I., and R. F. Freeman (1969), Rapid Annealing in *n*-Type Silicon Following Pulsed 10 MeV Electron Irradiation, *J. Appl. Phys.,* **40,** 2570.

57 Arizumi, T., T. Wada, and A. Yoshida (1965), Anomalous Junction Capacitance in Tunnel-Diodes, *Japan J. Appl. Phys.,* **4,** 415.

58 Arkad'eva, E. N. (1963), The Effect of Capture Levels on the Kinetics of Impurity Photoconductivity in Semiconductors, *Sov. Phys. Solid State,* **4,** 2233.

59 Arkad'eva, E. N., L. G. Paritskii, and S. M. Ryvkin (1962), A Long-Wave Photoelectric Method for Probing Local Levels in Semiconductors, *Sov. Phys. Solid State,* **4,** 1157.

60 Aronov, D. A., and R. G. Shamasov (1967), Effect of Trapping Centers on Photoconductivity of Semiconductors in the Presence of Radiative Interband Recombination, *Sov. Phys. Solid State,* **8,** 1318.

61 Arseni, K. A. (1969), Diffusion and Solubility of Gold in Indium Phosphide, *Sov. Phys.—Semicond.,* **2,** 1464.

62 Arseni, K. A. (1969), The Diffusion and Solubility of Copper in Indium Phosphide, *Sov. Phys. Solid State,* **10,** 2263.

63 Arseni, K. A., and B. I. Boltaks (1969), Diffusion and Solubility of Silver in Indium Phosphide, *Sov. Phys. Solid State,* **10,** 2190.

64 Arthur, J. R. (1967), Absorption and Desorption of O_2 on GaAs(III) Surfaces, *J. Appl. Phys.,* **38,** 4023.

65 Asano, S., K. Omori, and T. Sumimoto (1968), Activator-Free Luminescence of Zn (S:Se) Phosphors, *Sov. Phys. Solid State,* **10,** 891.

66 Ascarelli, G. (1962), Recombination of Electrons and Donors in *n*-Type Germanium, *Phys. Rev.,* **124,** 1321.

67 Ascarelli, G., and S. Rodriguez (1962), Recombination of Electrons and Donors in *n*-Type Germanium, *Phys. Rev.* **B127,** 167.

68 Ashkinadze, B. M., S. M. Ryvkin, and I. D. Yaroshetskii (1969), Thermal and Impact Ionization of Excitons in GaP in the Case of Two-Photon Excitation, *Sov. Phys.—Semicond.,* **3,** 455.

69 Ashkinadze, B. M., A. A. Patrin, and I. D. Yaroshetskii (1972), Absorption of Light by Nonequilibrium Carriers and Recombination in Silicon at High Injection Levels, *Sov. Phys.—Semicond.,* **5,** 1471.

70 Ashley, K. L. (1963), Investigation of the Effects of Space-Charge on the Conduction Mechanisms of Double-Injection in Semi-Insulators, Ph.D. Thesis, Electrical Engineering Department, Carnegie-Mellon University, Pittsburgh, Pennsylvania.

71 Ashley, K. L., and A. G. Milnes (1964), Double Injection in Deep-Lying Impurity Semiconductors, *J. Appl. Phys.,* **35,** 369.

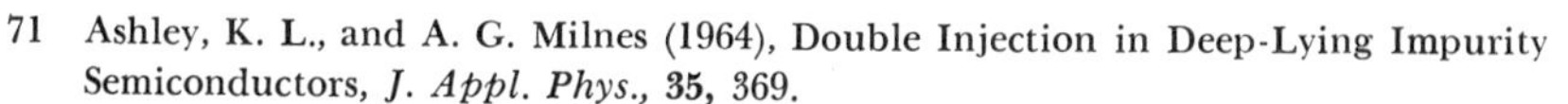

72 Ashley, K. L., W. E. Ham, and R. T. Brown (1970), Investigation of the Electron Statistics of the Neutral Acceptor Level of Gold in Germanium, *J. Appl. Phys.,* **41,** 3840.

73 Ashley, K. L., V. Jayakumur, and R. T. Brown (1971a), Degeneracy Ratio of Positive to Neutral States of Gold in Germanium, *J. Appl. Phys.,* **42,** 1240.

74 Ashley, K. L., R. T. Brown, and T. A. Blaske (1971b), Optical Phonon Scattering $\mu_{\mathrm{H}}\mu/_{\mathrm{D}}$, and Statistics of *p*-Type Zinc-Doped Silicon, *J. Appl. Phys.,* **42,** 2482.

75 Assour, J. M. (1972), Identification of Chemical Constituents of Defects in Silicon, *J. Electrochem. Soc.,* **119,** 1270.

76 Astaf'ev, N. I., M. I. Barnik, and B. I. Beglov (1971), Photo-Impact Ionization of Impurity Centers in Germanium, *Sov. Phys.—Semicond.,* **4,** 1240.

77 Astrov, Yu. A., and A. A. Kastal'skii (1972), Breakdown of Shallow Donors in Pure *n*-Type Germanium, *Sov. Phys.—Semicond.,* **5,** 1111.

78 Atalla, M. M., and D. Kahng (1965), Varactor Diode with Concentration of Deep-Lying Impurities and Enabling Circuitry, U.S. Patent No. 3,176,151.

79 Atalla, M. M., et al (1966), *One Micron Photodetectors,* Final Report #DA 44-009 AMC-262 (T), Hewlett-Packard, California.

80 Aubrey, J. E., W. Gubler, T. Henningsen, and S. H. Koenig (1963), Piezoresistance and Piezo-Hall effect in *n*-Type Silicon, *Phys. Rev.,* **130,** 1667.

81 Aukerman, L. W., M. F. Millea, and M. McColl (1967), Diffusion Lengths of Electrons and Holes in GaAs, *J. Appl. Phys.,* **38,** 685.

82 Avak'yants, G. M., and V. M. Arutyunyan (1970), Influence of the Impact Ionization of Trapping Levels on the Current-Voltage Characteristic of a Diode Operating under Double Injection Conditions, *Sov. Phys.—Semicond.,* **3,** 814.

83 Avak'yants, G. M., Z. N. Adamyan, R. S. Barsegyan, and S. A. Tarumyan (1971), Influence of Temperature on Oscillations in S-Type Diodes Made of Cadmium-Compensated Silicon, *Sov. Phys.—Semicond.,* **5,** 717.

84 Aven, M., and J. S. Prenner (1967), *Physics and Chemistry of II–VI Compounds,* Wiley, New York.

85 Aven, M., R. B. Hall, and W. Garwacki (1968), *p-n* Junctions as Artificial Diffusion Barriers for Native Defects, *Appl. Phys. Lett.,* **13,** 292.

86 Aver'yanova, M. M., and I. V. Varlamov (1969), Some Properties of *n-p-n* Structures Based on Gold-Doped Germanium, *Sov. Phys.—Semicond.,* **3,** 775.

87 Aver'yanova, M. M., I. V̇. Varlamov, V. A. Petrusevich, É. A. Poltoratskii, I. A. Sondaevskaya, and V. I. Stafeev (1971), Photoelectric Properties of Semiconductor Structures made of Gold-Doped Germanium, *Sov. Phys.—Semicond.,* **5,** 198.

88 Azhdarov, G. Kh., and V. I. Tagirov (1971), Investigation of the Energy Levels of Nickel in Germanium-Silicon Alloys, *Sov. Phys.—Semicond.,* **5,** 975.

89 Bachrach, R. Z., and O. G. Lorimor (1972), Measurement of the Extrinsic Room-Temperature Minority Carrier Lifetime in GaP, *J. Appl. Phys.,* **43,** 500.

90 Backenstoss, G. (1957), Conductivity Mobilities of Electrons and Holes in Heavily Doped Silicon, *Phys. Rev.,* **108,** 1416.

91 Bacon, D. D. (1959), Treatment of Semiconductive Material, U.S. Patent No. 2,871,111.

92 Badalov, A. Z. (1970), Photoconductivity of Gold Doped *n*-Type Silicon, *Sov. Phys.—Semicond.,* **3,** 1435.

93 Badalov, A. Z., and V. B. Shuman (1970), Diffusion of Cu in *n*-Type Si, *Sov. Phys.—Semicond.,* **3,** 1137.

94 Badenko, L. A. (1964), Investigation of the Behavior of Indium and Antimony Impurities in Germanium by the Electrical Diffusion Method, *Sov. Phys. Solid State,* **6,** 762.

95 Baev, I. A., and E. G. Valyashko (1966), An Investigation of the Distribution of Inhomogeneous Regions in Semiconductors, *Sov. Phys. Solid State,* **7,** 2093.

96 Bagaev, V. S. (1970), Electro-Optical Effect in Gallium Arsenide and Cadmium Telluride, *Sov. Phys.—Semicond.,* **3,** 1418.

97 Bagaev, V. S., Yu. N. Berozashvili, B. M. Vul, E. I. Zavaritskaya, and A. P. Shotov (1964), Mechanism of Radiative Recombination in Gallium Arsenide, *Sov. Phys. Solid State,* **6,** 959.

98 Bagaev, V. S., Yu. N. Berozashvili, B. M. Vul, E. I. Zavaritskaya, L. V. Keldysh, and A. P. Shotov (1964), Concerning the Energy Level Spectrum of Heavily Doped Gallium Arsenide, *Sov. Phys. Solid State,* **6,** 1093.

99 Bagaev, V. S., Yu. N. Berozashvili, and B. M. Vul (1968), Mobile Electrical Domains in Semiinsulating Gallium Arsenide, *Sov. Phys.—Semicond.,* **2,** 700.

100 Baicker, J. A., and P. H. Fang (1972), Transient Conductivity of Silicon, *J. Appl. Phys.,* **43,** 125.

101 Bailey, L. G. (1966), Preparation and Properties of Silicon Telluride, *J. Phys. Chem. Solids,* **27,** 1593.

102 Bailey, R. F., and T. G. Mills (1969), Diffusion Parameters of Platinum in Silicon, *Semiconductor Silicon,* Electrochemical Society, New York, p. 481.

103 Bailey, R. F., and T. G. Mills (1969), Diffusion Parameters of Platinum in Silicon, *J. Electrochem. Soc.,* **116,** 147C.

104 Bakanowski, A. E., and J. H. Forster (1960), Electrical Properties of Gold-Doped Diffused Silicon Computer Diodes, *Bell Syst. Tech. J.,* **39,** 87.

105 Baker, J. A. (1969), Oxygen and Carbon Content of Czochralski Silicon Crystals, *J. Electrochem. Soc.,* **116,** 149C.

106 Baker, J. A., T. N. Tucker, N. E. Moyer, and R. C. Buschert (1968), Effect of Carbon on the Lattice Parameter of Silicon, *J. Appl. Phys.,* **39,** 4365.

107 Balk, P. (1971), Surface Properties of Oxidized Germanium-Doped Silicon, *J. Electrochem. Soc.,* **118,** 494.

108 Balkanski, M., E. da Silva, and W. Nazarewicz (1965), *Spectroscopy of Point Defects, U.S. Gov. Res. Dev. Rep. Index,* **40,** April 5, 1965, 121(A), Document AD 611 194, National Technical Information Service, Springfield, Virginia.

109 Baltensperger, W. (1953), On Conduction in Impurity Bands, *Phil. Mag.,* **44,** 1355.

110 Baranenkov, A. I., and V. V. Osipov (1969), Current-Voltage Characteristics of Long Diodes Made of Compensated Semiconductors, *Sov. Phys.—Semicond.,* **3,** 30.

111 Baranenkov, A. I., and V. V. Osipov (1971), Injection Breakdown of Compensated Semiconductors, *Sov. Phys.—Semicond.,* **5,** 740.

112 Baranowski, J. M., J. W. Allen, and G. L. Pearson (1968), Absorption Spectrum of Nickel in Gallium Phosphide, *Phys. Rev.,* **167,** 758.

113 Baranskii, P. I., and E. A. Glushkov (1970), Negative Magnetoresistance of *n*-Type Gallium Arsenide, *Sov. Phys.—Semicond.,* **3,** 1169.

114 Barber, H. D. (1967), Effective Mass and Intrinsic Concentration in Silicon, *Solid-State Electron.,* **10,** 1039.

115 Bardsley, W. (1960), The Electrical Effects of Dislocations in Semiconductors, *Progress in Semiconductors,* Vol. 4, Wiley, New York, p. 155.

116 Barker, R. E., Jr. (1966), Some Comments on the Photo-Magnetoelectric Effect in Silicon, *Phys. Rev.,* **149,** 663.

117 Barnett, A. M. (1966), Filamentary Injection Currents in Semi-Insulating Silicon, Ph.D. Thesis, Electrical Engineering Department, Carnegie-Mellon University, Pittsburgh, Pennsylvania.

118 Barnett, A. M. (1968), Light-Emitting Devices Utilizing Current Filaments in Semi-Insulating GaAs, *International Symposium on Gallium Arsenide, Dallas, Texas, 1968,* Institute of Physics, London.

119 Barnett, A. M. (1969), Current Filaments in Semiconductors, *IBM J. Res. Dev.,* **13,** 522.

120 Barnett, A. M., and H. A. Jensen (1968), Observation of Current Filaments in Semi-Insulating GaAs, *Appl. Phys. Lett.,* **12,** 341.

121 Barnett, A. M., and A. G. Milnes (1966), Filamentary Injection in Semi-Insulating Silicon, *J. Appl. Phys.,* **37,** 4215.

122 Barnik, M. I., B. I. Beglov, D. A. Romanychev, and Yu. S. Kharionovskii (1971), Photoelectric Properties of Cobalt-Doped Germanium, *Sov. Phys.—Semicond.,* **5,** 87.

123 Baron, R. (1965), Effects of Diffusion on Double Injection in Insulators, *Phys. Rev.,* **137,** A272.

124 Baron, R., and J. W. Mayer (1970), *Double Injection in Semiconductors,* Academic Press, New York.

125 Baron, R., O. J. Marsh, and J. W. Mayer (1966), Transient Response of Double Injection in a Semiconductor of Finite Cross Section, *J. Appl. Phys.,* **37,** 2614.

126 Baron, R., G. A. Shifrin, and O. J. Marsh (1969), Electrical Behavior of Group III and V Implanted Dopants in Silicon, *J. Appl. Phys.,* **40,** 3702.

127 Barraud, A. (1963), Le comportement de l'arsersiure de gallium de haute resistivite soumis a des champs electrical intenses, *C. R. Acad. Sci. Paris,* **256,** 3632.

128 Barraud, A. (1967), *Current Instabilities in High-Resistivity Gallium Arsenide Produced by Strong Electric Field,* Ph.D. Thesis, Document N 69-18810-3/26, National Technical Information Service, Springfield, Virginia.

129 Barrera, J. S. (1967), Fluctuation Phenomena in Double Injection Devices Under Space Charge and Recombination Limited Conditions, Ph.D. Thesis, Electrical Engineering Department, Carnegie-Mellon University, Pittsburgh, Pennsylvania.

130 Barry, M. L., and P. Olofsen (1969), Doped Oxides as Diffusion Sources, *J. Electrochem. Soc.,* **116,** 854.

131 Bartelink, D. J., and D. L. Scharfetter (1969), Avalanche Shock Fronts in *p-n* Junctions, *Appl. Phys. Lett.,* **14,** 320.

132 Bartholomew, C. Y. (1967), Recombination Centers in Silicon Transistor Emitter-Base Junctions, *IEEE Trans. Electron Devices,* **ED-14,** 452.

133 Baryshev, N. S., E. E. Vdovkina, A. P. Martynovich, I. M. Nesmelova, N. Tsitsina, and I. I. Aver'yanov (1967), Deep-Lying Energy Levels in Indium Antimonide, *Sov. Phys. Solid State,* **8,** 1800.

134 Baryshev, N. S., B. I. Beglov, and K. Ya. Shtivel'man (1967), Splitting of the Ground Level of Double Acceptors in Semiconductors, *Sov. Phys. Solid State,* **9,** 1463.

135 Basetskii, V. Ya., N. A. Goryunova, O. G. Peskov, and G. F. Kholuyanov (1969), Electroluminesence and Photoluminescence of Gallium Phosphide Doped with Zinc and Boron, *Sov. Phys.—Semicond.,* **2,** 856.

136 Basinski, J., and R. Olivier (1966), Ionization Energy and Impurity Band Conduction of Shallow Donors in *n*-Gallium Arsenide, *Can. J. Phys.,* **45,** 119.

137 Bassett, R. J., and C. A. Hogarth (1972), Thermally-Stimulated Current From the Gold Acceptor Trapping Level in Silicon, *Int. J. Electron.,* **33,** 217.

138 Batavin, V. V. (1967), Influence of Oxygen in Silicon on the Motion of Dislocations Generated by Diffusion, *Sov. Phys. Solid State,* **8,** 2478.

139 Batavin, V. V., G. V. Popova, and L. A. Batavina (1967), Effect of Oxygen in Silicon on Light Emission from *p-n* Junctions, *Sov. Phys. Solid State,* **8,** 2005.

140 Batavin, V. V., V. M. Mikhaélyan, G. V. Popova, and V. N. Fedorenko (1972), Influence of Copper Impurity on the Carrier Mobility in Epitaxial Gallium Arsenide Films, *Sov. Phys.—Semicond.,* **6,** 67.

141 Batra, I. P., and H. Seki (1970), Photocurrent Due to Pulse Illumination in the Presence of Trapping, *J. Appl. Phys.,* **41,** 3409.

142 Batra, I. P., K. K. Kanazawa, B. H. Schechtman, and H. Seki (1971), Charge-Carrier Dynamics Following Pulsed Photoinjection, *J. Appl. Phys.,* **42,** 1124.

143 Baynham, A. C., and A. D. Boardman (1971), *Plasma Effects in Semiconductors: Helicon and Alfvén Waves,* Taylor and Francis, London, and Barnes and Noble, New York.

144 Bazhenov, V. K., and N. N. Solov'ev (1972), Deep Transition-Metal Centers in Gallium Arsenide, *Sov. Phys.—Semicond.,* **5,** 1589.

145 Bazhenov, V. K., V. A. Presnov, and S. P. Fedotov (1968), Paramagnetic Resonance in Fe-Doped Gallium Arsenide, *Sov. Phys. Solid State,* **10,** 205.

146 Bazhenov, V. K., V. A. Presnov, and S. P. Fedotov (1971), Effect of Electric Fields in the Paramagnetic Resonance of Deep Centers, *Sov. Phys.—Semicond.,* **4,** 2038.

147 Beattie, A. R. (1962), Quantum Efficiency in InSb, *J. Phys. Chem. Solids,* **24,** 1049.

148 Beattie, A. R., and R. W. Cunningham (1962), Large-Signal Photomagnetoelectric Effect, *Phys. Rev.,* **125,** 533.

149 Bebb, H. B. (1969), Application of the Quantum-Defect Method to Optical Transitions Involving Deep Effective-Mass-Like Impurities in Semiconductors, *Phys. Rev.,* **185,** 1116.

150 Bebb, H. B. (1972), Comments on Radiative Capture by Impurities in Semiconductors, *Phys. Rev.,* **5,** 4201.

151 Bebb, H. B., and R. A. Chapman (1967), Application of Quantum Defects Techniques to Photoionization of Impurities, *J. Phys. Chem. Solids,* **28,** 2087.

152 Bebb, H. B., and R. A. Chapman (1971), Theory of Deep Impurity Centers in Semiconductors, *Proceedings of the 3rd International Conference on Photoconductivity, Stanford University, 1969,* E. M. Pell, Ed., Pergamon Press, New York, p. 245.

153 Beer, A. C., R. D. Baxter, and F. J. Reid (1965), *High-Mobility Low-Melting-Point Group III-V Compound Semiconductors,* Document AD 624 562, National Information Service, Springfield, Virginia.

154 Beglov, B. I., Yu. S. Kharionovskii, and S. G. Yudin (1969), Hopping Photoconductivity of Gold-Doped Germanium, *Sov. Phys.—Semicond.,* **3,** 242.

155 Belen'kii, G. L., I. B. Ermolovich, N. B. Luk'yanchikova, and M. K. Sheinkman (1968), Determination of the Radiative Yield of the Electron Capture by Sensitizing Recombination Centers in CdSe, *Sov. Phys. Solid State,* **9,** 2626.

156 Beleznay, F., and G. Pataki (1966), Remarks on the Recombination of Electrons and Donors in n-Type Germanium, *Phys. Status Solidi.,* **13,** 499.

157 Beloglazov, A. V., R. Bindemann, V. M. Grachev, and A. E. Yunovich (1970), Structure of the Radiation Spectra of Gallium Phosphide Associated with Zn–O Impurity Complexes, *Sov. Phys.—Semicond.,* **3,** 1352.

158 Belozerskii, G. N., I. A. Gusev, Yu. A. Nemilov, and A. V. Shvedchikov (1967), Study of the Behavior of Impurity Atoms in Diatomic Crystals of InSb and GaSb, *Sov. Phys. Solid State,* **8,** 1680.

159 Belyaev, A. D., and V. G. Malogolovets (1964), Effective Cross Section for Electron Capture by Negative Iron Ions in Germanium, *Sov. Phys. Solid State,* **5,** 2229.

160 Belyaev, A. D., and S. S. Malogolovets (1965), Temperature Dependence of the Cross Section for the Capture of Holes by Impurity Centers in Germanium, *Sov. Phys. Solid State,* **7,** 1529.

161 Belyaev, A. D., and E. G. Miselyuk (1965), Electron Recombination at Negative Ions of Nickel in Germanium, *Sov. Phys. Solid State,* **6,** 2101.

162 Belyaev, Yu. I., and V. A. Zhidkov (1961), The Diffusion of Beryllium in Germanium, *Sov. Phys. Solid State,* **3,** 133.

163 Bemski, G. (1958a), Method of Improving the Minority Carrier Lifetime in a Single Crystal Silicon Body, U.S. Patent No. 2,827,436.

164 Bemski, G. (1958b), Recombination Properties of Gold in Silicon, *Phys. Rev.,* **111,** 1515.

165 Bemski, G. (1958c), Recombination in Semiconductors, *Proc. IRE,* **46,** 990.

166 Berkeliev, A., and K. Durdyev (1972), Quenching of the Photoconductivity of High-Resistivity *n*-Type InP, *Sov. Phys.—Semicond.,* **6,** 458.

167 Berkovskii, F. M., and R. S. Kasymova (1967), Effects of Charge Transfer of Deep Impurity Levels During Electric Injection, *Sov. Phys. Solid State,* **8,** 1580.

168 Berkovskii, F. M., and S. M. Ryvkin (1963), Effect of Optical Charge-Exchange in Impurity Centers on the Photo-EMF Kinetics in Germanium, *Sov. Phys. Solid State,* **5,** 278.

169 Berkovskii, F. M., S. M. Ryvkin, and N. B. Strokan (1961), Effect of Trapping Levels on the Decay of Current Through *p-n* Junction, *Sov. Phys. Solid State,* **3,** 169.

170 Berkovskii, F. M., S. M. Ryvkin, and N. B. Strokan (1962), Influence of Trapping Levels on Current Relaxation in *p-n* Junction Devices, *Sov. Phys. Solid State,* **3,** 2566.

171 Berkovskii, F. M., R. S. Kasymova, and S. M. Ryvkin (1963), Sensitization of Photodiodes by Optical Charge Exchange Among Impurities, *Sov. Phys. Solid State,* **5,** 382.

172 Berman, L. S. (1971), Barrier Capacitance of an Alloyed Diode with Deep Impurities Which Exchange Electrons with the Conduction and Valence Bands, *Sov. Phys.—Semicond.,* **4,** 1295.

173 Berman, L. S. (1971), Barrier Capacitance of an Alloyed Diode With Deep Impurities That Exchange Electrons With Conduction and Valence Bands (Transient Processes), *Sov. Phys.—Semicond.,* **5,** 592.

174 Berman, L. S. (1972), Barrier Capacitance of Silicon Diodes With Gold Impurities in the Base, *Sov. Phys.—Semicond.,* **6,** 285.

175 Berman, P. A. (1972), Summary of Results of JPL Lithium-doped Solar Cell Development Program, *Conference Record of Ninth IEEE Photovoltaic Specialists Conference,* IEEE, New York, p. 281.

176 Besfamil'naya, V. A. (1969), Field Dependence of Oscillations of the Impurity Photoconductivity of *p*-Type Germanium, *Sov. Phys.—Semicond.,* **3,** 380.

177 Besfamil'naya, V. A., and V. V. Ostroborodova (1965), Recombination Properties of Shallow Levels of Gold and Copper in *p*-Type Germanium, Determined from the Noise Spectrum, *Sov. Phys. Solid State,* **6,** 3005.

178 Bestfamil'naya, V. A., and V. V. Ostroborodova (1969a), Recombination Properties of Impurity Centers in *p*-Type Ge, *Sov. Phys.—Semicond.,* **3,** 15.

179 Besfamil'naya, V. A., and V. V. Ostroborodova (1969b), Influence of an Electric Field on the Recombination and Scattering of Carriers in *p*-Type Germanium with Deep Impurities, *Sov. Phys.—Semicond.,* **3,** 110.

180 Besfamil'naya, V. A., V. V. Ostroborodova, and L. M. Shlita (1969), Recombination of Holes at Shallow Acceptors in *p*-Type Germanium, *Sov. Phys.—Semicond.,* **2,** 888.

181 Bess, L. (1957), Possible Mechanism for Radiationless Recombination in Semiconductors, *Phys. Rev.,* **105,** 1469.

182 Bhargava, R. N. (1969), Negative Resistance in GaP Electroluminescent Diodes, *Appl. Phys. Lett.,* **14,** 193.

183 Bhargava, R. N., C. Michel, W. L. Lupatkin, R. L. Bronnes, and S. K. Kurtz (1972), Mg-O Complexes in GaP—A Yellow Diode, *Appl. Phys. Lett.,* **20,** 227.

184 Bhide, V. G., N. V. Bhat, and K. R. Rambhad (1969), X-Ray Spectroscopic Study of Zinc Selenide, *J. Appl. Phys.*, **39,** 4744.

185 Billig, E. (1959), Semiconductor Activated with Dissociated Ammonia, U.S. Patent No. 2,887,453.

186 Bisio, G. R., and A. E. Chiabrera (1970), Recombination Waves in Au-Doped Si, *Appl. Phys. Lett.*, **16,** 181.

187 Björkqvist, K., B. Domeij, L. Eriksson, G. Fladda, A. Fontell, and J. W. Mayer (1968), Lattice Location of Dopant Elements Implanted into Ge, *Appl. Phys. Lett.*, **13,** 379.

188 Black, J., and P. Lublin (1964), Electrical Measurements and X-Ray Lattice Parameter Measurements of GaAs Doped with Se, Te, Zn, and Cd, and the Stress Effects of these Elements as Diffusants in GaAs, *J. Appl. Phys.*, **35,** 2462.

189 Blagosklonskaya, L. E., E. M. Gershenzon, Yu. P. Ladyzhinskii, and A. P. Popova (1969), Electron Scattering by Neutral Acceptors in Semiconductors, *Sov. Phys. Solid State,* **10,** 2374.

190 Blakemore, J. S. (1956), Photoconductivity in Indium-Doped Silicon, *Can. J. Phys.*, **34,** 938.

191 Blakemore, J. S. (1962), *Semiconductor Statistics,* Pergamon Press, New York.

192 Blakemore, J. S. (1967), Radiative Capture by Impurities in Semiconductors, *Phys. Rev.*, **163,** 809.

193 Blakemore, J. S., and K. C. Nomura (1960), Influence of Transverse Modes on Photoconductive Decay in Filaments, *J. Appl. Phys.*, **31,** 753.

194 Blakemore, J. S., and C. E. Sarver (1968), Photoconductivity Associated with Indium Acceptors in Silicon, *Phys. Rev.*, **173,** 767.

195 Blakeslee, A. E., and J. E. Lewis (1969), Copper Contamination at the Substrate-Epitaxial Layer Interface in GaAs, Abstract No. 121, Electrochemical Society Spring Meeting, May 1969.

196 Blanc, J., and L. R. Weisberg (1961), Energy-Level Model for High-Resistivity Gallium Arsenide, *Nature,* **192,** 155.

197 Blanc, J., R. H. Bube, and H. E. MacDonald (1961), Properties of High-Resistivity Gallium Arsenide Compensated with Diffused Copper, *J. Appl. Phys.*, **32,** 1666.

198 Blanc, J., R. H. Bube, and L. R. Weisberg (1964), Behavior of Lattice Defects in GaAs, *J. Phys. Chem. Solids,* **25,** 225.

199 Blashku, A. I., B. I. Boltaks, T. D. Dzhafarov, and F. P. Kesamanly (1971), Interaction Between Zinc and Tellurium in Gallium Arsenide, *Sov. Phys.—Semicond.*, **5,** 664.

200 Blätte, M., W. Schairer, and F. Willmann (1970), Photoluminescence of Ag-Doped *p*-Type GaAs, *Solid-State Commun.*, **8,** 1265.

201 Blinov, L. M. (1965), Cross Sections for the Capture of Holes by Mercury Ions in Germanium, *Sov. Phys. Solid State,* **7,** 738.

202 Blinov, L. M., E. A. Bobrova, V. S. Vavilov, and G. N. Galkin (1968), Recombination of Nonequilibrium Carriers in Silicon in the Case of High Photoexcitation Levels, *Sov. Phys. Solid State,* **9,** 2537.

203 Blocker, T. G., R. H. Cox, and T. E. Hasty (1970), Interpretation of Anomalous Layers at GaAs n^+ n^- Step Junctions, *Solid-State Commun.*, **8,** 1313.

204 Blouke, M. M., and R. L. Williams (1972), Gain Saturation in High-Resistivity Si:B Photoconductors, *Appl. Phys. Lett.*, **20**, 25.

205 Boblylev, B. A., and A. F. Kravchenko (1970), Photoacoustic Spectra of High-Resistivity Gallium Arsenide, *Sov. Phys.—Semicond.*, **3**, 1126.

206 Boer, K. W. (1968), Light-Induced Modulation of Absorption (LIMA) for Defect Level Analysis, *Bull. Am. Phys. Soc.*, **13**, 727.

207 Boer, K. W. (1971), Determination of Field-Dependent Carrier Density and Mobility in Photoconductors Using High-Field Domains, *Proceedings of the 3rd International Conference on Photoconductivity, Stanford University, 1969*, E. M. Pell, Ed., Pergamon Press, New York.

208 Boer, K. W., and G. Dohler (1969), Influence of Boundary Conditions on High-Field Domains in Gunn Diodes, *Phys. Rev.*, **186**, 793.

209 Boer, K. W., S. Oberlander, and J. Voigt (1958), Zur Theorie der Glow-Kurven, *Z. Naturforsch*, **13a**, 544.

210 Boer, K. W., A. S. Esbitt, and W. M. Kaufman (1966), Evaporated and Recrystallized CdS Layers, *J. Appl. Phys.*, **37**, 2664.

211 Boer, K. W., J. W. Feitknecht, and D. G. Kannenberg (1966), Properties of Recrystallized Evaporated CdS Layers, *Phys. Status Solidi*, **16**, 697.

212 Boer, K. W., G. Dohler, G. A. Dussell, and P. Voss (1968), Experimental Determination of Changes in Conductivity with Electric Field, Using a Stationary High-Field Domain Analysis, *Phys. Rev.*, **169**, *Abstr.*, A16.

213 Boer, K. W., G. A. Dussel, and P. Voss (1969), Experimental Evidence for a Reduction of the Work Function of Blocking Gold Contacts with Increasing Photocurrents in CdS, *Phys. Rev.*, **179**, 703.

214 Bogoroditskii, N. P., A. N. Pikhtin, and D. A. Yas'kov (1968), Electroluminescence of Gallium Phosphide Prepared by the Zone Melting Method, *Sov. Phys. Solid State*, **9**, 2297.

215 Bohn, P. P. (1970), A Nonsaturating Velocity-Field Approximation for Improved Invariant Domain Analysis of Gunn Effect Devices, *Proc. IEEE*, **58**, 1397.

216 Boiko, I. I., E. I. Rashba, and A. P. Trofimenko (1960), Thermally Stimulated Conductivity in Semiconductors, *Sov. Phys. Solid State*, **2**, 99.

217 Boltaks, B. I. (1963), *Diffusion in Semiconductors*, Academic Press, New York.

218 Boltaks, B. I., and G. S. Kulikov (1965), Diffusion of Silver on the Surface of Silicon, *Sov. Phys. Solid State*, **6**, 1519.

219 Boltaks, B. I., and S. I. Rembeza (1967), Diffusion and Electrical Transport of Zinc in Indium Arsenide, *Sov. Phys. Solid State*, **8**, 2117.

220 Boltaks, B. I., and S. I. Rembeza (1969), Effect of Dislocations on the Decomposition of Solid Solutions of Elements of the Copper Group in Indium Arsenide, *Sov. Phys. Solid State*, **10**, 2477.

221 Boltaks, B. I., and H. Shih-yin (1961), Diffusion, Solubility and the Effect of Silver Impurities on Electrical Properties of Silicon, *Sov. Phys. Solid State*, **2**, 2383.

222 Boltaks, B. I., and F. S. Shishiyanu (1964), Diffusion and Solubility of Silver in Gallium Arsenide, *Sov. Phys. Solid State*, **5**, 1680.

223 Boltaks, B. I., and V. I. Sokolov (1964), Diffusion of Gold in Indium Antimonide, *Sov. Phys. Solid State*, **6**, 600.

224 Boltaks, B. I., G. S. Kulikov, and R. Sh. Malkovich (1960), The Effect of Gold on the Electrical Properties of Silicon, *Sov. Phys. Solid State,* **2,** 167.

225 Boltaks, B. I., G. S. Kulikov, and R. Sh. Malkovich (1960), Electrical Transport of Gold in Silicon, *Sov. Phys. Solid State,* **2,** 2134.

226 Boltaks, B. I., V. I. Dzhafarov, V. I. Sokolov, and F. S. Shishiyanu (1964), Diffusion and Electrical Transport of Zinc in Gallium Arsenide, *Sov. Phys. Solid State,* **6,** 1181.

227 Boltaks, B. I., S. Ya. Ksendzov, and S. I. Rembeza (1969), Concentration Dependence of the Diffusion Coefficient of Zinc in Indium Arsenide, *Sov. Phys. Solid State,* **10,** 2186.

228 Bonch-Bruevich, V. L. (1966), Concerning the Theory of Generation-Recombination Noise in Semiconductors, *Sov. Phys. Solid State,* **7,** 1728.

229 Bonch-Bruevich, V. L. (1969), Recombination Self-Oscillations, *Sov. Phys.—Semicond.,* **3,** 305.

230 Bonch-Bruevich, V. L., and A. A. Drugova (1966), The Theory of Radiative Recombination via Impurity Centers in Homogeneous Semiconductors, *Sov. Phys. Solid State,* **7,** 2491.

231 Bonch-Bruevich, V. L., and V. B. Glasko (1962), The Theory of "Cascade" Carrier Recombination in Homopolar Semiconductors, *Sov. Phys. Solid State,* **4,** 371.

232 Bonch-Bruevich, V. L., and S. G. Kalashnikov (1965), The Possibility of Recombination Instability in Semiconductors, *Sov. Phys. Solid State,* **7,** 599.

233 Bonch-Bruevich, V. L., and Sh. M. Kogan (1965), The Formation of Domains in Semiconductors with Negative Differential Resistance, *Sov. Phys. Solid State,* **7,** 15.

234 Bonch-Bruevich, V. L., and E. G. Landsberg (1968), Recombination Mechanism, *Phys. Status Solid,* **29,** 9.

235 Bonch-Bruevich, V. L., and E. B. Sokolova (1964), Concerning a Possible Recombination Mechanism, *Sov. Phys. Solid State,* **5,** 1986.

236 Bonch-Bruevich, V. L., A. V. Burlakov, and I. N. Taganov (1968), Interaction of Luminescence Centers in the Electroluminescence of ZnS:Cu, Al Phosphors, *Sov. Phys. Solid State,* **9,** 2769.

237 Bonsels, W., and J. L. Lambert (1969), The Carbon Content of Semiconductor Silicon, *Semiconductor Silicon,* The Electrochemical Society, New York, p. 89.

238 Booth, A. H. (1954), Calculation of Electron Trap Depths from Thermoluminescence Maxima *Can. J. Chem.,* **32,** 214.

239 Borghi, L., P. DeStefano, and P. Mascheretti (1970), Photoconductivity of Neutron-Irradiated Gallium Arsenide, *J. Appl. Phys.,* **41,** 4665.

240 Borkovskaya, O. Yu., V. V. Zuev, V. I. Lyashenko, and V. L. Shirdich (1971), Photoelectric Phenomena in Surface Layers of Silicon, *Sov. Phys.—Semicond.,* **4,** 1056.

241 Borrello, S. R., and H. Levinstein (1962), Preparation and Properties of Mercury-Doped Germanium, *J. Appl. Phys.,* **33,** 2947.

242 Bortfield, D. P., B. J. Curtis, and H. Meier (1972), Electrical Properties of Carbon-Doped Gallium Phosphide, *J. Appl. Phys.,* **43,** 1293.

243 Bower, R. W., R. Baron, J. W. Mayer, and O. J. Marsh (1966), Deep (1–10 μm) Penetration of Ion-Implanted Donors in Silicon, *Appl. Phys. Lett.,* **9,** 203.

244 Bowman, D. L. (1967), Photoconductivity and Photo-Hall Measurements on High-Resistivity GaP, *J. Appl. Phys.,* **38,** 568.

245 Bräunlich, P. (1967), Comment on the Initial-Rise Method for Determining Trap Depths, *J. Appl. Phys.,* **38,** 2516.

246 Bräunlich, P., and P. Kelly (1970), II. Correlations between Thermoluminescence and Thermally Stimulated Conductivity, *Phys. Rev.* **B1,** Part 1, 1596.

247 Braunstein, R. (1963), Lattice Vibration Spectra of Germanium–Silicon Alloys, *Phys. Rev.,* **130,** 879.

248 Bray, R. (1969), A Perspective on Acoustoelectric Instabilities, *IBM J. Res. Dev.,* **13,** 487.

249 Brehm, G. E., and G. L. Pearson (1970), Gamma Radiation Damage in GaAs, *Bull. Am. Phys. Soc.,* **15,** 398.

250 Brehm, G. E., and G. L. Pearson (1972), Gamma-Radiation Damage in Epitaxial Gallium Arsenide, *J. Appl. Phys.,* **43,** 569.

251 Brodin, M. S., Yu. P. Gnatenko, M. V. Kurik, and V. M. Matlak (1970), Influence of Impurities on the Optical Properties of CdTe. I. Group III and VI Impurities, *Sov. Phys.—Semicond.,* **3,** 835.

252 Brodovoi, V. A., and N. Z. Derikot (1972), Properties of GaAs:Cu in Strong Electric Fields, *Sov. Phys. Semicond.,* **6,** 237.

253 Brodovoi, V. A., and N. I. Kolesnik (1971), Photoconductivity of Cu-Doped GaAs in Strong Electric Fields, *Sov. Phys.—Semicond.,* **4,** 1770.

254 Brooks, H. (1955), Electrical Properties of Germanium and Silicon, *Advances in Electronics and Electron Physics,* Vol. 7, Academic Press, New York.

255 Broser, I., and R. Broser-Warminsky (1955), Statistisch-Kinetische Theorie der Lumineszenz und elektrischen Leitfahigkeit Von Storstellenhalbleitern, *Ann. Phys.,* **16,** 361.

256 Brotherton, S. D. (1971), Electrical Properties of Gold at the Silicon-Dielectric Interface, *J. Appl. Phys.,* **42,** 2085.

257 Brown, J. M., and A. G. Jordan (1966), Injection and Transportation of Added Carriers in Silicon at Liquid-Helium Temperatures, *J. Appl. Phys.,* **37,** 337.

258 Brown, R. A. (1966), Effect of Impurity Conduction on Electron Recombination in Germanium and Silicon at Low Temperatures, *Phys. Rev.,* **148,** 974.

259 Brown, R. A., and S. Rodriguez (1962), Low-Temperature Recombination of Electrons and Donors in *n*-Type Germanium and Silicon, *Phys. Rev.,* **153,** 890.

260 Brown, W. J. Jr., and J. S. Blakemore (1972), Transport and Photoelectrical Properties of Gallium Arsenide Containing Deep Acceptors, *J. Appl. Phys.,* **43,** 2242.

261 Brüchner, B. (1971). Electrical Properties of Gold-Doped Silicon, *Phys, Status Solidi,* **A4,** 685.

262 Bube, R. H. (1955), Photoconductivity and Crystal Imperfections in CdS Crystals, *J. Chem. Phys.,* **23,** 18.

263 Bube, R. H. (1960), *Photoconductivity of Solids,* Wiley, New York.

264 Bube, R. H., and E. L. Lind (1958), Photoconductivity of Zinc Selenide Crystals and a Correlation of Donor and Acceptor Levels in II–VI Photoconductors, *Phys. Rev.,* **110,** 1040.

265 Bube, R. H., and H. E. MacDonald (1962), Temperature Dependence of Photo-Hall Effects in High-Resistivity GaAs, *Phys. Rev.,* **128,** 2062.

266 Bube, R. H., H. E. MacDonald, and J. Blanc (1961), Photo-Hall Effects in Photoconductors, *J. Phys. Chem. Solids,* **22,** 173.

267 Bube, R. H., G. A. Dussel, C. T. Ho, and L. D. Miller (1966), Determination of Electron Trapping Parameters, *J. Appl. Phys.,* **37,** 21.

268 Buckingham, M. J., and E. A. Faulkner (1969), On the Theory of Logarithmic Silicon Diodes, *Radio. Electron. Eng.,* **38,** 33.

269 Bullis, W. M. (1966), Properties of Gold in Silicon, *Solid-State Electron.,* **9,** 143.

270 Bullis, W. M. (1968), *Measurement of Carrier Lifetime in Semiconductors: An Annotated Bibliography Covering the Period 1949–1967,* Document AD 674 627 National Technical Information Service, Springfield, Virginia.

271 Bullis, W. M., and V. Harrap (1964), Properties of Germanium in Indium Antimonide, *Physics of Semiconductors, Proceedings of the 7th International Conference, Paris, 1964,* Academic Press, New York and London, Dunod, Editeur, Paris.

272 Bullough, R., and R. C. Newman (1963), The Interaction of Impurities with Dislocations in Silicon and Germanium, *Progress in Semiconductors,* Vol. 7, Wiley, New York, p. 99.

273 Bulman, P. J., G. S. Hobson, and B. C. Taylor (1972), *Transferred Electron Devices,* Academic Press, New York and London.

274 Bulthuis, K. (1968), Anomalous Penetration of Gallium and Indium Implanted in Silicon, *Phys. Lett. A,* **27,** 193.

275 Burdiyan, I. I., S. B. Mal'tsev, I. F. Mironov, and Yu. G. Shreter (1972), Photoluminescence of Silicon-Doped Gallium Antimonide, *Sov. Phys.—Semicond.,* **5,** 1734.

276 Burger, R. M., and R. P. Donovan (1967), *Fundamentals of Silicon Integrated Device Technology;* Vol. 1. *Oxidation, Diffusion and Epitaxy,* Prentice-Hall, Englewood Cliffs, New Jersey.

277 Burgiel, J. C., and H. J. Braun (1969), Electroabsorption by Substitutional Copper Impurities in GaAs, *J. Appl. Phys.,* **40,** 2583.

278 Burmeister, R. A., and D. A. Stevenson (1967), Electrical Properties of n-Type CdSe, *Phys. Status Solidi.,* **24,** 683.

279 Burnham, R. D., P. D. Dapkus, N. Holonyak, Jr., and J. A. Rossi (1969), Laser Transition to Band Edge or to Impurity States in GaAs:Ge, *Appl. Phys. Lett.,* **14,** 190.

280 Burstein, E., J. W. Davisson, E. E. Bell, W. J. Turner, and H. G. Lipson (1954), Infrared Photoconductivity Due to Neutral Impurities in Germanium, *Phys. Rev.,* **93,** 65.

281 Burstein, E., B. Henvis, G. Picus, and R. Wallis (1956), Absorption Spectra of Impurities in Silicon-I Group III Acceptors, *J. Phys. Chem. Solids,* **1,** 65.

282 Burton, J. A. (1954), Impurity Centers in Ge and Si, *Physica,* **XX,** 845.

283 Burton, L. C., and A. H. Madjid (1969), Coulomb Screening in Intrinsic Medium-Gap Semiconductors and the Electrical Conductivity of Silicon at Elevated Temperatures, *Phys. Rev.,* **185,** 1127.

284 Buryak, E. V., S. A. Kaufman, and K. M. Kulikov (1963), Cross Section for the Capture of Holes by Singly Charged Gold Ions in Germanium, *Sov. Phys. Solid State,* **5,** 249.

285 Butcher, P. N., and W. Fawcett (1966), Stable Domain Propagation in the Gunn Effect, *Br. J. Appl. Phys.,* **17,** 1425.

286 Butcher, P. N., W. Fawcett, and C. Hilsum (1966), A Simple Analysis of Stable Domain Propagation in the Gunn Effect, *Br. J. Appl. Phys.,* **17,** 841.

287 Butuzov, V. E., and B. V. Kornilov (1969), Degeneracy Factor of the Deep Level of Zinc in Silicon, *Sov. Phys.—Semicond.,* **3,** 769.

288 Byer, N. E. (1970), Electroluminescence in Amphoteric Silicon-Doped GaAs Diodes. I. Steady State Response, *J. Appl. Phys.,* **41,** 1597.

289 Byer, N. E. (1970), Electroluminescence in Amphoteric Silicon-Doped GaAs Diodes. II. Transient Response, *J. Appl. Phys.,* **41,** 1602.

290 Bykovskii, Yu. A. (1968), Negative Resistance of Gold-Doped p-n_i-n Silicon Diodes, *Sov. Phys.—Semicond.,* **1,** 1295.

291 Bykovskii, Yu. A., K. N. Vinogradov, V. V. Zuev, and Yu. P. Kozyrev (1970), Negative Photoconductivity of Gold-Doped Silicon, *Sov. Phys.—Semicond.,* **3,** 933.

292 Bykovskii, Yu. A., K. N. Vinogradov, and V. V. Zuev (1970), Characteristic Features of the n-Type Negative Resistance and of the Negative Photoconductivity of Gold-Doped n-Type Si Samples, *Sov. Phys.—Semicond.,* **3,** 935.

293 Bykovskii, Yu. A., V. F. Elesin, and V. V. Zuev (1970), Negative Differential Conductivity in a Semiconductor Containing Attractive Impurity Centers, *Sov. Phys. —Semicond.,* **3,** 1442.

294 Cagnina, S. F. (1969), Enhanced Gold Solubility Effect in Heavily n-Type Silicon, *J. Electrochem. Soc.,* **116,** 498.

295 Carballés, J. C., D. Diguet, and J. Lebailly (1968), Electrical Properties of Solution Grown GaAs Layers, *International Symposium on Gallium Arsenide, Dallas, Texas, 1968,* Institute of Physics, London

296 Carchano, H., and C. Jund (1970), Electrical Properties of Silicon Doped with Platinum, *Solid-State Electron.,* **13,** 83.

297 Cardon, F. (1963), Impact Ionization of Centers in Solids by Electrons Accelerated by an Electric Field, *Phys. Status Solidi,* **3,** 339.

298 Cardona, M., K. L. Shaklee, and F. H. Pollak (1966), Electroreflectance at a Semiconductor-Electrolyte Interface, *Phys. Rev. Lett.,* **17,** A12.

299 Carlson, A. W. (1969), Negative Resistance in Current-Voltage Curves of Avalanche Diodes, *Proc. IEEE,* **57,** 351.

300 Carlson, R. O. (1956), Properties of Silicon Doped with Manganese, *Phys. Rev.,* **104,** 937.

301 Carlson, R. O. (1957), Double-Acceptor Behavior of Zinc in Silicon, *Phys. Rev.,* **108,** 1390.

302 Carlson, R. O., R. N. Hall, and E. M. Pell (1959), Sulfur in Silicon, *J. Phys. Chem. Solids,* **8,** 81.

303 Carter, J. R., Jr. (1966), Effect of Electron Energy on Defect Introduction in Silicon, *J. Phys. Chem. Solids,* **27,** 913.

304 Carter, W. E., H. K. Gummel, and B. R. Chawla (1972), Interpretation of Capacitance vs. Voltage Measurements of p-n Junctions, *Solid-State Electron.,* **15,** 195.

305 Casey, H. C., Jr., and G. L. Pearson (1964), Rare Earths in Covalent Semiconductors: the Thulium-Gallium Arsenide System, *J. Appl. Phys.*, **35,** 3401.

306 Casey, H. C., and F. A. Trumbore (1970), Single Crystal Electroluminescent Materials, *Mater. Sci. Eng.*, **6,** 69.

307 Casey, H. C., F. Ermanis, and K. B. Wolfstirn (1969), Variation of Electrical Properties with Zn Concentration in GaP, *J. Appl. Phys.*, **40,** 2945.

308 Casey, H. C., Jr., F. Ermanis, L. C. Luther, L. R. Dawson, and H. W. Verleur (1971), Electrical Properties of Sn-Doped GaP., *J. Appl. Phys.*, **42,** 2130.

309 Castagné, R., and A. Vapaille (1971), Description of the SiO_2—Si Interface Properties by Means of Very Low Frequency MOS Capacitance Measurements, *Surf. Sci.*, **28,** 157.

310 Castrucci, P. P., E. G. Grochowski, G. D. O'Rourke, and R. Plimley (1970), Resistivity Changes Caused by Gold Diffusion in Epitaxial *n-p-n* Transistors, *IEEE Trans. Electron Devices*, **ED-17,** 170.

311 Caywood, J. M., and C. A. Mead (1969), Origin of Field-Dependent Collection Efficiency in Contact-Limited Photoconductors, *Appl. Phys. Lett.*, **15,** 14.

312 Cecchi, R., A. Loria, M. Martini, and G. Ottaviani (1968), Electrons and Holes Drift Velocity in Silicon at Very Low Temperature, *Solid-State Commun.*, **6,** 727.

313 Chaikovskii, I. A., and V. A. Kovarskii (1968), Recombination Mechanism of Carrier Thermalization in Extrinsic Semiconductors, *Sov. Phys.—Semicond.*, **1,** 1389.

314 Champness, C. H. (1956), The Statistics of Divalent Impurity Centers in a Semiconductor, *Proc. Phys. Soc. (London)*, **B69,** 1335.

315 Chang, C. M. (1964), *Optical Absorption and Photoluminescence of Doped GaAs and* $(In_xGa_{1-x})As$, Stanford University Report, SU SEL 64-031, Document AD 442 491, National Technical Information Service, Springfield, Virginia.

316 Chang, D. M., and J. L. Moll (1966), Direct Observation of the Drift Velocity as a Function of the Electric Field in Gallium Arsenide, *Appl. Phys. Lett.*, **9,** 283.

317 Chang, Y. F. (1966), Capacitance of *p-n* Junctions: Space-Charge Capacitance, *J. Appl. Phys.*, **37,** 2337.

318 Chapnin, V. A. (1969), Electrical Properties of CdTe Containing Multiply Charged Acceptors, *Sov. Phys.—Semicond.*, **3,** 481.

319 Chattopadhyaya, S. K., and V. K. Mathur (1969), Normal and Anomalous Dember Effect, *J. Appl. Phys.*, **40,** 1930.

320 Chattopadhyaya, S. K., and V. K. Mathur (1971), Photomagnetoelectric Effect in Graded Band-Gap Semiconductors, *Phys. Rev.*, **B3,** 3390.

321 Chemeresyuk, G. G., and V. V. Serdyuk (1969), Short-Wavelength Quenching of the Longitudinal Photoconductivity of Cadmium Selenide Single Crystals, *Sov. Phys.—Semicond.*, **3,** 336.

322 Chen, R. (1969), Glow Curves with General Order Kinetics, *J. Electrochem. Soc.*, **116,** 1254.

323 Cheng, L. J., and M. L. Swanson (1970), Photoconductivity in Neutron-Irradiated *p*-Type Si, *J. Appl. Phys.*, **41,** 2627.

324 Cheng, Y. C. (1970), A Proposal Concerning the Nature of Interface States in Si/SiO_2, *Surf. Sci.*, **20,** 434.

325 Cheng, Y. C. (1970), A Model for Interface States in Silicon/Silicon Dioxide Structure. *Surf. Sci.,* **23,** 432.

326 Cherki, M., and A. H. Kalma (1970), Photoconductivity Studies of Defects in *p*-Type Silicon: Boron Interstitial and Aluminum Interstitial Defects, *Phys. Rev.* **B1,** 647.

327 Cheroff, G., J. Heer, and S. Triebwasser (1961), Trapping and Recombination Measurement by a Light Modulation Technique, *J. Phys. Chem. Solids,* **22,** 51.

328 Cherry, R. J., and J. W. Allen (1962), Excited States of Copper in Semiconductors, *Proceedings of the International Conference on the Physics of Semiconductors,* Institute of Physics, London, 1962.

329 Chiabrera, A. (1972) Semiconductors' Characterization: Kinetics of One Energy-Level Recombination Centers and Surface States, *Solid-State Electron.,* **15,** 277.

330 Chikovani, R. I., and Ya. E. Pokrovskii (1967), Determination of the Photoionization Cross Section of Negatively Charged Indium Atoms in Silicon, *Sov. Phys. Solid State,* **8,** 1856.

331 Chirkin, L. K. (1966), Negative Differential Resistance of Silicon Carbide Crystals, *Sov. Phys. Solid State,* **7,** 1787.

332 Choo, S. C. (1970), Carrier Lifetimes in Semiconductors with Two Interacting or Two Independent Recombination Levels, *Phys. Rev.* **B1,** 687.

333 Choo, S. C. (1971), Space-Charge Recombination in a Forward-Biased Diffused *p-n* Junction, *Solid-State Electron.,* **14,** 1201.

334 Choo, S. C., and E. M. Heasell (1962), Technique for the Measurement of Short Carrier Lifetimes, *Rev. Sci. Instrum.,* **33,** 1331.

335 Choyke, W. J., D. W. Feldman, and L. Patrick (1970), Donor-Acceptor Pair Spectrum in Cubic SiC, *Bull. Am. Phys. Soc.,* **15,** 279.

336 Chu, J. L., G. Persky, and S. M. Sze (1972) Thermionic Injection and Space-Charge-Limited Current in Reach-Through p^+np^+ Structures, *J. Appl. Phys.,* **43,** 3510.

337 Chua, W. B., and K. Rose (1970), Electrical Properties of High-Resistivity Nickel-Doped Silicon, *J. Appl. Phys.,* **41,** 2644.

338 Chuenkov, V. A. (1967), Theory of Impact Ionization in Ionic Semiconductors, *Sov. Phys. Solid State,* **9,** 35.

339 Chuenkov, V. A. (1968), Impact Ionization of Impurity Atoms in Semiconductors *Sov. Phys.—Semicond.,* **2,** 292.

340 Chynoweth, A. G., and K. G. McKay (1956), Photon Emission from Avalanche Breakdown in Silicon, *Phys. Rev.,* **102,** 369.

341 Cleland, J. W., and J. H. Crawford, Jr. (1970), Radiation Effects in Tellurium-Doped Germanium, *Phys. Rev.* **B1,** 713.

342 Clemetson, W. J. (1969), Self-Scanned Optical Transducer Using Acoustic Domains in CdS, *IEEE Trans. Electron Devices,* **ED-16,** 1080.

343 Coerver, L. E. (1969), New Method of Measurement of Diode Junction Parameters, *IEEE Trans. Electron Devices,* **ED-16,** 1082.

344 Cohen, M. E., and P. T. Landsberg (1966), Effect of Compensation on Breakdown Fields in Homogeneous Semiconductors, *Phys. Rev. Lett.,* **17,** A14; also (1967), *Phys. Rev.,* **154,** 683.

345 Cohen, M. M., and F. D. Bedard (1968), Electrical Properties of Single-Crystal Gallium Phosphide Doped with Zinc, *J. Appl. Phys.,* **39,** 75.

346 Coldren, L. A. (1971), Trapping Model for InSb Thin Films, *Appl. Phys. Lett.,* **18,** 319.

347 Coleman, P. D., R. C. Eden, and J. N. Weaver (1964), Mixing and Detection of Coherent Light in a Bulk Photoconductor, *IEEE Trans. Electron Devices,* **ED-11,** 488.

348 Collet, M. C. (1970), Recombination-Generation Centers Caused by 60° Dislocations in Silicon, *J. Electrochem. Soc.,* **117,** 259.

349 Collins, C. B. (1959), Silicon Current Controlling Devices, U.S. Patent No. 2,871,330.

350 Collins, C. B., and R. O. Carlson (1957), Properties of Silicon Doped with Iron or Copper, *Phys. Rev.,* **108,** 1409.

351 Collins, C. B., R. O. Carlson, and C. J. Gallagher (1957), Properties of Gold-Doped Silicon, *Phys. Rev.,* **105,** 1168.

352 Collins, D. R. (1968), The Effect of Gold on the Properties of the Si–SiO_2 System, *J. Appl. Phys.,* **39,** 4133.

353 Collins, D. R., D. K. Schroder, and C. T. Sah (1966), Gold Diffusivities in SiO_2 and Si Using the MOS Structure, *Appl. Phys. Lett.,* **8,** 323.

354 Comas, J. and E. A. Wolicki (1970), Argon Content in (III) Silicon for Sputtering Energies below 200eV, *J. Electrochem. Soc.,* **116,** 1197.

355 Comizzoli, R. B., and P. J. Warter, Jr. (1969), Hole-Trap Depletion-Layer Formation in CdS, *Phys. Rev.,* **180,** 767.

356 Constantinescu, C. R., and I. Petrescu-Prahova (1967), Acceptor Behavior of Germanium in Gallium Arsenide, *J. Phys. Chem. Solids,* **28,** 2397.

357 Conti, M., and A. Panchieri (1970), Electrical Properties of Platinum in Silicon, *J. Electrochem. Soc.,* **117,** 100C.

358 Conwell, E. M. (1956), Impurity Band Conduction in Germanium and Silicon, *Phys. Rev.,* **103,** 51.

359 Conwell, E. M. (1958), Properties of Silicon and Germanium: II, *Proc. IRE,* **46,** 1281.

360 Conwell, E. M. (1966), Electron-Hole Generation in GaAs, *Appl. Phys. Lett.,* **9,** 383.

361 Conwell, E. M. (1970), Negative Differential Conductivity, *Phys. Today,* **June 1970,** 35.

362 Conwell, E. M., and M. O. Vassel (1966), High-Field Distribution Function in GaAs, *IEEE Trans. Electron Devices,* **ED-13,** 22.

363 Cooper, J. A. Jr., E. R. Ward, and R. J. Schwartz (1972), Surface States and Insulator Traps at the Si_3N_4-GaAs Interface, *Solid-State Electron.,* **15,** 1219.

364 Copeland, J. A. (1966), Stable Space Charge Layers in Two-Valley Semiconductors, *J. Appl. Phys.,* **37,** 3602.

365 Copeland, J. A. (1968), Growth of Two- and Three-Dimensional Space Charge from Negative Differential Resistivity, *J. Appl. Phys.,* **39,** 5101.

366 Copeland, J. A. (1969), A Technique for Directly Plotting the Inverse Doping Profile of Semiconductor Wafers, *IEEE Trans. Electron Devices,* **ED-16,** 445.

367 Copeland, J. A. (1971), Semiconductor Impurity Analysis from Low-Frequency Noise Spectra, *IEEE Trans. Electron Devices,* **ED-18,** 50.

368 Corbett, J. W. (1966), *Electron Radiation Damage in Semiconductors and Metals,* Academic Press, New York.

369 Corbett, J. W. and G. D. Watkins, Ed., (1971), *Radiation Effects in Semiconductors,* Gordon and Breach, London, New York and Paris.

370 Corbett, J. W., R. S. McDonald, and G. D. Watkins (1964), The Configuration and Diffusion of Isolated Oxygen in Silicon and Germanium, *J. Phys. Chem. Solids,* **25,** 873.

371 Corbett, J. W., G. D. Watkins, and R. S. McDonald (1964), New Oxygen Infrared Bands in Annealed Irradiated Silicon, *Phys. Rev.,* **135A,** 1381.

372 Costato, M., and S. Scavo (1968), Hot-Electron Variable Effective Mass in Silicon, *Nuovo Cimento, B,* **56,** 343.

373 Cowell, T. A. T., and J. Woods (1967), The Evaluation of Thermally Stimulated Current Curves, *Br. J. Appl. Phys.,* **18,** 1045.

374 Craford, M. G., A. H. Herzog, N. Holonyak, Jr., and D. L. Keune (1970), Long Lifetime (Laser) States in *p*-Type Si-Doped GaAs, *J. Appl. Phys.,* **41,** 2648.

375 Crandall, R. S. (1968), Electrical Conduction in *n*-Type Cadmium Sulfide at Low Temperatures, *Phys. Rev.,* **169,** 577.

376 Crandall, R. S. (1971), Impurity Photoconductivity and Impact Ionization of Shallow States in Semiconductors, *J. Appl. Phys.,* **42,** 3933.

377 Cresswell, M. W. (1969), Analytical Treatment of Bulk and Surface Recombination in Minority Carrier Mobility Measurement in *p*-Si, *J. Appl. Phys.,* **40,** 2578.

378 Crowder, B. L., and F. F. Morehead, Jr. (1969), Annealing Characteristics of *n*-Type Dopants in Ion-Implanted Silicon, *Appl. Phys. Lett.,* **14,** 313.

379 Crowder, B. L., F. I. Morehead, and P. R. Wagner (1966), Efficient Injection Electroluminescence in ZnTe by Avalanche Breakdown, *Appl. Phys. Lett.,* **8,** 148.

380 Crowell, C. R., and K. Nakano (1972), Deep Level Impurity Effects on the Frequency Dependence of Schottky Barrier Capacitance, *Solid State Electron.,* **15,** 605.

381 Csavinszky, P. (1963), Corrections to the Ground State Energies of Shallow Donors in Silicon and Germanium, *J. Phys. Chem. Solids,* **24,** 1003.

382 Cunningham, R. W. (1963), Apparatus for the Measurement of the Photomagnetoelectric Effect, *Rev. Sci. Instrum.,* **34,** 83.

383 Cunningham, R. W., E. E. Harp, and W. M. Bullis (1962), Deep Acceptor Levels in Indium Antimonide, *Proceedings of the International Conference on the Physics of Semiconductors,* Institute of Physics, London, p. 732.

384 Curby, R. C., and D. K. Ferry (1971), Mobility of Hot Electrons in *n*-Type InAs, *Phys. Rev.* **B3,** 3379.

385 Curtis, O. L., Jr. (1968), Steady-State Photoconductivity in the Presence of Traps, *Phys. Rev.,* **172,** 773.

386 Cusano, D. A. (1964), Radiative Recombination from GaAs Directly Excited by Electron Beams, *Solid-State Commun.,* **2,** 353.

387 Cuthbert, J. D. (1971), Auger Electron Conductivity in Silicon, *J. Appl. Phys.,* **42,** 747.

388 Cuthbert, J. D. (1971), Luminescence and Free Carrier Decay Times in Semiconductors Containing Isoelectronic Traps, *J. Appl. Phys.*, **42,** 739.

389 Cuthbert, J. D., and R. A. Stern (1968), Free-Carrier Decay Times in Semiconductors Containing Donors and Isoelectronic Traps, *Bull. Am. Phys. Soc.*, **13,** 1475.

390 Cuthbert, J. D., C. H. Henry, and P. J. Dean (1968), Temperature-Dependent Radiative Recombination Mechanisms in GaP (Zn, O) and GaP (Cd, O), *Phys. Rev.*, **170,** 739.

391 Cuttriss, D. B. (1961), Relation between Surface Concentration and Average Conductivity in Diffused Layers in Germanium, *Bell Syst. Tech. J.*, **40,** 509.

392 Daley, D. F., and K. A. Pickar (1969), Electron Paramagnetic Resonance in Ion-Implanted Silicon, *Appl. Phys. Lett.*, **15,** 267.

393 Dapkus, P. D., et al. (1970), Spontaneous and Stimulated Carrier Lifetime (77° K) in a High-Purity, Surface-Free GaAs Epitaxial Layer, *J. Appl. Phys.*, **41,** 4194.

394 Das, P. (1970), Explanation of Current Oscillations in Cadmium Sulphide with Optically Polished Parallel End Surfaces, *Appl. Phys. Lett.*, **16,** 165.

395 Das, P., and A. J. Steckl (1970), Current Oscillations in Cadmium Sulphide with Optically Polished Parallel Surfaces, *Appl. Phys. Lett.*, **16,** 163.

396 Dascalu, D. (1968), Space-Charge Waves and High-Frequency Negative Resistance of SCL Diodes, *Int. J. Electron.*, **25,** 301.

397 Dashevskii, M. Ya., V. S. Ivleva, L. Ya. Krol', I. N. Kurilenko, L. B. Litvak-Gorskaya, R. S. Mitrofanova, and E. Yu. Fridlyand (1971), Investigation of the Behavior of Manganese in Indium Antimonide, *Sov. Phys.—Semicond.*, **5,** 757.

398 Davey, J. E., and T. Pankey (1969), Structural and Optical Evaluation of Vacuum-Deposited GaP Films, *J. Appl. Phys.*, **40,** 212.

399 Davis, E. A., and W. D. Compton (1965), Compensation Dependence of Impurity Conduction in Antimony-Doped Germanium, *Phys. Rev.*, **140,** A2183.

400 Davis, W. D. (1959), Lifetime and Capture Cross Sections in Gold-Doped Silicon, *Phys. Rev.*, **114,** 1006.

401 Davison, S. G., and J. D. Levine (1970), Surface States, in *Solid State Phys.*, Vol. 25, H. Ehrenreich, F. Seitz and D. Turnbull, Eds., Academic Press, New York and London, pp. 1-149.

402 Davydov, A. B. (1969), Effect of Electron-Electron Collisions on the Mobility of "Warm" Electrons in n-Type Ge and n-Type Si at 78°K, *Sov. Phys.—Semicond.*, **2,** 1239.

403 Dawber, P. G., and R. J. Elliott (1963), Theory of Optical Absorption by Vibrations of Defects in Silicon, *Proc. Phys. Soc.*, **81,** 453.

404 Dawber, P. G., and R. J. Elliott (1963), The Vibrations of an Atom of Different Mass in a Cubic Crystal, *Proc. Roy. Soc.*, **A273,** 222.

405 Dean, P. J. (1970), Recombination Processes Associated with "Deep States" in Gallium Phosphide, *J. Lumin.*, **1, 2,** 398.

406 Dean, P. J. (1971), Isoelectronic Trap Li-Li-O in GaP, *Phys. Rev.*, **B4,** 2596.

407 Dean, P. J., and R. A. Faulkner (1969), The Nitrogen Isoelectronic Trap in Phosphorus-Rich Gallium Arsenide Phosphide, *Appl. Phys. Lett.*, **14,** 210.

408 Dean, P. J., and R. A. Faulkner (1969), Zeeman Effect and Crystal-Field Splitting of Excitons Bound to Isoelectronic Bismuth in Gallium Phosphide, *Phys. Rev.*, **185,** 1064.

409 Dean, P. J., and C. H. Henry (1968), Electron-Capture ("Internal") Luminescence From The Oxygen Donor in Gallium Phosphide, *Phys. Rev.,* **176,** 928.

410 Dean, P. J., C. J. Frosch, and C. H. Henry (1968), Optical Properties of the Group IV Elements Carbon and Silicon in Gallium Phosphide, *J. Appl. Phys.,* **39,** 5631.

411 Dean, P. J., J. D. Cuthbert, R. A. Faulkner, and E. Schönherr (1969), The Isoelectronic Trap Antimony in Gallium Phosphide, *Bull. Am. Phys. Soc.,* **14,** 395.

412 Dean, P. J., J. D. Cuthbert, and R. T. Lynch (1969), Interimpurity Recombinations Involving the Isoelectronic Trap Bismuth in Gallium Phosphide, *Phys. Rev.,* **179,** 754.

413 Dean, P. J., G. Kaminsky, and R. B. Zetterstrom (1969), Low-Level Interband Absorption in Phosphorus-Rich Gallium Arsenide-Phosphide, *Phys. Rev.,* **181,** 1149.

414 Dean, P. J., E. A. Schönherr, and R. B. Zetterstrom (1970), Pair Spectra Involving Shallow Acceptor Mg in GaP, *J. Appl. Phys.,* **41,** 3475.

415 Dean, P. J., R. A. Faulkner, and S. Kimura (1971), Optical Properties of the Donor Tin in Gallium Phosphide, *Phys. Rev.,* **B2,** 4062.

416 Dean, P. J., R. A. Faulkner, S. Kimura, and M. Ilegems (1971), Optical Properties of Excitons Bound to Neutral Acceptors in GaP, *Phys. Rev.* **B4,** 1926.

417 Dean, R. H. (1968), Transient Double Injection in Germanium, *Appl. Phys. Lett.,* **13,** 164.

418 Dean, R. H. (1969), Transient Double Injection in Trap-Free Semiconductors, *J. Appl. Phys.,* **40,** 585.

419 Dearnaley, G., and D. C. Northrop (1966), *Semiconductor Counters for Nuclear Radiations,* 2nd ed., Wiley, New York.

420 de Biasi, R. S., and S. S. Yee (1972), Current Oscillations in n-Type Ge at Low Temperatures, *J. Appl. Phys.,* **43,** 609.

421 de Biasi, R. S., and S. S. Yee (1970), A Bulk Germanium LSA-Mode Oscillator, *Proc. IEEE,* **58,** 1404.

422 de Kock, A. J. R. (1970), Vacancy Clusters in Dislocation-Free Silicon, *Appl. Phys. Lett.,* **16,** 100.

423 de Kock, A. J. R. (1971), The Elimination of Vacancy-Cluster Formation in Dislocation-Free Silicon Carbide, *J. Electrochem. Soc.,* **118,** 1851.

424 Deuling, H. J. (1970), Double Injection Currents in Long p-i-n Diodes with One Trapping Level, *J. Appl. Phys.,* **41,** 2179.

425 Devaux, P., and M. Schott (1967), Thermally Stimulated Currents without Optical Excitation, Application to Copper Phtalocyanine, *Phys. Status Solidi,* **20,** 301.

426 Dewald, J. F. (1960), The Charge and Potential Distributions at the Zinc Oxide Electrode, *Bell Syst. Tech. J.,* **39,** 615.

427 Dick, C. L., and B. Ancker-Johnson (1969), An Explanation of Impact Ionization in InSb, *Bull. Am. Phys. Soc.,* **14,** 746.

428 Dick, C. L., and B. Ancker-Johnson (1971), Type S Bulk Negative Differential Conductivity in n-InSb, *Appl. Phys. Lett.,* **18,** 124.

429 DiDomenico, M., Jr., J. M. Dishman, and K. P. Sinha (1971), Free-to-Bound and Bound Exciton Transitions at Isoelectronic Impurities, GaP (Zn,O) *Phys. Rev.,* **B4,** 1270.

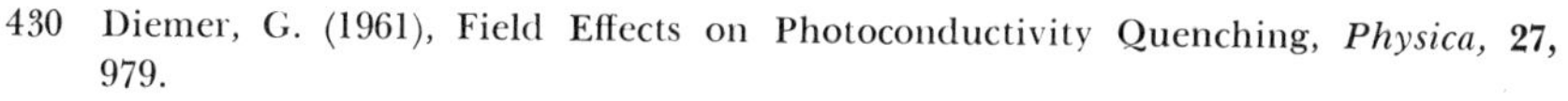

430 Diemer, G. (1961), Field Effects on Photoconductivity Quenching, *Physica,* **27,** 979.

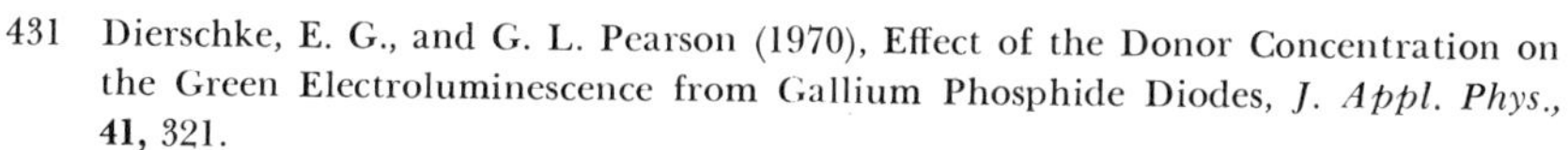

431 Dierschke, E. G., and G. L. Pearson (1970), Effect of the Donor Concentration on the Green Electroluminescence from Gallium Phosphide Diodes, *J. Appl. Phys.,* **41,** 321.

432 DiLorenzo, J. V. (1971), Analysis of Impurity Distribution in Homoepitaxial n on n^+ Films of GaAs. II., *J. Electrochem. Soc.,* **118,** 1645.

433 DiLorenzo, J. V., and G. E. Moore, Jr., (1971), Effects of the $AsCl_3$ Mole Fraction on the Incorporation of Germanium, Silicon, Selenium and Sulfur into Vapor Grown Epitaxial Layers of GaAs, *J. Electrochem. Soc.,* **118,** 1823.

434 DiLorenzo, J. V., R. B. Marcus, and R. Lewis (1971), Analysis of Impurity Distribution in Homoepitaxial n on n^+ Films of GaAs Which Contain High-Resistivity Regions, *J. Appl. Phys.,* **42,** 729.

435 Dinan, J. H., L. K. Galbraith, and T. E. Fischer (1971), Electronic Properties of Clean Cleaved [110] GaAs Surfaces, *Surf. Sci.,* **26,** 587.

436 Dingle, R. (1969), Radiative Lifetimes of Donor-Acceptor Pairs in *p*-Type Gallium Arsenide, *Phys. Rev.,* **184,** 788.

437 Dingle, R., and K. F. J. Rodgers (1969), Radiative Lifetimes in Gallium Arsenide, *Appl. Phys. Lett.,* **14,** 183.

438 Dionne, G. (1968), Nature of Stacking-Fault Defects in Epitaxial Silicon Layers, *J. Appl. Phys.,* **39,** 2940.

439 Dishman, J. M. (1970), Radiative and Non-Radiative Recombination at Oxygen in GaP (Zn,O), *Bull. Am. Phys. Soc.,* **15,** 348.

440 Dishman, J. M. (1972), Characterization of Dominant Recombination Centers in Semiconductors from the Temperature Dependence of Luminescence Excitation Spectra: GaP (Zn, O), *Phys. Rev.,* **B5,** 2258.

441 Dishman, J. M., and M. Di Domenico, Jr., (1971), Optical Absorption by Impurities in *p*-Type Gallium Phosphide, *Phys. Rev.,* **B4,** 2621.

442 Dishman, J. M., D. F. Daly, and W. P. Knox (1972), Deep Hole Traps in *n*-Type Liquid Encapsulated Czochralski GaP, *J. Appl. Phys.,* **43,** 4693.

443 Distler, G. I., and S. A. Kobzareva (1966), Direct Observation of Active Centers on Surfaces of Semiconducting Crystals, *Sov. Phys. Solid State,* **7,** 1977.

444 Dittfeld, H. J., and J. Voigt (1963), Vergleichende Untersuchungen von Lietfahigkeitsglowkurven an CdS, *Phys. Status Solidi,* **3,** 1941.

445 Dixon, J. R., and D. P. Enright (1959), Effect of Heat Treatment Upon the Electrical Properties of Indium Arsenide, *J. Appl. Phys.,* **30,** 753.

446 Djerassi, H., J. Merlo-Flores, and J. Messier (1966), Effect of ^{60}Co γ-Rays on High-Resistivity *p*-Type Si, *J. Appl. Phys.,* **37,** 4510.

447 Dmitriev, A. G., D. N. Nasledov, and B. V. Tsarenkov (1972), Impedance of Silicon-Doped *p-n* Junctions in GaAs, *Sov. Phys.—Semicond.,* **5,** 1829.

448 Dmitruk, N. L., V. A. Zuev, V. I. Lyashenko, and A. K. Tereshchenko (1970), Photoelectric Effects in the Surface Region of Gallium Arsenide, *Sov. Phys.—Semicond.,* **4,** 555.

449 Dneprovskaya, T. S., V. M. Stuchebnikov, and A. E. Yunovich (1969), Tunnel Effects in *p-n* Junctions in Gallium Antimonide. I. Electrical Properties, *Sov. Phys.—Semicond.,* **3,** 580.

450 Dobrego, V. P. (1970), Hopping Photoconductivity in Germanium with Two Types of Shallow Impurity, *Sov. Phys.—Semicond.*, **3,** 1400.

451 Dobrego, V. P., and S. M. Ryvkin (1962), Negative Photoconductivity in Germanium at the Temperature of Liquid Helium, *Sov. Phys. Solid State,* **4,** 402.

452 Dobrego, V. P., and S. M. Ryvkin (1964), Photoconductivity by a Hopping Process and Impurity Recombination, *Sov. Phys. Solid State,* **6,** 928.

453 Dobrego, V. P., and I. S. Shlimak (1969), Nonradiative Interimpurity Recombination in Germanium at 4.2°K, *Sov. Phys.—Semicond.*, **3,** 137.

454 Dobrego, V. P., Ya. Oksman, S. M. Ryvkin, and V. N. Smirnov (1965), Hopping Conductivity and Photoelectric Effect in Germanium, *Sov. Phys. Solid State,* **6,** 2275.

455 Dobrego, V. P., S. M. Ryvkin, and A. L. Shkol'nik (1965), Interimpurity Recombination in Gallium Arsenide, *Sov. Phys. Solid State,* **7,** 671.

456 Dobrego, V. P., S. M. Ryvkin, and I. S. Shlimak (1967), Influence of Uniaxial Compression on the Interimpurity Radiative Recombination in Germanium, *Sov. Phys. Solid State,* **9,** 1131.

457 Dobrego, V. P., S. M. Ryvkin, and I. S. Shlimak (1967), Radiative Interimpurity Recombination, *Sov. Phys. Solid State,* **8,** 1689.

458 Dobroval'skii, V. N., and F. W. An (1968), Negative Photoconductivity of InSb in a Magnetic Field, *Sov. Phys.—Semicond.*, **3,** 669.

459 Dobrovol'skii, V. N., and O. V. Tretyak (1969), Impedance of High-Resistivity Gallium Arsenide Samples with Nonlinear Current-Voltage Characteristics, *Sov. Phys.—Semicond.*, **3,** 57.

460 Dobson, P. S., and J. D. Filby (1968), Precipitation of Boron in Silicon, *J. Cryst. Growth,* **3, 4,** 209.

461 Doerbeck, F. H., E. E. Harp, and H. A. Strack (1968), GaAs High-Temperature Devices and Circuits, *International Symposium on Gallium Arsenide, Dallas, Texas, 1968,* Institute of Physics, London.

462 Döhler, G. (1967), On the Field Emission of Minority Carriers in Photoconductors, *Phys. Status Solidi,* **19,** 555.

463 Dolocan, V. (1969), Effects of Spatial Dependence of Recombination Centers on the *I-V* Characteristics of *p-n* Junctions, *J. Appl. Phys.*, **40,** 4095.

464 Domanevskii, D. S., and V. D. Tkachev (1971), Impurity Cathodoluminescence of Semi-Insulating Gallium Arsenide, *Sov. Phys.—Semicond.*, **4,** 1790.

465 Doo, V. Y., D. R. Nichols, and G. A. Silvey (1966), Preparation and Properties of Pyrolytic Silicon Nitride, *J. Electrochem. Soc.*, **113,** 1279.

466 Dorman, P. W. (1968), Domain Properties in GaAs Oscillating at kHz Frequencies, *Proc. IEEE,* **56,** 372.

467 Dorward, R. C., and J. S. Kirkaldy (1968), Complex Formation in the Ge–As–Cu System, *J. Appl. Phys.*, **39,** 4022.

468 Dorward, R. C., and J. S. Kirkaldy (1968), Effect of Grain-Boundaries on the Solubility of Copper in Silicon, *J. Mater. Sci.*, **3,** 502.

469 Dorward, R. C., and J. S. Kirkaldy (1969), Solubility of Gold in *p*-Type Silicon, *J. Electrochem. Soc.*, **116,** 1284.

470 Driver, M. C., and G. T. Wright (1963), Thermal Release of Trapped Space Charge in Solids, *Proc. Phys. Soc. (London),* **81,** 141.

471 Dubrovskaya, N. S., S. S. Meskin, N. F. Nedel'skii, V. N. Ravich, V. I. Sobolev, and B. V. Tsarenkov (1969), High-Efficiency Electroluminescence of Epitaxial GaAs *p-n* Structures, *Sov. Phys.—Semicond.*, **2,** 1525.

472 Duc Cuong, N., and J. Blair (1966), Impurity Photovoltaic Effect in Cadmium Sulfide, *J. Appl. Phys.*, **37,** 1660.

473 Dudenkova, A. V., and V. V. Nikitin (1967), Lifetime in Single Crystals of Gallium Arsenide, *Sov. Phys. Solid State,* **8,** 2432.

474 Dudenkova, A. V., and V. V. Nikitin (1967), Utilization of a Laser to Measure the Lifetime of Excess Carriers in Gallium Arsenide Single Crystals, *Sov. Phys. Solid State,* **9,** 664.

475 Dumke, W. P. (1963), Interband Transitions and Maser Action, *Phys. Rev.*, **127,** 1559.

476 Dumke, W. P. (1964), Theory of the Negative Resistance in *p-i-n* Diodes, *Physics of Semiconductors, Proceedings of the 7th International Conference, Paris, 1964,* Academic Press, New York and London, Dunod, Editeur, Paris, p. 611.

477 Dunlap, W. C., Jr. (1958), Germanium Current Controlling Devices, U.S. Patent No. 2,860,218.

478 Dunlap, W. C., Jr. (1954), Properties of Zinc-, Copper-, Platinum-Doped Germanium, *Phys. Rev.*, **96,** 40.

479 Dunlap, W. C., Jr. (1956), Impurities in Germanium, *Progress in Semiconductors,* Vol. 2, Wiley, New York.

480 Dushkin, V. A., B. Ya. Kovarskii, V. I. Murygin, V. V. Rakitin, and V. I. Stafeev (1971), Noise Oscillations of the Currents in *S*-Type Diodes Made of Cadmium Compensated Silicon., *Sov. Phys.—Semicond.*, **4,** 1076.

481 Dushkin, V. A., G. A. Egiazaryan, V. I. Murygin, and V. I. Stafeev, (1971), Noise Properties of Diode Structures Made of Chromium-Compensated Gallium Arsenide, *Sov. Phys.—Semicond.*, **4,** 1504.

482 Dussel, G. A., and R. H. Bube (1966a), Further Considerations on a Theory of Superlinearity in CdS and Related Materials, *J. Appl. Phys.*, **37,** 13.

483 Dussel, G. A., and R. H. Bube (1966b), Validity of Steady-State Photoconductivity Lifetime under Nonsteady-State Conditions, *J. Appl. Phys.*, **37,** 934.

484 Dussel, G. A., and R. H. Bube (1966c), Electric Field Effects in Trapping Processes, *J. Appl. Phys.*, **37,** 2797.

485 Dussel, G. A., and R. H. Bube (1966d), Theory of Thermally Stimulated Conductivity in a Previously Photoexcited Crystal, *Phys. Rev. Abstr.*, **17,** A13.

486 Dzhafarov, T. D. (1971), Neutron-Radiation Defects in Oxygen- and Carbon-Doped Silicon, *Sov. Phys.—Semicond.*, **5,** 697.

487 Ebner, G. C., and P. E. Gray (1966), Static V-I Relationships in Transistors at High Injection Levels, *IEEE Trans. Electron Devices,* **ED-13,** 692.

488 Edmond, J. T. (1959), The Behavior of Some Impurities in III–V Compounds, *Proc. Phys. Soc. (London),* **73,** 622.

489 EerNisse, E. P. (1971), Accurate Capacitance Calculations for *p-n* Junctions Containing Traps, *Appl. Phys. Lett.*, **18,** 183.

490 Egiazaryan, G. A., V. I. Gontar', and V. S. Rubin (1970), Some Investigations of *S*-Type Diodes Made of Semi-Insulating Gallium Arsenide, *Sov. Phys.—Semicond.*, **3,** 1389.

491 Eliseev, P. G., and S. G. Kalashnikov (1963), Recombination Properties of Nickel in Germanium, *Sov. Phys. Solid State,* **5,** 233.

492 Eliseev, P. G., I. Ismailov, A. B. Ormont and A. E. Yunovich (1967), Spontaneous and Coherent Photoluminescence of Indium Phosphide, *Sov. Phys.—Semicond.,* **3,** 799.

493 Eliseev, P. G., I. Ismailov, and L. I. Mikhailina (1970), Spontaneous Radiative Recombination in InP *p-n* Junctions in the Case of Small Currents, *Sov. Phys. Solid State,* **8,** 2703.

494 Elistratov, A. M., and P. R. Kamadzhiev (1961), Study of the Decomposition of a Supersaturated Solid Solution of Copper in Germanium, *Sov. Phys. Solid State,* **2,** 2621.

495 Elistratov, A. M., and P. R. Kamadzhiev (1963), X-Ray Investigation of the Decomposition of the Supersaturated Solid Solution of Ni in Ge, *Sov. Phys. Solid State,* **4,** 2557.

496 Elkin, E. L., and G. D. Watkins (1968), Defects in Irradiated Silicon: Electron Paramagnetic Resonance and Electron-Nuclear Double Resonance of the Arsenic- and Antimony-Vacancy Pairs, *Phys. Rev.,* **174,** A23.

497 Emel'yanenko, O. V., and D. N. Nasledov (1970), Magnetoresistance in Lightly Doped *n*-Type GaAs at Low Temperatures, *Sov. Phys.—Semicond.,* **3,** 1356.

498 Emel'yaneko, O. V., T. S. Lagunova, and D. N. Nasledov (1961), Impurity Zones in *p* and *n*-Type Gallium Arsenide Crystals, *Sov. Phys. Solid State,* **3,** 144.

499 Emel'yanenko, O. V., T. S. Lagunova, D. N. Nasledov, and G. N. Talalakin (1965), Formation and Properties of an Impurity Band in *n*-Type GaAs, *Sov. Phys. Solid State,* **7,** 1063.

500 Enck, R. C., and A. Honig (1969), Radiative Spectra from Shallow Donor-Acceptor Electron Transfer in Silicon, *Phys. Rev.,* **177,** 1182.

501 Engeler, W., H. Levinstein, and C. Stannard, Jr. (1961), Photoconductivity in *p*-Type Indium Antimonide with Deep Acceptor Impurities, *J. Phys. Chem. Solids,* **22,** 249.

502 Epstein, A. S., and J. F. Caldwell (1964), Degradation in Zinc-Doped GaAs Tunnel Diodes, *J. Appl. Phys.,* **35,** 2481.

503 Eriksson, L., J. A. Davies, N. G. E. Johansson, and J. W. Mayer (1969), Implantation and Annealing Behavior of Group III and V Dopants in Silicon as Studied by the Channeling Technique, *J. Appl. Phys.,* **40,** 842.

504 Ermanis, F., H. C. Casey, and K. B. Wolfstirn (1968), Thermal Ionization Energies of Cd and Zn in GaP, *J. Appl. Phys.,* **39,** 4856.

505 Ermolovich, I. B., and A. V. Lyubchenko (1968), Mechanism of the Green Edge Luminescence in CdS Single Crystals and the Parameters of Luminescence Centers, *Sov. Phys.—Semicond.,* **2,** 1364.

506 Ermolovich, I. B., N. B. Luk'yanchikova, and M. K. Sheinkman (1968), Determination of the Radiative Capture Yield of Electrons at Sensitizing Recombination Centers in CdS, *Sov. Phys. Solid State,* **9,** 2280.

507 Esina, N. P., and T. V. Ushkova (1970), Electroluminescence of *p-n* Junctions in Indium Arsenide, *Sov. Phys.—Semicond.,* **3,** 1140.

508 Evans, D. A., and P. T. Landsberg (1963), Recombination Statistics for Auger Effects with Applications to *p-n* Junctions, *Solid-State Electron.,* **6,** 169.

509 Evans, D. A., and P. T. Landsberg (1964), *Theory of the Decay of Excess Carrier Concentrations in Semiconductors*, Document 608 454, National Technical Information Service, Springfield, Virginia.

510 Fahrner, W., and A. Goetzberger (1972), Determination of Deep Energy Levels in Si by MOS Techniques, *Appl. Phys. Lett.*, **21**, 329.

511 Fairbairn, W. M. (1971), The Effect of Bulk Impurities on Localized Surface States, *Surf. Sci.*, **25**, 587.

512 Fairfield, J. M., and B. V. Gokhale (1965), Gold as a Recombination Center in Silicon, *Solid-State Electron.*, **8**, 685.

513 Fairfield, J. M., and B. V. Gokhale (1966), Control of Diffused Diode Recovery Time through Gold Doping, *Solid-State Electron.*, **9**, 905.

514 Fairfield, J. M., and B. J. Masters (1966), Self-Diffusion in Silicon, *Bull. Am. Phys. Soc.*, Ser. 2, **11**, 739.

515 Fairfield, J. M., and G. H. Schwuttke (1966), Precipitation Effects in Diffused Transistor Structures, *J. Appl. Phys.*, **37**, 1536.

516 Fang, P. H., and P. Iles (1969), Impurity Effects on Annealing of Radiation Defects in Silicon, *Appl. Phys. Lett.*, **14**, 131.

517 Fang, P. H., and T. Tanaka (1971), Interference Effect on Annealing Temperature of *A* and *E* Centers in Silicon, *J. Appl. Phys.*, **42**, 5333.

518 Fang, P. H., Y. M. Liu, J. R. Carter, Jr., and R. G. Downing (1968), Interaction of Lithium with Oxygen and Defects in Silicon, *Appl. Phys. Lett.*, **12**, 57.

519 Fang, P. H., H. Tarko, P. J. Drevinsky, and P. Iles (1970), Impurity Effects on Annealing of Radiation Defects in *p*-Type Silicon, *Appl. Phys. Lett.*, **17**, 426.

520 Fazakas, A. B., A. Friedman, and V. Novacu (1969), Carrier Pulse Propagation under Double Injection Conditions, *J. Appl. Phys.*, **40**, 764.

521 Feder, R., and T. Light (1968), Precision Thermal Expansion Measurements of Semi-Insulating GaAs, *J. Appl. Phys.*, **39**, 4870.

522 Feher, G. (1959), Electron Spin Resonance Experiments on Donors in Silicon, I. Electronic Structure of Donors by the Electron Nuclear Double Resonance Technique, *Phys. Rev.*, **114**, 1219.

523 Fekete, D. (1968), Closed-Form Solution for a Symmetrically Doped Steady-State *p-n* Junction, *J. Appl. Phys.*, **39**, 4263.

524 Feldman, J. M., and K. M. Hergenrother (1971), Radiative Recombination at Isoelectronic Point Defects in Ge, *J. Appl. Phys.*, **42**, 5563.

525 Feldman, L. C., and B. R. Appleton (1969), Unidirectional Channeling and Blocking: A New Technique for Defect Studies, *Appl. Phys. Lett.*, **15**, 305.

526 Fenimore, E., T. Mortka, and J. C. Corelli (1972), Radiation-Induced Extrinsic Photoconductivity in Li-Doped Si, *J. Appl. Phys.*, **43**, 1062.

527 Ferman, J. W. (1968), Nucleation and Growth of Li Precipitates in Si, *J. Appl. Phys.*, **39**, 3771.

528 Ferro, A. P., and S. K. Ghandhi (1970), Observations of Current Filaments in Chromium-Doped GaAs, *Appl. Phys. Lett.*, **16**, 196.

529 Ferro, A. P., and S. K. Ghandi (1970), Thermally Induced Oscillations and Negative Resistance in GaAs Double-Injection Devices, *Appl. Phys. Lett.*, **17**, 183.

530 Ferro, A. P., and S. K. Ghandhi (1971), Properties of Gallium Arsenide Double-Injection Devices, *J. Appl. Phys.*, **42**, 4015.

531 Ferry, D. K., and H. Heinrich (1968), Effect of Magnetic Fields on Impact Ionization Rates and Instabilities in InSb, *Phys. Rev. Abstr.*, **169,** A15.

532 Fertin, J. L., J. Lebailly, and E. Deyris (1966), Recent Progress in the Preparation of Very Pure Gallium Arsenide, *International Symposium on Gallium Arsenide, Dallas, Texas, 1968,* Institute of Physics, London, p. 46.

533 Fischer, H. (1961), Millimicrosecond Light Source with Increased Brightness, *J. Opt. Soc. Am.*, **51,** 543.

534 Fischler, S. (1962a), Vapor Growth and Doping of Silicon Crystals with Tellurium as Carrier, *Metallurgy of Advanced Electronic Materials,* Interscience, New York, p. 272.

535 Fischler, S. (1962b), Correlation between Maximum Solid Solubility and Distribution Coefficient for Impurities in Ge and Si, *J. Appl. Phys.*, **33,** 1615.

536 Fischler, S., and R. F. Brebrick (1962), Maximum Solubility of Impurities in Solid Ge and Si, *Bull. Am. Phys. Soc.*, **7,** 234.

537 Fisher, P., R. L. Jones, A. Onton, and A. K. Ramdas (1966), A Piezo-Spectroscopic Determination of Symmetries of Acceptor States in Silicon and Germanium, *Proceedings of the International Conference on the Physics of Semiconductors, Kyoto, Japan, 1966*; also *J. Phys. Soc. Japan, Suppl.*, **21,** 224.

538 Fistul', V. I. (1965), Determination of the Deep Copper Level in GaAs by a Tunnel Spectroscopy Method, *Sov. Phys. Solid State,* **6,** 2999.

539 Fleming, P. S. (1968), GaAs Domain Mode Frequency Memory, *Proc. IEEE,* **56,** 2082.

540 Fleming, W. J., and J. E. Rowe (1971), Emission of Microwave Noise Radiation from InSb, *J. Appl. Phys.*, **42,** 435.

541 Fogels, E. A., and C. A. T. Salama (1971), Characterization of Surface States at the Si-SiO_2 Interface Using the Quasi-Static Technique, *J. Electrochem. Soc.*, **118,** 2002.

542 Folland, N. O. (1970), Shapes of Two-Phonon Recombination Peaks in Silicon, *Phys. Rev.* **B1,** 1648.

543 Fomin, V. G., M. G. Mil'vidskii, and E. V. Solov'eva (1967), Effect of Structural Defects on Some Electrical Properties of Germanium Doped with Gold and Antimony, *Sov. Phys. Solid State,* **8,** 1795.

544 Forbes, L., and C. T. Sah (1969), Application of the Distributed Equilibrium Equivalent Circuit Model to Semiconductor Junctions, *IEEE Trans. Electron Devices,* **ED-16,** 1036.

545 Foster, J. E., and J. M. Swartz (1970), Electrical Characteristics of the Silicon Nitride-Gallium Arsenide Interface, *J. Electrochem. Soc.*, **117,** 1410.

546 Foster, L. M., and J. Scardefield (1969), Oxygen Doping of Solution-Grown GaP, *J. Electrochem. Soc.*, **116,** 494.

547 Foster, L. M., and J. Scardefield (1969), Anomalous Electrical Properties of Solution Grown, *p*-Type GaP, *Appl. Phys. Lett.*, **14,** 25.

548 Fowler, A. B. (1961), Photo-Hall Effect in CdSe Sintered Photoconductors, *J. Phys. Chem. Solids,* **22,** 181.

549 Foyt, A. G., J. P. Donnelly, and W. T. Lindley (1969), Efficient Doping of GaAs by Se^+ Ion Implantation, *Bull. Am. Phys. Soc.*, **14,** 395.

550 Foyt, A. G., W. T. Lindley, and J. P. Donnelly (1970), *n-p* Junction Photodetectors in InSb Fabricated by Proton Bombardment, *Appl. Phys. Lett.*, **16**, 335.

551 Frank, R. I., and J. G. Simmons (1967), Space-Charge Effects on Emission-Limited Current Flow in Insulators, *J. Appl. Phys.*, **38**, 832.

552 Frankl, D. R. (1967), *Electrical Properties of Semiconductor Surfaces*, Pergamon Press, New York.

553 Fredricks, W. J., F. J. Keneshea, D. S. Bloom, and A. B. Scott (1963), *Investigation of Solid State Electrolyte*, Document AD 407809, National Technical Information Service, Springfield, Virginia.

554 Frenkel, J. (1938), On the Theory of Electric Breakdown of Dielectrics and Electronic Semiconductors, *Tech. Phys. USSR*, **5**, 685; *Phys. Rev.*, **54**, 647.

555 Fritzsche, H. (1955), Electrical Properties of Germanium Semiconductors at Low Temperatures, *Phys. Rev.*, **99**, 406.

556 Fritzsche, H. (1958), Resistivity and Hall Coefficient of Antimony-Doped Germanium at Low Temperatures, *J. Phys. Chem. Solids*, **6**, 69.

557 Fu, H. S., and C. T. Sah (1969), Lumped Model Analysis of the Low-Frequency Generation Noise in Gold-Doped Silicon Junction-Gate Field-Effect Transistors, *Solid-State Electron.*, **12**, 605.

558 Fuller, C. S., and F. H. Doleiden (1958), Interactions between Oxygen and Acceptor Elements in Silicon, *J. Appl. Phys.*, **29**, 1264.

559 Fuller, C. S., and F. J. Morin (1957), Diffusion and Electrical Behavior of Zinc in Silicon, *Phys. Rev.*, **105**, 379.

560 Fuller, C. S., and K. B. Wolfstirn (1962), Diffusion, Solubility, and Electrical Behavior of Li in GaAs Single Crystals, *J. Appl. Phys.*, **33**, 2507.

561 Fuller, C. S., and K. B. Wolfstirn (1965), Investigation of Quenched-in Defects in Ge and Si by Means of ^{64}Cu, *J. Phys. Chem. Solids*, **26**, 1463.

562 Fuller, C. S., and K. B. Wolfstirn (1966a), Formation of Vacancy-Donor Associates during Cu and Ni Precipitation in Ge, *J. Phys. Chem. Solids*, **27**, 1431.

563 Fuller, C. S., and K. B. Wolfstirn (1966b), Cu-Doubling Effect in Gallium Arsenide, *J. Phys. Chem. Solids*, **27**, 1889.

564 Fuller, C. S., K. B. Wolfstirn, and H. W. Allison (1967), Hall-Effect Levels Produced in Te-Doped GaAs Crystals by Cu Diffusion, *J. Appl. Phys.*, **38**, 2873.

565 Furukawa, Y. (1967), Trap Levels in Gallium Arsenide, *Japan. J. Appl. Phys.*, **6**, 675.

566 Furukawa, Y., K. Kajiyama, and T. Aoki (1966), Copper Diffused Gallium Arsenide *p-n* Junctions, *Japan. J. Appl. Phys.*, **5**, 39.

567 Furukawa, Y., K. Kajiyama, Y. Seki, and K. Sugame (1967), Trap Density in Epitaxially Grown GaAs, *Japan. J. Appl. Phys.*, **6**, 413.

568 Furusho, K. (1964), Study on Precipitates in Oxygen-Doped Silicon Single Crystals by X-Ray Diffraction Micrography, *Japan. J. Appl. Phys.*, **3**, 203.

569 Galavanov, V. V., and V. G. Oding (1969), Influence of Compensation of a Deep Level on the Electrical Properties of *p*-Type InSb, *Sov. Phys.—Semicond.*, **3**, 238.

570 Galavanov, V. V., S. G. Metreveli, N. V. Siukaev, and S. P. Starasel'tseva (1969), The Electrical Properties of *p*-Type InP, *Sov. Phys.—Semicond.*, **3**, 94.

571 Galbraith, L. K., and T. E. Fischer (1972), Temperature- and Illumination-Dependence of the Work Function of Gallium Arsenide, *Surf. Sci.,* **30,** 185.

572 Galkin, G. N., E. L. Nolle, and V. S. Vavilov (1962), Recombination Levels in *p*-Silicon Created by High-Temperature Heat Treatment, *Sov. Phys. Solid State,* **33,** 1708.

573 Galkin, G. N., F. F. Kharakhorin, and E. V. Shatkovskii (1971), Recombination of Nonequilibrium Carriers in Indium Arsenide at High Excitation Levels, *Sov. Phys.—Semicond.,* **5,** 387.

574 Galkina, T. I., N. A. Penin, and V. A. Rassushin (1967), Determination of the Energy Position of the Cadmium Acceptor Level in Indium Arsenide, *Sov. Phys. Solid State,* **8,** 1990.

575 Gansauge, P., and W. Hoffmeister (1966), Radiotracer Measurements of Copper Contamination in GaAs from Quartz, *Solid-State Electron.,* **9,** 89.

576 Garbuny, M., T. P. Vogl, and J. R. Hansen (1957), Method for the Generation of Very Fast Light Pulses, *Rev. Sci. Instrum.,* **28,** 826.

577 Gardner, E. E., G. H. Schwuttke, and H. De Angelis (1966), Effects of High-Energy Nitrogen Bombardment on Single-Crystal Silicon, *Bull. Am. Phys. Soc., Ser. 2,* **11,** 739.

578 Gardner, W. A., G. A. Marlor, and R. H. Bube (1968), Gain Studies in Photoconductors, *J. Appl. Phys.,* **39,** 4869.

579 Garin, B. M. and V. I. Stafeev (1972), Effect of Illumination on the Static Current-Voltage Characteristic of a Long Diode, *Sov. Phys.—Semicond.,* **6,** 63.

580 Garlick, G. F. J., and A. F. Gibson (1948), The Electron Trap Mechanism of Luminescence in Sulphide and Silicate Phosphors, *Proc. Phys. Soc. (London),* **60,** 574.

581 Gaylord, T. K., P. L. Shah, and T. A. Rabson (1969), Gunn Effect Bibliography Supplement, *IEEE Trans. Electron Devices,* **ED-16,** 490.

582 Gebauhr, W. C. J. (1969), Trace Analysis in Semiconductor Silicon. A Comparison of Methods, *Semiconductor Silicon,* The Electrochemical Society, New York, p. 517; also *J. Electrochem. Soc.,* **116,** 148C.

583 George, E. V., and G. Bekefi (1969), Effects of Contacts on the Emission from Indium Antimonide, *Appl. Phys. Lett.,* **15,** 33.

584 Gerasimenko, N. N., Yu. V. Loburets, G. F. Polyakov, and L. S. Smirnov (1966), Investigation of Cathodoluminescence of Cadmium Sulfide, *Sov. Phys. Solid State,* **8,** 1165.

585 Gerasimov, A. B., and S. M. Ryvkin (1965), Current Oscillation in Germanium Containing Radiation Defects, *Sov. Phys. Solid State,* **7,** 526.

586 Gerasimov, A. B., B. M. Konovalenko, I. M. Kotina, and Kh. F. Umarova (1967), Concerning the Kinetics of Impurity Photoconductivity in Silicon Containing Radiation Defects, *Sov. Phys. Solid State,* **8,** 2390.

587 Gerasimov, A. B., B. M. Konovalenko, S. M. Ryvkin, Kh. F. Umarova, and I. D. Yaroshetskii (1967), Photoelectret States in Silicon Containing Radiation Defects, *Sov. Phys. Solid State,* **8,** 2581.

588 Gerlach, E. (1969), Localized Levels Near Semiconductor Surfaces, *Surf. Sci.,* **13,** 446.

589 Gershanov, V. Yu. (1966), Removal of Gold from Silicon by the Gettering Method, Uch. Zap. Kabardino-Balkar Gos. Univ., No. 31, 37-9.

590 Gershenzon, E. M., and L. B. Litvak-Gorskaya (1963), Investigation of the Saturation of the Drift Velocity of Hot Carriers in *p*-Type Germanium, *Sov. Phys. Solid State,* **5,** 1074.

591 Gershenzon, M., F. A. Trumbore, R. M. Mikulyak, and M. Kowalchik (1965), Radiative Recombination between Deep-Donor-Acceptor Pairs in GaP, *J. Appl. Phys.,* **36,** 1528.

592 Ghandhi, S. K., and F. L. Thiel (1969), The Properties of Nickel in Silicon, *Proc. IEEE,* **57,** 1484.

593 Ghandhi, S. K., K. E. Mortenson, and J. N. Park 1965), Impact Ionization in Cobalt-Doped Silicon, *Proc. IEEE,* **53,** 635.

594 Ghandhi, S. K., K. E. Mortenson, and J. N. Park (1966), Impact Ionization Devices, *IEEE Trans. Electron Devices,* **ED-13,** 515.

595 Ghoshtagore, R. N. (1969), Diffusion of Nickel in Amorphous Silicon Dioxide and Silicon Nitride Films, *J. Appl. Phys.,* **40,** 4374.

596 Ghoshtagore, R. N. (1971), Dopant Diffusion in Silicon III. Acceptors, *Phys. Rev.* **B3,** 2507.

597 Gibbons, J. F., and J. L. Moll (1965), The Doping of Semiconductors by Ion Bombardment, *Nucl. Instrum. Methods,* **38,** 165.

598 Gibbons, J. F., and V. K. G. Reddi (1962), Electrical Breakdown Phenomena in Gold-Doped Silicon, *IRE Prof. Group Electron Devices Trans.,* **ED-9,** 511.

599 Gibbons, P. E. (1969), Problems Concerning the Spatial Distribution of Deep Impurities in Semiconductors, *Solid-State Electron.,* **12,** 989.

600 Gibbons, P. E., and N. G. Blamires (1965), The Design and Application of Lithium Drift Silicon Diodes as Nuclear Radiation Detectors, *J. Sci. Instrum.,* **42,** 862.

601 Gilmer, T. E., R. K. Franks, and R. J. Bell (1965), An Optical Study of Lithium and Lithium–Oxygen Complexes as Donor Impurities in Silicon, *J. Phys. Chem. Solids,* **26,** 1195.

602 Ginodman, V. B., P. S. Gladkov, B. G. Zhurkin, and B. V. Kornilov (1972), EPR of Zinc Atoms in *p*-Type Silicon, *Sov. Phys.—Semicond.,* **5,** 1930.

603 Gippius, A. A., and V. S. Vavilov (1963), Radiative Recombination at Dislocations in Germanium, *Sov. Phys. Solid State,* **4,** 1777.

604 Gippius, A. A., and V. S. Vavilov (1965), Radiative Recombination at Radiation Structure Defects in Germanium, *Sov. Phys. Solid State,* **7,** 515.

605 Glazov, V. M., and V. S. Zemskov (1968), *Physicochemical Principles of Semiconductor Doping,* Israel Program for Scientific Translations, Jerusalem.

606 Glicksman, M. (1961), Instabilities of a Cylindrical Electron Hole Plasma in a Magnetic Field, *Phys. Rev.,* **124,** 1655.

607 Glicksman, M. (1962), Using Instability Characteristics of Semiconductor Plasmas, *Electronics,* **March 9, 1962,** 56.

608 Glicksman, M. (1969), Summary of Microwave Emission from InSb: Gross Features and Possible Explanations, *IBM J. Res. Dev.,* **13,** 626.

609 Glinchuk, K. D., and N. M. Litovchenko (1964a), Recombination of Current Carriers at Zinc Atoms in *n*-Type Silicon, *Sov. Phys. Solid State,* **5,** 2197.

610 Glinchuk, K. D., and N. M. Litovchenko (1964b), Trapping of Charge Carriers in Heat-Treated Silicon, *Sov. Phys. Solid State,* **5,** 2305.

611 Glinchuk, K. D., and N. M. Litovchenko (1965), Activation of Impurity Centers in Silicon, *Sov. Phys. Solid State,* **6,** 2963.

612 Glinchuk, K. D., and N. M. Litovchenko (1967), Quenching of Photoconductivity in Silicon, *Sov. Phys. Solid State,* **8,** 2011.

613 Glinchuk, K. D., and E. G. Miselyuk (1963), Cross Sections for the Capture of Electrons by Negatively Charged Atoms of the Deep Impurity Centers in Germanium, *Sov. Phys. Solid State,* **4,** 2684.

614 Glinchuk, K. D., E. G. Miselyuk, and N. N. Fortunatova (1960), A Study of the State of Silver and Gold Local Levels in Germanium, *Sov. Phys. Solid State,* **1,** 1234.

615 Glinchuk, K. D., N. M. Litovchenko, and E. G. Miselyuk (1963), Capture and Trapping of Electrons by Positively Charged Tellurium Atoms in Germanium, *Sov. Phys. Solid State,* **5,** 690.

616 Glinchuk, K. D., A. D. Denisova, and N. M. Litovchenko (1964), Recombination of Charge Carriers on Zinc Atoms in *p*-Type Silicon, *Sov. Phys. Solid State,* **5,** 1412.

617 Glinchuk, K. D., A. D. Denisova, and N. M. Litovchenko (1966a), The Photoconductivity of Silicon Doped with Deep Impurities, *Sov. Phys. Solid State,* **7,** 2963.

618 Glinchuk, K. D., N. M. Litovchenko, and V. A. Novikova (1966b), Trapping of Carriers in Plastically Deformed Silicon, *Sov. Phys. Solid State,* **8,** 777.

619 Glinchuk, K. D., N. M. Litovchenko, and L. F. Linnik (1972), Recombination of Electrons and Holes at Deep Impurity Centers in Germanium Excited by Laser Radiation, *Sov. Phys.—Semicond.,* **5,** 2088.

620 Glodeanu, A. (1967), Helium-Like Impurities in Semiconductors, *Phys. Status Solidi,* **19,** K43.

621 Gloriozova, R. I., M. I. Iglitsyn, and L. I. Kolesnik (1969), Energy Spectrum of Deep Centers in High-Resistivity Gallium Phosphide, *Sov. Phys.—Semicond.,* **3,** 790.

622 Gobrecht, H., and D. Hofmann (1966), Spectroscopy of Traps by Fractional Glow Techniques, *J. Phys. Chem. Solids,* **27,** 509.

623 Godik, É. É. (1966), Influence of an Electric Field on the Capture and Scattering of Holes in Boron-Doped Silicon, *Sov. Phys. Solid State,* **8,** 1228.

624 Godik, É. É., and Ya. E. Pokrovskii (1965), Hole Capture Coefficients of Indium and Boron Atoms in Silicon, *Sov. Phys. Solid State,* **6,** 1870.

625 Goetzberger, A., and S. M. Sze (1969), Metal Insulator Semiconductor MIS Physics, *Applied Solid State Science,* Vol. 1, R. Wolfe, Ed., Academic Press, New York, p. 154.

626 Gold, L. (1965), Role of Traps in a Steady-State Space Charge, *J. Appl. Phys.,* **36,** 2731.

627 Goldberg, C. (1969), Electric Current in a Semiconductor Space-Charge Region, *J. Appl. Phys.,* **40,** 4612.

628 Gol'dberg, Yu. A., and B. V. Tsarenkov (1972), Investigation of the Relaxation of the Capacitance of Surface-Barrier Structures and Determination of the Parameters of Impurity Atoms in *n*-Type GaAs, *Sov. Phys.—Semicond.,* **5,** 1553.

629 Goldstein, B., and S. S. Perlman (1966), Electrical and Optical Properties of High-Resistivity Gallium Phosphide, *Phys. Rev.,* **148,** 715.

630 Golikova, A. A., and A. G. Orlov (1964), Mobility of Holes in Germanium Doped with Aluminum and Indium, *Sov. Phys. Solid State,* **5,** 1393.

631 Goncharov, L. A., V. V. Emtsev, T. V. Mashovets, and S. M. Ryvkin (1972), Divacancy-Donor Complexes in Oxygen-Free Gamma-Irradiated Germanium, *Sov. Phys.—Semicond.,* **6,** 369.

632 Gontar', V. M., G. A. Egiazaryan, V. S. Rubin, V. I. Murygin, and V. I. Stafeev (1970), Some Properties of Diode Structures made of Semi-Insulating Gallium Arsenide, *Sov. Phys.—Semicond.,* **3,** 1460.

633 Gontar', V. M., G. A. Egiazaryan, V. S. Rubin, V. I. Murygin, and V. I. Stafeev (1971), Negative Differential Conductance in Illuminated High-Resistivity Gallium Arsenide, *Sov. Phys.—Semicond.,* **5,** 939.

634 Gorban', I. S., and V. M. Kosarev (1967), Radiative and Nonradiative Recombination of Minority Carriers in Gallium Phosphide, *Sov. Phys. Solid State,* **8,** 2967.

635 Gorban', I. S., Yu. M. Suleimanov, and Yu. M. Shvaidak (1969), Radiative Recombination at Deep Impurity States in Silicon Carbide, *Sov. Phys.—Semicond.,* **3,** 101.

636 Gorelenok, A. T., B. V. Tsarenkov, and N. G. Chiabrishvili (1971), Temperature Dependence of the Impurity Photoluminescence of Cr-Doped GaAs, *Sov. Phys.—Semicond.,* **5,** 95.

637 Goucher, F. S. (1951), Measurement of Hole Diffusion in *n*-Type Germanium, *Phys. Rev.,* **81,** 475.

638 Gräfe, W. (1971), Extension of the Surface Recombination Model and Unique Determination of the Recombination Energy by Means of Reverse Current Measurements, *Phys. Status Solidi* (A), 4, 655.

639 Gramberg, G. (1971), Temperature Dependence of Space Charge Capacitance of Silicon Carbide Diodes, *Solid State Electron.,* **14,** 1067.

640 Gray, P. E., and R. B. Adler (1965), A Simple Method for Determining the Impurity Distribution Near a *p-n* Junction, *IEEE Trans. Electron Devices,* **ED-12,** 475.

641 Grebene, A. B. (1968), Comments on Auger Recombination in Semiconductors, *J. Appl. Phys.,* **39,** 4866.

642 Green, D., R. H. Glaenzer, A. G. Jordan, and A. J. Noreika (1968), Observation of Parallel Arrays of Pure Edge Dislocations in Silicon, *J. Appl. Phys.,* **39,** 2937.

643 Green, M. A., V. A. K. Temple, and J. Shewchun (1972), A Simplified Computational Treatment of Recombination Centers in the Transmission Line Equivalent Circuit Model of a Semiconductor, *Solid-State Electron.,* **15,** 1027.

644 Gregory, B. L., and A. G. Jordan (1964), Experimental Investigations of Single Injection in Compensated Silicon at Low Temperatures, *Phys. Rev.,* **134,** A378.

645 Grekhov, I. V., and Yu. N. Serezhkin (1972), Avalanche Breakdown in Silicon With a High Concentration of Oxygen, *Sov. Phys.—Semicond.,* **5,** 1902.

646 Grenning, D. A., and A. H. Herzog (1968), Dislocations and Their Relation to Irregularities in Zinc-Diffused GaAsP *p-n* Junctions, *J. Appl. Phys.,* **39,** 2783.

647 Gribnikov, Z. S. (1965), Theory of Current Injection in "Long" Diodes, *Sov. Phys. Solid State,* **7,** 191.

648 Gribnikov, Z. S., and V. I. Mel'nikov (1969), Electron-Hole Scattering in Semiconductors at High Injection Levels, *Sov. Phys.—Semicond.,* **2,** 1133.

649 Grimmeiss, H. G., and G. Olofsson (1969), Charge-Carrier Capture and Its Effect on Transition Capacitance in GaP–Cu Diodes, *J. Appl. Phys.,* **40,** 2526.

650 Grinberg, A. A., and N. I. Kramer (1967), Photoionization of Shallow Impurity Levels in Semiconductors with Phonon Participation, *Sov. Phys. Solid State,* **8,** 2675.

651 Grinberg, A. A., and N. I. Kramer (1968), Scattering of Light by Charged Centers in Semiconductors, *Sov. Phys. Solid State,* **10,** 458.

652 Grinberg, A. A., L. G. Paritskii, and S. M. Ryvkin (1960), The Effect of Trapping Levels in Semiconductors on Steady State Photoconductivity and Nonequilibrium Carrier Lifetime, *Sov. Phys. Solid State,* **7,** 1403.

653 Grinberg, I. S., P. I. Knigin, and Yu. S. Korolev (1972), Oscillations of the Current in Cadmium-Compensated *n*-Type Silicon, *Sov. Phys.—Semiconduct.,* **6,** 354.

654 Grinshtein, P. M., B. A. Sakharov, and V. I. Fistul' (1969), Kinetics of the Formation of Complexes in Heavily Doped Germanium, *Sov. Phys.—Semicond.,* **2,** 1142.

655 Gross, C., G. Gaetano, T. N. Tucker, and J. A. Baker (1972), Comparison of Infrared and Activation Analysis Results in Determining the Oxygen and Carbon Content in Silicon, *J. Electrochem. Soc.,* **119,** 926.

656 Gross, E. F., and L. G. Suslina (1966), Radiation Spectrum of Donor-Acceptor Pairs in Zinc Sulfide Crystals, *Sov. Phys. Solid State,* **8,** 696.

657 Grossweiner, L. I. (1953), A Note on the Analysis of First-Order Glow Curves, *J. Appl. Phys.,* **24,** 1306.

658 Guétin, P. (1969), Interaction between a Light Beam and a Gunn Oscillator Near the Fundamental Edge of GaAs, *J. Appl. Phys.,* **40,** 4114.

659 Guichar, G. M., C. Sebenne, F. Proix, and M. Balkanski (1972), First.Order Stark Effect in Phosphorus-Doped Silicon from Photoconductivity on Impurity Levels, *Phys. Rev.,* **B5,** 422.

660 Guion, W. G., and D. K. Ferry (1971), Bulk Negative Differential Conductivity in a Near Cathode Region in *n*-Type Germanium, *J. Appl. Phys.,* **42,** 2502.

661 Gulamova, M. A., I. Z. Karimova, and P. I. Knigin (1971), Cadmium Levels in Silicon, *Sov. Phys.—Semicond.* **5,** 687.

662 Gulbransen, E. A., K. F. Andrew, and F. A. Brassart (1966), Oxidation of Silicon at High Temperatures and Low Pressure under Flow Conditions and the Vapor Pressure of Silicon, *J. Electrochem. Soc.,* **113,** 834.

663 Gulyaev, Yu. V. (1961), Contribution to the Statistics of Recombination of Electrons and Holes in Impurity Centers in Semiconductors, *Sov. Phys. Solid State,* **3,** 279.

664 Gulyaev, Yu. V. (1962), On the Theory of Current Carrier Recombination at Dislocations in Semiconductors, *Sov. Phys. Solid State,* **4,** 941.

665 Gulyaev, Yu. V., and V. V. Proklov (1968), Acoustoelectric Effect and the Amplification of Sound by a Current in a Semiconductor Containing Traps, *Sov. Phys. —Semicond.,* **1,** 1245.

666 Gulyaev, Yu. V., V. S. Ivleva, and M. I. Iglitsyn (1967), Lifetime of Excess Charge Carriers in InSb Single Crystals Containing Ge and Au Impurities, *Sov. Phys. Solid State,* **8,** 1972.

667 Gunn, J. B. (1963), Microwave Oscillations of Current in III–V Semiconductors, *Solid-State Commun.,* **1,** 88.

668 Gunn, J. B. (1969), A Topological Theory of Domain Velocity in Semiconductors, *IBM J. Res. Dev.,* **13,** 591.

669 Gupta, D. C., and J. Y. Chan (1972), Direct Measurement of Impurity Distribution in Semiconductive Materials, *J. Appl. Phys.,* **43,** 515.

670 Gurevich, L. É., and É. R. Gasanov (1970), Theory of Spontaneous Current Oscillations in Germanium-Type Crystals Doped with Gold, *Sov. Phys.—Semicond.,* **3,** 1008.

671 Gurevich, L. É., and É. R. Gasanov (1971), Amplitude of Spontaneous Oscillations in a Semiconductor Containing Deep Traps, *Sov. Phys.—Semicond.,* **4,** 1718.

672 Gurevich, L. É., I. V. Ioffe, and A. M. Kovnatskii (1967), Excitation of Oscillations in Semiconductors in the Case of Strong Inhomogeneity of the Current Density, *Sov. Phys. Solid State,* **8,** 2309.

673 Gusev, I. A., and A. N. Murin (1964), Zinc Diffusion in Indium Antimonide, *Sov. Phys. Solid State,* **6,** 932.

674 Gusev, I. A., L. I. Molkanov, and A. N. Murin (1964), Diffusion of Certain Rare Earth Elements into Germanium, *Sov. Phys. Solid State,* **6,** 980.

675 Gusev, V. M., V. I. Kurrinyi, I. I. Kruglov, I. V. Ryzhikov, B. V. Sestroretskii, and Yu. A. Sin'kov (1967), Influence of the Degree of Ionic Doping of the p-Type and n-Type Regions on the Current-Voltage and Modulation Characteristics of a Silicon p-i-n Diode, *Sov. Phys.—Semicond.,* **3,** 759.

676 Gutkin, A. A., M. B. Kagan, D. N. Nasledov, B. A. Kholev, and T. A. Shaposhnikova (1971), Photocapacitance Effect in GaAs p-n Junctions With Deep Impurity Levels, *Sov. Phys.—Semicond.,* **5,** 1006.

677 Gutkin, A. A., M. B. Kagan, A. A. Lebedev, B. A. Kholev, and T. A. Shaposhnikova (1972), Nonadditive Photoconductivity of GaAs p-i-n Structures Under Combined Excitation Conditions, *Sov. Phys.—Semicond.,* **6,** 204.

678 Guts, V. V., A. S. Kirichuk, and L. A. Kosyachenko (1969), Dependence of the Electroluminescence Brightness and Current on the Reverse Bias Applied to SiC Diodes, *Sov. Phys.—Semicond.,* **3,** 331.

679 Guttler, G., and H. J. Queisser (1969), Photovoltaic Effect of Gold in Silicon, *J. Appl. Phys.,* **40,** 4994.

680 Haake, C. H. (1957), Critical Comment on a Method for Determining Electron Trap Depths, *J. Opt. Soc. Am.,* **47,** 649.

681 Haas, C. (1960), The Diffusion of Oxygen in Silicon and Germanium, *J. Phys. Chem. Solids,* **15,** 108.

682 Habegger, M. A., and H. Y. Fan (1964), Oscillatory Intrinsic Photoconductivity of GaSb and InSb, *Phys. Rev. Lett.,* **12,** 99.

683 Hackler, W., and C. Kikuchi (1965), Effect of Radiation-Induced Dislocations on Lithium Mobility in Silicon, *Bull. Am. Phys. Soc.,* **10,** 600.

684 Hadamovsky, H. F. (1967), Distribution of Copper and Gold in Float-Zoned Single Crystal Silicon, *Kristall und Technik.,* **2,** 415.

685 Haering, R. R., and E. N. Adams (1960), Theory and Application of Thermally Stimulated Currents in Photoconductors, *Phys. Rev.,* **117,** 451.

686 Hagenlocher, A. K. (1968), Electric Field Distribution in a n^+-π-n^+ Silicon Structure in the Presence of Space-Charge-Limited Currents, *Bull. Am. Phys. Soc.,* **13,** 1676.

687 Hagenlocher, A. K. (1969), Carrier Profile in Nickel-Doped Silicon under Space-Charge-Limited Current Conditions, *Bull. Am. Phys. Soc.,* **14,** 114.

688 Hagenlocher, A. K., and W. T. Chen (1969), Space-Charge-Limited Current Instabilities in n^+-π-n^+ Silicon Diodes, *IBM J. Res. Dev.,* **13** 533.

689 Haisty, R. W. (1967), Effect of Illumination Time on Thermally Stimulated Currents in Semi-Insulating GaAs, *Appl. Phys. Lett.,* **10,** 31.

690 Haisty, R. W., and G. R. Cronin (1964), A Comparison of Doping Effects of Transition Elements in Gallium Arsenide, *Physics of Semiconductors, Proceedings of the 7th International Conference, Paris, 1964,* Academic Press, New York and London, Dunod, Editeur, Paris, p. 1161.

691 Haisty, R. W., E. W. Mehal, and R. Stratton (1962), Preparation and Characterization of High-Resistivity GaAs, *J. Phys. Chem. Solids,* **23,** 829.

692 Hakki, B. W. (1971), Theory of Luminescent Efficiency of Ternary Semiconductors, *J. Appl. Phys.,* **42,** 4981.

693 Hakki, B. W., and R. W. Dixon (1969), Phonon Frequency Spectra of Traveling Acoustoelectric Domains in CdS, *Appl. Phys. Lett.,* **14,** 185.

694 Hall, R. B., and H. H. Woodbury (1968), The Diffusion and Solubility of Phosphorus in CdTe and CdSe, *J. Appl. Phys.,* **39,** 5361.

695 Hall, R. N. (1957), Variation of the Distribution Coefficient and Solid Solubility with Temperature, *J. Phys. Chem. Solids,* **3,** 63.

696 Hall, R. N. (1959), Recombination Processes in Semiconductors, *Proc. IEEE.,* **106B,** 923.

697 Hall, R. N., and J. H. Racette (1964), Diffusion and Solubility of Copper in Extrinsic and Intrinsic Germanium, Silicon, and Gallium Arsenide, *J. Appl. Phys.,* **35,** 379.

698 Halperin, A., and A. A. Braner (1960), Evaluation of Thermal Activation Energies from Glow Curves, *Phys. Rev.,* **117,** 408.

699 Ham, W. E. (1972), Surface Charge Effects on the Resistivity and Hall Coefficient of Thin Silicon on Sapphire Films, *Appl. Phys. Lett.,* **21,** 440.

700 Ham, W. E., and K. L. Ashley (1970), Direct Observation of the Ashley-Milnes Three-Halves Power Law Potential Versus Distance in the Prebreakdown Square Law Region of Gold-Doped Germanium Double-Injection Diodes, *Appl. Phys. Lett.,* **16,** 273.

701 Ham, W. E., and K. L. Ashley (1972), Prebreakdown Behavior of Long Gold-Doped Germanium n^+pp^+ Diodes, *J. Appl. Phys.,* **43,** 149.

702 Hannay, N. B. (Ed.) (1959), *Semiconductors* Reinhold, New York.

703 Hansen, M., and K. Anderko (1958), *Constitution of Binary Alloys,* 2nd ed., McGraw-Hill, New York.

704 Hara, T., and I. Akasaki (1968), Electrical Properties of Sulfur-Doped Gallium Phosphide, *J. Appl. Phys.,* **39,** 285.

705 Harper, F. E., and B. W. Hakki (1969), Time-Decay Characteristics for the Green Emission from a GaP *p-n* Junction, *J. Appl. Phys.*, **40,** 672.

706 Harper, F. E., S. Strassler, and B. W. Hakki (1968), Time-Decay Characteristics for the Red Emission from GaP, *J. Appl. Phys.*, **39,** 3661.

707 Harper, J. G., H. E. Matthews, and R. H. Bube (1970), Two-Carrier Photothermoelectric Effects in GaAs, *J. Appl. Phys.*, **41,** 3182.

708 Harris, J. S., Y. Nannichi, and G. L. Pearson (1969), Ohmic Contacts to Solution-Grown Gallium Arsenide, *J. Appl. Phys.*, **40,** 4575.

709 Hartke, J. L. (1968), The Three-Dimensional Poole-Frenkel Effect, *J. Appl. Phys.*, **39,** 4871.

710 Hasegawa, F. (1972), Origin of the High Resistance Region at the Epitaxial GaAs Layer-Substrate Interface, *J. Electrochem. Soc.*, **119,** 930.

711 Hasiguti, R. R. (1968), *Lattice Defects in Semiconductors*, Pennsylvania State University Press, University Park, Pennsylvania.

712 Hattori, T., and H. Kanzaki (1968), Preparation and Characteristics of Ge:Cu Detector, *Japan J. Appl. Phys.*, **7,** 1541.

713 Hayakawa, H., M. Kikuchi, and Y. Abe (1966), Continuous Current Oscillation in GaAs Caused by Acoustoelectric Effect, *Japan. J. Appl. Phys.*, **5,** 734.

714 Haydl, W. H., and C. F. Quate (1965), Microwave Emission from *n*-Type Cadmium Sulphide, *Appl. Phys. Lett.*, **7,** 45.

715 Haydl, W. H., K. Harker and C. F. Quate (1967), Current Oscillations in Piezoelectric Semiconductors, *J. Appl. Phys.*, **38,** 4295.

716 Haynes, J. R., and W. Shockley (1951), The Mobility and Life of Injected Holes and Electrons in Germanium, *Phys. Rev.*, **81,** 835

717 Haynes, J. R., and W. C. Westphal (1956), Radiation Resulting from Recombination of Holes and Electrons in Silicon, *Phys. Rev.*, **101,** 1676.

718 Heine, V. (1965), Theory of Surface States, *Phys. Rev.*, **138,** A1689.

719 Hemment, P. T. F., and P. R. C. Stevens (1969), Study of the Anisotrophy of Radiation Damage Rates in *n*-Type Silicon, *J. Appl. Phys.*, **40,** 4893.

720 Henderson, H. T., and K. L. Ashley (1969a), A Negative Resistance Diode Based upon Double Injection in Thallium-Doped Silicon, *Proc. IEEE*, **57,** 1677.

721 Henderson, H. T., and K. L. Ashley (1969b), Space-Charge-Limited Current in Neutron-Irradiated Silicon with Evidence of the Complete Lampert Triangle, *Phys. Rev.*, **186,** 811.

722 Henderson, H. T., K. L. Ashley, and M. K. L. Shen (1972), Third Side of the Lampert Triangle: Evidence of Traps-Filled-Limit Single-Carrier Injection, *Phys. Rev.*, **B6,** 4079.

723 Henderson, R. C. (1972), Silicon Cleaning with Hydrogen Peroxide Solutions: A High Energy Electron Diffraction and Auger Electron Spectroscopy Study, *J. Electrochem. Soc.*, **119,** 772.

724 Henisch, H. K. (1962), *Electroluminescence*, Pergamon Press, New York.

725 Henry, C. H., P. J. Dean, and J. D. Cuthbert (1968), New Red Pair Luminescence from GaP, *Phys. Rev.*, **166,** 754.

726 Henzler, M. (1971), The Origin of Surface States, *Sur. Sci.*, **25,** 650.

727 Hergenrother, K. M., and J. M. Feldman (1969), Radiative Recombination in Tin-Doped Germanium, *J. Appl. Phys.*, **40,** 2323.

728 Herring, C., and R. G. Breckenridge (Eds.) (1956), *Photoconductivity Conference at Atlantic City 1954,* Wiley, New York.

729 Hervouet, C., J. Lebailly, P. L. Hugon, and R. Veilex (1965), Current Oscillations in GaAs under Acoustic Amplification Conditions, *Solid-State Commun.,* **3,** 413.

730 Herzog, A. H., W. O. Groves, and M. G. Craford (1969), Electroluminescence of Diffused $GaAs_{1-x}P_x$ Diodes with Low Donor Concentrations, *J. Appl. Phys.,* **40,** 1830.

731 Hesse, K., and H. Strack (1972), On the Frequency Dependence of GaAs Schottky Barrier Capacitances, *Solid State Electron.,* **15,** 767.

732 Hewes, R. A. (1968), Recombination Lifetimes in Gamma-Irradiated Silicon, *J. Appl. Phys.,* **39,** 4106.

733 Higuchi, H., and H. Tamura (1965), Measurement of the Lifetime of Minority Carriers in Semiconductors with a Scanning Electron Microscope, *Japan. J. Appl. Phys.,* **4,** 316.

734 Hill, D. E. (1970), Activation Energy of Holes in Zn-Doped GaAs, *J. Appl. Phys.,* **41,** 1815.

735 Hilsum, C. (1959), Some Effects of Copper as an Impurity in Indium Arsenide, *Proc. Phys. Soc. (London),* **73,** 685.

736 Hilsum, C., and B. R. Holeman (1961), Carrier Lifetime in GaAs, *Proceedings of the International Conference on the Physics of Semiconductors, Prague,* Academic Press, New York and London, p. 962.

737 Hilsum, C., and A. C. Rose-Innes (1961), *Semiconducting III–V Compounds,* Pergamon Press, New York.

738 Hirabayashi, K. (1971), The Effect of Adsorbed Water on the Surface Photovoltage on GaAs, *Surf. Sci.,* **82,** 627.

739 Hiramatsu, M., and S. Okazaki (1969), Double Injection in Trap-Free Silicon at Low Operating Points, *J. Appl. Phys.,* **40,** 5312.

740 Hiramatsu, M., M. Kusaka, and S. Okazaki (1970), Double Injection in Trap-Free Silicon at High Operating Points, *Japan. J. Appl. Phys.,* **9,** 854.

741 Ho, L. T., and A. K. Ramdas (1972), Excitation Spectra and Piezospectroscopic Effects of Magnesium Donors in Silicon, *Phys. Rev.,* **B5,** 462.

742 Hodgkinson, R. J. (1954), Thermal Acceptors in Germanium, *Physica,* **XX,** 1.

743 Hoffman, A. (1961), Physikalische Methoden zur Störstellenanalyse beim Silizium, *Halbeiterprobleme,* Vol. 5, Vieweg and Sohn, Braunschweig, p. 152.

744 Hofman, D., J. A. Lely, and J. Volger (1957), The Dielectric Constant of SiC, *Physica,* **XXIII,** 236.

745 Holeman, B. R., and C. Hilsum (1961), Photoconductivity in Semi-Insulating Gallium Arsenide, *J. Phys. Chem. Solids,* **22,** 19.

746 Holland, M. G., and W. Paul (1962), Effect of Pressure on the Energy Levels of Impurities in Semiconductors. I. Arsenic, Indium, and Aluminum in Silicon, *Phys. Rev.,* **128,** 30.

747 Holland, M. G., and W. Paul (1962), Effect of Pressure on the Energy Levels of Impurities in Semiconductors. III. Gold in Germanium, *Phys. Rev.,* **128,** 43.

748 Holonyak, N., Jr. (1962), Double Injection Diodes and Related DI Phenomena in Semiconductors, *Proc. IRE,* **50,** 2421.

749 Holonyak, N., Jr., and S. F. Bevacqua (1963), Oscillations in Semiconductors Due to Deep Levels, *Appl. Phys. Lett.,* **2,** 71.

750 Homma, K., Y. Kobayashi, and T. Fukami (1972), Effect of Magnetic Field on Current Filament in Gold-Doped Silicon, *Appl. Phys. Lett.,* **21,** 154.

751 Hoogenstraaten, W. (1958), Electron Traps in ZnS Phosphors, *Philips Res. Rep.,* **13,** 515.

752 Hori, J. (1962), Observations on Electroluminescence in SiC Single Crystals, *Japan. J. Appl. Phys.,* **1,** 262.

753 Hornbeck, J. A., and J. R. Haynes (1955), Trapping of Minority Carriers in Silicon, I. *p*-Type Silicon, *Phys. Rev.,* **97,** 311.

754 Hovel, H. J., and J. J. Urgell (1971), Switching and Memory Characteristics of ZnSe-Ge Heterojunctions, *J. Appl. Phys.,* **42,** 5076.

755 Hoyt, P. L., and R. W. Haisty (1966), The Preparation of Epitaxial Semi-Insulating Gallium Arsenide by Iron Doping, *J. Electrochem. Soc.,* **113,** 296.

756 Hrostowski, H. J., and R. H. Kaiser (1958), Infrared Spectra of Heat Treatment Centers in Silicon, *Phys. Rev. Lett.,* **1,** 199.

757 Hu, S. M. (1966), Properties of Amorphous Silicon Nitride Films, *J. Electrochem. Soc.,* **113,** 693.

758 Hu, S. M., and M. R. Poponiak (1972), Habit and Morphology of Copper Precipitates in Silicon, *J. Appl. Phys.,* **43,** 2067.

759 Hu, S. M., and S. Schmidt (1968), Interactions in Sequential Diffusion Processes in Semiconductors, *J. Appl. Phys.* **39,** 4272.

760 Hughes, F. D., and R. J. Tree (1970), Warm Electron Effects in GaAs at Low Temperatures, *J. Phys. C: Solid-State Phys.,* **3,** 1943.

761 Hughes, W. E. (1968), Current Controlled Negative Resistance in *n*-Type Gallium Arsenide, *Proc. IEEE,* **56,** 1715.

762 Hulme, K. F., and J. B. Mullin (1962), Indium Antimonide—A Review of Its Preparation, Properties and Device Applications, *Solid-State Electron.,* **5,** 211.

763 Hung, C. S., and J. R. Gliessman (1954), Resistivity and Hall Effect of Germanium at Low Temperatures, *Phys. Rev.,* **96,** 1226.

764 Hunsperger, R. G., and O. J. Marsh (1969), Electrical Properties of Zinc and Cadmium Ion Implanted Layers in Gallium Arsenide, *J. Electrochem. Soc.,* **116,** 488.

765 Hunsperger, R. G., and E. D. Wolf (1971), Anneal Behavior of Cd Ion Implanted GaAs, *J. Electrochem. Soc.,* **118,** 1847.

766 Hunsperger, R. G., O. J. Marsh, and C. A. Mead (1958), The Presence of Deep Levels in Ion Implanted Junctions, *Appl. Phys. Lett.,* **13,** 295.

767 Hunsperger, R. G., R. G. Wilson, and D. M. Jamba (1972), Mg and Be Ion Implanted GaAs, *J. Appl. Phys.,* **43,** 1318.

768 Huth, F. (1970), Deep Acceptor Levels in Cd-Doped GaAs, *Phys. Status Solidi* (A), **1,** K37.

769 Hutson, A. R., and D. L. White (1962), Elastic Wave Propagation in Piezoelectric Semiconductors, *J. Appl. Phys.,* **33,** 40.

770 Hwang, C. J. (1968a), Effect of Heat Treatment on the 1.370 eV Photoluminescence Emission Band in Zn-Doped GaAs, *J. Appl. Phys.,* **39,** 4307.

771 Hwang, C. J. (1968b), Photoluminescence Study of Defect Formation During Copper Diffusion in Zn-Doped GaAs, *J. Appl. Phys.,* **39,** 4313.

772 Hwang, C. J. (1968c), Photoluminescence Study of Thermal Conversion in GaAs Grown from Silica Boats, *J. Appl. Phys.,* **39,** 5347.

773 Hwang, C. J. (1969a), Evidence for Luminescence Involving Arsenic-Vacancy-Acceptor Centers in *p*-Type GaAs, *Phys. Rev.,* **180,** 827.

774 Hwang, C. J. (1969b), Optical Properties of *n*-Type GaAs. I. Determination of Hole Diffusion Length from Optical Absorption and Photoluminescence Measurements, *J. Appl. Phys.,* **40,** 3731.

775 Hwang, C. J. (1969c), Optical Properties of *n*-Type GaAs. II. Formation of Efficient Hole Traps During Annealing in Te-Doped GaAs, *J. Appl. Phys.,* **40,** 4584.

776 Hwang, C. J. (1969d), Optical Properties of *n*-Type GaAs. III. Relative Band-Edge Recombination Efficiency of Si–Te–Doped Crystals Before and After Heat Treatment, *J. Appl. Phys.,* **40,** 4591.

777 Hwang, C. J. (1971), Excitation and Doping Dependences of Electron Diffusion Length in GaAs Junction Lasers, *J. Appl. Phys.,* **42,** 757.

778 Hwang, C. J., and J. C. Dyment (1968), Correlation of GaAs Junction Laser Thresholds with Photoluminescence Measurements, *International Symposium on Gallium Arsenide, Dallas, Texas, 1968,* Institute of Physics, London.

779 Hwang, W., and K. C. Kao (1972), A Unified Approach to the Theory of Current Injection in Solids with Traps Uniformly and Non-Uniformly Distributed in Space and in Energy, and Size Effects in Anthracene Films, *Solid-State Electron.,* **15,** 523.

780 Ibragimov, N. I., M. G. Shakhtakhtinskii, and A. A. Kuliev (1963), The Influence of an Electric Field on Diffusion of Thallium in Germanium Single Crystals, *Sov. Phys. Solid State,* **5,** 632.

781 Ibragimov, V. Yu., N. M. Kolchanova, D. N. Nasledov, and G. N. Talakin (1972), Aspects of the Negative Photoconductivity in Gallium Arsenide Crystals, *Sov. Phys.—Semicond.,* **6,** 42.

782 Iglitsyn, M. I., and Yu. A. Kontsevoi (1961), Determination of the Physical Parameters of Recombination Centers Created by Copper in Germanium, *Sov. Phys. Solid State,* **2,** 1039.

783 Iglitsyn, M. I., and E. V. Solov'eva (1969), Determination of Ionization Energy of Cadmium in Indium Arsenide, *Sov. Phys.—Semicond.,* **2,** 884.

784 Iglitsyn, M. I., E. M. Kistova, Y. N. Maslov, and E. S. Yurova (1969), Some Properties of the Ternary Compound $GaAs_{1-x}P_x$ Doped with Selenium, *Sov. Phys.—Semicond.,* **3,** 19.

785 Ignatkina, R. S., I. A. Kucherenko, S. S. Meskin, V. N. Ravich, B. V. Tsarenkov, and E. G. Shevchenko (1969), Photoluminescence of Epitaxial GaP, *Sov. Phys.—Semicond.,* **2,** 1057.

786 Ignatkov, V. D., and V. E. Kosenko (1961), Evaporation of Germanium in Tellurium Vapor, *Sov. Phys. Solid State,* **3,** 65.

787 Ignatkov, V. D., and V. E. Kosenko (1962), Diffusion of Tellurium in Germanium, *Sov. Phys. Solid State,* **4,** 1193.

788 Iida, M., and M. Kimata (1968), Multiple Negative Resistance in Ge Diode, *Japan. J. Appl. Phys.,* **7,** 188.

789 Iida, M., and M. Kimata (1968), A Simplified Theory of Negative Resistance Diode, *Japan. J. Appl. Phys.,* **7,** 1078.

790 Iizuka, T. (1968), Investigation of Microprecipitates in Highly Te-Doped GaAs Crystals, *Japan. J. Appl. Phys.,* **7,** 490.

791 Iizuka, T., M. Kikuchi, and K. Kanasaki (1963), Hexagonal Platelets Observed in Nickel Diffused Silicon, *Japan. J. Appl. Phys.,* **2,** 309.

792 Ikizli, M. N., A. I. Ivashchenko, S. N. Nasledov, and S. V. Slobodchikov (1971), Electrical Properties and Electroluminescence of *p-n* Junctions in Gold- and Silver-Doped GaP, *Sov. Phys.—Semicond.,* **5,** 403.

793 Ikizli, M. N., D. N. Nasledov, and S. V. Slobodchikov (1971), Influence of Temperature and Illumination on the Current-Voltage Characteristics of GaP: Au and GaP: Ag Diodes, *Sov. Phys.—Semicond.,* **5,** 1042.

794 Ikoma, T., and B. Jeppson (1972), Determination of Hole and Electron Traps in LPE GaAs from Capacitance Measurements, *Proceedings of the Symposium on GaAs and Related Compounds, Boulder, Colorado, 1972,* Physical Society, London.

795 Ilgems, M., and W. C. O'Mara (1972), Diffusion of Beryllium into Gallium Phosphide, *J. Appl. Phys.,* **43,** 1190.

796 Im, H. B., H. E. Matthews, and R. H. Bube (1970a), Evidence for Photochemical Changes in Traps in CdS Crystals, *J. Appl. Phys.,* **41,** 2581.

797 Im, H. B., H. E. Matthews, and R. H. Bube (1970b), Evidence for Hole Traps in CdS Crystals, *J. Appl. Phys.,* **41,** 2751.

798 Imenkov, A. N., L. M. Kogan, M. M. Kozlov, S. S. Meskin, D. N. Nasledov, and B. V. Tsarenkov (1966), Influence of Impurities on the Recombination Radiation Spectra of Gallium Arsenide, *Sov. Phys. Solid State,* **7,** 2519.

799 Imenkov, A. N., M. M. Kozlov, D. N. Nasledov, and B. V. Tsarenkov (1967), Kinetics of Recombination of Excess Carriers in GaAs *p-n* Junctions, *Sov. Phys. Solid State,* **8,** 1669.

800 Ing, S. W., Jr., and G. C. Gerhard (1965), A High Gain Silicon Photodetector, *Proc. IEEE,* **53,** 1714.

801 Inoguchi, T., and T. Tanaka (1966), Photo-Effect on the Barrier Capacitance of *p-n* Junction, *Japan. J. Appl. Phys.,* **5,** 456.

802 Inoue, T., and M. Ohyama (1970), On the Temperature Dependence of Mixed Conduction in Cr-Doped GaAs, *Solid-State Commun.,* **8,** 1309.

803 Ioffe, A. V. (1968), Effect of Various Impurities on the Thermal Conductivity of the Germanium Lattice, *Sov. Phys. Solid State,* **9,** 1914.

804 Ioffe, I. V., and A. M. Kovnatskii (1966), Explanation of the Observed Field-Distribution Oscillations in Germanium with an Injecting Contact, *Sov. Phys. Solid State,* **7,** 2035.

805 Ipatova, I. P., R. F. Kazarinov, and A. V. Subashiev (1966), The Faraday Effect for "Hot" Electrons in Germanium and Silicon, *Sov. Phys. Solid State,* **7,** 1714.

806 Irmler, H. (1958), Tiefe Energieniveaus in Silicium, *Z. Naturforsch.,* **13a,** 557.

807 Irvin, J. C. (1962), Resistivity of Bulk Silicon and of Diffused Layers in Silicon, *Bell Syst. Tech. J.,* **41,** 387.

808 Ishiguro, T., S. Nitta, A. Hotta, and T. Tanaka (1965), Current Oscillation in Tellurium at High Electric Field, *Japan. J. Appl. Phys.*, **4**, 703.

809 Ishikawa, R., and T. Mitsuma (1966), Luminescent Properties of CdTe Diode, *Japan. J. Appl. Phys.*, **5**, 1171.

810 Ismailov, I. M., D. N. Nasledov, M. A. Sipovskaya, and Yu. S. Smetannikova (1969), Negative Photoconductivity of n-Type InSb, *Sov. Phys.—Semicond.* **3**, 467.

811 Ismailov, I. M., D. N. Nasledov, M. A. Sipovskaya, and Yu. S. Smetannikova (1970), Negative Photoconductivity of p-Type InSb at Low Temperature, *Sov. Phys.—Semicond.*, **3**, 1154.

812 Ismailov, I. M., S. N. Krivonogov, S. N. Nasledov, M. A. Sipovskaya, and Yu. S. Smetannikova (1971), The Features of Negative Impurity Photoconductivity in n-Type InSb, *Sov. Phys.—Semicond.*, **5**, 779.

813 Ivakhno, V. N., and D. N. Nasledov (1965), Dependence of Quantum Yield upon Photon Energy for p-n Junctions in InSb, *Sov. Phys. Solid State*, **6**, 1651.

814 Ivanov, I. L., and S. M. Ryvkin (1958), Occurrence of Current Oscillations in Specimens of Germanium Placed in an Electric Field and a Longitudinal Magnetic Field, *J. Tech. Phys. (USSR)*, **28**, 774.

815 Ivanov, M. A. (1968), Coherent Effects in Crystals Containing Electron Impurity Centers, *Sov. Phys. Solid State*, **9**, 1731.

816 Ivanov, V. G. (1967), Recombination in High-Resistivity Silicon, *Sov. Phys. Solid State*, **8**, 1306.

817 Ivanov, Yu. L. (1963), The Cross Sections for Capture of a Photon and an Electron by the Third Level of Copper in Germanium, *Sov. Phys. Solid State*, **4**, 1665.

818 Ivanov, Yu. L. (1963), Temperature-Dependence of the Activation Energy of the Second and Third Levels of Copper in Germanium, *Sov. Phys. Solid State*, **5**, 888.

819 Ivanov, Yu. L., and S. M. Ryvkin (1962), Optically Induced Charge Exchange in Impurity Centers and the Kinetics of Extrinsic Photoconductivity, *Sov. Phys. Solid State*, **4**, 1089.

820 Ivanova, E. A., D. N. Nasledov, and B. V. Tsarenkov (1964), Carrier Lifetime in the Space-Charge Layer of GaAs p-n Junctions, *Sov. Phys. Solid State*, **6**, 604.

821 Ivey, H. F. (1963), *Electroluminescence and Related Effects*, Academic Press, New York.

822 Iwauchi, S., and T. Tanaka (1968), The Effects of Traps in the Semiconductor on the Characteristics of MOS Transistors, *Japan. J. Appl. Phys.*, **7**, 1237.

823 Jackson, J. A., J. R. Szedon, and T. A. Temofonte (1972), An Effect of Organic Electron Donors and Acceptors on a Real Silicon Surface, *J. Electrochem. Soc.*, **119**, 1424.

824 Jayakumar, V., and K. L. Ashley (1972), Photoconductivity in Germanium Doped With Gold, *J. Appl. Phys.*, **43**, 4030.

825 Jayson, J. S. (1972), Analysis of Recombination of Excitons Bound to Deep Neutral Donors and Acceptors, *Phys. Rev.*, **B6**, 2372.

826 Jayson, J. S., and R. Z. Bachrach (1971), Response-Time Measurements of Exciton and Pair Radiative Recombination Associated with the Zn-O Isoelectronic Complex in GaP, 4 to 100°K, *Phys. Rev.* **B4,** 477.

827 Jayson, J. S., R. Z. Bachrach, P. D. Dapkus, and N. E. Schumaker (1972), Evaluation of the Zn–O Complex and Oxygen-Donor Electron-Capture Cross Section in *p*-Type GaP: Limits on the Quantum Efficiency of Red-Emitting (Zn–O)-Doped Material, *Phys. Rev.,* **B6,** 2357.

828 Jeffers, W. Q., and C. J. Johnson (1968), Spectral Response of the Ge:Ga Photoconductive Detector, *Appl. Opt.,* **7,** 1859.

829 Jensen, H. C., and R. J. Cashman (1954), Distribution of Trapping Levels in CdSe, *Phys. Rev.,* **96,** 798.

830 Jervis, T. R. (1970), Mobility Effects Due to Dilute Multilevel Impurities in GaAs, *J. Appl. Phys.,* **41,** 3551.

831 Johnson, L. (1960), *Germanium and Indium Antimonide Infrared Detectors,* Document AD 234 116, National Technical Information Service, Springfield, Virginia.

832 Johnson, L., and H. Levinstein (1960), Infrared Properties of Gold in Germanium, *Phys. Rev.,* **117,** 1191.

833 Johnson, R. T., Jr. (1968), Fast Neutron Irradiation Effects in CdS, *J. Appl. Phys.,* **39,** 3517.

834 Jonscher, A. K. (1957a), Drift of Minority Carriers in the Presence of Trapping, *Proc. Phys. Soc. (London),* **LXX,** 223.

835 Jonscher, A. K. (1957b), Diffusion of Minority Carriers in the Presence of Trapping, *Proc. Phys. Soc. (London),* **LXX,** 230.

836 Jordan, A. S. (1970), The Solid Solubility Isotherms of Zn in GaP and GaAs, Electrochemical Society Extended Abstract No. 145, Spring Meeting, Los Angeles, May 1970, p. 372.

837 Joshi, M. L., and S. Dash (1966), Distribution and Precipitation of Gold in Phosphorus-Diffused Silicon, *J. Appl. Phys.,* **37,** 2453.

838 Joyce, B. D., and E. W. Williams (1971), The Preparation and Photoluminescent Properties of High Purity Vapour Grown InP Layers, *Proceedings of the 3rd International Symposium on Gallium Arsenide and Related Compounds, Aachen, 1971,* The Institute of Physics, London, p. 97.

839 Kachlishvili, Z. S. (1968), Theory of Low-Temperature Breakdown in Extra-Pure Germanium, *Sov. Phys.—Semicond.,* **2,** 478.

840 Kagan, M. S., and S. G. Kalashnikov (1966), Electrical Instability in Germanium Due to Hot Electron Recombination on Repulsive Centers, *Proceedings of the International Conference on Physics of Semiconductors, Kyoto, 1966, J. Phys. Soc. Japan,* Suppl., **21,** 537.

841 Kagan, M. B., and T. L. Lyubashevskaya (1971), Influence of Recombination Processes in the Space-Charge Region on the Photoconductivity Spectrum of a Heterostructure, *Sov. Phys.—Semicond.,* **4,** 1217.

842 Kagan, M. S., S. G. Kalashnikov, V. A. Kemarskii, and E. G. Landsberg (1972), Influence of Optical Charge Exchange Between Impurity Levels on the Nonlinearity of the Current-Voltage Characteristics of Germanium, *Sov. Phys.—Semicond.,* **5,** 1780.

843 Kahng, D. (1968), Electroluminescence of Rare-Earth and Transition Metal Molecules in II–VI Compounds Via Impact Excitation, *Appl. Phys. Lett.*, **13**, 210.

844 Kaiser, W. (1957), Electrical and Optical Properties of Heat-Treated Silicon, *Phys. Rev.*, **105**, 1751.

845 Kaiser, W., and G. H. Wheatley (1959), Hot Electrons and Carrier Multiplication in Silicon at Low Temperature, *Phys. Rev. Lett.*, **3**, 334.

846 Kaiser, W., P. H. Keck, and C. F. Lange (1956), Infrared Absorption and Oxygen Content in Silicon and Germanium, *Phys. Rev.*, **101**, 1264.

847 Kaiser, W., H. L. Frisch, and H. Reiss (1958), Mechanism of the Formation of Donor States in Heat-Treated Silicon, *Phys. Rev.*, **112**, 1546.

848 Kalashnikov, S. G., and K. Konstantinesku (1960), A Possible Method of Determining the Ratio of Capture Cross Sections of Recombination Centers in Semiconductors, *Sov. Phys. Solid State*, **1**, 1612.

849 Kalashnikov, S. G., and A. K. Mednikov (1961), The Effect of a Copper-Impurity Atmosphere at Dislocations in Germanium on Recombination, *Sov. Phys. Solid State*, **2**, 1847.

850 Kalashnikov, S. G., and A. I. Morozov (1961), Investigation of Trapping at Copper Atoms in Germanium, *Sov. Phys. Solid State*, **2**, 2505.

851 Kalashnikov, S. G., and K. P. Tissen (1959), Recombination of Electrons and Holes at Nickel Atoms in Germanium, *Sov. Phys. Solid State*, **1**, 491.

852 Kalashnikov, S. G., and K. P. Tissen (1960), The Cross Section of Capture of Electrons and Holes by Atoms of Nickel in Germanium, *Sov. Phys. Solid State*, **1**, 1603.

853 Kalashnikov, S. G., and K. P. Tissen (1961), Capture and Recombination at Multielectron Capture Centers in Semiconductors, *Sov. Phys. Solid State*, **2**, 2443.

854 Kalashnikov, S. G., A. I. Morozov, and B. A. Stankovskii (1967), Effect of Traps on the Amplification of Ultrasound on Cadmium Sulfide, *Sov. Phys. Solid State*, **9**, 670.

855 Kalashnikov, S. G., G. S. Pado, V. I. Pustovoit, and E. F. Tokarev (1970), Electrothermal Instability in Cadmium Selenide Single Crystal, *Sov. Phys.—Semicond.*, **3**, 864.

856 Kalashnikov, S. G., R. G. Maev, and V. I. Pustovoit (1971), Theory of Thermal Quenching of the Photoconductivity, *Sov. Phys.—Semicond.*, **5**, 456.

857 Kalikstein, K., B. Kramer, and S. Gelfman (1968), Investigation of Photoconductivity in ZnS and CdS Phosphors with Microwave Methods, *J. Appl. Phys.*, **39**, 4252.

858 Kal'nin, A. A., V. V. Pasynkov, Yu. M. Tairov, and D. A. Yas'kov (1963), *Investigation of p-Junctions Produced on the Basis of Silicon Carbide Alloyed with Beryllium*, Document AD 678 302, National Technical Information Service, Springfield, Virginia.

859 Kal'nin, A. A., Yu. M. Tairov, and D. A. Yas'kov (1966), Luminescence of Silicon Carbide Containing Beryllium Impurity, *Sov. Phys. Solid State*, **8**, 755.

860 Kal'nin, A. A., V. V. Pasynkov, Yu. M. Tairov, and D. A. Yas'kov (1967), Photoluminescence of Silicon Carbide with Beryllium Impurity, *Sov. Phys. Solid State*, **8**, 2381.

861 Kalyuzhnaya, G. A., Ya. A. Oksman, V. N. Smirnov, and Yu. V. Shmartsev (1964), Investigation of the Photoconductivity of GaP by the Contactless Method, *Sov. Phys. Solid State,* **6,** 915.

862 Kanazawa, K. K., and I. P. Batra (1972), Deep-Trapping Kinematics, *J. Appl. Phys.,* **43,** 1845.

863 Kanda, Y., Y. Kato and K. Narito (1972), Effects of Impurity Compensation on Electrical Properties of *p*-Type Germanium Under Hydrostatic Pressure, *J. Appl. Phys.,* **43,** 490.

864 Kar, S., and W. E. Dahlke (1972), Interface States in MOS Structures with 20-40 Å Thick SiO_2 Films on Nondegenerate Si, *Solid-State Electron.,* **15,** 221.

865 Karageorgii-Alkalaev, P. M., and A. Yu Leiderman (1968), Theory of Kinetic Oscillations in Semiconductors, *Sov. Phys. Solid State,* **9,** 1691.

866 Karakushan, E. I., and V. I. Stafeev (1961), Magnetodiodes, *Sov. Phys. Solid State,* **3,** 493.

867 Karkhanin, Yu. I., and O. V. Tretyak (1966), Photocurrent Oscillations in High-Resistance GaAs, *Sov. Phys. Solid State,* **7,** 2787.

868 Karpenko, V. P., P. G. Kasherininov, and O. A. Matveev (1972), Effect of Charge Exchange Between Deep Centers Near a *p-n* Junction on the Properties of Germanium Gamma-Ray Detectors, *Sov. Phys.—Semicond.,* **6,** 93.

869 Karpova, I. V., and Ya. E. Pokrovskii (1965), Radiative Capture of Carriers by Neutral Indium and Antimony Atoms in Germanium, *Sov. Phys. Solid State,* **6,** 2905.

870 Karpova, I. V., V. G. Alekeseeva, and S. G. Kalashnikov (1962), Recombination Properties of Gold in *n*-Germanium, *Sov. Phys. Solid State,* **4,** 461.

871 Kasamanyan, Z. A. (1969), Effect of an Impurity on the Energy Spectrum of Electrons in Semiconductors, *Sov. Phys.—Semicond.,* **2,** 1425.

872 Kasherininov, P. G., V. P. Karpenko, and O. A. Matveev (1971), Photoelectret Effect in Semiconductors, *Sov. Phys.—Semicond.,* **5,** 51.

873 Kasperovich, N. S., B. M. Nikolaev, and Ya. A. Oksman (1967), The Photodielectric Effect in Compensated Germanium, *Sov. Phys. Solid State,* **9,** 343.

874 Kastal'skii, A. A., T. Risbaev, I. M. Fishman, and Yu. G. Shreter (1972), Role of Double Acceptors in Radiative Recombination in *n*-Type GaSb, *Sov. Phys.—Semicond.,* **5,** 1391.

875 Kato, H., and S. Takayanagi (1963), Diffusion of Indium in Cadmium Telluride, *Japan. J. Appl. Phys.,* **2,** 250.

876 Kato, H., M. Yokozawa, and S. Takayanagi (1965), Diffusion of Se in CdTe, *Japan. J. Appl. Phys.,* **4,** 1019.

877 Katrich, G. A., O. G. Sarbei, and D. T. Tarashchenko (1965), Surface Conductivity of Gold-Doped Germanium, *Sov. Phys. Solid State,* **7,** 1091.

878 Katsitadze, A. P., Z. S. Kachlishvili, and V. A. Morozova (1971), Influence of an Electric Field on Recombination in Ge Doped with Deep Impurities, *Sov. Phys.—Semicond.,* **4,** 1486.

879 Kaufman, S. A., K. M. Kulkov, N. P. Likhtman, and N. Sh. Khaikin (1965), Some Characteristics of the Photoconductivity Kinetics in High-Resistivity *n*-Type Germanium Samples Doped with Gold, *Sov. Phys. Solid State,* **7,** 668.

880 Kaufman, S. A. N. Sh. Khaikin, and G. T. Yakovleva (1969), Influence of Injecting Contacts on the Impurity Photoconductivity Kinetics, *Sov. Phys.—Semicond.*, **3,** 485.

881 Kavalyauskene, G. S., and V. S. Rimkyavichyus (1969), Thermally Stimulated Capacitor Discharge Method, *Sov. Phys.—Semicond.*, **3,** 378.

882 Kazarinov, R. F., and V. V. Osipov (1969), Injection Breakdown in a Long Diode, *Sov. Phys.—Semicond.*, **3,** 690.

883 Kazarinov, R. F., V. I. Stafeev, and R. A. Suris (1967a), Negative Differential Resistance of Germanium with Deep Levels During the Injection of Minority Carriers, *Sov. Phys.—Semicond.*, **1,** 1078.

884 Kazarinov, R. F., V. I. Stafeev, and R. A. Suris (1967b), Impedance and Transient Processes in Germanium Diodes with Deep Levels, *Sov. Phys.—Semicond.*, **1,** 1084.

885 Kazarinov, R. F., R. A. Suris, and V. P. Kholodnov (1970), Lux-Ampere Characteristic of a Germanium *p-n-p* Phototransistor with a Gold-Compensated Base, *Sov. Phys.—Semicond.*, **3,** 1363.

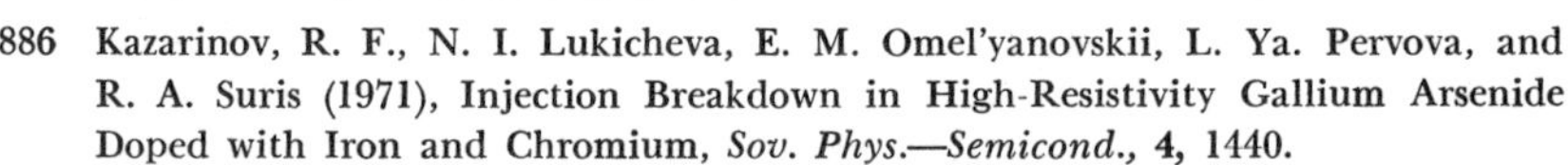

886 Kazarinov, R. F., N. I. Lukicheva, E. M. Omel'yanovskii, L. Ya. Pervova, and R. A. Suris (1971), Injection Breakdown in High-Resistivity Gallium Arsenide Doped with Iron and Chromium, *Sov. Phys.—Semicond.*, **4,** 1440.

887 Kazarinov, R. F., R. A. Suris, and A. V. Shenkerman (1971), Electrical Properties of Junctions Formed by Impurities with Deep Levels, *Sov. Phys.—Semicond.*, **4,** 1794.

888 Kazarinov, R. F., R. A. Suris, and B. I. Fuks (1972), "Thermal-Current" Instability in Compensated Semiconductors, *Sov. Phys.—Semicond.*, **6,** 500.

889 Keating, P. N. (1961), Thermally Stimulated Emission and Conductivity Peaks in the Case of Temperature Dependent Trapping Cross Sections, *Proc. Phys. Soc. (London)*, **78,** 1408.

890 Keating, P. N. (1964), Effect of Shallow Trapping and the Thermal-Equilibrium Recombination Center Occupancy on Double-Injection Currents in Insulators, *Phys. Rev.*, **135,** A1407.

891 Kendall, D. L. (1968), Diffusion, *Semiconductors and Semimetals*, Vol. 4, Academic Press, New York, p. 163.

892 Kendall, D. L., and D. B. De Vries (1969), Diffusion in Silicon, *Semiconductor Silicon*, The Electrochemical Society, New York, p. 358.

893 Kenigsberg, N. L., and A. N. Chernets (1969), Preparation and Properties of Thin Films of ZnO, *Sov. Phys. Solid State*, **10,** 2235.

894 Kennedy, D. I., E. S. Koteles, and W. A. Webb (1969), Voltage Dependence of Electroluminescence from GaP Diodes Prepared by Liquid Epitaxial Techniques, *J. Appl. Phys.*, **40,** 875.

895 Kennedy, D. P. (1969), On the Ambipolar Diffusion of Impurities into Silicon, *Proc. IEEE*, **57,** 1203.

896 Kern, W., and J. P. White (1970), Interface Properties of Chemically Vapor Deposited Silica Films on Gallium Arsenide, *RCA Rev.*, **31,** 771.

897 Kesamanly, F. P., E. E. Klotyn'sh, T. S. Lagunova and D. N. Nasledov (1964), The Impurity Zone in *n*-InP Crystals, *Sov. Phys. Solid State*, **6,** 741.

898 Kessler, H. K., and N. N. Winogradoff (1966), Surface Aspects of the Thermal Degradation of GaAs *p-n* Junction Lasers and Tunnel Diodes, *IEEE Trans. Electron Devices,* **ED-13,** 688.

899 Keune, D. L., *et al.* (1971), Optical Phase-Shift Measurement of Carrier Decay Times (77°K) on Lightly Doped Double-Surface and Surface-Free Epitaxial GaAs, *J. Appl. Phys.,* **42,** 204.

900 Keyes, R. J. (1960), *Photo and Thermal Effects in Compensated Zinc-Doped Germanium,* Document AD 245 671, National Technical Information Service, Springfield, Virginia.

901 Kharlamova, T. E., and D. A. Tairova (1964), Effect of Radiation from Radioisotopes on the Characteristics of Silicon Carbide *p-n* Junctions, *Sov. Phys. Solid State,* **6,** 503.

902 Kholodar', G. A., and V. L. Vinetskii (1966), Self-Compensation of Conductivity in Semiconductors, *Sov. Phys. Solid State,* **8,** 676.

903 Kholuyanov, G. F. (1966), The Roles of Boron, Nitrogen and Gallium in the Electroluminescence of Silicon Carbide *p-n* Junctions, *Sov. Phys. Solid State,* **7,** 2620.

904 Khukhryanskii, Yu. P. (1964), Solubility of Indium in Germanium at 350–550°C, *Sov. Phys. Solid State,* **6,** 1222.

905 Kichigin, D. A., and V. P. Lobachev (1966), Negative Conductance in Nickel-Doped Germanium, *Sov. Phys. Solid State,* **8,** 200.

906 Kiess, H. (1967), Theoretical Considerations Concerning Saturation of Photocurrents, *J. Phys. Chem. Solids,* **82,** 1473.

907 Kikuchi, Y., N. Chubachi, and K. Iinuma (1967), Temperature Dependence of Electron Drift Mobility for Ultrasonic Amplification in Cadmium Sulfide in Relation to Electron Trapping Effects, *Japan. J. Appl. Phys.,* **6,** 1251.

908 Kimata, M., and M. Iida (1968), Current Flow in Negative Resistance Diode, *Japan. J. Appl. Phys.,* **7,** 177.

909 Kimata, M., and K. Kani (1965), A Simple Model for Negative Resistance by Impact Ionization of Deep Level Impurities, *Japan. J. Appl. Phys.,* **4,** 737.

910 Kingston, R. H., Ed. (1957), *Semiconductor Surface Physics,* University of Pennsylvania Press, Philadelphia.

911 Kirov, K., and V. Zhelev (1965), Study of the Effects of an Electric Field on Trap Filling in CdS by the use of Thermally Stimulated Currents, *Phys. Status Solidi,* **8,** 431.

912 Klaassen, F. M. (1962), Carrier Density Fluctuations in a Two-Level Impurity Semiconductor, *Phys. Status Solidi,* **2,** 299.

913 Klaassen, F. M., J. Blok, N. G. Booy, and F. J. de Hoog (1961), Generation-Recombination Noise in Various Photoconductive Semiconductors, *J. Phys. Chem. Solids,* **22,** 391.

914 Klawiter, W. F., and S. Nakai (1970), Calculation of the Conductivity of Impurity Band Conduction in Semiconductors, *Bull. Am. Phys. Soc.,* **15,** 304.

915 Klein, W. (1969), Photoemission from Cesium Oxide Covered GaInAs, *J. Appl. Phys.,* **40,** 4384.

916 Kleinfelder, W. J. (1967), *Properties of Ion-Implanted Boron Nitrogen and Phosphorus in Single-Crystal Silicon,* Stanford University Electronics Laboratories, Report K701-1, SU-SEL-67-015.

917 Klimenko, A. P., and Yu. O. Tkhorik (1965), *Investigating Recombination on Nickel Atoms in p-Germanium at High Injection Levels,* Document AD 621 005, National Technical Information Service, Springfield, Virginia.

918 Klimka, L. A., and K. D. Glinchuk (1970), Recombination of Hot Holes at Nickel Ions in Germanium, *Sov. Phys.—Semicond.*, **4,** 1361.

919 Klimka, L. A., S. P. Kal'venas, and Yu. K. Pozhela (1971), Temperature Dependence of the Capture Coefficient of Hot Electrons in Ni-Doped *n*-Type Ge, *Sov. Phys.—Semicond.,* **5,** 891.

920 Knab, O. D., A. I. Petrov, V. D. Frolov, V. I. Shveikin, and I. A. Shmerkin (1972), Decay of Photoluminescence Emitted by Gallium Arsenide, *Sov. Phys.—Semicond.,* **5,** 1429.

921 Knepper, R. W. (1969), Electrical Characteristics and Fluctuation Phenomena in Semiconductor Double Injection Devices, Ph.D. Thesis, Carnegie-Mellon University, Pittsburgh, Pennsylvania.

922 Knepper, R. W., and A. G. Jordan (1972a), Double Injection in *p-π-n* Silicon Devices, *Solid-State Electron.,* **15,** 45.

923 Knepper, R. W., and A. G. Jordan (1972b), Electrical Fluctuations in Silicon Double Injection Devices, *Solid State Electron.*, **15,** 59.

924 Kodera, H. (1963), Diffusion Coefficients of Impurities in Silicon Melt, *Japan J. Appl. Phys.,* **2,** 212.

925 Kodera H. (1964), Solid Solubility of Gold in Germanium, *Japan J. Appl. Phys.,* **3,** 369.

926 Koenig, S. H. (1959), Hot and Warm Electrons—a Review, *J. Phys. Chem. Solids,* **8,** 227.

927 Koenig, S. H., and G. R. Gunther-Mohr (1957), The Low-Temperature Electrical Conductivity of *n*-Type Germanium, *J. Phys. Chem. Solids,* **2,** 268.

928 Kogan, L. M., S. S. Meskin, and A. Ya. Goikhman (1964), Diffusion of Cadmium and Zinc into Gallium Arsenide from the Gaseous Phase, *Sov. Phys. Solid State,* **6,** 882.

929 Kogan, Sh. M., T. M. Lifshits, and V. I. Sidorov (1965), Photoconductivity Resulting from Optical Transitions between Impurity Centers, *Sov. Phys. Solid State,* **6,** 2636.

930 Kohn, W. (1957), Shallow Impurity States in Silicon and Germanium, *Solid State Phys.,* **5,** 257.

931 Kolchanova, N. M., and G. N. Talalakin (1966), Anomalous Behavior of the Mobility in Gallium Arsenide Doped with Oxygen, *Sov. Phys. Solid State,* **7,** 2522.

932 Kolchanova, N. M., D. N. Nasledov, and G. N. Talalakin (1970), Properties of Gallium Arsenide Doped with Iron and Nickel, *Sov. Phys.—Semicond.,* **4,** 106.

933 Kolchanova, N. M., D. N. Nasledov, M. A. Mirdzhalilova, and V. Yu. Ibragimov (1970), Impurity Photoconductivity of Gallium Arsenide Crystals, *Sov. Phys.—Semicond.,* **4,** 294.

934 Kolesnik, L. I. (1962), Recombination at Dislocation Lines in Germanium, *Sov. Phys. Solid State,* **4,** 1066.

935 Kolomeev, M. P., and N. Sh. Khaikin (1972), Investigation of the Kinetics of the Impurity Photoconductivity in High-Resistivity Compensated Semiconductors, *Sov. Phys.—Semicond.,* **6,** 324.

936 Konopleva, R. F., and S. R. Novikov (1969), Influence of Copper Impurity and of Dislocations on the Formation of Defects in Germanium and Silicon under Fast-Neutron Irradiation, *Sov. Phys—Semicond.*, **2**, 1080.

937 Konopleva, R. F., S. R. Novikov, and E. E. Rubinova (1966), Defects in Silicon Produced by Irradiation with Fast Neutrons at 77°K, *Sov. Phys. Solid State*, **8**, 264.

938 Konopleva, R. F., S. R. Novikov, E. E. Rubinova, Yu. A. Zaporozchenko, V. N. Pokrovskii, and L. N. Nikityuk (1970), Radiation Defects in Germanium Irradiated with High Frequency Protons, *Sov. Phys.—Semicond.*, **3**, 948.

939 Konstantinesku, K. (1960), Determination of the Ratio of the Electron and Hole Capture Cross Sections of Copper Atoms in Germanium, *Sov. Phys. Solid State*, **1**, 1615.

940 Konstantinov, O. V., and V. I. Perel' (1965), Recombination Waves in Semiconductors, *Sov. Phys. Solid State*, **6**, 2691.

941 Konstantinov, O. V., V. I. Perel', and B. V. Tsarenkov (1968), Calculations of the Conditions for the Excitation of Recombination Waves in Germanium Containing Compensating Impurities, *Sov. Phys. Solid State*, **10**, 686.

942 Kornilov, B. V. (1963), Determination of Effective Capture Cross Sections of Majority Carriers by Copper Atoms in Germanium by the Nonsteady-State Extrinsic Photoconductivity, *Sov. Phys. Solid State*, **4**, 1771.

943 Kornilov, B. V. (1964a), Absorption in Silicon Doped with Zinc, *Sov. Phys. Solid State*, **5**, 2420.

944 Kornilov, B. V. (1964b), Selective Insensitivity of a *p-n* Junction in Deep-Lying Zinc Levels in Silicon, *Sov. Phys. Solid State*, **5**, 2441.

945 Kornilov, B. V. (1964c), Phenomenon of Impulse Ionization of a Deep Zinc Level in *p*-Type Silicon, *Sov. Phys. Solid State*, **6**, 268.

946 Kornilov, B. V. (1965a), Optically Induced Charge Exchange in Zinc Impurity Levels in Silicon, *Sov. Phys. Solid State*, **6**, 2982.

947 Kornilov, B. V. (1965b), The Phenomenon of the Generation of Oscillations in an *n*-Type Germanium Plate Doped with Copper, *Sov. Phys. Solid State*, **7**, 267.

948 Kornilov, B. V. (1965c), Carrier Recombination at Zinc Atoms in *n*-Silicon, *Sov. Phys. Solid State*, **7**, 1446.

949 Kornilov, B. V. (1966a), Determination of Effective Capture Cross Section for Holes by Singly Negatively Charged Zinc Atoms in *p*-Silicon by the Double Injection Method, *Sov. Phys.—Semicond.*, **7**, 2794.

950 Kornilov, B. V. (1966b), Recombination of Carriers at Zinc Atoms in *p*-Type Silicon, *Sov. Phys. Solid State*, **8**, 157.

951 Kornilov, B. V. (1969), The Problem of the Existence of Low-Ionization Energy Levels of Zinc in Silicon, *Sov. Phys.—Semicond.*, **2**, 1480.

952 Kornilov, B. V., and A. V. Anfimov (1967a), Nonlinear Effects in *n*-Type Silicon Compensated with Zinc, *Sov. Phys.—Semicond.*, **1**, 279.

953 Kornilov, B. V., and A. V. Anfimov (1967b), Nonsinusoidal Current Oscillations in *n*-Type Silicon Compensated with Zinc, *Sov. Phys. Solid State*, **8**, 2742.

954 Kornilov, B. V., and S. E. Gorskii (1968), Effective Cross Section for the Capture of a Hole by a Doubly Charged Negative Zinc Atom in Silicon, *Sov. Phys.—Semicond.*, **2**, 216.

955 Kornilov, B. V., V. A. Vil'kotskii, G. V. Aleksandrova, G. N. Tereshko, and T. P. Tsarevskaya (1971), Compensation, by Deep-Level Impurities of High-Resistivity Epitaxial Films of Gallium Arsenide, *Sov. Phys.—Semicond.*, **5,** 119.

956 Korol'kov, V. I., and V. N. Romanenko (1964), Concentration Dependence of the Segregation Coefficients of Some Group III and V Impurities in Germanium, *Sov. Phys. Solid State,* **5,** 2130.

957 Korsun', V. M., and A. M. Nemchenko (1967), Electroluminescence in the Case of Cu Diffusion in ZnS Crystals, *Sov. Phys. Solid State,* **8,** 2988.

958 Korsunskaya, N. E., I. V. Markevich and M. K. Sheinkman (1968), Formation of New Local Centers in CdS Single Crystals in the Presence of Free Electrons and Holes, *Sov. Phys. Solid State,* **10,** 409.

959 Korzo, V. F., P. S. Kireev, and G. A. Lyashchenko (1969), Band Structure of Si-SiO_2, *Sov. Phys.—Semicond.,* **2,** 1545.

960 Kosenko, V. E. (1960), Diffusion and Solubility of Cadmium in Germanium, *Sov. Phys. Solid State,* **1,** 1481.

961 Kosenko, V. E. (1962), "Slow" Diffusion of Silver into Germanium, *Sov. Phys. Solid State,* **4,** 42.

962 Koshenko, V. E., and L. A. Khomenko (1962), Diffusion of Silver at the Surface of Germanium, *Sov. Phys. Solid State,* **3,** 2166.

963 Kotina, I. M., N. E. Mazurik, S. R. Novikov, and A. Kh. Khusainov (1969), Effect of "Impurity" Radiation on the Capacitance of a *p-n* Junction, *Sov. Phys.—Semicond.,* **3,** 319.

964 Kotina, I. M., S. R. Novikov, N. E. Mazurik, and A. Kh. Khusainov (1969) Transient Photo-EMF Due to the Ionization of Impurity Centers in a Space-Charge Layer, *Sov. Phys.—Semicond.,* **3,** 362.

965 Koval', Yu. P., V. N. Mordkovich, E. M. Temper, and V. A. Kharchenko (1972), Behavior of Oxygen in Irradiated Silicon, *Sov. Phys.—Semicond.,* **5,** 2061.

966 Kovalevskaya, G. G., E. E. Klotyn'sh, D. N. Nasledov, and S. V. Slobodchikov (1967), Some Electrical and Photoelectric Properties of InP Doped with Copper, *Sov. Phys. Solid State,* **8,** 1922.

967 Kovtonyuk, N. F. 1965), Recombination-Gradient Photo-EMF in Semiconductors, *Sov. Phys. Solid State,* **7,** 1243.

968 Kovtunyuk, N. F., and Yu. I. Gokhfel'd (1969), Negative Differential Conductance in *n*-Type Germanium, *Sov. Phys.—Semicond.,* **2,** 1412.

969 Kowalchik, M., A. S. Jordan, and M. H. Read (1972), Coprecipitation of Ga_2O_3 in the Liquid-Phase Epitaxial Growth of GaP, *J. Electrochem. Soc.,* **119,** 756.

970 Kozlov, Yu. I., and R. Sh. Malkovich (1967), Decomposition of a Solid Solution of Gold in Silicon, *Sov. Phys. Solid State,* **9,** 384.

971 Kozlovskaya, V. M., and R. N. Rubinshtein (1962), Calculation Solubility and Vapor Pressure for Semiconductor-Dopant Systems, *Sov. Phys. Solid State,* **3,** 2434.

972 Krag, W. E., and H. J. Zeiger (1962), Infrared Absorption Spectrum of Sulfur-Doped Silicon, *Phys. Rev. Lett.,* **8,** 485.

973 Krag, W. E., W. H. Kleiner, H. J. Zeiger, and S. Fischler (1966), Effects of Strain on Infrared Absorption Spectra of Sulfur-Doped Silicon, *Bull. Am. Phys. Soc.,* **11,** 205.

974 Krag, W. E., W. H. Kleiner, H. J. Zeiger, and S. Fischler (1966), Sulfur Donors in Silcon: Infrared Transitions and Effects of Calibrated Uniaxial Stress, *Proceedings of the International Conference on the Physics of Semiconductors, Kyoto, 1966, J. Phys. Soc. Japan, Suppl.*, **21**, 230.

975 Kremen, A. (1968), Copper Donors in Silicon, *Czech. J. Phys.*, **18**, 937.

976 Kressel, H., and F. Z. Hawrylo (1970), Ionization Energy of Mg and Be Acceptors in GaAs, *J. Appl. Phys.*, **41**, 1865.

977 Kressel, H., and H. Nelson (1969), Electrical and Optical Properties of n-Type Si-Compensated GaAs Prepared by Liquid-Phase Epitaxy, *J. Appl. Phys.*, **40**, 3720.

978 Kressel, H., and H. Von Philipsborn (1969), Properties of Si-Doped GaAs Prepared by Vapor Phase Growth, Abstract No. 61, Electrochemical Society Spring Meeting, May 1969.

979 Kressel, H., and H. Von Philipsborn (1970), Electrical and Optical Properties of Vapor-Grown GaAs:Si, *J. Appl. Phys.*, **41**, 2244.

980 Kressel, H., J. U. Dunse, H. Nelson, F. Z. Hawrylo (1968a), Luminescence in Silicon-Doped GaAs Grown by Liquid-Phase Epitaxy, *J. Appl. Phys.*, **39**, 2006.

981 Kressel, H., F. Z. Hawrylo, and P. LeFur (1968b), Luminescence Due to Ge Acceptors in GaAs, *J. Appl. Phys.*, **39**, 4059.

982 Kressel, H., F. Z. Hawrylo, and M. S. Abrahams (1968c), Observations Concerning Radiative Efficiency and Deep-Level Luminescence in n-Type GaAs Prepared by Liquid-Phase Epitaxy, *J. Appl. Phys.*, **39**, 5139.

983 Kressel, H., H. Nelson, and F. Z. Hawrylo (1968d), New Deep Level Luminescence in GaAs:Sn, *J. Appl. Phys.*, **39**, 5647.

984 Kressel, H., F. Z. Hawrylo, and N. Almeleh (1969a), Properties of Efficient Silicon-Compensated $Al_xGa_{1-x}As$ Electroluminescent Diodes, *J. Appl. Phys.*, **40**, 2248.

985 Kressel, H., H. Nelson, and F. Z. Hawrylo (1969b), Comment on "Local-Mode Absorption and Defects in Compensated Silicon-Doped Gallium Arsenide" by W. G. Spitzer and W. Allred, *J. Appl. Phys.*, **40**, 3069.

986 Kressel, H., F. H. Nicholl, F. Z. Hawrylo, and H. F. Lockwood (1970), Luminescence in Indirect Bandgap $Al_xGa_{1-x}As$, *J. Appl. Phys.*, **41**, 4692.

987 Krivov, M. A., E. V. Malisova and G. S. Shishkova (1966), *Electric Properties of Gallium Arsenide with Admixture of Gold*, Document AD 638 456, National Technical Information Service, Springfield, Virginia.

988 Krivov, M. A., E. V. Malisova, and É. N. Mel'chenko (1970), Investigation of the Behavior of Gold in Gallium Arsenide, *Sov. Phys.—Semicond.*, **4**, 693.

989 Kroemer, H. (1966), Nonlinear Space-Charge Domain Dynamics in a Semiconductor with Negative Differential Mobility, *IEEE Trans. Electron Devices*, **ED-13**, 27.

990 Kroko, L. J. (1966), Studies of SiC, Ph.D. Thesis, Carnegie-Mellon University, Pittsburgh, Pennsylvania.

991 Kruse, P. W., L. D. McGlaughlin, and R. D. McQuistan (1962), *Elements of Infrared Technology*, Wiley, New York.

992 Kryukova, I. V., and V. S. Vavilov (1964), Orientation Dependence of the Formation of Radiation Defects in n-Type Silicon, *Sov. Phys. Solid State*, **2**, 266.

993 Kucherenko, I. V. (1965), Mobility of Holes and Electrical Breakdown in Silicon at Low Temperatures, *Sov. Phys. Solid State*, **7**, 818.

994 Kuczynski, G. C., K. R. Iyer, and C. W. Allen (1972), Effect of Light on the Mobility of Dislocations in Germanium, *J. Appl. Phys.*, **43,** 1337.

995 Kuhn-Kuhnenfeld, F. (1972), Selective Photoetching of Gallium Arsenide, *J. Electrochem. Soc.*, **119,** 1063.

996 Kukimoto, H., C. H. Henry, and G. L. Miller (1972), Photocapacitance Studies of Deep Double Electron-Trap Oxygen in Gallium Phosphide, *Appl. Phys. Lett.*, **21,** 251.

997 Kulshreshtha, A. P., and V. A. Goryunov (1966), Calculation of Thermally Stimulated Currents, *Sov. Phys. Solid State*, **8,** 1540.

998 Kulshreshtha, A. P., and A. E. Yunovich (1966), Some Electrical Properties of Semi-Insulating GaAs, *Sov. Phys. Solid State*, **7,** 2058.

999 Kuno, H. J., J. R. Collard and A. R. Gobat (1969), Avalanche Breakdown Voltage of GaAs p^+-n-n^+ Diode Structures, *Appl. Phys. Lett.,* **14,** 343.

1000 Kuranova, L. A., Yu. M. Degot', Yu. R. Nosov (1970), A Semiconductor Diode with a Microsecond Reverse-Resistance Recovery Time, *Sov. Phys.—Semicond.*, **3,** 1175.

1001 Kurbatov, L. N., N. N. Mochalkin, A. D. Britov, and E. M. Omel'yanovskii (1969), Radiative Transitions between Deep Acceptors and the Valence Band in Gallium Arsenide, *Sov. Phys.—Semicond.*, **3,** 528.

1002 Kurilo, P. M., E. Seitov, and M. I. Khitren' (1971), Influence of Heat Treatment on the Electrical Properties of n-Type Silicon Heavily Doped with Oxygen, *Sov. Phys.—Semicond.*, **4,** 1953.

1003 Kurnick, S. W. (1959), *High Temperature Broad Band Semiconductors,* Document AD 244 912, National Technical Information Service, Springfield, Virginia.

1004 Kurnick, S. W., and R. N. Zitter (1956), Photoconductive and Photoelectromagnetic Effects in InSb, *J. Appl. Phys.*, **27,** 278.

1005 Kurokawa, K. (1967), Transient Behavior of High-Field Domains in Bulk Semiconductors, *Proc. IEEE,* **55,** 1615.

1006 Kurova, I. A., and S. G. Kalashnikov (1964), Electrical Instability in Germanium, *Sov. Phys. Solid State*, **5,** 2359.

1007 Kurova, I. A., and N. N. Ormont (1965), The Impurity Photoconductivity Spectra of Gold-Doped Germanium at Low Temperatures, *Sov. Phys. Solid State*, **6,** 2970.

1008 Kurova, I. A., and N. N. Ormont (1967), Impurity Photoconductivity Spectra of Silicon at Low Temperatures, *Sov. Phys. Solid State*, **8,** 2264.

1009 Kurova, I. A., and V. V. Ostroborodova (1965), Kinetics of the Low-Temperature Impurity Photoconductivity in p-Type Germanium Containing Gold, *Sov. Phys. Solid State*, **7,** 547.

1010 Kurova, I. A., and N. D. Tyapkina (1961), Electrical Conductivity of Germanium with Lithium Impurity at Low Temperatures, *Sov. Phys. Solid State*, **2,** 2761.

1011 Kurova, I. A., S. G. Kalashnikov, and N. D. Tyapkina (1962), Kinetics of Impurity Photoconductivity in n-Type Germanium Containing Gold, *Sov. Phys. Solid State*, **4,** 1104.

1012 Kurova, I. A., V. V. Ostroborodova, and N. N. Ormont (1965), Low-Temperature Voltage Sensitivity of Gold-Doped p-Type Germanium, *Sov. Phys. Solid State*, **7,** 755.

1013 Kurova, I. A., M. Vrana, and V. S. Vavilov (1967), Observation of the Movement of Electric Domains in n-Ge Having a Partially Compensated Upper Au Level, *Sov. Phys. Solid State*, **8,** 1892.

1014 Kurova, I. A., M. Vrana, and P. Berndt (1969), Some Investigations of Current-Voltage Characteristics of Gold-Doped n-Type Germanium, *Sov. Phys.—Semicond.*, **2,** 1530.

1015 Kurskii, Yu. A. (1964), Thermal Trapping of Electrons by a Neutral Center in Semiconductors, *Sov. Phys. Solid State*, **6,** 1162.

1016 Kustov, V. G., and V. P. Orlov (1970), Diffusion Length of Minority Carriers in Gallium Arsenide, *Sov. Phys.—Semicond.*, **3,** 1457.

1017 Kusuya, I. A., and S. Kiode (1958), A Theory of Impurity Conduction. II, *J. Phys. Soc. Japan*, **13,** 1287.

1018 Kuznetsova, E. M. (1964), Uniqueness of Characteristic Times for Uniform Excitation and for Charge Carrier Injection, *Sov. Phys. Solid State*, **6,** 1213.

1019 Lacey, S. D., L. N. Large, and D. R. Wright (1969), Cathodoluminescence of Oxygen-Implanted Zinc-Doped Gallium Phosphide, *Electron. Lett.*, **5,** 203.

1020 Ladany, I. (1971), Electroluminescence Characteristics and Efficiency of GaAs:Si Diodes, *J. Appl. Phys.*, **42,** 654.

1021 Ladany, I., and S. H. McFarlane, III (1969), Comparison of Liquid-Encapsulated and Solution-Grown Substrates for Efficient GaP Diodes, *J. Appl. Phys.*, **40,** 4984.

1022 Lagowski, J., C. L. Balestra, and H. C. Gatos (1972), Electronic Characteristics of "Real" CdS Surfaces, *Surf. Sci.*, **29,** 213.

1023 Lambert, J. L. (1971), Gold Diffusion and Bulk Vacancy Generation in Silicon, Electrochemical Society Extended Abstracts, 139th National Meeting, p. 197.

1024 Lambert, J. L., and M. Reese (1968), The Gettering of Gold and Copper from Silicon, *Solid State Electron.*, **2,** 1055.

1025 Lambert, L. M. (1962), Impact Ionization of Impurities in Heavily Compensated Germanium, *J. Phys. Chem. Solids*, **23,** 1481.

1026 Lampert, M. A. (1956), Simplified Theory of Space-Charge-Limited Currents in an Insulator with Traps, *Phys. Rev.*, **103,** 1648.

1027 Lampert, M. A. (1962), Double Injection in Insulators, *Phys. Rev.*, **125,** 126.

1028 Lampert, M. A. (1964), Volume-Controlled Current Injection in Insulators, *Rep. Prog. Phys.*, **27,** 329.

1029 Lampert, M. A., and P. Mark (1970), *Current Injection in Solids*, Academic Press, New York.

1030 Lampert, M. A., and A. Rose (1959), Transient Behavior of the Ohmic Contact, *Phys. Rev.*, **113,** 1236.

1031 Lampert, M. A., and A. Rose (1961), Volume-Controlled, Two-Carrier Currents in Solids: The Injected Plasma Case, *Phys. Rev.*, **121,** 26.

1032 Lancaster, G. (1967), *Electron Spin Resonance in Semiconductors*, Plenum Press, New York.

1033 Landsberg, E. G., and S. G. Kalashnikov (1961), Capture Cross Sections for Electrons in Manganese-Doped Germanium Atoms, *Sov. Phys. Solid State*, **3,** 1137.

1034 Landsberg, E. G., and S. G. Kalashnikov (1963), Properties of Manganese Recombination Centers in Germanium, *Sov. Phys. Solid State*, **5,** 777.

1035 Landsberg, P. T., and A. R. Beattie (1959), Auger Effect in Semiconductors, *J. Phys. Chem. Solids*, **8,** 73.

1036 Landsberg, P. T., *et al.* (1963), *Auger Effects Involving Recombination Centers,* Document AD 434 598 (also see AD 606 682), National Technical Information Service, Springfield, Virginia.

1037 Landsberg, P. T., D. A. Evans, and C. Rhys-Roberts (1964), Auger Recombination and Impact Ionization Involving Traps in Semiconductors, *Proc. Phys. Soc.,* **84,** 915.

1038 Langer, D. W., Y. S. Park, and R. N. Euwema (1966), Phonon Coupling in Edge Emission and Photoconductivity of CdSe, CdS, and Cd (Se_xS_{1-x}), *Phys. Rev.,* **152,** 788.

1039 Lappo, M. T., and V. D. Tkachev (1972), Excitation Spectrum of new Acceptor-Type Radiation Defects in Silicon, *Sov. Phys.—Semicond.,* **5,** 1411.

1040 Larrabee, R. D., and M. C. Steel (1960), The Oscillistor—A New Type of Semiconductor Oscillator, *J. Appl. Phys.,* **31,** 1519.

1041 Lashkarev, V. E. (1963), Certain Properties of the Quasi-Polar Photoconductivity of Semiconductors, *Sov. Phys. Solid State,* **5,** 303.

1042 Lashkarev, V. E. (1963), Effect of an Additional Fast Recombination Center on the Phenomenological Photocurrent Yield, *Sov. Phys. Solid State,* **5,** 309.

1043 Lashkarev, V. E., A. V. Lyubchenko, and M. K. Sheinkman (1965), Combined Investigation of the Kinetics of Recombination and Infrared Photocurrent Quenching Processes in Cadmium Sulfide, *Sov. Phys. Solid State,* **7,** 1388.

1044 Lashkarev, V. E., V. K. Malyutenko, and V. A. Romanov (1966), A Method for Determining the Lifetime of Excess Carriers in Monopolar Photoconductors, *Sov. Phys. Solid State,* **8,** 51.

1045 Lashkarev, G. V., A. I. Dmitriev, G. A. Sukach, and V. A. Shershel' (1972), Magnetoresistance of Germanium Doped with Neodymium and Europium, *Sov. Phys.—Semicond.,* **5,** 1808.

1046 Latshaw, G. L., G. D. Sprouse, P. B. Russell, G. M. Kalvius, and S. S. Hanna (1968), Implantation of ^{57}Fe into Si and Ge, *Bull, Am. Phys. Soc.,* **13,** 1649.

1047 Lax, M. (1959), Giant Traps, *J. Phys. Chem. Solids,* **8,** 66.

1048 Lax, M. (1960), Cascade Capture of Electrons in Solids, *Phys. Rev.,* **119,** 1502.

1049 Leadon, R., and J. A. Naber (1969), Recombination Lifetime in High-Purity Silicon at Low Temperatures, *J. Appl. Phys.,* **40,** 2633.

1050 Lebedev, A. A., and A. T. Mamadalimov (1972), Thermal and Field Quenching of the Photoconductivity in Ni-Doped Silicon, *Sov. Phys.—Semicond.,* **6,** 96.

1051 Lebedev, A. A., and N. A. Sultanov (1969), Zinc-Compensated Silicon Diodes with Negative Resistance, *Sov. Phys.—Semicond.,* **2,** 1295.

1052 Lebedev, A. A., and N. A. Sultanov (1969), Cadmium-Doped Silicon Diodes with Negative Resistance, *Sov. Phys.—Semicond.,* **2,** 1543.

1053 Lebedev, A. A., and N. A. Sultanov (1969), Anomalous Current-Voltage Characteristics of Silicon Diodes with a Negative Resistance, *Sov. Phys.—Semicond.,* **3,** 108.

1054 Lebedev, A. A., and N. A. Sultanov (1969), Negative Resistance Diodes Made of Silicon Doped with Mercury, *Sov. Phys.—Semicond.,* **3,** 276.

1055 Lebedev, A. A., and N. A. Sultanov (1969), Platinum-Doped Silicon Diodes Exhibiting Negative Resistance, *Sov. Phys.—Semicond.,* **3,** 530.

1056 Lebedev, A. A., and N. A. Sultanov (1971), Some Properties of Chromium-Doped Silicon, *Sov. Phys.—Semicond.*, **4,** 1900.

1057 Lebedev, A. A., A. T. Mamadalimov, and N. A. Sultanov (1971), Investigation of Sulfur-Doped Silicon Diodes Exhibiting S-Type Negative Resistance, *Sov. Phys.—Semicond.*, **5,** 17.

1058 Lebedev, A. A., N. A. Sultanov, and V. M. Tuchkevich (1971), N-Type Negative Resistance and Photoconductivity of Sulfur-Doped *p*-Type Si, *Sov. Phys.—Semicond.*, **5,** 25.

1059 Lebedev, A. A., A. T. Mamadalimov, and N. A. Sultanov (1971), Negative Photoeffect in Diodes Made of Zinc-Doped *p*-Type Silicon Crystals, *Sov. Phys.—Semicond.*, **5,** 773.

1060 Lebedev, A. A., A. T. Mamadalimov, and N. A. Sultanov (1972), Principal Parameters of Diodes Made of Nickel-Doped Silicon, *Sov. Phys.—Semicond.*, **5,** 1990.

1061 Lee, D. B. (1969), Anisotropic Etching of Silicon, *J. Appl. Phys.*, **40,** 4569.

1062 Lee, D. H. (1970), Temperature Dependence of Double Injection in a Long Silicon $p^+\pi n^+$ Structure, *J. Appl. Phys.*, **41,** 3467.

1063 Lee, C. C., and H. Y. Fan (1972), Two Photon Absorption and Photoconductivity in GaAs and InP, *Appl. Phys. Lett.*, **20,** 18.

1064 Lee, D. H., and M. A. Nicolet (1969), Thermal Noise in Double Injection, *Phys. Rev.*, **184,** 806.

1065 Lee, H.-S., and S. M. Sze (1970), Silicon *p-i-n* Photodetector Using Internal Reflection Method, *IEEE Trans. Electron Devices*, **ED-17,** 342.

1066 Leenov, D., and R. G., Stewart (1968), A Proposed Method for Rapid Determination of Doping Profiles in Semiconductor Layers, *Proc. IEEE*, **56,** 2095.

1067 Leiderman, A. Yu. (1961), On the Negative Photo-Effect in a Metal-Semiconductor Contact, *Izv. An Uzb. SSR, Fiz.-Mat. Nauk*, **1,** 54.

1068 Leiderman, A. Yu. (1970), Influence of Trapping Centers on Current Characteristics of a Semiconductor pnn^+ Structure, *Sov. Phys.—Semicond.*, **3,** 1251.

1069 Lemke, H. (1965), Zur Problem der Lebensdauermessung mit der Doppelimpulsmethode bei Silizium-Einkristallen, *Phys. Status Solidi*, **12,** 115.

1070 Lemke, H. (1966a), Zur Zeitabhängigkeit raumladungsbegrenzter Injektionsströme in Halbleitern, *Phys. Status Solidi*, **16,** 413.

1071 Lemke, H. (1966b), Zeitabhängigkeit Raumladungsbegrenzter Injektionsströme in Fe-Dotiertem p-Silizium, *Phys. Status Solidi*, **16,** 427.

1072 Leung, P. C., L. H. Skolnik, W. P. Allred, and W. G. Spitzer (1972), Infrared Absorption Study of Li-Diffused Mg-Doped GaAs, *J. Appl. Phys.*, **43,** 4096.

1073 Levinstein, H. (1959), Impurity Photoconductivity in Germanium, *Proc. IRE*, **47,** 1478.

1074 Levitt, R. S., and A. Honig (1961), Low Temperature Electron Trapping Lifetimes and Extrinsic Photoconductivity in *n*-Type Silicon Doped with Shallow Impurities, *J. Phys. Chem. Solids*, **22,** 269.

1075 Li, S. S., and C. I. Huang (1972), Investigation of the Recombination and Trapping Processes of Photodetected Carriers in Semi-Insulating Cr-Doped GaAs Using PME and PC Methods, *J. Appl. Phys.*, **43,** 1757.

1076 Li, S. S., and H. F. Tseng (1971), Study of Room Temperature Photomagnetoelectric and Photoconductive Effects in Au-Doped Silicon, *Phys. Rev.* **B4,** 490.

1077 Liao, J. H. (1970), Noise in InSb Double-Injection Space-Charge-Limited Diodes, *Electron. Lett.*, **6**, 175.

1078 Lifshits, T. M., F. Ya. Nad', and V. I. Sidorov (1967), Impurity Photoconductivity of Germanium Doped with Antimony, Arsenic, Boron or Indium, *Sov. Phys. Solid State*, **8**, 2567.

1079 Lilly, A. C., L. I. Stewart, and R. M. Henderson (1970), Thermally Stimulated Currents in Mylar High-Field Low-Temperature Case, *J. Appl. Phys.*, **41**, 2001.

1080 Lindquist, P. F., and R. H. Bube (1970), Photocapacitance Effects at Cu_2S–CdS Heterojunctions, *Bull. Am. Phys. Soc.*, **15**, 1591.

1081 Lindström, J. L. (1971), Flash X-Ray Irradiation of *p-n* Junctions: A Method to Measure Minority Carrier Lifetimes, Diffusion Constants and Generation Constants, *Solid-State Electron.*, **14**, 827.

1082 Litovchenko, V. G., and V. P. Kovbasyuk (1969), Criterion of "Heavy" Doping for Deep Level Centers and the Nature of the Fast Surface Trap Spectrum in Silicon, *Sov. Phys.—Semicond.*, **3**, 733.

1083 Litovchenko, V. G., A. P. Gorban', and V. P. Kovbasyuk (1965), Investigation of the Effect of Photocarrier Trapping at the Surface of Silicon, *Sov. Phys. Solid State*, **7**, 449.

1084 Liu, S. T. (1967), Thermal Noise in Space-Charge-Limited Solid State Diodes, *Solid-State Electron.*, **10**, 253.

1085 Liu, Y-Z., J. L. Moll, and W. E. Spicer (1969), Effects of Heat Cleaning on the Photoemission Properties of GaAs Surfaces, *Appl. Phys. Lett.*, **14**, 275.

1086 Loebner, E. E., *et al.* (1966), *One Micron Photodector*, Final Technical Report, Hewlett-Packard Laboratories, Palo Alto, California.

1087 Loescher, D. H., J. W. Allen, and G. L. Pearson (1966), The Application of Crystal Field Theory to the Electrical Properties of Co Impurities in GaP, *Proceedings of the International Conference on the Physics of Semiconductors, Kyoto, 1966, J. Phys. Soc. Japan, Suppl.* **21**, 239.

1088 Loewenstein, M., and A. Honig (1966), Photoexcited Electron Capture by Ionized and Neutral Shallow Impurities in Silicon at Liquid-Helium Temperatures, *Phys. Rev.*, **144**, 781.

1089 Logan, M. A. (1961), An AC Bridge for Semiconductor Resistivity Measurements Using a Four-Point Probe, *Bell Syst. Tech. J.*, **XL**, 885.

1090 Logan, R. A., and A. J. Peters (1959), Diffusion of Oxygen in Silicon, *J. Appl. Phys.*, **30**, 1627.

1091 Logan, R. A., J. M. Rowell, and F. A. Trumbore (1964), Phonon Spectra of Ge–Si Alloys, *Phys. Rev.*, **136**, A1751.

1092 Logan, R. M. (1971), Analysis of Heat Treatment and Formation of Gallium-Vacancy-Tellurium Complexes in GaAs, *J. Phys. Chem. Solids*, **32**, 1755.

1093 Long, D., and J. Myers (1959), Ionized-Impurity Scattering Mobility of Electrons in Silicon, *Phys. Rev.*, **115**, 1107.

1094 Long, D. (1960), Scattering of Conduction Electrons by Lattice Vibrations in Silicon, *Phys. Rev.*, **120**, 2024.

1095 Long, D. (1968), *Energy Bands in Semiconductors*, Interscience, New York.

1096 Longini, R. L., and R. F. Greene (1955), Ionization Interaction between Impurities in Semiconductors and Insulators, *Phys. Rev.*, **102**, 993.

1097 Lorimor, O. G., and W. G. Spitzer (1966), Local Mode Absorption in Compensated Silicon-Doped Gallium Arsenide, *J. Appl. Phys.,* **37,** 3687.

1098 Lorimor, O. G., and M. E. Weiner (1972), *p*-Type Dopants for GaP Green Light Emitting Diodes, *J. Electrochem. Soc.*, **119,** 1576.

1099 Lucovsky, G. (1965), On the Photoionization of Deep Impurity Centers in Semiconductors, *Solid-State Commun.*, **3,** 299.

1100 Lucovsky, G. (1966), Optical Absorption Associated with Deep Semiconductor Impurity Centers, *Bull. Am. Phys. Soc.*, **11,** 206.

1101 Ludwig, G. W., and R. L. Watters (1956), Drift and Conductivity Mobility in Silicon, *Phys. Rev.*, **101,** 1699.

1102 Ludwig, G. W., and H. H. Woodbury (1961), Interactions between Impurities in Silicon, *International Conference on Semiconductor Physics, Prague, 1960,* Academic Press, New York and London.

1103 Lukicheva, N. I., O. V. Pelevin, and L. Ya. Pervova (1971), Carrier Lifetimes in High-Resistivity Iron-Doped Gallium Arsenide, *Sov. Phys.—Semicond.,* **5,** 169.

1104 Luk'yanchikova, N. B., and M. K. Sheinkman (1967), Photocurrent Noise and Superlinearity of Lux-Ampere Characteristics of CdS and CdSe Single Crystals, *Sov. Phys. Solid State,* **8,** 2398.

1105 L'vova, E. Yu. (1964), On the Influence of Cadmium on the Probability of Electron Recombination in Germanium, *Sov. Phys. Solid State,* **5,** 2239

1106 Lyubchenko, A. V., M. K. Sheinkman, V. A. Brodovoi, and N. M. Krolevets (1968), Determination of the Parameters of Recombination Centers in Copper-Doped Gallium Arsenide, *Sov. Phys.—Semicond.*, **2,** 406.

1107 MacMillan, H. F., and R. H. Bube (1970), Long Lifetime Hole Traps in CdTe, *Bull. Am. Phys. Soc.*, **15,** 1616.

1108 McCaldin, J. O., and J. W. Mayer (1970), Donor Behavior in Indium-Alloyed Silicon, *Appl. Phys. Lett.*, **17,** 365.

1109 McCaldin, J. O., M. J. Little, and A. E. Widmer (1965), The Solubility of Sodium in Silicon, *J. Phys. Chem. Solids*, **26,** 1119.

1110 McCombs, A. E., Jr. (1971), Impact Ionization of Deep Impurities in Silicon, Ph.D. Thesis, Carnegie-Mellon University, Pittsburgh, Pennsylvania. [See also *Int. J. Electron.,* **32,** 361 (1972)].

1111 McCombs, A. E., Jr., and A. G. Milnes (1968), Calculation of Drift Velocity in Silicon at High Electric Fields, *Int. J. Electron.*, **24,** 573.

1112 McCumber, D. E., and A. G. Chynoweth (1966), Theory of Negative-Conductance Amplification and of Gunn Instabilities in "Two Valley" Semiconductors, *IEEE Trans. Electron Devices,* **ED-13,** 4.

1113 McGroddy, J. C. (1970), Current Oscillations and Negative Differential Conductivity in n-Type Germanium, *IEEE Trans. Electron Devices,* **ED-17,** 207.

1114 McIntyre, R. J. (1970), Comparison of Photomultipliers and Avalanche Photodiodes for Laser Applications, *IEEE Trans. Electron Devices,* **ED-17,** 347.

1115 McIrvine, E. C. (1969), *Phenomenology of Impurity Conduction in Semiconductors,* Document AD 240 541, National Technical Information Service, Springfield, Virginia.

1116 McKay, K. G. (1948), Electron Bombardment Conductivity in Diamond, *Phys. Rev.*, **74,** 1606.

1117 McKay, K. G. (1954), Avalanche Breakdown in Silicon, *Phys. Rev.*, **94,** 877.

1118 McKelvey, J. P. (1966), *Solid State and Semiconductor Physics,* Harper Row, New York.

1119 McWhorter, A. L., and R. H. Rediker (1959), The Cryosar—A New Low-Temperature Computer Component, *Proc. IRE,* **47,** 1207.

1120 Madelung, O. (1964), *Physics of III–V Compounds,* Wiley, New York.

1121 Magee, T. J., A. M. Hermann, and R. J. Deck (1972), Observation of Defect Annealing in Neutron-Irradiated Silicon by Space-Charge-Limited Current, *Phys. Rev.*, **B5,** 3364.

1122 Maher, A. T., B. G. Streetman, and N. Holonyak (1969), Infrared Detection Properties of Zn-Doped Si *p-i-n* Diodes, *IEEE Trans. Electron Devices,* **ED-16,** 963.

1123 Makarevich, A. I., and L. Yu. Raines (1969), Relative Influence of *A* and *E* Centers on Changes in the Electrical Properties of Silicon During Irradiation, *Sov. Phys.—Semicond.*, **2,** 985.

1124 Makovskii, L. L. (1971), Influence of Trapping of Carriers on the Kinetics of Charge Collection in Semiconductor Particle Detectors, *Sov. Phys—Semicond.*, **4,** 1351.

1125 Makovskii, L. L., N. B. Strokan, S. M. Ryvkin, and A. Kh. Khusainov (1970), Influence of the Contact Potential Difference on the Width of a Space-Charge Region in a Metal-Semiconductor System in the Presence of a Compensating Impurity, *Sov. Phys.—Semicond.*, **3,** 1205.

1126 Malkovich, R. Sh. (1968), Changes in the Concentration of Gold in Silicon, *Sov. Phys. Solid State,* **9,** 1676.

1127 Malkovich, R. Sh., and N. A. Alimbarashvili (1963), The Effect of an Electric Field on the Diffusion of Zinc on Silicon, *Sov. Phys. Solid State,* **4,** 1725.

1128 Mannami, M., K. Izumi, T. Hayashida, and M. Nakagawa (1964), Segregation of Copper Supersaturated Solid Solution in Germanium, *Japan J. Appl. Phys.*, **3,** 508.

1129 Mantena, N. R., and E. E. Loebner (1971), Noise in Zinc-Doped Silicon Photoconductors, *Proceedings of the Third International Conference on Photoconductivity, Stanford University, 1969,* E. M. Pell, Ed., Pergamon Press, New York, p. 53.

1130 Many, A., and G. Rakavy (1962), Theory of Transient Space-Charge-Limited Currents in Solids in the Presence of Trapping, *Phys. Rev.*, **126,** 1980.

1131 Many, A., S. Z. Weisz and M. Simhony (1962), Space-Charge-Limited Currents in Iodine Single Crystals, *Phys. Rev.*, **126,** 1989.

1132 Margoninski, Y., and Y. Walzer (1966), Unique Determination of the Parameters of Surface Recombination Centers in Semiconductors, *Phys. Rev. Lett.*, **17,** A12.

1133 Marlor, G. A., and J. Woods (1965), Space-Charge-Limited Currents and Electron Traps in CdS Crystals, *Br. J. Appl. Phys.*, **16,** 1449.

1134 Marsh, O. J., R. Baron, and J. W. Mayer (1965), Effect of Light on Population of Recombination Centers in Structures Exhibiting Double Injection, *Appl. Phys. Lett.*, **7,** 120.

1135 Marsh, O. J., R. Baron, G. A. Shifrin, and J. W. Mayer (1968), The Electrical Behavior of Implanted Bismuth in Silicon, *Appl. Phys. Lett.*, **13,** 199.

1136 Martens, M., L. Mehrkam, and F. Williams (1969), Theory of Donor-Acceptor Dipairs in Semiconductors, *Phys. Rev.*, **186,** 757.

1137 Martin, J. A. (1969), Activation Analysis and Autoradiography of Silicon, *J. Electrochem. Soc.*, **116,** 149C.

1138 Maruska, H. P., and J. J. Tietjen (1969), The Preparation and Properties of Vapor-Deposited Single Crystalline GaN, *Appl. Phys. Lett.*, **15,** 327.

1139 Mason, H. J., Jr., and J. S. Blakemore (1972), Spectral Dependence of Photoconductivity in Indium-Doped Silicon, *J. Appl. Phys.*, **43,** 2810.

1140 Masters, B. J., and J. M. Fairfield (1966), Silicon Self-Diffusion, *Appl. Phys. Lett.*, **8,** 280.

1141 Masters, B. J., and J. M. Fairfield (1969), Arsenic Isoconcentration Diffusion Studies in Silicon, *J. Appl. Phys.*, **40,** 2390.

1142 Mataré, H. F., and C. W. Laakso (1969), Space-Charge Domains at Dislocation Sites, *J. Appl. Phys.*, **40,** 476.

1143 Mathur, M. P., and N. Pearlman (1969), Phonon Scattering by Neutral Donors in Germanium, *Phys. Rev.*, **180,** 833.

1144 Mattis, R. L., W. E. Phillips, and W. M. Bullis (1968), *Measurement and Interpretation of Carrier Lifetime in Silicon and Germanium*, Document AD 671 552, National Technical Information Service, Springfield, Virginia.

1145 Matukura, Y. (1959), Effects of Heat Treatment upon the Electrical Properties of Silicon, *J. Phys. Soc. Japan*, **14,** 918.

1146 Matukura, Y. (1963), Grain Boundary States in Silicon and Germanium, *Japan. J. Appl. Phys.*, **2,** 91.

1147 Matveenko, Yu. A., V. I. Murygin, and V. S. Rubin (1969), Some Properties of Nickel-Doped Gallium Arsenide, *Sov. Phys.—Semicond.*, **2,** 1536.

1148 Mayburg, S. (1961), Direct Recombination in GaAs and Some Consequences in Transistor Design, *Solid-State Electron.*, **2,** 195.

1149 Mayer, J. W., R. Baron, and O. J. Marsh (1965), Observation of Negative Resistance in Long Silicon $p\pi n$ Diodes, *Appl. Phys. Lett.*, **6,** 38.

1150 Mayer, J. W., R. Baron, and O. J. Marsh (1965), Observation of Double Injection in Long Silicon *p-i-n* Structures, *Phys. Rev.*, **137,** A286.

1151 Mayer, J. W., O. Meyer, N. G. E. Johansson, and S. T. Picraux (1969), Lattice Location of Group I_A, II_B and VI_B Elements Implanted in Silicon, *Bull. Am. Phys. Soc.*, **14,** 1159.

1152 Mayer, J. W., M. Martini, K. R. Zanio, and I. L. Fowler (1970), Influence of Trapping and Detrapping Effects in Si(Li), Ge(Li) and CdTe Detectors, *IEEE Trans. Nucl. Sci.*, **NS-17,** 3.

1153 Mazurczyk, V. J., and H. Y. Fan (1970), Spectral Oscillation of Impurity Photoconductivity and the Photo-Hall Effect in *p*-Type InSb, *Phys. Rev.*, **B1,** 4037.

1154 Mednikov, A. K. (1960), Extraction of Nickel from Germanium, *Sov. Phys. Solid State*, **1,** 1704.

1155 Meilikhov, E. Z. (1965), Determination of the Parameters of Trapping and Recombination Levels in CdS, *Sov. Phys. Solid State*, **7,** 1228.

1156 Meilikhov, E. Z. (1966), Photodielectric Effect and Negative Photoconductivity in Ge at 10^{10}-cps Frequency, *Sov. Phys. Solid State*, **8,** 428.

1157 Melehy, M. A., and W. Shockley (1961), Response of a *p-n* Junction to a Linearly Decreasing Current, *IRE Trans. Electron Devices,* **ED-8,** 135.

1158 Melngailis, I., and A. G. Milnes (1961), The Cryosistor—A Field-Controlled Impact Ionization Switch, *Proc. IRE,* **49,** 1616.

1159 Melngailis, I., and A. G. Milnes (1962), Filamentary Impact Ionization in Compensated Germanium at 4.2°K, *J. Appl. Phys.,* **33,** 995.

1160 Melngailis, I., and P. E. Tannenwald (1969), Far Infrared and Submillimeter Impact Ionization Modulator, *Proc. IEE,* **57,** 806.

1161 Mendel, E., and K. Yang (1969), Polishing of Silicon by the Cupric Ion Process, *Proc. IEEE,* **57,** 1476.

1162 Merz, J. L., R. A. Faulkner, and P. J. Dean (1969), Excitonic Molecule Bound to the Isoelectronic Nitrogen Trap in GaP, *Phys. Rev.,* **188,** 1228.

1163 Meskin, S. S., V. N. Ravich, and B. V. Tsarenkov (1969), Control of the Radiation Spectrum of *p-n* GaP Structures, *Sov. Phys—Semicond.,* **2,** 1157.

1164 Messenger, R. A., and J. S. Blakemore (1971), Photoabsorption Cross Section for Silicon Doped with Indium, *Phys. Rev.* **B4,** 1873.

1165 Mette, H. L., and A. Boatright (1965), Photomagnetoelectric Effect in Thin *p*-Type Silicon Crystals, *Phys. Rev.,* **140,** A919.

1166 Mezei, F., and A. Zawadowski (1971), Kinematic Change in Conduction-Electron Density of State, Due to Impurity Scattering. II. Problem of an Impurity Layer and Tunneling Anomalies, *Phys. Rev.* **B3,** 3127.

1167 Migitaka, M. (1962), Silicon Negative Resistance Diodes, *Japan. J. Appl. Phys.,* **1,** 189.

1168 Migitaka, M. (1965), Anomalous Properties of Silicon Recrystallized Layers Containing Indium Atoms, *J. Appl. Phys.,* **36,** 2139.

1169 Mikhailov, G. B., and Yu. N. Nikolaev (1972), Structure and Capacitance of Diffused *p-n* Junctions With Deep Impurity Centers, *Sov. Phys.—Semicond.,* **6,** 375.

1170 Mikoshiba, N. (1962), Strong-Field Magnetoresistance of Impurity Conduction in *n*-Type Germanium, *Phys. Rev.,* **127,** 1962.

1171 Miles, H. M. (1969), Extrinsic Photoconductivity from Edge Dislocations in Germanium, *J. Appl. Phys.,* **40,** 2720.

1172 Milevskii, L. S. (1961), A Mechanism for the Introduction of Recombination Centers in Germanium and Silicon During Low-Temperature Quenching, *Sov. Phys. Solid State,* **2,** 1980.

1173 Milevskii, L. S. (1962), A Study of the Charge Carrier Lifetime on Different Stages of Saturation of Dislocation Impurity Atmospheres by Copper Atoms, *Sov. Phys. Solid State,* **4,** 311.

1174 Milevskii, L. S. (1962), The Effect of Heat Treatment on the Current Carrier Lifetime in Silicon Doped with Copper, *Sov. Phys. Solid State,* **4,** 606.

1175 Millea, M. F. (1966), The Effect of Heavy Doping on the Diffusion of Impurities in Silicon, *J. Phys. Chem. Solids,* **27,** 315.

1176 Millea, M. F., and L. W. Aukerman (1966), *The Effect of Electron Lifetime on the Electroluminescence of Diffused GaAs Diodes,* Document AD 803 903, National Technical Information Service, Springfield, Virginia.

1177 Miller, A., and E. Abrahams (1960), Impurity Conduction at Low Concentrations, *Phys. Rev.*, **120,** 745.

1178 Miller, S. L. (1955), Avalanche Breakdown in Germanium, *Phys. Rev.*, **99,** 1234.

1179 Milnes, A. G., and D. L. Feucht (1972), *Heterojunctions and Metal-Semiconductor Junctions*, Academic Press, New York and London.

1180 Milton, A. F. (1970), Frequency Dependence of the Debye Length in Compensated Extrinsic Photoconductors, *Appl. Phys. Lett.*, **16,** 285.

1181 Milton, A. F., and M. M. Blouke (1971), Sweepout and Dielectric Relaxation in Compensated Extrinsic Photoconductors, *Phys. Rev.* **B3,** 4312.

1182 Mil'vidskii, M. G., V. B. Osvenskii, and T. G. Yugova (1966), Dislocation Decoration in Gallium Arsenide Crystals, *Sov. Phys. Solid State*, **7,** 2791.

1183 Mil'vidskii, M. G., V. B. Osvenskii, and T. G. Yugova (1969), Some Peculiarities of the Decomposition of Supersaturated Solid Solutions of Copper in Gallium Arsenide, *Sov. Phys. Solid State*, **10,** 2144.

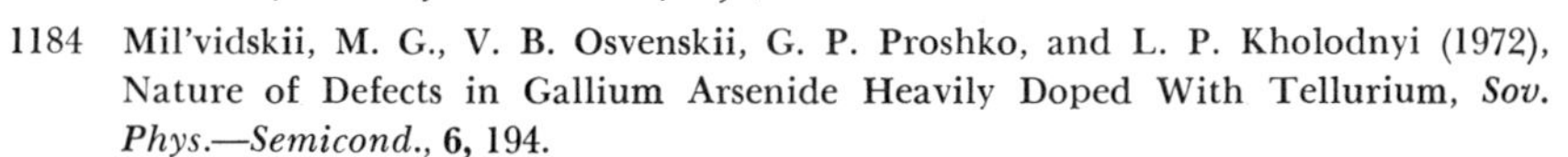

1184 Mil'vidskii, M. G., V. B. Osvenskii, G. P. Proshko, and L. P. Kholodnyi (1972), Nature of Defects in Gallium Arsenide Heavily Doped With Tellurium, *Sov. Phys.—Semicond.*, **6,** 194.

1185 Minden, H. T. (1969), Recent Advances in Gallium Arsenide Materials Technology, *Solid-State Tech.*, **12,** 25.

1186 Mirdzhalilova, M. A., and L. G. Paritskii (1967), Thermally Stimulated EMF of a *p-n* Junction, *Sov. Phys. Solid State*, **8,** 2468.

1187 Mirianashvili, Sh. M., and D. I. Nanobashvili (1971), Impact Ionization in Compensated High-Resistivity *p*-Type InSb, *Sov. Phys.—Semicond.*, **4,** 1606.

1188 Mitchell, I. V., J. W. Mayer, J. K. Kung, and W. G. Spitzer (1971), Investigation of Te-Doped GaAs Annealing Effects by Optical- and Channeling-Effect Measurements, *J. Appl. Phys.*, **42,** 3982.

1189 Miyauchi, T., H. Sonomura, and N. Yamamoto (1969), Electrical Properties of Gallium Phosphide, *Japan. J. Appl. Phys.*, **6,** 1409.

1190 Mogab, C. J., and W. D. Kingery (1968), Preparation and Properties of Noncrystalline Silicon Carbide Films, *J. Appl. Phys.*, **39,** 3640.

1191 Montillo, F., and P. Balk (1971), High-Temperature Annealing of Oxidized Silicon Surfaces, *J. Electrochem. Soc.*, **118,** 1463.

1192 Moore, A. R. (1968), Direct Observation of Acoustoelectric Domains, *Appl. Phys. Lett.*, **13,** 126.

1193 Moore, J. S., C. M. Penchina, N. Holonyak, M. D. Sirkis, and T. Yamada (1966), Electrical Oscillations in Silicon Compensated with Deep Levels, *J. Appl. Phys.*, **37,** 2009.

1194 Moore, J. S., N. Holonyak, Jr., M. D. Sirkis, and M. M. Blouke (1967), Space-Charge Recombination Oscillations in Silicon, *Appl. Phys. Lett.*, **10,** 58.

1195 Moore, J. S., M. C. P. Chang, and C. M. Penchina (1971), Energy Levels in Cobalt Compensated Silicon, *Bull. Am. Phys. Soc.*, **16,** 141.

1196 Moore, M. J. (1969), A Deep Donor Level in *n*-Type Silicon Carbide, *J. Electrochem. Soc.*, **116,** 109.

1197 Moore, W. J. (1970), Magnetic Field Effects on the Excitation Spectra of Group II Double Acceptors in Germanium, *Bull. Am. Phys. Soc.*, **15,** 279.

1198 Mordkovich, V. N. (1963), The Effects of Oxygen on the Electrical Properties of *n*-Type Silicon, *Sov. Phys. Solid State,* **4,** 2662.

1199 Mordkovich, V. N. (1964), The Influence of Oxygen on the Conductivity of Silicon, *Sov. Phys. Solid State,* **6,** 654.

1200 Mordkovich, V. N. (1965), Effect of Oxygen on Recombination in Silicon, *Sov. Phys. Solid State,* **6,** 1716.

1201 Morgan, T. N., M. Pilkuln, and H. Rupprecht (1965), Effect of Deep Levels on the Optical and Electrical Properties of Copper Doped GaAs *p-n* Junctions, *Phys. Rev.,* **138A,** A1551.

1202 Morgan, T. N., B. Welber, and R. N. Bhargava (1968), Optical Properties of Cd–O and Zn–O Complexes in GaP, *Phys. Rev.,* **166,** 751.

1203 Morgan, T. N., T. S. Plaskett, and G. D. Pettit (1969), Pair Spectra Involving Si Donors in GaP, *Phys. Rev.,* **180,** 845; also *Bull. Am. Phys. Soc.* (1969), **14,** 396.

1204 Moriizumi, T., and K. Takahashi (1969), Si- and Ge-Doped GaAs *p-n* Junctions, *Japan. J. Appl. Phys.,* **8,** 348.

1205 Morin, F. J., J. P. Maita, R. G. Shulman, and N. B. Hannay (1954), Impurity Levels in Silicon, *Phys. Rev.,* **96,** 833.

1206 Morozov, A. I., and S. G. Kalashnikov (1962), Trapping Effects in Zinc-Doped Germanium, *Sov. Phys. Solid State,* **3,** 2520.

1207 Mort, J. (1968), Transient Photoconductivity in Trigonal Selenium Single Crystals, *J. Appl. Phys.,* **39,** 3543.

1208 Mort, J. (1971), Electronic Transport in Amorphous Silicon Films: Comments, *Phys. Rev.,* **B3,** 3576.

1209 Morton, G. A. (1965), Infrared Detectors, *RCA Rev.,* **26,** 3.

1210 Moss, T. S. (1961), Optical Absorption Edge in GaAs and Its Dependence on Electric Field, *J. Appl. Phys., Suppl.,* **32,** 2136.

1211 Mott, N. F. (1956), On the Transition to Metallic Conduction in Semiconductors, *Can. J. Phys.,* **34,** 1356.

1212 Mott, N. F., and R. W. Gurney (1940), *Electronic Processes in Ionic Crystals,* Oxford University Press, London and New York, p. 83.

1213 Mott, N. F., and W. D. Twose (1961), The Theory of Impurity Conduction, *Adv. Phys.,* **10,** 107.

1214 Muller, M. W. (1968), Current Filaments in Avalanching *p-i-n* Diodes, *Appl. Phys. Lett.,* **12,** 218.

1215 Muller, M. W., and H. Guckel (1968), Negative Resistance and Filamentary Currents in Avalanching Silicon p^+-i-n^+ Junctions, *IEEE Trans. Electron Devices,* **ED-15,** 560.

1216 Mullin, J. B., A. Royle, and B. W. Straughan (1971), The Preparation and Electrical Properties of InP Crystals Grown by Liquid Encapsulation, *Proceedings of the 3rd International Symposium on Gallium Arsenide and Related Compounds, Aachen, 1971,* The Institute of Physics, London, p. 49.

1217 Munakata, C., and H. Todokoro (1966), A Method of Measuring Lifetime for Minority Carriers Induced by an Electron Beam in Germanium, *Japan. J. Appl. Phys.,* **5,** 249.

1218 Muravakii, B. S., V. S. Gusakov, N. G. Kruzhilina, and A. G. Shued (1966), Current Oscillations in Compensated Germanium and Silicon, *Sov. Phys. Solid State,* **7,** 2748.

1219 Murygin, V. I., and V. S. Rubin (1969), Resonance and Generation of Oscillations in Nickel-Doped Gallium Arsenide *S*-Type Diodes, *Sov. Phys—Semicond.*, **3,** 765.

1220 Murygin, V. I., and V. S. Rubin (1970), Some Investigations of the Electrical Properties and Injection Conductivity of High-Resistivity Nickel-Doped Gallium Arsenide, *Sov. Phys.—Semicond.*, **3,** 810.

1221 Murygin, V. I., and V. S. Rubin (1970), Electroluminescence of *S*-Type Diodes Made of Iron- and Nickel-Doped Gallium Arsenide, *Sov. Plays.—Semicond.*, **3,** 917.

1222 Murzin, V. N., A. I. Demeshina, and L. M. Umarov (1969), Long-Wavelength Infrared Absorption Spectra of *p*-Type InSb, *Sov. Phys.—Semicond.*, **3,** 367.

1223 Murzin, V. N., A. I. Demeshina, and L. M. Umarov (1972), Far-Infrared Absorption Spectra and the Energy Levels of Shallow Acceptor Impurities in InSb, *Sov. Phys.—Semicond.*, **6,** 419.

1224 Musabelov, T. Yu., and V. B. Sandomirskii (1965), Impedance of a Dielectric Diode with Traps, *Sov. Phys. Solid State*, **7,** 1366.

1225 Muss, D. R. (1955), Capacitance Measurements on Alloyed Indium–Germanium Junction Diodes, *J. Appl. Phys.*, **26,** 1514.

1226 Nagae, M. (1958), On the Recombination Proccsses by the Auger Effect, *Prog. Theor. Phys.*, **19,** 339.

1227 Nagae, M. (1958), On the Auger Effect in the Deep Traps in Si, *Prog. Theor. Phys.*, **19,** 341.

1228 Nakamura, M., T. Kato, and N. Oi (1968), A Study of Gettering Effect of Metallic Impurities in Silicon, *Japan. J. Appl. Phys.*, **7,** 512.

1229 Nakano, K., and C. R. Crowell (1970), Deep Level Impurity Effects on the Frequency Dependence of Schottky Barrier Capacitance, *Bull. Am. Phys. Soc.*, **15,** 1614.

1230 Nakano, T., and T. Oku (1967), Temperature Dependence of Recombination Lifetime in Gallium Arsenide Electroluminescent Diodes, *Japan. J. Appl. Phys.*, **6,** 1212.

1231 Nakashima, K., and Y. Inuishi (1968), Recombination Centers in γ-Ray Irradiated Boron-Doped *p*-Type Si, *Japan. J. Appl. Phys.*, **7,** 965.

1232 Nakhmanson, R. S. (1964), Theory of Surface Capacitance, *Sov. Phys. Solid State*, **6,** 859.

1233 Nakhmanson, R. S. (1966), Inclusion of Recombination within the Space-Charge Region Near the Surface of a Semiconductor, *Sov. Phys. Solid State*, **7,** 2776.

1234 Nasledov, D. N., and V. V. Negreskul (1970), Radiative Recombination in Lightly and Heavily Doped *p*-Type GaAs, *Sov. Phys.—Semicond.*, **3,** 1012.

1235 Nathan, M. I., and W. Paul (1962), Effect of Pressure on the Energy Levels of Impurities in Semiconductors, II. Gold in Silicon, *Phys. Rev.*, **128,** 38.

1236 Neizvestnyi, I. G., S. V. Pokrovskaya, and A. V. Rzhanov (1972), Influence of Certain Treatments on Surface Recombination on Germanium, *Sov. Phys.—Semicond.*, **6,** 281.

1237 Nes, E., and J. Washburn (1971), Precipitation in High-Purity Silicon Single Crystals, *J. Appl. Phys.*, **42,** 3562.

1238 Nes, E., and J. Washburn (1972), Precipitate Colonies in Silicon, *J. Appl. Phys.*, **43,** 2005.

1239 Neudorfer, M. L., and S. S. Yee (1969), Computer Analysis of Current Instabilities in Piezoelectric Semiconductors, *IEEE Trans. Electron Devices,* **ED-16,** 1069.

1240 Neuringer, L. J., and W. Bernard (1961a), Generation-Recombination Noise and Capture Cross-Sections in *p*-Type Gold-Doped Germanium, *J. Phys. Chem. Solids,* **22,** 385.

1241 Neuringer, L. J., and W. Bernard (1961b), Generation-Recombination Noise in *p*-Type Gold-Doped Germanium, *Phys. Rev. Lett.,* **6,** 455.

1242 Newhouse, V. L., E. S. Furgason, and T. D. Ellis (1969), A New Negative Magneto-Resistance Effect in Germanium, *Appl. Phys. Lett.,* **14,** 308.

1243 Newman, R. (1955), Optical Properties of Indium-Doped Silicon, *Phys. Rev.,* **99,** 465.

1244 Newman, R., and W. W. Tyler (1959), Photoconductivity in Germanium, *Solid State Physics, Vol. 8* (Ed. F. Seitz and D. Turnbull), Academic Press, New York, p. 49.

1245 Newman, R. C., and J. Wakefield (1961), The Diffusivity of Carbon in Silicon, *J. Phys. Chem. Solids,* **19,** 230.

1246 Newman, R. C., and J. B. Willis (1965), Vibration Absorption of Carbon in Silicon, *J. Phys. Chem. Solids,* **26,** 373.

1247 Nicholas, K. H., and J. Woods (1964), The Evaluation of Electron Trapping Parameters from Conductivity Glow Curves in Cadmium Sulphide, *Br. J. Appl. Phys.,* **15,** 783.

1248 Nicolet, M. A. (1966), Unipolar Space-Charge-Limited Current in Solids with Nonuniform Spacial Distribution of Shallow Traps, *J. Appl. Phys.,* **37,** 4224.

1249 Nikolaeva, E. A., and V. N. Lozovskii (1967), Temperature Dependence of the Distribution Coefficient of Silver in Silicon, *Sov. Phys.—Semicond.,* **1,** 381.

1250 Ning, T. H., and C. T. Sah (1972), Multivalley Effective-Mass Approximation for Donor States in Silicon. I. Shallow-Level Group-V Impurities, *Phys. Rev.,* **B4,** 3468.

1251 Nolle, É. L. (1967), Recombination through Exciton States in Semiconductors, *Sov. Phys. Solid State,* **9,** 90.

1252 Nolle, É. L. (1969), Kinetics of Recombination through Exciton States in Semiconductors, *Sov. Phys.—Semicond.,* **2,** 1397.

1253 Nolle, É. L., V. M. Malovetskaya, and V. S. Vavilov (1962), The Effect of Oxygen on the Lifetime of Nonequilibrium Carriers in *p*-Type Silicon, *Sov. Phys. Solid State,* **4,** 1010.

1254 Northrop, D. C., P. R. Thornton, and K. E. Trezise (1964), Electrical Transients in High-Resistivity Gallium Arsenide, *Solid-State Electron.,* **7, 17.**

1255 Norton, P., and H. Levinstein (1972), Recombination Cross Section for Holes at a Singly Ionized Copper Impurity in Germanium, *Phys. Rev.,* **B6,** 489.

1256 Novototskii-Vlasov, Yu. F. (1965), Role of Water in the Formation of Dominant Surface Recombination Centers on Germanium, *Sov. Phys. Solid State,* **7,** 870.

1257 Nozaki, T., Y. Yatsuragi, and N. Akiyama (1970), Concentration and Behavior of Carbon in Semiconductor Silicon, *J. Electrochem. Soc.,* **117,** 156C.

1258 Nuese, C. J., J. J. Gannon, R. H. Dean, H. F. Gossenberger, and R. E. Enstrom (1972), GaAs Vapor-Grown Bipolar Transistors, *Solid-State Electron.,* **15,** 81.

1259 Nuyts, W., and R. van Overstraeten (1971), Influence of the Incomplete Ionization of Impurities on the Capacitance of *p-n* Junctions, *J. Appl. Phys.*, **42**, 3988.

1260 Nygren, S. F., and G. L. Pearson (1969), Zinc Diffusion into Gallium Phosphide under High and Low Phosphorus Overpressure, *J. Electrochem. Soc.*, **11**, 648.

1261 Oda, J., and Y. Matukura (1967), Uniaxial Stress Effect of Ge *p-n* Junctions Doped with Copper, *Japan. J. Appl. Phys.*, **6**, 411.

1262 Ogawa, T. (1966), Influence of Metal Impurities on Breakdown Characteristics of High-Voltage Silicon n^+p Junctions, *Japan. J. Appl. Phys.*, **5**, 145.

1263 Okada, M., and M. Kikuchi (1964), Observations of Impurity Precipitation in CdSe Single Crystals, *Japan. J. Appl. Phys.*, **3**, 362.

1264 Okazaki, S., and M. Hiramatsu (1968), Experimental Investigation of Double Injection in *p*-Type Silicon, *Japan. J. Appl. Phys.*, **7**, 557.

1265 Oksman, Ya. A., and G. A. Sizova (1969), Hopping Conduction Measured Using an Alternating Current, *Sov. Phys.—Semicond.*, **2**, 1512.

1266 Oksman, Ya. A., and O. M. Smirnov (1969), High-Temperature Photodielectric Effect in Semiconductors, *Sov. Phys.—Semicond.*, **3**, 440.

1267 Oldham, W. G., and S. S. Naik (1972), Admittance of *p-n* Junctions Containing Traps, *Solid-State Electron.*, **15**, 1085.

1268 Omel'yanovskii, E. M., L. Ya. Pervova, E. P. Rashevskaya, N. N. Solov'ev, and V. I. Fistul' (1970), Deep Levels in Semiinsulating Iron-Doped Gallium Arsenide, *Sov. Phys.—Semicond.*, **4**, 316.

1269 Onton, A. (1969), Optical Absorption Due to Excitation of Electrons Bound to Si and S in GaP, *Phys. Rev.*, **186**, 786.

1270 Onton, A., P. Fisher, and A. K. Ramdas (1967), Spectroscopic Investigation of Group-III Acceptor States in Silicon, *Phys. Rev.*, **163**, 686.

1271 Ormont, A. B., E. A. Poltoratskii, and A. E. Yunovich (1966), Radiation Recombination in *p-n* Junctions of Gallium Arsenide in the Case of Weak Currents, *Sov. Phys. Solid State*, **8**, 431.

1272 Osipov, V. V. (1968), Current-Voltage Characteristics of a Four-Layer Structure with Deep Impurity Levels, *Sov. Phys.—Semicond.*, **2**, 184.

1273 Osipov, V. V., and V. A. Kholodnov (1971), Theory of Diodes With Radiative and Nonradiative Impurity Recombination, *Sov. Phys.—Semicond.*, **4**, 1932.

1274 Osipov, V. V., and V. I. Stafeev (1968), Theory of Long Diodes with Negative Resistance, *Sov. Phys.—Semicond.*, **1**, 1486.

1275 Osipov, V. V., and V. A. Kholodnov (1972), Negative Resistance of Compensated Semiconductor Diodes Under Double Injection Conditions, *Sov. Phys.—Semicond.*, **5**, 1218.

1276 Osipov, V. V., and V. A. Kholodnov (1972), Special Features of Current-Voltage Characteristics of Diodes Made of Overcompensated Semiconductors, *Sov. Phys.—Semicond.*, **6**, 381.

1277 Ostroborodova, V. V. (1965), Degeneracy Factors of Impurity Levels and the Analysis of Electrical Properties of Germanium Doped with Gold, *Sov. Phys. Solid State*, **7**, 484.

1278 Ostroborodova, V. V., and P. Dias (1970), Ionization Energy of Zinc in Gallium Phosphide, *Sov. Phys.—Semicond.*, **3**, 1319.

1279 Ostroborodova, V. V., and S. V. Ivanova (1965), Problem of the Ionization Energy and the Degeneracy Factor of the Lower Level of Nickel in Germanium, *Sov. Phys. Solid State,* **6,** 2787.

1280 Ostroborodova, V. V., and L. A. Kandidova (1970), Determination of the Characteristics of Impurity Centers in Semi-Insulating Gallium Arsenide, *Sov. Phys.—Semicond.,* **4,** 892.

1281 Pado, G. S., V. I. Pustovoit, and E. F. Tokarev (1968), Electrical Domain Formation in CdSe Single Crystals by Joule Heating, *Sov. Phys. Solid State,* **10,** 1374.

1282 Pajot, B. (1967), Structure Fine de l'Absorption a 9μ des Groupements Si_2O dans le Silicium a Basse Temperature, *J. Phys. Chem. Solids,* **28,** 73.

* 1283 Palei, V. M., I. M. Vikulin, and N. G. Tonkoshkur (1972), Influence of Trapping Levels on Microplasma Turn-on Delay Time, *Sov. Phys.—Semicond.,* **6,** 74.

1284 Panish, M. B., and H. G. Casey, Jr. (1969), Temperature Dependence of the Energy Gap in GaAs and GaP, *J. Appl. Phys.,* **40,** 163.

1285 Pankey, T., Jr., and J. E. Davey (1970), Effects of Neutron Irradiation on the Optical Properties of Thin Films and Bulk GaAs and GaP, *J. Appl. Phys.,* **41,** 697.

1286 Pankove, J. I. (1971), *Optical Processes in Semiconductors,* Prentice Hall, Englewood Cliffs, New Jersey.

1287 Panousis, P. T. (1968), Charge Storage Effects in *p-i-n* Diodes Containing a Large Trap Density, *IEEE Trans. Electron Devices,* **ED-15,** 688.

1288 Panousis, P. T., R. H. Krambeck, and W. C. Johnson (1969), Field-Enhanced Emission and Constant Field Domains in Semiconductors with Deep Traps, *Appl. Phys. Lett.,* **15,** 79.

1289 Panteleev, V. A., and N. E. Rudoi (1971), Influence of the Level of Doping on the Concentration of Vacancies in Silicon, *Sov. Phys.—Semicond.,* **4,** 1160.

1290 Paramonova, R. A., and A. F. Plotnikov (1963), Some Problems in the Kinetics of Impurity Photoconductivity of Copper-Doped Germanium, *Sov. Phys. Solid State,* **4,** 2077.

1291 Paramonova, R. A., and A. V. Rzhanov (1963), Bulk Recombination in Copper-Doped Germanium Crystals, *Sov. Phys. Solid State,* **4,** 1335.

1292 Parillo, L. C., and W. C. Johnson (1972), Acceptor State of Gold in Silicon-Resolution of an Anomaly, *Appl. Phys. Lett.,* **20,** 104.

1293 Paritskii, L. G. (1964), Dependence of Steady-State Impurity Photoconductivity and Lifetime in a Semiconductor on the Fermi Level, *Sov. Phys. Solid State,* **6,** 845.

1294 Paritskii, L. G., and S. M. Ryvkin (1962), Effect of Trapping Levels on the Relaxation of Photoconductivity in CdS Monocrystals, *Sov. Phys. Solid State,* **3,** 511.

1295 Paritskii, L. G., O. M. Sreseli, B. E. Yakovlev, and V. B. Yarzhembitskii (1971), Negative Photoconductivity of Copper-Doped Germanium, *Sov. Phys.—Semicond.,* **5,** 673.

1296 Park, Y. S., and C. H. Chung (1971), Type Conversion and *p-n* Junction Formation in Lithium-Ion-Implanted ZnSe, *Appl. Phys. Lett.,* **18,** 99.

1297 Parmenter, R. H., and W. Ruppel (1959), Two Carrier Space-Charge-Limited Current in a Trap-Free Insulator, *J. Appl. Phys.,* **30,** 1548.

1298 Pasechnik, Yu. A., and O. V. Snitko (1966), Photoconductivity of Silicon in the Infrared Region of the Spectrum, Due to Surface Electron States, *Sov. Phys. Solid State*, **7,** 2759.

1299 Pataki, G. (1968), Effect of Domain Shape on the Stable Domain Propagation in Hot Electron Semiconductors, *IX International Conference on the Physics of Semiconductors, Moscow,* Nauka, Leningrad.

1300 Pataki, G. (1968), Velocity and Current Determination for Steadily Travelling Domains in Semiconductors, I, Recombination Instability, II, Watkins-Gunn Effect, *Acta Phys. Acad. Sci. Hung.*, **24,** 119 and 377.

1301 Pataki, G. (1969), Effects of Traps on Steadily Travelling Domains in Semiconductors, *Acta Phys. Acad. Sci. Hung.*, **27,** 421.

1302 Patrick, L., and W. J. Choyke (1959), Impurity Bands and Electroluminescence in SiC *p-n* Junctions, *J. Appl. Phys.,* **30,** 236.

1303 Pauling, L. (1960), *The Nature of the Chemical Bond,* Cornell University Press, Ithaca, New York, p. 247.

1304 Pavlichenko, V. I., I. V. Ryzhikov, T. G. Kmita, P. M. Karageorgii-Alkalaev, and A. Yu. Leiderman (1966), Electroluminescence of Silicon Carbide Diodes, *Sov. Phys. Solid State,* **8,** 984.

1305 Pavlov, P. V., D. I. Tetel'baum, E. I. Zorin, and V. I. Alekseev (1967), Distribution of Implanted Atoms and Radiation Defects in the Ion Bombardment of Silicon (Monte Carlo Calculation Method), *Sov. Phys. Solid State,* **8,** 2141.

1306 Peart, R. F. (1966), Self-Diffusion in Intrinsic Silicon, *Phys. Status Solidi,* **15,** K119.

1307 Peka, G. P., and Yu. I. Karkhanin (1972), Energy Spectrum of Deep Levels and the Mechanism of Radiative Recombination in GaAs:Cr, *Sov. Phys—Semicond.,* **6,** 261.

1308 Peka, G. P., V. A. Brodovoi, and L. I. Gorshkov (1972), Field Control of the Intensity of Radiative Recombination Emitted by GaAs:Cu Under Carrier Heating Conditions, *Sov. Phys.—Semicond.,* **5,** 1592.

1309 Pell, E. M. (1971), *Proceedings of the 3rd International Conference on Photoconductivity, Stanford University 1969,* Pergamon, New York.

1310 Penchina, C. M., J. S. Moore, and N. Holonyak, Jr. (1966), Energy Levels and Negative Photoconductivity in Cobalt-Doped Silicon, *Phys. Rev.,* **143,** 634.

1311 Penin, N. A., T. I. Galkina, and N. D. Tyapkina (1969), Radiative Capture of Electrons by Negative Ions and Neutral Atoms of Beryllium in Germanium, *Sov. Phys.—Semicond.,* **3,** 236.

1312 Penning, P. (1956), Annealing of Germanium Supersaturated with Nickel, *Phys. Rev.,* **102,** 1414.

1313 Perel', V. I., and A. L. Éfros (1968), Capacitance of a *p-n* Junction with Deep Impurities, *Sov. Phys.—Semicond.,* **1,** 1403.

1314 Pervova, L. Ya., N. N. Solov'ev, V. I. Fistul', and O. V. Pelevin (1969), Possibility of Constructing a Fast Switch from an Iron-Doped Gallium Arsenide Crystal, *Sov. Phys.—Semicond.,* **3,** 254.

1315 Peters, D. W. (1969), Small-Signal Impedance of Gold-Doped p^+-n-n^+ Silicon Diodes, *IEEE Trans. Electron Devices,* **ED-16,** 828.

1316 Peters, D. W., and M. Shipley (1968), Forward Transient Characteristics of Gold-Doped Silicon p^+-n-n^+ Diodes, *IEEE Trans. Electron Devices*, **ED-15,** 852.

1317 Petritz, R. L. (1959), Fundamentals of Infrared Detectors, *Proc. IRE*, **47,** 1458.

1318 Phillips, J. C. (1970), Dielectric Theory of Impurity Binding Energies. I. Group-V Donors in Si and Ge, *Phys. Rev.* **B1,** 1540.

1319 Phillips, J. C. (1970), Dielectric Theory of Impurity Binding Energies. II. Donor and Isoelectric Impurities in GaP, *Phys. Rev.* **B1,** 1545.

1320 Pickard, P. S., and M. V. Davis (1970), Analysis of Electron Trapping in Alumina Using Thermally Stimulated Electrical Currents, *J. Appl. Phys.*, **41,** 2636.

1321 Picraux, S. T., N. G. E. Johansson, and J. W. Mayer (1969), Hall Measurement on Cd and Te Implanted Silicon, *J. Electrochem. Soc.*, **116,** 147C.

1322 Picus, G. S. (1961), Recombination Processes in Far Infrared Photoconductors, *J. Phys. Chem. Solids*, **22,** 159.

1323 Picus, G. S., and C. J. Buczek (1968), *Far Infrared Laser Receiver Investigation*, Document AD 833 267, National Technical Information Service, Springfield, Virginia.

1324 Pikhtin, A. N., D. A. Yas'kov, Omar A. H. Omar (1971), Impurity and Intervalley Photoconductivity of Gallium Phosphide, *Sov. Phys.—Semicond.*, **4,** 1274.

1325 Pilkuhn, M. H. (1969), Avalanche Breakdown in GaP, *J. Appl. Phys.*, **40,** 3162.

1326 Pines, B. Ya., and I. P. Grebennik (1963), An Electron-Diffraction Study of Heterodiffusion in the Ge–Si System, *Sov. Phys. Cryst.*, **8,** 11.

1327 Pinson, W. E., and R. Bray (1964), Experimental Determination of the Energy Distribution Functions and Analysis of the Energy-Loss Mechanisms of Hot Carriers in p-Type Germanium, *Phys. Rev. A.*, **136,** 1449.

1328 Plotnikov, A. F., V. D. Tkachev, and V. S. Vavilov (1963), Remanent Impurity Photoconductivity Spectra of Silicon Single Crystals, *Sov. Phys. Solid State*, **4,** 2616.

1329 Poehler, T. O. (1971), Magnetic Freezeout and Impact Ionization in GaAs, *Phys. Rev.* **B4,** 1223.

1330 Poehler, T. O., and D. Abraham (1964), Electric Field Excitation of Electrons from Shallow Traps in CdSe Thin-Film Triodes, *J. Appl. Phys.*, **35,** 2452.

1331 Poirier, R., and J. Olivier (1969), Hot Electron Emission from Silicon into Silicon Dioxide by Surface Avalanche, *Appl. Phys. Lett.*, **15,** 364.

1332 Pokrovskii, Ya. E., and K. I. Svistunova (1961), Investigation of Recombination in Silicon Alloyed with Gallium, Indium and Antimony, *Sov. Phys. Solid State*, **3,** 551.

1333 Pokrovskii, Ya. E., and K. I. Svistunova (1962), Radiation Capture of Electrons by Indium Atoms in Silicon, *Sov. Phys. Solid State*, **3,** 2058.

1334 Pokrovskii, Ya. E., and K. I. Svistunova (1964a), Some Peculiarities of Radiative Trapping of Electrons by Indium and Gallium Atoms in Silicon, *Sov. Phys. Solid State*, **5,** 1373.

1335 Pokrovskii, Ya. E., and K. I. Svistunova (1964b), Radiative Capture of Carriers by Impurity Atoms in Silicon aand Germanium, *Sov. Phys. Solid State*, **6,** 13.

1336 Pokrovskii, Ya. E., and K. I. Svistunova (1965a), Impurity Recombination Radiation of Diodes Made of Indium-Doped n-Type Silicon, *Sov. Phys. Solid State*, **7,** 1275.

1337 Pokrovskii, Ya. E., and K. I. Svistunova (1965b), Impurity and Interimpurity Radiative Recombination in Silicon, *Sov. Phys. Solid State,* **7,** 1478.

1338 Pokrovskii, Ya. E., and K. I. Svistunova (1969), On the Problem of Electron Capture Coefficients of the Group III Acceptors in Silicon, *Phys. Status Solidi,* **33,** 517.

1339 Pollak, M. (1962), Some Aspects of Non-Steady-State Conduction in Bands and Hopping Processes, *Proceedings of the International Conference on the Physics of Semiconductors,* Institute of Physics, London.

1340 Pollak, M. (1964), Approximations for the AC Impurity Hopping Conduction, *Phys. Rev.,* **133,** A564.

1341 Pollak, M., and T. H. Geballe (1961), Low-Frequency Conductivity Due to Hopping Processes in Silicon, *Phys. Rev.,* **122,** 1742.

1342 Pollak, M., and D. H. Watt (1969), Study of Impurity Distributions (Mainly Lithium in Silicon) Using AC Hopping Conduction, *Phys. Rev.,* **140,** A87.

1343 Poltoratskii, E. A., and V. M. Stuchebnikov (1966), Diffusion of Beryllium in Gallium Arsenide, *Sov. Phys. Solid State,* **8,** 770.

1344 Polyakov, Yu. I., and B. V. Petukhov (1971), Corrections to the Hydrogen Model of Donors in Semiconductors, *Sov. Phys.—Semicond.,* **5,** 692.

1345 Pomerantz, D. I. (1972), Effects of Grown-In and Process-Induced Defects in Single Crystal Silicon, *J. Electrochem. Soc.,* **119,** 255.

1346 Pontinen, R. E., and T. M. Sanders, Jr. (1965), *Electron Spin Resonance Studies of Shallow Donors in Germanium,* Document AD 616 788, National Technical Information Service, Springfield, Virginia.

1347 Pope, M., and N. E. Geacintov (1969), Excitement in Excitons, *Ind. Res.,* **September 1969,** 68.

1348 Potter, R. M., J. M. Blank, and A. Addamiano (1969), Silicon Carbide Light-Emitting Diodes, *J. Appl. Phys.,* **40,** 2253.

1349 Potts, H. R., and G. L. Pearson (1966), Annealing and Arsenic Overpressure Experiments on Defects in Gallium Arsenide, *J. Appl. Phys.,* **37,** 2098.

1350 Pratt, B., and F. Friedman (1966), Diffusion of Lithium into Ge and Si, *J. Appl. Phys.,* **37,** 1893.

1351 Preier, H. (1968), Electron and Hole-Capture Coefficient of Indium in Silicon at Low Temperatures, *J. Appl. Phys.,* **39,** 194.

1352 Pritchard, R. L. (1959), Transition Capacitance of *p-n* Junctions, *Semicond. Prod.,* **2,** 31.

1353 Proklov, V. V., E. E. Godik, and Ya. E. Pokrovskii (1965), Generation-Recombination Noise in Silicon Doped with Boron and Indium, *Sov. Phys. Solid State,* **7,** 263.

1354 Pruniaux, B. R., and A. C. Adams (1972), Dependence of Barrier Height of Metal Semiconductor Contact (Au-GaAs) on Thickness of Semiconductor Surface Layer, *J. Appl. Phys.,* **43,** 1980.

1355 Ptashchenko, A. A., V. V. Serdyuk, and I. A. Kuz'menko (1966), Infrared Quenching of the Impurity Photoconductivity in Cadmium Sulfide, *Sov. Phys. Solid State,* **8,** 1291.

1356 Pumper, E. Ya., and I. V. Prostoserdova (1964), Diffusion of Zinc in Indium Antimonide, *Sov. Phys. Solid State,* **6,** 692.

1357 Pustovoit, A. K. (1971), Antimony-Copper Interimpurity Recombination in Germanium, *Sov. Phys.—Semicond.*, **5,** 784.

1358 Pustovoit, A. K. (1972), Negative Resistance in the Low-Temperature Breakdown of Compensated *n*-Type Ge, *Sov. Phys.—Semicond.*, **5,** 1535.

1359 Putley, E. H. (1960), *The Hall Effect and Related Phenomena*, Butterworths, London.

1360 Quat, V. T., and M. A. Nicolet (1972), Electron Trapping in Neutron-Irradiated Silicon Studied by Space-Charge-Limited Current, *J. Appl. Phys.*, **43,** 2755.

1361 Quist, T. M. (1968), Copper-Doped Germanium Detectors, *Proc. IEEE,* **56,** 1212.

1362 Raag, V., and R. E. Berlin (1968), A Silicon-Germanium Solar Thermoelectric Generator, *Energy Convers.*, **8,** **161.**

1363 Rado, W. A., W. J. Johnson, and R. L. Crawley (1972), Effect of Aluminum on the Amphoteric Behavior of Silicon in $Ga_{1-x}Al_xAs$, *J. Appl. Phys.,* **43,** 2763.

1364 Rai-Choudhury, P., and D. K. Schroder (1972), Lifetimes and Diode Characteristics in Epitaxial Silicon, *J. Electrochem. Soc.*, **119,** 1580.

1365 Rai-Choudhury, P., A. J. Noreika, and M. I. Theodore (1969), Carbon in Expitaxial Silicon, *J. Electrochem. Soc.*, **116,** 97.

1366 Randall, J. T., and M. H. F. Wilkins (1945), Phosphorescence and Electron Traps, *Proc. Roy. Soc.*, **184A,** 365.

1367 Rao-Sahib, T. S., and D. B. Wittry (1969), Measurement of Diffusion Lengths in *p*-Type Gallium Arsenide by Electron Beam Excitation, *J. Appl. Phys.*, **40,** 3745.

1368 Rashevskaya, E. P., and V. I. Fistul' (1967), Infrared Absorption Spectra of Gallium Arsenide Heavily Doped with Sulfur, Selenium, and Tellurium, *Sov. Phys. Solid State,* **9,** 1443.

1369 Rashevskaya, E. P., and V. I. Fistul' (1968), Infrared Absorption of Gallium Arsenide, Doped with Group VI Impurities, *Sov. Phys. Solid State,* **9,** 2849.

1370 Rau, J. W., and C. R. Kannewurf (1966), Intrinsic Absorption and Photoconductivity in Single Crystal $SiTe_2$, *J. Phys. Chem. Solids,* **27,** 1097.

1371 Ravich, Yu. I. (1961), Determining the Characteristic Parameters of Minority Carriers in Semiconductors from Measurements of Photoconductivity and the Photomagnetic Effect, *Sov. Phys. Solid State,* **3,** 1162.

1372 Reddi, V. G. K. (1963), *Electrical Ionization Phenomenon in Gold-Doped Silicon,* Stanford University Solid State Electronics Laboratory, Technical Report No. 4712–1.

1373 Reddi, V. G. K. (1965), Tunable High-Pass Filter Characteristics of a Special MOS Transistor, *IEEE Trans. Electron Devices,* **ED-12,** 581.

1374 Redfield, D., and J. P. Wittke (1970), Luminescent Properties of Energy Band Tail States, *Bull. Am. Phys. Soc.*, **15,** 255.

1375 Reiss, H., and C. S. Fuller (1956), Influence of Holes and Electrons on the Solubility of Lithium in Boron-Doped Silicon, *Trans. AIME J. Met.*, **206,** 276.

1376 Reiss, H., C. S. Fuller, and F. J. Morin (1956), Chemical Interaction Among Defects in Germanium and Silicon, *Bell Syst. Tech. J.*, **35,** 535.

1377 Rembeza, S. I. (1969), Diffusion of Au in InP Single Crystals, *Sov. Phys.—Semicond.*, **3,** 519.

1378 Reuter, H., and K. Hübner (1971), Impact Ionization in Crossed Fields in Semiconductors, *Phys. Rev.*, **B4,** 2575.

1379 Revesz, A. G., J. Reynolds, and J. Lindmayer (1971), New Aspects of Failure Mechanism in Germanium Tunnel Diodes, *Solid State Electron.*, **14,** 1137.

1380 Rhzanov, A. V. (1962), Recombination Statistics for Carrier Trapping by Excited States of Recombination Centers, *Sov. Phys. Solid State*, **3,** 2680.

1381 Riccius, H. D. (1968), Infrared Lattice Vibrations of Zinc Selenide and Zinc Telluride, *J. Appl. Phys.*, **39,** 4381.

1382 Richman, D. (1964), *"Rare-Earth Doping of GaAs", Synthesis and Characterization of Electronically Active Materials,* RCA Laboratories Report, Document AD 452 788, National Technical Information Service, Springfield, Virginia.

1383 Richman, P. (1968), The Effect of Gold Doping upon the Characteristics of MOS Field-Effect Transistors with Applied Substrate Voltage, *Proc. IEEE*, **56,** 774.

1384 Ridley, B. K. (1963), Specific Negative Resistance in Solids, *Proc. Phys. Soc. (London)*, **82,** 954.

1385 Ridley, B. K. (1965), Propagation of Space-Charge Waves in a Conductor Exhibiting a Differential Negative Resistance, *Proc. Phys. Soc. (London)*, **86,** 637.

1386 Ridley, B. K. (1966), The Influence of Traps on the Watkins-Gunn Effect, *Br. J. Appl. Phys.*, **17,** 595.

1387 Ridley, B. K., and R. G. Pratt (1965), Hot Electrons and Negative Resistance at 20°K in n-Type Germanium Containing Au− Centers, *J. Phys. Chem. Solids*, **26,** 21.

1388 Ridley, B. K., and T. B. Watkins (1961), The Dependence of Capture Rate on Electric Field and the Possibility of Negative Resistance in Semiconductors, *Proc. Phys. Soc.*, **78,** 710.

1389 Ridley, B. K., and T. B. Watkins (1961), Negative Resistance and High Electric Field Capture Rates in Semiconductors, *J. Phys. Chem. Solids*, **22,** 155.

1390 Ridley, B. K., and P. H. Wisbey (1967), Non-Linear Theory of Electrical Domains in the Presence of Trapping, *Br. J. Appl. Phys.*, **18,** 761.

1391 Ripper, J. E. (1968), Theory of Q-Switching and Time Delays in GaAs Junction Lasers, *International Symposium on Gallium Arsenide, Dallas, Texas, 1968,* Institute of Physics, London.

1392 Roberts, G. G., and F. W. Schmidlin (1969), Study of Localized Levels in Semi-Insulators by Combined Measurements of Thermally Activated Ohmic and Space-Charge-Limited Conduction, *Phys. Rev.*, **180,** 785.

1393 Roberts, G. I., and C. R. Crowell (1970), Capacitance Energy Level Spectroscopy of Deep-Lying Semiconductor Impurities Using Schottky Barriers, *J. Appl. Phys.*, **41,** 1767.

1394 Robertson, J. B. (1968), Effect of Beryllium on Oxygen in Silicon, *Bull. Am. Phys. Soc.*, **13,** 1475.

1395 Robertson, J. B., and R. K. Franks (1968), Beryllium as an Acceptor in Silicon, *Solid-State Commun.*, **6,** 825.

1396 Rodriguez, S., and T. D. Schultz (1969), Effects of Resonant Phonon Interactions on Shapes of Impurity Absorption Lines, *Phys. Rev.*, **178,** 1252.

1397 Rodriguez, V., and M.-A. Nicolet (1969), Drift Velocity of Electrons in Silicon at High Electric Fields from 4.2 to 300°K, *J. Appl. Phys.*, **40,** 496.

1398 Rogachev, A. A., and S. M. Ryvkin (1962), Temperature Dependence of the Radiative Recombination Cross Section in Germanium, *Sov. Phys. Solid State*, **4,** 1233.

1399 Rogachev, A. A., and S. M. Ryvkin (1965), Effect of Screening on the Recombination Cross Sections in the Presence of a Coulomb Barrier, *Sov. Phys. Solid State,* **6,** 3003.

1400 Rogachev, A. A., and N. I. Sablina (1966), Disappearance of Impurity Levels in Germanium at High Injection Levels, *Sov. Phys. Solid State,* **8,** 146.

1401 Romanov, O. V., P. P. Konorov, and T. A. Kotova (1969), Influence of Atomic Hydrogen on Electrical Properties of the Surface of Germanium, *Sov. Phys.—Semicond.,* **3,** 98.

1402 Romanov, O. V., Yu. N. Demashov, and A. D. Andreev (1971), Influence of the Adsorption of Fe, Co and Ni Ions on the Surface Properties of Germanium, *Sov. Phys.—Semicond.,* **4,** 1132.

1403 Romanychev, D. A. (1966), Concerning Negative Photoconductivity in Germanium, *Sov. Phys. Solid State,* **8,** 1315.

1404 Ronen, R. S., and P. H. Robinson (1972), Hydrogen Chloride and Chlorine Gettering: An Effective Technique for Improving Performance of Silicon Devices, *J. Electrochem. Soc.,* **119,** 747.

1405 Rose, A. (1955a), Recombination Processes in Insulators and Semiconductors, *Phys. Rev.,* **97,** 322.

1406 Rose, A. (1955b), Space-Charge-Limited Currents in Solids, *Phys. Rev.,* **97,** 1538.

1407 Rose, A. (1964), Comparative Anatomy of Models for Double Injection of Electrons and Holes into Solids, *J. Appl. Phys.,* **35,** 2664.

1408 Rosental, A., and A. Kalda (1972), Third Side of the Lampert Triangle in Fitting Experimental Data, *Phys. Rev.,* **B6,** 4077.

1409 Rosenzweig, W. (1962), Diffusion Length Measurement by Means of Ionizing Radiation, *Bell Syst. Tech. J.,* **41,** 1573.

1410 Rosenzweig, W., W. H. Hackett, Jr., and J. S. Jayson (1969), Kinetics of Red Luminescence in GaP, *J. Appl. Phys.,* **40,** 4477.

1411 Rosier, L. L., and C. T. Sah (1971), Photoionization of Electrons at Sulfur Centers in Silicon, *J. Appl. Phys.,* **42,** 4000.

1412 Ross, E. C., and G. Warfield (1970), Superlinear Current Density vs. Electric Field in *p*-Type Silicon-on-Sapphire, *J. Appl. Phys.,* **41,** 2657.

1413 Rossi, J. A., N. Holonyak, Jr., P. D. Dapkus, J. B. McNeely, and F. V. Williams (1969), Laser-Recombination Transition in *p*-Type GaAs, *Appl. Phys. Lett.,* **15,** 109.

1414 Rossi, J. A., N. Holonyak, Jr., P. D. Dapkus, F. V. Williams, and J. W. Burd (1969), Laser Transitions in *p*-Type GaAs: Si, *J. Appl. Phys.,* **40,** 3289.

1415 Rossiter, E. L., and G. Warfield (1971), Transient Space-Charge-Limited Currents in Amorphous Selenium Thin Films, *J. Appl. Phys.,* **42,** 2627.

1416 Rosztoczy, F. E., and K. B. Wolfstirn (1971), Distribution Coefficient of Germanium in Gallium Arsenide Crystals Grown from Gallium Solution, *J. Appl. Phys.,* **42,** 426.

1417 Rosztoczy, F. E., and K. B. Wolfstirn (1970), Germanium-Doped Gallium Arsenide *J. Appl. Phys.,* **41,** 264.

1418 Rosztoczy, F. E., G. A. Antypas, and C. J. Casau (1971), Distribution Coefficients of Ge, Sn and Te in InP Grown by Liquid Phase Epitaxy, *Proceedings of the 3rd*

International Symposium on Gallium Arsenide and Related Compounds, Aachen, 1971, The Institute of Physics, London, p. 88.

1419 Rubinova, E. E., S. R. Novikov, and R. F. Konopleva (1970), Oscillations of the Photocurrent in Ge with Radiation Defects, *Sov. Phys.—Semicond.*, **3,** 1294.

1420 Ruch, J. C., and G. S. Kino (1968), Transport Properties of GaAs, *Phys. Rev.*, **174,** 921.

1421 Rupprecht, G. (1961), The Capture of Holes in Deep-Lying Acceptor Levels in Gold-Doped Germanium, *J. Phys. Chem. Solids*, **22,** 255.

1422 Rupprecht, G. (1963), Investigation of Surface States on Semiconductors by the Pulsed Field Effect, *Am. N. Y. Acad. Sci.*, **101,** 960.

1423 Ryabinkin, Yu. S. (1965), The Effect of Electric Field Strength on Space-Charge-Limited Current in Dielectrics and Semiconductors, *Sov. Phys. Solid State*, **6,** 2382.

1424 Ryabokon', V. N., and K. K. Svidzinskii (1972), Substitutional Acceptor Impurities in Semiconductors, *Sov. Phys.—Semicond.*, **5,** 1623.

1425 Ryan, R. D. (1969), The Gate Currents of Junction Field-Effect Transistors at Low Temperatures, *Proc. IEEE*, **57,** 1125.

1426 Ryan, R. D., and J. E. Eberhardt (1972), Hole Diffusion Length in High Purity *n* GaAs, *Solid-State Electron.*, **15,** 865.

1427 Rybka, V., M. Yoseli, and M. Aoki (1962), Diffusion of Silver in Gallium Arsenide, *J. Phys. Soc. Japan*, **17,** 1812.

1428 Ryvkin, S. M. (1961), Kinetics of Impurity Photoconductivity, *J. Phys. Chem. Solids*, **22,** 5.

1429 Ryvkin, S. M. (1964a), Photoelectric Effects Associated with Recharging of Impurity Centers, *Physics of Semiconductors, Proceedings of the 7th International Conference, Paris, 1964*, Academic Press, New York and London, Dunod, Editeur, Paris, p. 919.

1430 Ryvkin, S. M. (1964b), *Photoelectric Effects in Semiconductors*, Consultants Bureau, New York.

1431 Ryvkin, S. M., R. Yu. Khansevarov, and I. D. Yaroshetskii (1962), A Study of Impurity Photoconductivity in Germanium Irradiated by X-Rays, *Sov. Phys. Solid State*, **3,** 2331.

1432 Ryvkin, S. M., A. A. Grinberg, and N. I. Kramer (1966), Indirect Optical Transitions in Semiconductors Involving Carrier Interaction, *Sov. Phys. Solid State*, **7,** 1766.

1433 Ryvkin, S. M., O. A. Matveev, A. N. Sakharov, N. B. Strokan, and A. Kh. Khusainov (1971), Use of Copper-Compensated Germanium in the Fabrication of Spectrometric Counters, *Sov. Phys.—Semicond.*, **4,** 1103.

1434 Sacks, H. K. (1970), Low Frequency Oscillations and Deep Impurities in High Resistance GaAs, Ph.D. Thesis, Department of Electrical Engineering, Carnegie-Mellon University, Pittsburgh, Pennsylvania.

1435 Sacks, H. K., and A. G. Milnes (1970), Low Frequency Oscillations in Semi-Insulating Gallium Arsenide, *Int. J. Electron.*, **28,** 565.

1436 Sacks, H. K., and A. G. Milnes (1971), An Experimental Study of High Field Moving Domains Produced by Deep Centers in Semi-Insulating GaAs, *Int. J. Electron.*, **30,** 49.

1437 Sah, C. T. (1967a), Optical Method for the Direct Determination of Carrier Capture Cross Sections and Defect Energy Levels in Semiconductors, *Bull. Am. Phys. Soc.*, **12,** 356.

1438 Sah, C. T. (1967b), The Equivalent Circuit Model in Solid State Electronics—Part I: The Single Energy Level Defects Centers, *Proc. IEEE*, **55,** 654.

1439 Sah, C. T. (1967c), The Equivalent Circuit Model in Solid State Electronics—Part II: The Multiple Energy Level Impurity Centers, *Proc. IEEE*, **55,** 672.

1440 Sah, C. T., and V. G. K. Reddi (1964), Frequency Dependence of the Reverse-Biased Capacitance of Gold-Doped Silicon p^+n Step Junctions, *IEEE Trans. Electron Devices*, **ED-11,** 345.

1441 Sah, C. T., and D. K. Schroder (1968), Effects of Multiply-Charged Gold Impurities on the Breakdown Voltage of Silicon *p-n* Junctions, *Appl. Phys. Lett.*, **12,** 141.

1442 Sah, C. T., and W. Shockley (1958), Electron-Hole Recombination Statistics in Semiconductors through Flaws with Many Charge Conditions, *Phys. Rev.*, **109,** 1103.

1443 Sah, C. T., and A. F. Tasch, Jr. (1967), Precise Determination of the Multiphonon and Photon Carrier Generation Properties Using the Impurity Photovoltaic Effect in Semiconductors, *Phys. Rev. Lett.*, **19,** 69.

1444 Sah, C. T., and L. D. Yau (1969), Observation of the Ideal Generation-Recombination Noise Spectrum and Spectra with Voltage Variable Relaxation Time in Gold-Doped Silicon, *Appl. Phys. Lett.*, **14,** 267.

1445 Sah, C. T., R. N. Noyce, and W. Shockley (1957), Carrier Generation and Recombination in *p-n* Junctions and *p-n* Junction Characteristics, *Proc. IEEE*, **45,** 1228.

1446 Sah, C. T., S. Y. Wu, and F. H. Hielscher (1966), The Effects of Fixed Bulk Charge on the Thermal Noise in Metal-Oxide-Semiconductor Transistors, *IEEE Trans. Electron Devices*, **ED-13,** 410.

1447 Sah, C. T., A. F. Tasch, Jr., and D. K. Schroder (1967), Recombination Properties of the Gold Acceptor Level in Silicon Using the Impurity Photovoltaic Effect, *Phys. Rev. Lett.*, **19,** 71.

1448 Sah, C. T., L. Forbes, L. I. Rosier, A. F. Tasch, Jr., and A. B. Tole (1969), Thermal Emission Rates of Carriers at Gold Centers in Silicon, *Appl. Phys. Lett.*, **15,** 145.

1449 Sah, C. T., L. I. Rosier, and L. Forbes (1969a), Low-Temperature High-Frequency Capacitance Measurements of Deep-and Shallow-Level Impurity Center Concentrations, *Appl. Phys. Lett.*, **15,** 161.

1450 Sah, C. T., L. I. Rosier, and L. Forbes (1969b), Direct Observation of the Multiplicity of Impurity Charge States in Semiconductors from Low-Temperature High-Frequency Photo-Capacitance, *Appl. Phys. Lett.*, **15,** 316.

1451 Sah, C. T., L. Forbes, and W. H. Chan (1970a), A New Model of Negative Photocurrent, *Bull. Am. Phys. Soc.*, **15,** 1590.

1452 Sah, C. T., L. Forbes, L. I. Rosier, and A. F. Tasch, Jr. (1970b), Thermal and Optical Emission and Capture Rates and Cross Sections of Electrons and Holes at Imperfection Centers in Semiconductors from Photo and Dark Junction Current and Capacitance Experiments, *Solid-State Electron.*, **13,** 759.

1453 Sah, C. T., W. W. Chan, H. S. Fu, and J. W. Walker (1972), Thermally Stimulated Capacity (TSCAP) in *p-n* Junctions, *Appl. Phys. Lett.*, **20,** 193.

1454 Saito, H., and M. Hirata (1963), Nature of Radiation Defects in Silicon Single Crystals, *Japan. J. Appl. Phys.*, **2,** 678.

1455 Sal'kov, E. A., and M. K. Sheinkman (1963), A Method for Determining the Recombination-Level Parameters in Unipolar Photoconductors, *Sov. Phys. Solid State*, **5,** 289.

1456 Sandiford, D. J. (1957), Carrier Lifetime in Semiconductors for Transient Conditions, *Phys. Rev.*, **105,** 524.

1457 Sansbury, J. D. (1969), Conductivity and Hall Mobility of Ion-Implanted Silicon in Semi-Insulating Gallium Arsenide, *Appl. Phys. Lett.*, **14,** 311.

1458 Sarver, C. E., and J. S. Blakemore (1968), Photoconductivity Associated with Indium Acceptors in Silicon, *Bull. Am. Phys. Soc.*, **13,** 497.

1459 Sasaki, W. (1966), Negative Magnetoresistance in the Metallic Impurity Conduction, Proceedings of the International Conference on the Physics of Semiconductors, Kyoto, Japan, 1966, *J. Phys. Soc. Japan, Suppl.*, **21,** 543.

1460 Sato, S., and C. T. Sah (1970), Measurements of the Thermal-Emission Rates of Electrons and Holes at the Gold Centers in Silicon Using the Small-Signal-Pulsed Field Effect, *J. Appl. Phys.*, **41,** 4175.

1461 Sato, S., S. Kawaji, and A. Kobayashi (1964), Deep Trap Levels in Silicon *p-n* Junctions, *Japan. J. Appl. Phys.*, **3,** 496.

1462 Sato, S., S. Kawaji, and A. Kobayashi (1965), Electrical Properties of Carrier Generation-Recombination Centers in Silicon *p-n* Junctions, *J. Appl. Phys.*, **36,** 3779.

1463 Saul, R. H. (1969), Effect of Etching on the Efficiency and Emission Pattern of Annealed GaP Electroluminescent Diodes, *J. Appl. Phys.*, **40,** 4978.

1464 Saul, R. H., and W. H. Hackett (1970), Distribution of Impurities in Zn, O-Doped GaP Liquid Phase Epitaxy Layers, *J. Electrochem. Soc.*, **117,** 921.

1465 Saul, R. H., and W. H. Hackett, Jr. (1972), On the Incorporation of Oxygen in GaP Liquid Phase Epitaxy Layers, *J. Electrochem. Soc.*, **119,** 542.

1466 Saul, R. H., J. Armstrong, and W. H. Hackett, Jr. (1969), GaP Red Electroluminescent Diodes with an External Quantum Efficiency of 7%, *Appl. Phys. Lett.*, **15,** 229.

1467 Saunders, A. F., and G. T. Wright (1970), Interface States in the Silicon/Silicon-Dioxide System Observed by Thermally Stimulated Charge Release, *Electron. Lett.*, **6,** 207.

1468 Saunders, I. J. (1967), The Relationship between Thermally Stimulated Luminescence and Thermally Stimulated Conductivity, *Br. J. Appl. Phys.*, **18,** 1219.

1469 Saunders, I. J. (1968), A Gallium Arsenide Double Injection Diode, *Solid-State Electron.*, **11,** 1165.

1470 Saunders, I. J., and R. H. Jewitt (1965), *Annual Report on Research Project R7-27, Thin Film Circuit Element,* Document AD 480 752, National Technical Information Service, Springfield, Virginia.

1471 Sawamoto, K., M. Takahashi, H. Tomishima, and H. Toyoda (1966), Studies of Surface States in Si Crystal by Pulsed Field Effect, *Japan J. Appl. Phys.*, **5,** 191.

1472 Schade, H. (1969), Studies of Electron Bombardment Damage in GaAs by Thermally Stimulated Conductivity Measurements, *J. Appl. Phys.*, **40,** 2613.

1473 Schade, H., and D. Herrick (1969), Determination of Deep Centers in Silicon by Thermally Stimulated Conductivity Measurements, *Solid-State Electron.*, **12,** 857.

1474 Schade, H. C., C. J. Nuese, and D. Herrick (1970), Defect Centers in $GaAs_{1-x}P_x$ Electroluminescent Diodes Due to High-Energy Electron Irradiation, *J. Appl. Phys.*, **41,** 3783.

1475 Schafft, H. A., and S. Needham (1968), *A Bibliography on Methods for the Measurements of Inhomogeneities in Semiconductors*, Document AD 671 524, National Technical Information Service, Springfield, Virginia.

1476 Schairer, W., and N. Stath (1972), Impact Ionization of Donors in Semiconductors as a Tool For Photoluminescence Investigations, *J. Appl. Phys.*, **43,** 447.

1477 Scharfetter, D. L., and T. E. Seidel (1969), Comments on Static Negative Resistance in Avalanching Silicon p^+-i-n^+ Junctions, *IEEE Trans. Electron Devices,* **ED-16,** 970.

1478 Scher, H., and M. Lax (1969), Self-Consistent Calculation of Diffusion for Impurity Hopping Conduction, *Bull. Am. Phys. Soc.*, **14,** 311.

1479 Schibli, E. G. (1972), Determination of Impurity Profiles in Presence of Deep Levels by the Second-Harmonic Method, *Solid-State Electron.*, **15,** 137.

1480 Schibli, E., and A. G. Milnes (1967a), Deep Impurities in Silicon, *Mater. Sci. Eng.*, **2,** 173.

1481 Schibli, E., and A. G. Milnes (1967b), Lifetime and Capture Cross-Section Studies of Deep Impurities in Silicon, *Mater. Sci. Eng.*, **2,** 229.

1482 Schibli, E., and A. G. Milnes (1967c), Injected Carrier Flow in a Semi-Insulator Containing a Density Gradient of a Deep Impurity, *Solid-State Electron.*, **10,** 97.

1483 Schibli, E., and A. G. Milnes (1968), Effects of Deep Impurities on n^+-p Junction Reverse-Biased Small-Signal Capacitance, *Solid-State Electron.*, **11,** 323.

1484 Schlegel, E. S., and G. L. Schnable (1969), Test Structures for Study of Surface Effects, *IEEE Trans. Electron Devices*, **ED-16,** 387.

1485 Schmidlin, F. W. (1970), Electrical Switching Device Based on Charge-Controlled Double Injection, *Phys. Rev.* **B1,** 1583.

1486 Schroder, D. K. (1968), Impedance and Other Properties of Gold-Doped Silicon p-n Junctions, *Diss. Abstr.*, **29,** 735-B.

1487 Schroder, D. K., and J. Guldberg (1971), Interpretation of Surface and Bulk Effects Using the Pulsed MIS Capacitor, *Solid-State Electron.*, **14,** 1285.

1488 Schroder, D. K., A. F. Tasch, Jr., and C.-T. Sah (1968), The Spatial Variation of the Charged Gold Concentration in Silicon p-n Step Junctions, *IEEE Trans. Electron Devices*, **ED-15,** 553.

1489 Schultz, M. L., and B. K. Ridley (1969), Trap-Controlled Bunching of Electrons in Acoustoelectric Domains, *Solid-State Commun.*, **7,** 1027.

1490 Schultz, M. L., W. E. Harty, and C. D. Rowley (1962), *Report*, RCA Laboratories, Princeton, New Jersey.

1491 Schwob, H. P., and I. Zschokke-Gränacher (1972), A New Approach to Double Injection, *Solid-State Electron.*, **15,** 271.

1492 Sclar, N., and E. Burstein (1957), Impact Ionization of Impurities in Germanium, *J. Phys. Chem. Solids*, **2,** 1.

1493 Scrifes, D. R., N. Holonyak, Jr., R. D. Burnham, and H. R. Zwicker (1971), Optical Phase Shift Measurement (77°K) of Carrier Decay Time in Direct GaAsP, *Solid-State Electron.*, **14,** 949.

1494 Scrifes, D. R., *et al* (1972), Stimulated Emission and Laser Operation (cw 77°K) of Direct and Indirect $GaAs_{1-x}P_x$ on Nitrogen Isoelectronic Trap Transitions, *J. Appl. Phys.*, **43,** 2368.

1495 Sedgwick, T. O. (1968), Dominant Surface Electronic Properties of SiO_2-Passivated Ge Surfaces as a Function of Various Annealing Treatments, *J. Appl. Phys.*, **39,** 5066.

1496 Sedgwick, T. O., and S. Krongelb (1970), Negative Surface Charge in Ge MOS Structures, *J. Electrochem. Soc.*, **117,** 1199.

1497 Seeger, A. (1965), *Festkorperprobleme, Bd. IV,* Vieweg, Braunschweig.

1498 Seki, H., and I. P. Batra (1971), Photocurrents Due to Pulse Illumination in the Presence of Trapping II, *J. Appl. Phys.*, **42,** 2407.

1499 Selway, P. R., and W. M. Nicolle (1969), Negative Resistance in Chromium-Doped GaAs *p-i-n-* Diodes, *J. Appl. Phys.*, **40,** 4087.

1500 Semak, D. G. (1967), Thermally Stimulated Current Method Under Photo-Electret Conditions, *Sov. Phys. Solid State,* **9,** 966.

1501 Senechal, R. R. (1968), Comment on the Photocurrent Determination of the Emission Rates of the Gold Acceptor Level in Silicon, *J. Appl. Phys.*, **39,** 3515.

1502 Senechal, R. R., and J. Basinski (1968), Capacitance of Junctions on Gold-Doped Silicon, *J. Appl. Phys.*, **39,** 3723.

1503 Shah, J., and Y. Yacoby (1968), Electrical and Optical Enhancement of Photoconductivity in Semiinsulating GaAs, *Phys. Rev.*, **174,** 932.

1504 Shaklee, F. S., J. B. Larkin, and D. L. Kendall (1969), Heat Treatments of Gold-Doped Silicon Diodes, *Proc. IEEE,* **57,** 1481.

1505 Sharma, B. L. (1970), *Diffusion in Semiconductors,* Trans Tech Publications, Clausthal-Zellenfeld, Germany.

1506 Shashkov, Yu. M., and N. Ya. Shushlebina (1964), Growth of Silicon Carbide from a Silicon Melt, *Sov. Phys. Solid State,* **6,** 1134.

1507 Shaw, D. (1962), Group IVB Impurities in Aluminium Antimonide, *Proc. Phys. Soc. (London),* **80,** 161.

1508 Sheinkman, M. K. (1964), On a Possible Recombination Mechanism at Multicharged Centers in Semiconductors, *Sov. Phys. Solid State,* **5,** 2035.

1509 Sheinkman, M. K. (1965), Possibility of Auger Recombination in Multiply Charged Centers in Germanium and Silicon, *Sov. Phys. Solid State,* **7,** 18.

1510 Sheinkman, M. K., and G. L. Belen'kii (1969), Radiative Recombination in Unactivated ZnSe Single Crystals, *Sov. Phys.—Semicond.*, **2,** 1360.

1511 Sheinkman, M. K., I. Ya. Gorodetskii, and I. B. Ermolovich (1966), Temperature Dependences of the Cross Sections for the Capture of Electrons by Recombination Centers in CdS and CdSe, *Sov. Phys. Solid State,* **7,** 2538.

1512 Sheinkman, M. K., I. B. Ermolovich, and G. L. Belen'kii (1968), Nature of Infrared Luminescence ($\lambda_m = 1.2\ \mu$) in CdSe Single Crystals and Its Relationship to Photoconductivity, *Sov. Phys. Solid State,* **10,** 1394.

1513 Shepherd, W. H., and J. A. Turner (1962), Iron-Boron Pairing in Silicon, *J. Phys. Chem. Solids,* **23,** 1697.

1514 Sher, A. H., and W. J. Keery (1972), Improved Infrared-Response Technique for Determining Impurity and Defect Levels in Semiconductors, *Appl. Phys. Lett.*, **20,** 120.

1515 Shifrin, G. A., O. J. Marsh, and J. W. Mayer (1968), Donor Behavior in Thallium Implanted Silicon, *Bull. Am. Phys. Soc.*, **13,** 376.

1516 Shih, K. K., and G. D. Pettit (1968), Properties of GaP Green-Light-Emitting Diodes Grown by Liquid-Phase Epitaxy, *J. Appl. Phys.*, **39,** 5025.

1517 Shih, K. K., J. M. Woodall, S. E. Blum, and L. M. Foster (1968), Efficient Green Electroluminescence from GaP *p-n* Junctions Grown by Liquid-Phase Epitaxy, *J. Appl. Phys.*, **39,** 2962.

1518 Shimizu, T. (1967), Electronic Structure of Impurity Levels in Semiconductors I, *Electr. Electron. Abstr.*, **70,** Abstract 2244.

1519 Shirafuji, J. (1966), Temperature Dependence of the Photoconductive Lifetime in *n*-Type Gallium Arsenide, *Japan J. Appl. Phys.*, **5,** 469.

1520 Shirafuji, J. (1968), Temperature Dependence of the Photoconductive Lifetime in *n*-Type Gallium Arsenide Diffused with Copper, *Japan. J. Appl. Phys.*, **7,** 1074.

1521 Shirshov, Yu. M., V. A. Tyagai, and O. V. Snitko (1969), Detection of Discrete Trapping Levels on the Surface of Cadmium Sulfide Single Crystals, *Sov. Phys.—Semicond.*, **3,** 89.

1522 Shishiyanu, F. S., and B. I. Boltaks (1966), Energy Levels of Ag and Au in GaAs, *Sov. Phys. Solid State,* **8,** 1053.

1523 Shklovskii, B. I., and I. S. Shlimak (1972), Hopping Conduction in Germanium, *Sov. Phys.—Semicond.*, **6,** 104.

1524 Shockley, W. (1950), *Electrons and Holes in Semiconductors,* D. Van Nostrand, Princeton, New Jersey.

1525 Shockley, W. (1958), Electrons, Holes and Traps, *Proc. IRE,* **46,** 973.

1526 Shockley, W., and J. T. Last (1957), Statistics of the Charge Distribution for a Localized Flaw in a Semiconductor, *Phys. Rev.*, **107,** 392.

1527 Shockley, W., and W. T. Read, Jr. (1952), Statistics of the Recombinations of Holes and Electrons, *Phys. Rev.*, **87,** 835.

1528 Shohno, K. (1965), Current Oscillations in the Forward Direction of Si-pin-Diodes, *Japan. J. Appl. Phys.*, **4,** 699.

1529 Shulman, R. G. (1957), Tight-Bonding Calculations of Acceptor Energies in Germanium and Silicon, *J. Phys. Chem. Solids,* **2,** 115.

1530 Shul'pina, I. L., and E. K. Kov'ev (1967), X-Ray Diffraction Investigation of Defects Generated in Silicon During Deep Diffusion of Phosphorus, *Sov. Phys. Solid State,* **9,** 791.

1531 Shul'pina, I. L., L. V. Lainer, M. G. Mil'vidskii, and E. P. Rashevskaya (1967), Investigation of Defects Generated During the Heat Treatment of Oxygen-Bearing Silicon, *Sov. Phys. Solid State,* **9,** 1291.

1532 Shul'pina, I. L., A. I. Zaslavskii, and T. T. Dedegkaev (1968), Investigation of Precipitates Formed in Silicon During Heat Treatment, *Sov. Phys. Solid State,* **10,** 1070.

1533 Shuman, V. B. (1967), Diffusion of Gold in Silicon, *Sov. Phys.—Semicond.*, **1,** 790.

1534 Sidorov, V. I. (1964), Low-Temperature Electrical Conductivity and Hall Effect in Zinc-Doped Germanium, *Sov. Phys. Solid State*, **5,** 2199.

1535 Sidorov, V. I., T. E. Shushko, and A. Ya. Shul'man (1967), Investigation of Optical Absorption in Germanium Doped with Zinc and Compensated with Antimony, *Sov. Phys. Solid State*, **8,** 1608.

1536 Simhony, M., and R. Williams (1969), Impact Ionization of Filled Traps in Cadmium Sulfide, *J. Appl. Phys.*, **40,** 691.

1537 Simmons, J. G., and G. W. Taylor (1971), Nonequilibrium Steady-State Statistics and Associated Effects for Insulators and Semiconductors Containing an Arbitrary Distribution of Traps, *Phys. Rev.* **B4,** 502.

1538 Simmons, J. G., and G. W. Taylor (1972), Dielectric-Relaxation Currents in Insulators, *Phys. Rev.*, **B5,** 553.

1539 Singhal, S. P. (1971), Isolated Interstitials in Silicon, *Phys. Rev.*, **B4,** 2497.

1540 Singhal, S. P. (1972), Isolated Interstitials in Silicon. II, *Phys. Rev.*, **B5,** 4203.

1541 Sirota, N. N., and F. P. Korshunov (1969), Effect of γ Radiation on the Parameters and Structure of Silicon p-n Junctions, *Sov. Phys.—Semicond.*, **2,** 1389.

1542 Sixou, P., and G. Nuzillat (1972), Observation of Gold Levels in Silicon by MOS Capacitance, *Solid-State Electron.*, **15,** 945.

1543 Sklensky, A. F., and R. H. Bube (1972), Photoelectronic Properties of Zinc Impurity in Silicon, *Phys. Rev.*, **B6,** 1328.

1544 Sliva, P. O., and R. Bray (1965), Oscillatory Current Behavior in GaSb and Its Relation to Spontaneous Generation and Amplification of Ultrasonic Flux, *Phys. Rev. Lett.*, **14,** 372.

1545 Smirnov, L. S., V. S. Vavilov, and N. N. Gerasimenko (1963), Kinetics of Recombination Radiation of Silicon, *Sov. Phys. Solid State*, **4,** 1927.

1546 Smirnov, V. N., and G. N. Talalakin (1969), Electroluminescence in n-Type GaAs Crystals and the Acoustoelectrical Current Instability, *Sov. Phys. Solid State*, **10,** 2209.

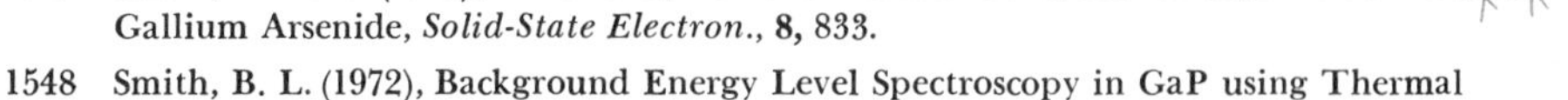

1547 Smith, A. W. (1965), Electro-Optic Observation of Space Charge Effects in Gallium Arsenide, *Solid-State Electron.*, **8,** 833.

1548 Smith, B. L. (1972), Background Energy Level Spectroscopy in GaP using Thermal Release of Trapped Space Charge in Schottky Barriers, *Appl. Phys. Lett.*, **21,** 350.

1549 Smith, E. F., and P. T. Landsberg (1966), Phonon Cascade Theory, *J. Phys. Chem. Solids*, **27,** 1727.

1550 Smith, G. E., and D. Kahng (1962), Negative Resistance of Silicon p-n Junctions at 4.2°K, *Solid-State Electron.*, **5,** 177.

1551 Smith, J. E., Jr., M. I Nathan, J. C. McGroddy, S. A. Porowski, and W. Paul (1969), Gunn Effect in n-Type InSb, *Appl. Phys. Lett.*, **15,** 242.

1552 Smith, P. C., and A. G. Milnes (1970), Solid Solubilities of Ag, Co and Pt in Silicon by Neutron Activation Analysis, Extended Abstracts of the 138th National Meeting of the Electrochemical Society; also *J. Electrochem. Soc.*, **117,** 260C (1970).

1553 Smith, P. C., and A. G. Milnes (1971), A Photoconductivity Decay Model Applied to Silver-Doped Silicon, *Int. J. Electron.*, **30,** 255.

1554 Smith, P. C., and A. G. Milnes (1972), Thermally Stimulated Current Measurements Applied to Silver-Doped Silicon, *Int. J. Electron.*, **32,** 697.

1555 Smith, R. A. (1958), *Semiconductors*, Cambridge University Press, London and New York, p. 62.

1556 Sohm, J. C. (1961), Ionisation par Choc des Impurites dans le Silicium, *J. Phys. Chem. Solids*, **18**, 181.

1557 Sokolov, V. I., and F. S. Shishiyanu (1964), Diffusion and Solubility of Gold in Gallium Arsenide, *Sov. Phys. Solid State*, **6**, 265.

1558 Sokolov, V. I., V. V. Makarov, E. N. Mokhov, G. A. Lomakina, and Yu. A. Vodakov (1969), Luminescence of Beryllium-Doped Silicon Carbide, *Sov. Phys. Solid State*, **10**, 2383.

1559 Sokolova, A. A., V. S. Vavilov, A. F. Plotnikov, and V. A. Chapnin (1969), Photoelectric Properties of Cadmium Telluride Containing Acceptors with Deep Levels, *Sov. Phys.—Semicond.*, **3**, 162.

1560 Sokolova, E. B. (1970), Cascade Capture of Carriers by a Linear Dislocation, *Sov. Phys.—Semicond.*, **3**, 1266.

1561 Solov'eva, E. V., and Yu. F. Lyutov (1971), Ionization Energy of Accidental Impurities in Gallium Arsenide, *Sov. Phys.—Semicond.*, **4**, 1545.

1562 Solov'eva, E. V., Yu. F. Lyutov, and V. M. Yakub (1971), Energy Spectrum of Centers in Undoped Gallium Arsenide, *Sov. Phys.—Semicond.*, **5**, 139.

1563 Sommers, H. S., Jr., and W. B. Teutsch (1964), Demodulation of Low-Level Broad-Band Optical Signals with Semiconductors: Part II—Analysis of the Photoconductive Detector, *Proc. IEEE*, **52**, 144.

1564 Sondaevskii, V. P., and V. I. Stafeev (1964), Injection in Semiconductors with Deep Impurity Levels, *Sov. Phys. Solid State*, **6**, 63.

1565 Sopryakov, V. I., V. D. Tkachev, A. V. Yukhnevich, and A. M. Yanchenko (1970), Characteristic Features of the Forward Branches of the Current-Voltage Characteristics of Silicon Diodes, *Sov. Phys.—Semicond.*, **3**, 1315.

1566 Spears, D. L., and R. Bray (1968), Optical Probing of Inhomogeneities in n-GaAs with Applications to the Acoustoelectric Instabilities, *J. Appl. Phys.*, **39**, 5093.

1567 Spitzer, W. G., and W. Allred (1968), Local-Mode Absorption and Defects in Compensated Silicon-Doped Gallium Arsenide, *J. Appl. Phys.*, **39**, 4999.

1568 Spitzer, W. G., and M. B. Panish (1969), Silicon-Doped Gallium Arsenide Grown from Gallium Solution: Silicon Site Distribution, *J. Appl. Phys.*, **40**, 4200.

1569 Spitzer, W. G., and M. Waldner (1965), Localized Mode Measurements of Boron- and Lithium-Doped Silicon, *J. Appl. Phys.*, **36**, 2450.

1570 Spitzer, W. G., T. E. Firle, M. Cutler, R. G. Shulman, and M. Becker (1955), Measurement of the Lifetime of Minority Carriers in Germanium, *J. Appl. Phys.*, **26**, 414.

1571 Spitzer, W. G., W. P. Allred, and S. E. Blum (1969), Vibrational Modes of Defects in GaP, *J. Appl. Phys.*, **40**, 2589.

1572 Sprokel, G. J. (1965), Interstitial-Substitutional Diffusion in a Finite Medium, Gold into Silicon, *J. Electrochem. Soc.*, **112**, 807.

1573 Sprokel, G. J., and J. M. Fairfield (1965), Diffusion of Gold into Silicon Crystals, *J. Electrochem. Soc.*, **112**, 200.

1574 Srour, J. R., and O. L. Curtis, Jr. (1970), Method for Determining Capture-Probability Temperature Dependences in Semiconductors, *J. Appl. Phys.*, **41**, 4200.

1575 Srour, J. R. and O. L. Curtis, Jr., (1972), Techniques for Obtaining Recombination-Center Parameters from Carrier Lifetime Studies, *J. Appl. Phys.*, **43**, 1782.

1576 Stafeev, V. I. (1959), Modulation of Diffusion Length as a New Principle of Operation of Semiconductor Devices, *Sov. Phys. Solid State*, **1**, 763.

1577 Stafeev, V. I. (1961), Forward Branch of the Current-Voltage Characteristics of a Nonsymmetric Diode, *Sov. Phys. Solid State*, **3**, 135.

1578 Stafeev, V. I. (1962), Photoconductivity in Semiconductor Diodes Induced by Carrier Lifetime Changes, *Sov. Phys. Solid State*, **3**, 1829.

1579 Stafeev, V. I. (1964), Investigation of Some Properties of Gold-Doped Germanium, *Sov. Phys. Solid State*, **5**, 2267.

1580 Staflin, T. (1965), Photo-Induced Infrared Absorption in Doped Silicon, *J. Phys. Chem. Solids*, **26**, 563.

1581 Starosel'tseva, S. P., V. S. Kulov, and S. G. Metreveli (1972), Indium Phosphide S-Type Diodes, *Sov. Phys.—Semicond.*, **5**, 1603.

1582 Statler, R. L., and B. J. Faraday (1967), Thermal Annealing of Silicon Solar Cells Irradiated by High-Energy Electrons, *Bull. Am. Phys. Soc.*, **12**, 205.

1583 Statz, H. (1963), Maximum Solid Solubility and Distribution Coefficient of Impurities in Germanium and Silicon, *J. Phys. Chem. Solids*, **24**, 699.

1584 Stein, H. J. (1966), Comparison of Neutron and Gamma-Ray Damage in n-Type Silicon, *J. Appl. Phys.*, **37**, 3382.

1585 Stein, H. J. (1967), Energy Dependence of Neutron Damage in Silicon, *J. Appl. Phys.*, **38**, 204.

1586 Stein, H. J. (1968), Electrical Studies of Neutron-Irradiated p-Type Silicon: Defect Structure and Annealing, *J. Appl. Phys.*, **39**, 5283.

1587 Stein, H. J., and W. Beezhold (1970), Localized Modes and Divacancy Absorption in Oxygen Ion Implanted Si, *Appl. Phys. Lett.*, **17**, 442.

1588 Stein, H. J., and R. Gereth (1968), Introduction Rates of Electrically Active Defects in n- and p-Type Silicon by Electron and Neutron Irradiation, *J. Appl. Phys.*, **39**, 2890.

1589 Stein, H. J., F. L. Vook, and J. A. Borders (1969), Direct Evidence of Divacancy Formation in Silicon by Ion Implantation, *Appl. Phys. Lett.*, **14**, 328.

1590 Sterkhov, V. A., V. A. Panteleev, and P. V. Pavlov (1967), Diffusion and Electrotransport of Indium and Silver Along Dislocations in Silicon, *Sov. Phys. Solid State*, **9**, 533.

1591 Stevenson, D. T., and R. J. Keyes (1955), Measurement of Carrier Lifetime in Germanium and Silicon, *J. Appl. Phys.*, **26**, 190.

1592 Stiles, P. J., L. L. Chang, L. Esaki, and R. Tsu (1970), High-Field Photoconductivity of Amorphous GeTe and GeSe Films, *Appl. Phys. Lett.*, **16**, 380.

1593 Stillman, G. E., C. M. Wolfe, I. Melngailis, C. D. Parker, P. E. Tannenwald, and J. O. Dimmock (1968), Far Infrared Photoconductivity in High-Purity Epitaxial GaAs, *Appl. Phys. Lett.*, **13**, 83.

1594 Stillman, G. E., C. M. Wolfe, and J. O. Dimmock (1969), Magnetospectroscopy of Shallow Donors in GaAs, *Solid-State Commun.*, **7**, 921.

1595 Stolen, R. H. (1969), Far-Infrared Absorption in High-Resistivity GaAs, *Appl. Phys. Lett.*, **15**, 74.

1596 Strack, H. A. (1966), Iron-Doped Gallium Arsenide Transistor, *International Symposium on Gallium Arsenide, Reading, U.K., 1966*, Institute of Physics and Physical Society, London, p. 206.

1597 Stratton, R. (1969), Carrier Heating or Cooling in a Strong Built-in Electric Field, *J. Appl. Phys.*, **40**, 4582.

1598 Streetman, B. G. (1966), Carrier Recombination and Trapping Effects in Transient Photoconductive Decay Measurements, *J. Appl. Phys.*, **37**, 3137.

1599 Streeman, B. G., and N. Holonyak, Jr. (1969), Current Oscillations in Deep-Level Doped Semiconductors, *IBM J. Res. Dev.*, **13**, 529.

1600 Streetman, B. G., M. M. Blouke, and N. Holonyak, Jr. (1967), Current Oscillations in Co-Doped Si *p-i-n* Structures, *Appl. Phys. Lett.*, **11**, 200.

1601 Streetman, B. G., N. Holonyak, Jr., V. H. Krone, and W. D. Compton (1969), Current Oscillations in Si *p-i-n* Devices After Irradiation with One MeV Electron, *Appl. Phys. Lett.*, **14**, 63.

1602 Stringfellow, G. B., and R. H. Bube (1968), Radiative Pair Transitions in *p*-Type ZnSe:Cu Crystals, *J. Appl. Phys.*, **39**, 3657.

1603 Su, J. L., Y. Nishi, and J. L. Moll (1971), A Deep Level in Zn-Doped Liquid Phase Epitaxial GaAs, *Solid-State Electron.*, **14**, 262.

1604 Suden, E. M., R. J. Camarata, and E. A. Jarmoc (1964), Negative Transfer Resistance Effect in *p*-Type Silicon, *Proc. IEEE*, **52**, 1051.

1605 Sugiyama, K. (1967), Recombination and Trapping Processes at Deep Centers in *n*-Type GaAs, *Japan. J. Appl. Phys.*, **6**, 601.

1606 Sumita, M. (1967), Photoconductive Effects of ZnO–Cu Crystals, *Japan. J. Appl. Phys.*, **6**, 418.

1607 Summers, C. J., R. Dingle, and D. E. Hill (1970), Far-Infrared Donor Absorption and Photoconductivity in Epitaxial *n*-Type GaAs, *Phys. Rev.*, **B1**, 1603.

1608 Sushkov, V. P., and E. B. Lyubyanitskaya (1968), The Properties of Gold-Doped *p*-Type Silicon, *Inorg. Mater.*, **4**, 419.

1609 Sushkov, V. P., and E. B. Lyubyanitskaya (1972), Radiative Recombination in Silicon-Doped *p*-Type GaAs, *Sov. Phys.—Semicond.*, **5**, 1974.

1610 Suto, K., and Jun-ichi Nishizawa (1972), Paramagnetic Resonance and Hall Coefficients in Fe-Doped *n*-Type GaP, *J. Appl. Phys.*, **43**, 2247.

1611 Suto, K., M. Aoki, M. Nakada, and S. Ibuki (1967), Photo-Induced Paramagnetic Resonance of Cr^+ in ZnSe, *J. Phys. Soc. Japan*, **22**, 1121.

1612 Suzuki, T. (1965), Microwave Emission and Low Frequency Instabilities in InSb, *Japan. J. Appl. Phys.*, **4**, 700.

1613 Svistunova, K. I. (1963), Decay of Nonequilibrium Conductivity in a Semiconductor Containing Two Types, *Sov. Phys. Solid State*, **5**, 82.

1614 Sworakowski, J. (1970), Space-Charge-Limited Currents in Solids with Nonuniform Spatial Trap Distribution, *J. Appl. Phys.*, **41**, 292.

1615 Sze, S. M., and J. C. Irvin (1968), Resistivity, Mobility and Impurity Levels in GaAs, Ge, and Si at 300°K, *Solid-State Electron.*, **11**, 599.

1616 Sze, S. M. (1969), *Physics of Semiconductor Devices*, Wiley, New York.

1617 Tacano, M., and S. Kataoka (1969), Continuous, Coherent Oscillation in *n*-InSb, *Appl. Phys. Lett.*, **15**, 345.

1618 Tacano, M., and S. Kataoka (1971), Mechanism of Continuous Coherent Oscillation in n-InSb, *J. Appl. Phys.*, **42,** 494.

1619 Taft, E. A., and R. O. Carlson (1969), Beryllium in Silicon, Paper 342 RNP, Electrochemical Society Meeting, New York, May 1969.

1620 Taft, E. A., and R. O. Carlson (1970), Beryllium as an Acceptor in Silicon, *J. Electrochem. Soc.*, **117,** 711.

1621 Taft, E. A., Jr., and F. H. Horn (1958), Silicon Semiconductor Devices, U.S. Patent No. 2,847,544.

1622 Tagirov, V. I., and A. A. Kuliev (1962), A Study of the Coefficient of Tantalum Distribution in Germanium During Germanium Crystallization, *Sov. Phys. Solid State,* **3,** 1944.

1623 Tairov, S. I., V. I. Tagirov, B. S. Alieva, and A. A. Kuliev (1968), Electrical Properties of Solid Solutions of Germanium and Silicon, *Inorg. Mater.*, **4,** 425.

1624 Takeshima, M. (1972), Auger Recombination in InAs, GaSb, InP and GaAs, *J. Appl. Phys.*, **43,** 4114.

1625 Tanaka, T., and Y. Inuishi (1964), Hall Effect Measurement of Radiation Damage and Annealing in Si, *J. Phys. Soc. Japan,* **19,** 167.

1626 Tansley, T. L. (1968), GaAs Based Photodetectors, International Symposium on Gallium Arsenide, Dallas, Texas, Institute of Physics, London (1968).

1627 Tao, T. F., and Y. Hsia (1969), Dopant Dependency of Fermi Level Location in Heavily Doped Silicon, *Appl. Phys. Lett.*, **13,** 291.

1628 Tarneja, K. S. (1970), Fabrication of High Speed Rectifiers through the Use of a Double-Epi Structure, *J. Electrochem. Soc.*, **117,** 101C.

1629 Tasch, A. F., and C. T. Sah (1967), Direct Optical Determination of Carrier Cross Sections at Gold Impurity Centers in Silicon Using Junction Depletion Region, *Bull. Am. Phys. Soc.*, **12,** 356.

1630 Tasch, A. F., Jr., and C. T. Sah (1970), Recombination-Generation and Optical Properties of Gold Acceptor in Silicon, *Phys. Rev.* **B1,** 800.

1631 Tauke, R. V., and B. J. Faraday (1966), Annealing Study in Electron-Irradiated n-Type Silicon, *J. Appl. Phys.*, **37,** 5009.

1632 Taylor, J. M. (1963), *Semiconductor Particle Detectors*, Butterworths, Washington, D.C.

1633 Taylor, R. J., and C. R. Westgate (1968), Dependence of Tunnel Diode Junction Capacitance on Trapping Levels, *Bull. Am. Phys. Soc.,* **13,** 708.

1634 Tefft, W. E., and K. R. Armstrong (1971), Shallow Donor Surface Impurity Levels in Si and Ge, *Surf. Sci.*, **24,** 535.

1635 Tefft, W. E., R. J. Bell, and H. V. Romero (1969), Some New Approaches to Shallow Impurity States, *Phys. Rev.*, **177,** 1194.

1636 Teich, M., R. J. Keyes, and R. H. Kingston (1966), Optimum Heterodyne Detection at 10.6-μm in Photoconductive Ge:Cu, *Appl. Phys. Lett.*, **9,** 357.

1637 Thiel, F. L., and S. K. Ghandhi (1970), Electronic Properties of Silicon Doped with Silver, *J. Appl. Phys.*, **41,** 254.

1638 Thomas, D. G. (1966), A Review of Radiative Recombination at Isoelectronic Donors and Acceptors, Proceedings of the International Conference on the Physics of Semiconductors, Kyoto, Japan, 1966, *J. Phys. Soc. Japan, Suppl.*, **21,** 265.

1639 Thomas, D. G. (Ed.) (1967), *II–VI Semiconducting Compounds,* Benjamin, New York.

1640 Thomas, D. G. (1968), Electroluminescence, *Phys. Today,* **21 February 1968,** 43.

1641 Thomas, D. G., M. Gershenzon, and F. A. Trumbore (1964), Pair Spectra and "Edge" Emission in Gallium Phosphide, *Phys. Rev.,* **133,** A269.

1642 Thomas, J. E., A. B. Stuber, and J. W. Dearlove (1972), X-Ray Study of Lithium Precipitation in Germanium, *J. Appl. Phys.,* **43,** 251.

1643 Thomas, R. N., and M. H. Francombe (1970), Influence of Impurities on Si(111) Surface Structures, *Appl. Phys. Lett.,* **17,** 180.

1644 Thompson, A. G., and J. E. Rowe (1969), Preparation and Optical Properties of $InAs_{1-x}P_x$ Alloys, *J. Appl. Phys.,* **40,** 3280.

1645 Thompson, Ch. (1964), The Effect of Deep Level Impurities on Depletion Layer Capacitance in Silicon Junctions, *Arch. Elektr. Uebertrag.,* **18,** 628.

1646 Thompson, D. A., H. D. Barber, and W. D. Mackintosh (1969), The Determination of Surface Contamination on Silicon by Large Angle Ion Scattering, *Appl. Phys. Lett.,* **14,** 102.

1647 Thurmond, C. D., and J. D. Struthers (1953), Equilibrium Thermochemistry of Solid and Liquid Alloys of Germanium and of Silicon. II. The Retrograde Solid Solubilities of Sb in Ge, Cu in Ge, and Cu in Si, *J. Phys. Chem.,* **57,** 831.

1648 Thurmond, C. D., F. A. Trumbore, and M. Kowalchik (1956), Germanium Solidus Curves, *J. Chem. Phys.,* **25,** 799.

1649 Tietjin, J. J., and D. Richman (1967), *II. Vapor-Phase Growth of High-Resistivity* $GaAs_{1-x}P_x$ *Alloys,* Document AD 815 548, National Technical Information Service, Springfield, Virginia.

1650 Timan, B. L., and V. M. Fesenko (1971), Influence of the Field Ionization Effect on the Current-Voltage Characteristic of a Dielectric Diode With a Nonuniform Distribution of Traps, *Sov. Phys.—Semicond.,* **5,** 486.

1651 Tissen, K. P. (1959), Extraction of Copper and Nickel from Germanium, *Sov. Phys. Solid State,* **1,** 1001.

1652 Title, R. S. (1969), EPR Study of Lithium-Diffused, Mn-Doped GaAs, *J. Appl. Phys.,* **40,** 4902.

1653 Tkach, Yu. Ya. (1972), Alternating-Current Thermally Stimulated Conductivity Method, *Sov. Phys.—Semicond.,* **6,** 451.

1654 Tkachev, V. D., and V. I. Urenev (1972), Interaction Between Simple Lattice Defects and Donor Impurity Atoms in Germanium, *Sov. Phys.—Semicond.,* **5,** 1324.

1655 Tkachev, V. D., and V. I. Urenev (1972), Influence of Oxygen on the Formation of Radiation Defects in Germanium, *Sov. Phys—Semicond.,* **5,** 1416.

1656 Tokumaru, Y. (1963), Properties of Silicon Doped with Nickel, *Japan. J. Appl. Phys.,* **2,** 542.

1657 Tokumaru, Y. (1969), Current Oscillations by Two Bulk Negative-Resistance Effects in Photoexcited GaAs, *Appl. Phys. Lett.,* **14,** 212.

1658 Tokumaru, Y. (1970), Low-Frequency Photocurrent Oscillations and Trapping Levels in High-Resistivity GaAs Doped with Oxygen, *Japan J. Appl. Phys.,* **9,** 95.

1659 Tokumaru, Y., and M. Kikuchi (1965), Switching Time and I–V Characteristics of Cu-Doped Ge Diodes, *Japan. J. Appl. Phys.,* **4,** 157.

1660 Tokumaru, Y., and M. Kikuchi (1968), Current Oscillation and Light Probe Measurement of High-Field Domain Velocity in Photo-Excited High-Resistivity GaAs, *Japan. J. Appl. Phys.*, **7**, 95.

1661 Tove, P. A., and L. G. Andersson (1973), Transient Space-Charge-Limited Current and Surface Recombination in Light-Pulse Excited Silicon, *Solid-State Electron.,* **16,** in press.

1662 Toyazawa, J. (1962), Theory of Localized Spins and Negative Magnetoresistance in the Metallic Impurity Conduction, *J. Phys. Soc. Japan*, **17**, 986.

1663 Traum, M. M., M. J. Zucker, and S. H. Charap (1969), Piezoresistance of Nickel-Doped Germanium, *Bull. Am. Phys. Soc.*, **14**, 352.

1664 Tretola, A. R., and J. C. Irvin (1968), Correlation of the Physical Location of Crystal Defects with Electrical Imperfections in GaAs *p-n* Junctions, *J. Appl. Phys.*, **39**, 3563.

1665 Tretyak, O. V. (1970), Instability of the Electric Current in Samples of High-Resistivity GaAs Compensated with Cr, *Sov. Phys.—Semicond.*, **4**, 517.

1666 Trifonov, V. I., and N. G. Yaremenko (1971), Deep Donor Level in *n*-Type InSb, *Sov. Phys.—Semicond.*, **5**, 839.

1667 Trifonov, E. D., A. S. Troshin, and E. E. Fradkin (1968), Electron-Phonon Interaction Model in Solid-State Laser Theory. Nonlinear Polarization of the Impurity Centers, *Sov. Phys. Solid State*, **9**, 1617.

1668 Trumbore, F. A. (1960), Solid Solubilities of Impurity Elements in Germanium and Silicon, *Bell Syst. Tech. J.*, **39**, 205.

1669 Trumbore, F. A., C. R. Isenberg, and E. M. Porbansky (1959), On the Temperature-Dependence of the Distribution Coefficient, *J. Phys. Chem. Solids*, **9**, 60.

1670 Tscholl, E. (1966), Effect of Ni on Thermally Stimulated Conductivity Spectra in CdS-Crystals, *Solid-State Commun.*, **4**, 87.

1671 Tseng, H. F., and S. S. Li (1972), Determination of Electron and Hole Capture Rates in Nickel-Doped Germanium Using Photomagnetoelectric and Photoconductive Methods, *Phys. Rev.*, **B6**, 3066.

1672 Tsertsvadze, A. A. (1969), Scattering of Carriers by Impurities in Strongly Compensated Semiconductors, *Sov. Phys.—Semicond.*, **3**, 346.

1673 Turner, C. W. (1968), Microwave Emission from Indium Antimonide Stimulated by Acoustic Wave Amplification, *J. Appl. Phys.*, **39**, 4246.

1674 Turner, J. A. (1966), Gallium Arsenide Field Effect Transistors, *International Symposium on Gallium Arsenide, Reading, U.K., 1966,* Institute of Physics and Physical Society, London, p. 213.

1675 Tyakina, N. D., and V. S. Vavilov (1965), Impurity Photoconductivity in *p*-Type Germanium Containing Beryllium, *Sov. Phys. Solid State*, **7**, 1010.

1676 Tyler, W. W. (1959), Deep Level Impurities in Germanium, *J. Phys. Chem. Solids,* **8**, 59.

1677 Tyler, W. W., and R. Newman (1959), Bistable Semiconductor Devices, U.S. Patent No. 2,871,377.

1678 Tyler, W. W., and R. Newman (1959), Germanium Current Controlling Devices, U.S. Patent No. 2,871,427.

1679 Tyler, W. W., and H. H. Woodbury (1959), Scattering of Carriers from Doubly Charged Impurity Sites in Germanium, *Phys. Rev.*, **102**, 647.

1680 Uchinokura, K., and S. Tanaka (1967), Relaxation Process of Ionized Impurity Pairs in Silicon, *Phys. Rev.*, **153,** 828.

1681 Ugai, Ya. A., and Yu. P. Khukhryanskii (1964), Solubility of Gallium in Bulk Germanium as Determined from Measurements of the Saturation Currents in *p-n* Junctions, *Sov. Phys. Solid State,* **6,** 1.

1682 Ukhanov, Yu, I. (1963), Faraday Effect in Silicon in the Infrared Region of Wavelengths, *Sov. Phys. Solid State,* **4,** 2010.

1683 Ukhanov, Yu. I. (1963), Determination of the Effective Electron Mass in GaAs Using the Faraday Effect, *Sov. Phys. Solid State,* **5,** 75.

1684 Vakulenko, O. V., and M. P. Lisitsa (1967), Absorption of Infrared Radiation by Free Carriers in *n*-Type GaAs, *Sov. Phys. Solid State,* **9,** 769.

1685 Valahas, T. M., J. S. Sochanski, and H. C. Gatos (1971), Electrical Characteristics of Gallium Arsenide "Real" Surfaces, *Surf. Sci.,* **26,** 41.

1686 Valov, Yu. A. (1969) An Attempt to Predict Thermodynamically the Behavior of Impurities in A^{III} B^{V} Semiconductors, *Sov. Phys.—Semicond.,* **3,** 534.

1687 van der Pauw, L. J. (1958), A Method of Measuring Specific Resistivity and Hall Effect of Discs of Arbitrary Shape, *Philips Res. Rep.,* **13,** 1.

1688 van der Ziel, A. (1954), *Noise,* Prentice-Hall, Englewood Cliffs, New Jersey.

1689 van der Ziel, A., and S. T. Hsu (1966), High-Frequency Admittance of Space-Charge-Limited Solid-State Diodes, *Proc. IEEE,* **54,** 1194.

1690 van Gelder, W., and E. H. Nicollian (1971), Silicon Impurity Distribution as Revealed by Pulsed MOS *C–V* Measurements, *J. Electrochem. Soc.,* **118,** 138.

1691 van Heerden, P. J. (1957), Primary Photocurrent in Cadmium Sulfide, *Phys. Rev.,* **106,** 468.

1692 van Opdorp, C., and J. Vrakking (1969), Avalanche Breakdown in Epitaxial SiC *p-n* Junctions, *J. Appl. Phys.,* **40,** 2320.

1693 van Roosbroeck, W. (1956), Theory of the Photomagnetoelectric Effect in Semiconductors, *Phys. Rev.,* **101,** 1713.

1694 van Roosbroeck, W. (1960), Theory of Current Carrier Transport and Photoconductivity in Semiconductors with Trapping, *Bell Syst. Tech. J.,* **39,** 515.

1695 van Vliet, K. M. (1958), Noise in Semiconductors and Photoconductors, *Proc. IRE,* **46,** 1004.

1696 Varlamov, I. V., and V. V. Osipov (1970), Filamentation of the Current in Multilayered Structures, *Sov. Phys.—Semicond.,* **3,** 803.

1697 Varlamov, I. V., and E. A. Poltoratskii (1969), Influence of Radial Carrier Currents on the Formation of Current-Voltage Characteristics of Diodes under Current Filamentation Conditions, *Sov. Phys.—Semicond.,* **3,** 259.

1698 Varlamov, I. V., and E. M. Shandarov (1970), Thermal Mechanism of Negative Differential Resistance and Formation of Current Filaments, *Sov. Phys.—Semicond.,* **3,** 1203.

1699 Varlamov, I. V., I. A. Sondaevskaya, and V. P. Sondaevskii (1967), Double Injection in Gold-Doped Silicon, *Sov. Phys.—Semicond.,* **1,** 375.

1700 Varlamov, I. V., I. A. Sondaevskaya, and V. P. Sondaevskii (1968), Possibility of Controlling the Photosensitivity of a Transistor by Illumination, *Sov. Phys.—Semicond.,* **2,** 342.

1701 Varlamov, I. V., V. V. Osipov, and E. A. Poltoratskii (1970), Current Filamentation in a Four-Layered Structure, *Sov. Phys.—Semicond.*, **3**, 978.

1702 Varlamov, I. V., E. A. Poltoratskii, and V. I. Stafeev (1970), A Method for Modulating the Diffusion Length of Minority Carriers, *Sov. Phys.—Semicond.*, **3**, 1181.

1703 Varshni, Y. P. (1967), Band-to-Band Radiative Recombination in Groups IV, VI and III-V Semiconductors (II), *Phys. Status Solidi*, **20**, 9.

1704 Vasil'ev, V. S., I. N. Kanevskii, and V. B. Osvenskii (1969), Measurement of the Diffusion Coefficient of Copper in Gallium Arsenide by an Ultrasonic Method, *Sov. Phys.—Semicond.*, **2**, 1495.

1705 Vas'kin, V. V., and V. A. Uskov (1968), Influence of Complex Formation on Impurity Diffusion in Semiconductors, *Sov. Phys. Solid State*, **10**, 985.

1706 Vavilov, V. S. (1965), *Effects of Radiation on Semiconductors*, Consultants Bureau, New York.

1707 Vavilov, V. S. (1965), Strahlungsdefekte in Halbleitern, *Phys. Status Solidi*, **11**, 447.

1708 Vavilov, V. S. (1965), The Nature and Energy Spectrum of Radiation Defects in Semiconductors, *Sov. Phys. USP*, **7**, 797.

1709 Vavilov, V. S. (1965), Interaction of Radiation Defects with Impurities, *Physics of Semiconductors, Proceedings of the 7th International Conference, Paris, 1964*, Academic Press, New York and London, Dunod, Editeur, Paris.

1710 Vavilov, V. S., E. N. Lotkova, and A. F. Plotnikov (1961), Photoconductivity and Infra-Red Absorption in Silicon Irradiated by Neutrons, *J. Phys. Chem. Solids*, **22**, 31.

1711 Vavilov, V. S., O. G. Koshelev, Y. P. Koval', and Y. G. Klyava (1967), A Study of Inter-Impurity Recombination between Phosphorus and Boron in Silicon, *Sov. Phys. Solid State*, **8**, 2770.

1712 Vavilov, V. S., I. A. Kurova, N. N. Ormont, and Yu. V. Petrov (1969), Current Oscillations Connected with the Acoustoelectric Effect in GaAs, *Sov. Phys.—Semicond.*, **3**, 212.

1713 Vavilov, V. S., A. F. Plotnikov, and V. É. Shubin (1972), Charge Exchange Between Traps in InSb Oxide in Au-InSb Oxide-InSb Structures, *Sov. Phys.—Semicond.*, **5**, 1799.

1714 Veinger, A. I. (1969), Electron Paramagnetic Resonance of Nitrogen Atoms in Various Poltypes of α-Silicon Carbide, *Sov. Phys.—Semicond.*, **3**, 52.

1715 Veinger, A. I., V. G. Ivanov, L. G. Paritskii, B. S. Ryvkin, and A. B. Fedortsov (1969), Characteristic Features of the Recombination of Hot Carriers in Copper-Doped Germanium, *Sov. Phys.—Semicond.*, **3**, 103.

1716 Veinger, A. I., and A. F. Shulekin (1969), Observation of Interimpurity Recombination in Silicon Carbide, *Sov. Phys.—Semicond.*, **3**, 113.

1717 Veinger, A. I., V. G. Ivanov, L. G. Paritskii, and S. M. Ryvkin (1969a), Recombination of Hot Electrons in Gold-Doped Silicon, *Sov. Phys.—Semicond.*, **2**, 1236.

1718 Veinger, A. I., V. G. Ivanov, L. G. Paritskii, B. S. Ryvkin, and A. B. Fedortsov (1969b), Characteristic Features of the Recombination of Hot Carriers in Copper-Doped Germanium, *Sov. Phys.—Semicond.*, **3**, 103.

1719 Venkateswaran, K., and D. J. Roulston (1972), Recombination Dependent Characteristics of Silicon p^+-n-n^+ Epitaxial Diodes, *Solid-State Electron.*, **15**, 311.

1720 Verleur, H. W., A. S. Barker, Jr., and C. N. Berglund (1968), Optical Properties of Pt and Platinum Silicide Films on Silicon Substrates, *Bull. Am. Phys. Soc.*, **13**, 456.

1721 Verwey, J. F. (1972), Hole Currents in Thermally Grown SiO_2, *J. Appl. Phys.*, **43**, 2273.

1722 Vick, G. L., and K. M. Whittle (1969), Solid Solubility and Diffusion Coefficients of Boron in Silicon, *J. Electrochem. Soc.*, **116**, 1142.

1723 Viehmann, W. (1969), Current Oscillations in Photoexcited Gallium-Arsenide, *Appl. Phys. Lett.*, **14**, 39.

1724 Vieland, L. J., and I. Kudman (1963), Behavior of Selenium in Gallium Arsenide, *J. Phys. Chem. Solids*, **24**, 437.

1725 Vilms, J., and W. E. Spicer (1965), Quantum Efficiency and Radiative Lifetime in *p*-Type Gallium Arsenide, *J. Appl. Phys.*, **36**, 2815.

1726 Vinetskii, R. M., and E. G. Miselyuk (1960), Determination of the Impurity Concentration in Germanium, *Sov. Phys. Solid State*, **2**, 60.

1727 Vinetskii, V. L., and G. A. Kholodar' (1966), Conductivity of Semiconductors Due to the Ionization of Thermal Defects of the Lattice ("Intrinsic-Defect" Conductivity), *Sov. Phys. Solid State*, **8**, 785.

1728 Vinetskii, V. L., and N. N. Kolychev (1963), Increase of the Lifetime of Nonequilibrium Electrons by the Introduction of Trapping Levels, *Sov.. Phys. Solid State*, **5**, 506.

1729 Vinetskii, V. L., I. D. Konozenko, and S. I. Shakhovtsova (1964), Nature of the Phenomenon Involved in the Generation of Photocurrent Pulses in Cadmium Selenide Crystals, *Sov. Phys. Solid State*, **5**, 1971.

1730 Vinogravoda, K. I., D. N. Nasledov, Yu. S. Smetannikova, and V. R. Felitsiant (1969), Double Inversion of the Hall Coefficient of *p*-Type InSb Doped with Fe and Ni, *Sov. Phys.—Semicond.*, **3**, 592.

1731 Violin, E. E., and G. F. Kholuyanov (1964), Recombination Radiation and Electrical Properties of Diffused *p-n* Junctions in SiC, *Sov. Phys. Solid State*, **6**, 465.

1732 Violin, E. E., and G. F. Kholuyanov (1964), On the Electroluminescence and Photoluminescence of Diffused *p-n* Junctions in SiC, *Sov. Phys. Solid State*, **6**, 1331.

1733 Violin, E. E., and G. F. Kholuyanov (1967), Extraction of Carriers by a *p-n* Junction Field and the Electroluminescence Mechanism of Silicon Carbide, *Sov. Phys. Solid State*, **8**, 2716.

1734 Vitovskii, N. A., and T. V. Mashovets (1964), Possibility of Precise Determination of the Activation Energy of Defect and Impurity Levels in Semiconductors, *Sov. Phys. Solid State*, **6**, 1297.

1735 Vogl, T. P., J. R. Hansen, and M. Garbuny (1960), Photoconductive Time Constants and Related Characteristics of *p*-Type Gold-Doped Germanium, *J. Opt. Soc. Am.*, **51**, 70.

1736 Vol'fson, A. A., S. M. Gorodetskii, and V. K. Subashiev (1965), A Study of Photoconductivity in Heavily Doped *p*-Silicon, *Sov. Phys. Solid State*, **7**, 53.

1737 Vol'fson, A. A., N. S. Zhdanovich, and V. K. Subashiev (1965), Quantum Yield of the Internal Photoeffect in *p*-Type Silicon Doped with Boron, *Sov. Phys. Solid State*, **6**, 2993.

1738 Vook, F. L., and H. J. Stein (1968), Infrared Absorption Bands in Carbon- and Oxygen-Doped Silicon, *Appl. Phys. Lett.*, **13**, 343.

1739 Vorob'ev, Yu. V., and O. V. Tretyak (1969), Investigation of the Influence of Electron Heating on the Instability of an Electric Current in High-Resistivity GaAs, *Sov. Phys.—Semicond.*, **3**, 515.

1740 Vorob'ev, Yu. V., Yu. I. Karkhanin, and O. V. Tretyak (1971), Investigation of Transient Electronic Processes in High-Resistivity GaAs, *Sov. Phys.—Semicond.*, **5**, 254.

1741 Vorobkalo, F. M., K. D. Glinchuk, and A. V. Prokhorovich (1969), Electroluminescence of *p*-Type GaAs Diodes, *Sov. Phys.—Semicond.*, **3**, 125.

1742 Vorobkalo, F. M., K. D. Glinchuk, and N. M. Litovchenko (1970), Investigation of the Photoconductivity Kinetics of Gallium Arsenide, *Sov. Phys.—Semicond.*, **4**, 717.

1743 Voronkov, V. V., G. I. Voronkova, and M. I. Iglitsyn (1970), Determination, Using the Hall Effect, of the Parameters of Energy Levels at Low Level Density, *Sov. Phys.—Semicond.*, **3**, 1449.

1744 Voronkov, V. V., G. I. Voronkova, and M. I. Iglitsyn (1972), Mobility and Concentration of Vacancies in Ge and Si, *Sov. Phys.—Semicond.*, **6**, 14.

1745 Voronkova, N. M., and D. N. Nasledov (1966), Concerning the Role of Trapping Levels in the Photoconductivity of GaAs, *Sov. Phys. Solid State*, **7**, 2050.

1746 Vrana, M., I. A. Kurova, and L. G. Mamsurova (1971), Influence of a Magnetic Field on the Recombination Domain Instability in Gold-Doped *n*-Type Ge, *Sov. Phys.—Semicond.*, **5**, 36.

1747 Vul, B. M. (1961), Impact Ionization and Tunnel Effect in Semiconductors, *Sov. Phys. Solid State*, **2**, 2631.

1748 Vul, B. M., E. I. Zavaritskaya, and A. P. Shotov (1964), Current-Voltage Characteristics of *p-n* Junctions in Heavily Doped Gallium Arsenide, *Sov. Phys. Solid State*, **6**, 1146.

1749 Vul', A. Yu., L. V. Golubev, T. A. Polyanskaya, and Yu. V. Shmartsev (1969), Structure of the Conduction Band of GaSb, *Sov. Phys.—Semicond.*, **3**, 256.

1750 Vul', A. Ya., L. V. Golubev, T. A. Polyanskaya, and Yu. V. Shmartsev (1969), Some Properties of Sulfur-Doped Gallium Antimonide, *Sov. Phys.—Semicond.*, **3**, 671.

1751 Vul', A. Ya., G. L. Bir, and Yu. V. Shmartsev (1971), Donor States of Sulfur in Gallium Antimonide, *Sov. Phys.—Semicond.*, **4**, 2005.

1752 Vul', A. Ya., L. V. Golubev, and Yu. V. Shmartsev (1971), Electrical Properties of Sulfur-Doped *n*-Type and *p*-Type Gallium Antimonide, *Sov. Phys.—Semicond.*, **5**, 1059.

1753 Wadhawa, R. P., and M. L. Sisodia (1967), Transit-Time Effects in the Presence of Traps in Space-Charge-Limited Solid-State Devices, *Int. J. Electron.*, **23**, 83.

1754 Wagener, J. L. (1964), Double-Injection Experiments in Semi-Insulating Diodes, Ph.D. Thesis, Carnegie-Mellon University, Pittsburgh, Pennsylvania.

1755 Wagener, J. L., and A. G. Milnes (1964), Post-Breakdown Conduction in Forward-Biased *p-i-n* Silicon Diodes, *Appl. Phys. Lett.*, **5**, 186.

1756 Wagener, J. L., and A. G. Milnes (1965), Double-Injection Experiments in Semi-Insulating Silicon Diodes, *Solid-State Electron.*, **8**, 495.

1757 Wakim, F. G. (1970), Photo-Current and Thermally Stimulated Current Excitation Spectra in Cubic ZnSe Crystals, *J. Appl. Phys.*, **41,** 835.

1758 Waldner, M., M. A. Hiller, and W. G. Spitzer (1965), Infrared Combination Mode Absorption in Lithium-Boron-Doped Silicon, *Phys. Rev. A,* **140,** 172.

1759 Walker, B. J., and J. N. Crouch, Jr. (1969), Internal Current Gain in Microwave-Biased Gallium-Doped Silicon 8–14-μm IR Detectors, *Proc. IEEE,* **57,** 2167.

1760 Walters, P. A. (1969), Step-Like Shifts in the Voltage-Current Characteristics of Power Thyristors and Their Effect on Transient Thermal Impedance Determinations, *IEEE Trans. Electron Devices,* **ED-16,** 497.

1761 Watanabe, H., and M. Wada (1964), Optical and Electrical Properties of ZnO Crystals, *Japan. J. Appl. Phys.*, **3,** 617.

1762 Watkins, G. D. (1965), A Review of EPR Studies in Irradiated Silicon, *Physics of Semiconductors, Proceedings of the 7th International Conference, Paris, 1964,* Academic Press, New York and London, Dunod, Editeur, Paris.

1763 Watkins, G. D., and J. W. Corbett (1964), Defects in Irradiated Silicon: Electron Paramagnetic Resonance and Electron-Nuclear Double Resonance of the Si–*E* Center, *Phys. Rev.*, **134,** A1359.

1764 Watkins, G. D., and R. P. Messmer (1970), An LCAO-MO Treatment for a Deep Level in a Semiconductor, *Proceedings of the 10th International Conference on the Physics of Semiconductors, MIT, Cambridge,* S. P. Keller, Ed., Pub. CONF 700801 USAEC, National Technical Information Service, Springfield, Virginia, p. 623.

1765 Watkins, T. B., and B. K. Ridley (1961), The Possibility of Negative Resistance Effects in Semiconductors, *Proc. Phys. Soc. (London),* **78,** 293.

1766 Weber, W. H. (1970), Double Injection in Long *p-i-n* Diodes with Deep Double-Acceptor Impurities, *Appl. Phys. Lett.*, **16,** 396.

1767 Weber, W. H., and G. W. Ford (1970), Double Injection in Semiconductors Heavily Doped with Deep Two-Level Traps, *Solid-State Electron.*, **13,** 1333.

1768 Weber, W. H., and G. W. Ford (1971), Space-Charge Recombination Oscillations in Double-Injection Structures, *Appl. Phys. Lett.*, **18,** 241.

1769 Weber, W. H., R. S. Elliott, and A. L. Cederquist (1971), Small-Signal Transient Double Injection in Semiconductors Heavily Doped with Deep Traps, *J. Appl. Phys.*, **42,** 2497.

1770 Weiner, M. E. (1972), Si Contamination in Open Flow Quartz Systems for the Growth of GaAs and GaP, *J. Electrochem. Soc.*, **119,** 496.

1771 Weiner, M. E., and A. S. Jordan (1972), Analysis of Doping Anomalies in GaAs by Means of a Silicon-Oxygen Complex Model, *J. Appl. Phys.*, **43,** 1767.

1772 Weisberg, L. R. (Ed.) (1961a), *High Temperature Semiconductor Research,* Report No. 5, Document AD 260 767, National Technical Information Service, Springfield, Virginia.

1773 Weisberg, L. R. (Ed.) (1961b), *High Temperature Semiconductor Research,* Report No. 6, Document AD 264 438, National Technical Information Service, Springfield, Virginia.

1774 Weisberg, L. R. (1968), Auger Recombination in GaAs, *J. Appl. Phys.*, **39,** 6096.

1775 Weisberg, L. R. (1971), Degradation Mechanisms in III–V Compound Electroluminescent Diodes, *Proceedings of the 3rd International Symposium on Gallium*

Arsenide and Related Compounds, Aachen, 1971, The Institute of Physics, London.

1776 Weisberg, L. R., and H. Schade (1968), A Technique for Trap Determinations in Low-Resistivity Semiconductors, *J. Appl. Phys.*, **39,** 5149.

1777 Weiser, K. (1962), Theory of Diffusion and Equilibrium Position of Interstitial Impurities in the Diamond Lattice, *Bull. Am. Phys. Soc.*, **7,** 234; also *Phys. Rev.*, **126,** 1427 (1962).

1778 Weiser, K., and R. S. Levitt (1964), Electroluminescent Gallium Arsenide Diodes with Negative Resistance, *J. Appl. Phys.*, **35,** 2431.

1779 Weiser, K., M. Drougard, and R. Fern (1967), Avalanching in GaAs p-π-p Structures, *J. Phys. Chem. Solids*, **28,** 171.

1780 Weltzin, R. D., R. A. Swalin, and T. E. Hutchinson (1965), Electron Microscopic Study of Precipitates and Defects in Germanium and Silicon, *Acta Metall.*, **13,** 115.

1781 Wenass, E. P., R. E. Leadon, J. A. Naber, and C. E. Mallon (1971), Recombination Lifetimes in Gold-Doped p-Type Silicon, *J. Appl. Phys.*, **42,** 2893.

1782 Wertheim, G. K. (1958), Transient Recombination of Excess Carriers in Semiconductors, *Phys. Rev.*, **109,** 1086.

1783 Wertheim, G. K., and W. M. Augustyniak (1956), Measurement in Short Carrier Lifetimes, *Rev. Sci. Instrum.*, **27,** 1062.

1784 Wertheim, G. K., A. Hausmann, and W. Sander (1971), *The Electronic Structure of Point Defects,* North Holland, Amsterdam, p. 224.

1785 Whan, R. E. (1966), Oxygen-Defect Complexes in Neutron-Irradiated Silicon, *J. Appl. Phys.*, **37,** 3378.

1786 Whan, R. E., and H. J. Stein (1963), Oxygen-Defect Complexes in Germanium, *Appl. Phys. Lett.*, **3,** 187.

1787 Whelan, J. M., J. D. Struthers, and J. A. Ditzenberger (1960), Distribution Coefficients of Various Impurities in Gallium Arsenide, *Properties of Elemental and Compound Semiconductors,* H. C. Gatos, Ed., Interscience, New York, p. 141.

1788 Whelen, F. G., and I. E. Distelhorst (1962), Mercury and Cadmium Improved IR Detectors, *Electronics,* **August 10, 1962,** 84.

1789 Whitaker, J. (1965), Electrical Properties of n-Type Aluminum Arsenide, *Solid-State Electron.*, **8,** 649.

1790 Wieder, H. H. (1971), A Review of the Electrical and Optical Properties of III-V Compound Semiconductor Films, *J. Vac. Sci. Technol.*, **8,** 210.

1791 Wilcox, W. R., and T. L. LaChapelle, Mechanism of Gold Diffusion into Silicon, *J. Appl. Phys.*, **35,** 240.

1792 Wiley, J. D., P. S. Peercy, and R. N. Dexter (1969), Helicons and Nonresonant Cyclotron Absorption in Semiconductors, I. InSb, *Phys. Rev.*, **181,** 1173.

1793 Willardson, R. K., and W. P. Allred (1966), Distribution Coefficients in Gallium Arsenide, *International Symposium on Gallium Arsenide, Reading, U.K.*, Institude of Physics and Physical Society, London.

1794 Willardson, R. K., and A. C. Beer (1966), *Semiconductors and Semimetals,* Vol. 2, *Physics of III–V Compounds*; Vol. 3, *Optical Properties of III–V Compounds,* Academic Press, New York.

1795 Willardson, R. K., and A. C. Beer (1970), *Semiconductors and Semimetals, Infrared Detectors,* Vol. 5, Academic Press, New York.

1796 Willardson, R. K., and H. L. Goering (Eds.) (1962), *Compound Semiconductors,* Vol. 1, *Preparation of III–V Compounds,* Reinhold, New York.

1797 Williams, E. W. (1968), Evidence for Self-Activated Luminescence in GaAs: The Gallium Vacancy-Donor Center, *Phys. Rev.,* **168,** 922.

1798 Williams, E. W. (1969), Electroreflectance of Impurities in GaAs: Manganese, Silicon and Cadmium, *Solid-State Commun.,* **7,** 541.

1799 Williams, F. (1969), Donor-Acceptor Pairs of Semiconductors, *Phys. Status Solidi,* **25,** 493.

1800 Williams, R. (1966), Determination of Deep Centers in Conducting Gallium Arsenide, *J. Appl. Phys.,* **37,** 3411.

1801 Williams, R., and A. Willis (1968), Electron Multiplication and Surface Charge on Zinc Oxide Single Crystals, *J. Appl. Phys.,* **39,** 3731.

1802 Williams, R. L. (1967), *High-Speed, Long-Wavelength Coherent Radiation Detectors,* Document AD 663 749, National Technical Information Service, Springfield, Virginia.

1803 Williams, R. L. (1969a), Anomalous Copper Distribution in and Near a Gallium-Diffused Layer in Germanium, *Bull. Am. Phys. Soc.,* **14,** 395.

1804 Williams, R. L. (1969b), Response Characteristics of Extrinsic Photoconductors, *J. Appl. Phys.,* **40,** 184.

1805 Williams, R. L. (1969c), Bipolar Photoeffect Arising from an Anomalous Copper Distribution in and Near a Gallium-Diffused Layer in Germanium, *J. Appl. Phys.,* **40,** 2932.

1806 Wilson, C. L., and S. J. Brient (1970), Comparison of Theoretical with Experimental *p-n* Junction Recombination Effects, *J. Appl. Phys.,* **41,** 4190.

1807 Wilson, P. G. (1967), Recombination in Silicon p-π-n Diodes, *Solid-State Electron.,* **10,** 145.

1808 Winteler, H. R., and A. Steinemann (1968), Correlation between Diffusion and Precipitation of Impurities in Dislocation Free GaAs, *International Symposium on Gallium Arsenide, Dallas, Texas, 1968,* Institute of Physics, London.

1809 Winter, J. M. (1966), Electron Spin Resonance of Impurities in Semiconductors, Proceedings of the International Conference on the Physics of Semiconductors, Kyoto, Japan, *Japan. I. Appl. Phys., Suppl.,* **21,** 1966.

1810 Wittry, D. B., and D. F. Kyser (1964), Use of Electron Probes in the Study of Recombination Radiation, *J. Appl. Phys.,* **35,** 2439.

1811 Wolfe, C. M., G. E. Stillman, and J. O. Dimmock (1970), Ionized Impurity Density in *n*-Type GaAs, *J. Appl. Phys.,* **41,** 504.

1812 Wolley, E. D., and R. Stickler (1967), Formation of Precipitates in Gold-Diffused Silicon, *J. Electrochem. Soc.,* **114,** 1287.

1813 Woodbury, H. H., and M. Aven (1968), Some Diffusion and Solubility Measurements of Cu in CdTe, *J. Appl. Phys.,* **39,** 5485.

1814 Woodbury, H. H., and G. W. Ludwig (1960), Vacancy Interactions in Silicon, *Phys. Rev. Lett.,* **5,** 96.

1815 Woodbury, H. H., and G. W. Ludwig (1960a), Spin Resonance of Transition Metals in Silicon, *Phys. Rev.,* **117,** 102.

1816 Woodbury, H. H., and G. W. Ludwig (1962), Spin Resonance of Pd and Pt in Silicon, *Phys. Rev.*, **126,** 466.

1817 Woodbury, H. H., and W. W. Tyler (1956), Recent Studies of Deep Impurity Levels in Germanium, *Bull. Am. Phys. Soc.*, **1,** 127.

1818 Woodbury, H. H., and W. W. Tyler (1955), Properties of Germanium Doped with Manganese, *Phys. Rev.*, **100,** 659.

1819 Woods, J., and K. H. Nicholas (1964), Photochemical Effects in CdS, *Br. J. Appl. Phys.*, **15,** 1361.

1820 Wright, G. T., and A. F. Ibrahim (1968), Single Injection, Double Injection and Negative Resistance in Gold-Doped High-Resistivity Silicon, *Electron. Lett.*, **4,** 597.

1821 Wright, H. C., and G. A. Allen (1966), Thermally Stimulated Current Analysis, *Br. J. Appl. Phys.*, **17,** 1181.

1822 Wright, H. C., R. E. Hunt, and G. A. Allen (1968), Crystal Efficiency for Injection Luminescence, *International Symposium on Gallium Arsenide, Dallas, Texas, 1968,* Institute of Physics, London.

1823 Wronski, C. R. (1970), Effects of Deep Centers on n-Type GaP Schottky Barriers, *J. Appl. Phys.*, **41,** 3805.

1824 Wyluda, B. J., and R. G. Shulman (1956), Copper in Germanium: Recombination Center and Trapping Center, *Phys. Rev.*, **102,** 1455.

1825 Wysocki, J. J. (1965), Thermally Stimulated Conductivity of Neutron-Irradiated Silicon, *J. Appl. Phys.*, **36,** 2968.

1826 Wysocki, J. J., P. Rappaport, E. Davison, R. Hand, and J. J. Loferski (1966), Lithium-Doped, Radiation-Resistant Silicon Solar Cells, *Appl. Phys. Lett.*, **9,** 44.

1827 Yamamoto, K., M. Yamada, and K. Abe (1970), Observations of Propagating Domain in Semiconductive CdS Using the Optical Probe Method, *J. Appl. Phys.*, **41,** 450.

1828 Yamamoto, S., S. T. Liu, and A. van der Ziel (1967), Thermal Noise in Ge p-ν-p SCL Diodes, *Appl. Phys. Lett.*, **11,** 140.

1829 Yamashita, A., M. Tanaka, T. Yamada, and I. Yamauchi (1966), Pressure-Sensitive Switching Diodes, *Proc. IEEE*, **54,** 317.

1830 Yamashita, J. (1961), Low-Temperature Electrical Breakdown in Germanium, *J. Phys. Soc. Japan*, **16,** 720.

1831 Yarbrough, D. W. (1968), Status of Diffusion Data in Binary Compound Semiconductors, *Solid-State Tech.*, **11, November 1968,** 23.

1832 Yau, L. D., and C. T. Sah (1972), Measurement of Trapped-Minority-Carrier Thermal Emission Rates from Au, Ag, and Co Traps in Silicon, *Appl. Phys. Lett.*, **21,** 157.

1833 Yeargan, J. R., and H. L. Taylor (1968), The Poole-Frenkel Effect with Compensation Present, *J. Appl. Phys.*, **39,** 5600.

1834 Yee, J. H. (1969), Calculations of Two-Photon Conductivity in Semiconductors, *Appl. Phys. Lett.*, **14,** 231.

1835 Yee, J. H., and G. A. Condas (1969), Photoluminescence Field Quenching and Breakdown of CdS at 4.2°K, *J. Appl. Phys.*, **40,** 2317.

1836 Yeh, T. H., and M. L. Joshi (1969), Strain Compensation in Silicon by Diffused Impurities, *J. Electrochem. Soc.*, **116,** 73.

1837 Yeh, T. H., S. M. Hu, and R. H. Kastl (1968), Diffusion of Tin in Silicon, *J. Appl. Phys.*, **39,** 4266.

1838 Yemel'yanenko, O. V., I. I. Klotyn'sh, and D. N. Nasledov (1968), *Galvanomagnetic and Thermomagnetic Properties of Gallium Arsenide After Diffusion of Copper,* Document AD 673 923, National Technical Information Service, Springfield, Virginia.

1839 Yokozawa, M., S. Otsoka, aand S. Takayanagi (1965), Distribution Coefficient and Electrical Behavior of In in CdTe Crystals, *Japan. J. Appl. Phys.*, **4,** 1018.

1840 Yoshida, M., and K. Furusho (1964), Behavior of Nickel as an Impurity in Silicon, *Japan. J. Appl. Phys.*, **3,** 521.

1841 Yoshida, M., Y. Yamaguchi, and H. Aoki (1963), Hexagonal Precipitates of Nickel in Silicon, *Japan. J. Appl. Phys.,* **2,** 305.

1842 Yu, P. W., and M. Glicksman (1972), Electrical Properties and Luminescence of Alloys of Gallium Phosphide and Zinc Selenide, *J. Appl. Phys.,* **43,** 4153.

1843 Yukhnevich, A. V. (1965), The Structure of the Spectrum of the Radiative Capture of Holes by *A*-Centers in Silicon, *Sov. Phys. Solid State,* **7,** 259.

1844 Yukhnevich, A. V., and V. D. Tkachev (1966), Radiative Recombination in Silicon Containing Radiation-Induced Defects, *Sov. Phys. Solid State,* **7,** 2746.

1845 Yukhnevich, A. V., and V. D. Tkachev (1966), Optical Analog of the Mossbauer Effect in Silicon, *Sov. Phys. Solid State,* **8,** 1004.

1846 Yukhnevich, A. V., V. D. Tkachev, and V. M. Lomako (1966), Concerning Impurity Radiative Recombination in Silicon Crystals, *Sov. Phys. Solid State,* **8,** 444.

1847 Yukhnevich, A. V., V. D. Tkachev, and M. V. Bortnik (1967), Annealing of Impurity Recombination Bands in Silicon Irradiated by γ-Rays, *Sov. Phys. Solid State,* **8,** 2571.

1848 Yunovich, A. E., P. G. Eliseev, I. A. Nakhodnova, A. B. Ormont, L. A. Osadchaya, and V. M. Stuchebnikov (1964), Radiative Recombination in GaAs *p-n* Junctions Formed by Beryllium Diffusion, *Sov. Phys. Solid State,* **6,** 1496.

1849 Yurova, E. S., E. V. Solov'eva, E. M. Kistova, L. I. D'yakonov. M. I. Iglitsyn, and M. N. Kevorkov (1972), "Self-Compensation" of Donors in Gallium Arsenide and in $GaAs_{1-x}P_x$ Solid Solutions, *Sov. Phys.—Semicond.,* **6,** 426.

1850 Zaikovskaya, M. A., O. A. Klimkova, and O. R. Niyazova (1971), Low-Temperature Doping of Semiconductors, *Sov. Phys.—Semicond.,* **5,** 802.

1851 Zakharova, G. A., M. A. Krivov, E. V. Malisova, and E. A. Popova (1972), Conditions and Mechanism of the Formation of Certain Impurity Centers in Copper-Doped Gallium Arsenide, *Sov. Phys.—Semicond.,* **6,** 171.

1852 Zanio, K. R. (1969), Improved CdTe for Gamma Detection Between −100°C and +100°C, *Appl. Phys. Lett.,* **14,** 56.

1853 Zanio, K. R. (1969), Characterization of Foreign Atoms and Native Defects in Single Crystals of Cadmium Telluride by High-Temperature Conductivity Measurements, *Appl. Phys. Lett.,* **15,** 260.

1854 Zanio, K. R., W. M. Akutagawa, and R. Kikuchi (1968), Transient Currents in Semi-Insulating CdTe Characteristics of Deep Traps, *J. Appl. Phys.,* **39,** 2818.

1855 Zavadskii, Yu. I., and B. V. Kornilov (1967), Slow Rise of the Photoresponse of n-Type Silicon Compensated with Zinc, *Sov. Phys.—Semicond.*, **1**, 1103.

1856 Zavadskii, Yu. I., and B. V. Kornilov (1970), Thermo-Oscillator: A New Functional Temperature-Sensitive Device, *Sov. Phys.—Semicond.*, **3**, 1211.

1857 Zavadskii, Yu. I., and B. V. Kornilov (1971), Effect of Light on Self-Oscillations of the Current in Zinc-Doped Silicon, *Sov. Phys.—Semicond.*, **4**, 1815.

1858 Zavadskii, Yu. I., and B. V. Kornilov (1971), Optical Absorption in Zinc-Doped n-Type Silicon, *Sov. Phys.—Semicond.*, **5**, 56.

1859 Zavaritskaya, E. I. (1961), Concerning Impact Ionization of Impurities in Germanium at Low Temperatures, *Sov. Phys. Solid State*, **3**, 1374.

1860 Zelevinskaya, V. M., and G. A. Kachurin (1972), Annealing of Gallium Arsenide Doped by Implantation of Group VI Ions, *Sov. Phys.—Semicond.*, **5**, 1455.

1861 Zelm, M., and C. E. Gaier (1970), Gold Concentration Profiles and Minority Carrier Lifetimes in *npn* Structures, Extended Abstracts, 138th National Meeting of the Electrochemical Society; also *J. Electrochem. Soc.*, **117**, 261C (1970).

1862 Zhang, H. I., and J. Callaway (1969), Energy-Band Structure and Optical Properties of GaSb, *Phys. Rev.*, **181**, 1163.

1863 Zhdanova, N. G. (1965), Electron Capture Coefficients of Negatively Charged Impurity Centers in Germanium, *Sov. Phys. Solid State*, **6**, 2024.

1864 Zhdanova, N. G., and V. G. Alekseeva (1963), Effect of Temperature on the Kinetics of Impurity Photoconductivity in n-Type Gold Doped Germanium, *Sov. Phys. Solid State*, **5**, 397.

1865 Zhdanova, N. G., and S. G. Kalashnikov (1964), Effect of Temperature on the Decay Kinetics of Extrinsic Photoconductivity in Copper-Doped Germanium, *Sov. Phys. Solid State*, **6**, 350.

1866 Zhdanova, N. G., M. S. Kagan, and S. G. Kalashnikov (1966a), Recombination of Hot Electrons at Repulsive Impurity Centers in Germanium, *Sov. Phys. Solid State*, **8**, 620.

1867 Zhdanova, N. G., M. S. Kagan, and S. G. Kalashnikov (1966b), Current Instability and Electrical Domains in Compensated Ge, *Sov. Phys. Solid State*, **8**, 632.

1868 Zhdanova, N. G., M. S. Kagan, S. G. Kalashnikov, and E. G. Landsberg (1972), Recombination of Hot Electrons Heated by a High-Frequency Electric Field, *Sov. Phys.—Semicond.*, **6**, 7.

1869 Zhumakulov, U. (1966), Dependence of the Effective Masses of Electrons and Holes on the Carrier Density in the Compounds GaAs, GaP and InAs, *Sov. Phys. Solid State*, **8**, 2476.

1870 Zhurkin, B. G., I. V. Kucherenko, and N. A. Penin (1967), Influence of Uniaxial Compression on Hopping Conduction in p-Type Si, *Sov. Phys. Solid State*, **8**, 2767.

1871 Zibuts, Yu. A., L. G. Paritskii, and S. M. Ryvkin (1964), Some Properties of Silicon Containing Mercury, Tungsten, Molybdenum and Platinum Impurities, *Sov. Phys. Solid State*, **5**, 2416.

1872 Zibuts, Yu. A., L. G. Paritskii, S. M. Ryvkin, and Zh. G. Dokholyan (1967), Photoelectric Properties of Silicon Doped with Copper, Tungsten, and Platinum, *Sov. Phys. Solid State*, **8**, 2041.

1873 Zijlstra, R. J. J., and A. Gisolf (1972), Thermal Noise in Double Injection Diodes Operating in the Insulator Regime, *Solid-State Electron.*, **15,** 877.

1874 Zitter, R. N. (1958), Role of Traps in the Photoelectromagnetic and Photoconductive Effects, *Phys. Rev.*, **112,** 852.

1875 Zohta, Y. (1970), A New Method for Determination of Deep-Level Impurity Centers in Semiconductors, *Appl. Phys. Lett.*, **17,** 284.

1876 Zohta, Y. (1972), Frequency Response of Gold Impurity Centers in the Depletion Layer of Reverse-Biased Silicon p^+n Junctions, *J. Appl. Phys.*, **43,** 1713.

1877 Zohta, Y., and Y. Ohmura (1972), Determination of the Spatial Distribution of Deep Centers from Capacitance Measurements of p-n Junctions, *Appl. Phys. Lett.*, **21,** 117.

1878 Zohta, Y., Y. Ohmura, and M. Kanazawa (1971), Shallow Donor State Produced by Proton Bombardment of Silicon, *Japan. J. Appl. Phys.*, **10,** 532.

1879 Zolataryov, V. F., D. G. Semak, and D. V. Chepur (1967), Thermally Stimulated Currents under Conditions of Persistent Internal Polarization, *Phys. Status Solidi*, **21,** 437.

1880 Zorin, E. I., P. V. Pavlov, and D. I. Tetel'baum (1968), Donor Properties of Nitrogen in Silicon, *Sov. Phys.—Semicond.*, **2,** 111.

1881 Zorin, E. I., P. V. Pavlov, .D I. Tetel'baum, and A. F. Khoklov (1972), Donor Properties and Diffusion of Potassium in Silicon, *Sov. Phys.—Semicond.*, **6,** 21.

1882 Zorin, E. I., P. V. Pavlov, D. I. Tetel'baum, and A. F. Khokhlov (1972), Difference Between the Properties of Sodium- and Potassium-Doped Silicon, *Sov. Phys. —Semicond.*, **6,** 344.

1883 Zoroglu, D. S., and I. C. Chang (1970), Spectra Investigation of Acoustoelectric Domains in n-GaAs by Microwave Emission Studies, *J. Appl Phys*, **41,** 2294

1884 Zotova, N. V., T. S. Lagunova, and D. N. Nasledov (1964), Negative Magnetoresistance of n-Type Indium Arsenide at Low Temperatures, *Sov. Phys. Solid State*, **5,** 2439.

1885 Zucker, J., and E. M. Conwell (1961), The Recombination of Hot Carriers in Germanium—Experimental, *J. Phys. Chem. Solids*, **22,** 141.

1886 Zucker, M. J., M. M. Traum, and S. H. Charap (1969), Piezoresistance of Si and Ge with Deep Acceptor Levels, *Bull. Am. Phys. Soc.*, **14,** 557.

1887 Zuev, V. A., and V. G. Litovchenko (1972), Photoeffect Spectra in Silicon at Different States of Bulk and Surface, I. Theoretical Analysis, *Surf. Sci.*, **32,** 365.

1888 Zuev, V. A., V. G. Litovchenko, and V. V. Antoshchuk (1972), Photoeffect Spectra in Silicon at Different States of Bulk and Surface, II. Experimental Results, *Surf. Sci.*, **32,** 377.

1889 Zuleeg, R., and V. Ranon (1969), Coupling of Light Emission and Negative Resistance in a GaAs Diode, *Appl. Phy. Lett.*, **15,** 168.

1890 Zwerdling, S., K. J. Button, B. Lax, and L. M. Roth (1960), Internal Impurity Levels in Semiconductors: Experiments in p-Type Silicon, *Phys. Rev. Lett.*, **4,** 173.

1891 Zwicker, H. R. (1972), Steady State Field Driven Conduction in Semiconductors with Trapping, PhD. Thesis, University of Illinois, Urbana, Illinois.

1892 Zwicker, H. R., B. G. Streetman, N. Holonyak, Jr., and A. M. Andrews (1970a), Double Injection in Au-Doped Si p-π-n Diodes, *Appl. Phys. Lett.*, **16,** 63.

1893 Zwicker, H. R., B. G. Streetman, N. Holonyak, Jr., and A. M. Andrews (1970b), Double Injection in Semiconductors with Multivalent Trapping Centers, *J. Appl. Phys.*, **41,** 4697.

1894 Zwicker, H. R., D. L. Keune, N. Holonyak, Jr., and R. D. Burnham (1971), Optical Phase Shift Measurement of Carrier Decay-Time on Thin Semiconductor Samples With Surface Losses, *Solid-State Electron.*, **14,** 1023.

Author Index

The numbers without parentheses show the pages of the text on which the author is mentioned. Numbers in parentheses refer to paper numbers in the bibliography.

Subject Index

The numbers without parentheses show the page of the text on which the subject matter is discussed. Numbers in parentheses, however, refer to paper numbers in the bibliography relating to the subject.